THE ROMANTIC
CONCEPTION OF LIFE

Flora, goddess of spring and flowers, oil painting (1790) by Angelika Kauffmann, friend of Goethe. (Private collection.)

The Romantic Conception of Life

•◆•◆•◆•

SCIENCE AND PHILOSOPHY IN THE AGE OF GOETHE

•◆•◆•◆•

Robert J. Richards

THE UNIVERSITY OF CHICAGO PRESS
CHICAGO AND LONDON

Science and Its Conceptual Foundations
Edited by David L. Hull

ROBERT J. RICHARDS is a professor of history, psychology, and philosophy at the University of Chicago. He is the author *Darwin and the Emergence of Evolutionary Theories of Mind and Behavior* (1987) and *The Meaning of Evolution: The Morphological Construction and Ideological Reconstruction of Darwin's Theory* (1992).

The University of Chicago Press, Chicago 60637
The University of Chicago Press, Ltd., London
© 2002 by The University of Chicago
All rights reserved. Published 2002
Printed in the United States of America

11 10 09 08 07 06 05 04 03 02   1 2 3 4 5

ISBN: 0-226-71210-9 (cloth)

Library of Congress Cataloging-in-Publication Data

Richards, Robert J.
    The romantic conception of life : science and philosophy in the age of
Goethe / Robert J. Richards.
        p.   cm. — (Science and its conceptual foundations)
Includes bibliographical references and index.
    ISBN 0-226-71210-9 (alk. paper)
    1. Romanticism—Germany.   2. German literature—18th century—
History and criticism.   3. German literature—19th century—History and
criticism.   4. Literature and science—Germany.   5. Philosophy—German.
I. Title.   II. Series.
    PT361 .R53   2002
    830.9'145—dc21

                                                                2002001769

For Barbara

All art should become science and all science art;
poetry and philosophy should be made one.

Friedrich Schlegel, *Kritische Fragmente*, 115

Contents

Illustrations xiii
Acknowledgments xv
Prologue xvii

1 Introduction: A Most Happy Encounter   1
    The Historical Meaning of Naturphilosophie *and Romantic Biology* 6

PART ONE
THE EARLY ROMANTIC MOVEMENT IN LITERATURE,
PHILOSOPHY, AND SCIENCE

2 The Early Romantic Movement   17
    Wilhelm and Friedrich Schlegel 23
    Novalis: The Romantic Personality 25
    Caroline Böhmer and the Mainz Revolution 36
    *The Schlegels in Jena: The Break with Schiller and the Politics
    of Romanticism* 45
    Fichte, the Philosopher of Freedom 59
    The Salons of Berlin 91
    Friedrich Schleiermacher: The Poetics and Erotics of Religion 94
    Friedrich Schlegel's Aesthetic Theory 105

3 Schelling: The Poetry of Nature   114
    Schelling's Early Life 116
    Naturphilosophie *128*
    Schelling in Jena 147
    Transcendental Idealism and Poetic Construction 151
    Schelling's Affair with Caroline and the Tragedy of Auguste 166
    Schelling's Identity Philosophy 176

4 Denouement: Farewell to Jena   193
    The Meaning of Romanticism 199

PART TWO
SCIENTIFIC FOUNDATIONS OF THE
ROMANTIC CONCEPTION OF LIFE

5 Early Theories of Development: Blumenbach and Kant   207
    *Embryology and Theories of Descent in the Seventeenth and
    Eighteenth Centuries* 211

Blumenbach's Theory of the Bildungstrieb   216
Kant's Theory of Biological Explanation   229

6   Kielmeyer and the Organic Powers of Nature   238
Lecture on Organic Forces   241
Theory of Species Origin and Transformation   246
Critique of Kant and the Idealists   248

7   Johann Christian Reil's Romantic Theories of Life and Mind, or
Rhapsodies on a Cat-Piano   252
Early Training and Practice   253
Lebenskraft   255
Studies of Mental Illness   261
The Romantic Movement in Halle   273
The Romantic Naturphilosoph   278
Final Years: War and Romance   282

8   Schelling's Dynamic Evolutionism   289
Biological Treatises   291
Critical Analysis of the Biological Theories of Contemporaries   292
Nature as a Dynamically Shifting Balance of Forces   294
Theory of Dynamic Evolution   298

9   Conclusion: Mechanism, Teleology, and Evolution   307
*Appendix: Theories of Irritability, Sensibility, and
Vital Force from Haller to Humboldt*   313

PART THREE
GOETHE, A GENIUS FOR POETRY,
MORPHOLOGY, AND WOMEN

10   The Erotic Authority of Nature   325
Growing Up in Frankfurt   330
University Education   334
The Law, Herder, and Lotte   339
The Weimar Councillor and the Frustrated Lover   355
The Science of Goethe's First Weimar Period   365
*The Unity of Biological Nature: Goethe's Discovery
of the Zwischenkiefer in Human Beings*   367
The Impact of Spinoza   376
Goethe's Italian Journey: Art, Nature, and the Female   382
Conclusion   404

11   Goethe's Scientific Revolution   407
Homecoming   409
The Foundations of Morphology   413

Friendship with Schiller and Induction into the Kantian Philosophy 421
The Science of Morphology 434
The Romantic Circle and Schelling 457
Zur Morphologie 471
*The Vertebral Theory of the Skull: Goethe's Dispute with Oken
and the Truth of Memory* 491

12 Conclusion: The History of a Life in Art and Science 503

PART FOUR
EPILOGUE

13 The Romantic Conception of Life 511

14 Darwin's Romantic Biology 514
The Romantic Movement 516
Darwin's Romantic Conception of Nature 522
Romantic Nature in the Origin of Species 526
Darwin's Theory of Morals 540
Conclusion 552

Bibliography 555
Index 573

Illustrations

	Frontispiece, Flora, goddess of flowers	ii
1.1.	The vertebrate archetype, from Carl Gustav Carus, *Von den Ur-Theilen des Knochen und Schalengerustes*	9
2.1.	Friedrich Schlegel	24
2.2.	Friedrich von Hardenberg (Novalis)	27
2.3.	Sophie von Kühn	29
2.4.	The freedom tree	41
2.5.	Caroline Michaelis Böhmer Schlegel Schelling	44
2.6.	August Wilhelm Schlegel	48
2.7	Johann Gottlieb Fichte	73
2.8.	Friedrich Schlegel	89
2.9.	Friedrich Daniel Schleiermacher	98
3.1.	Friedrich Wilhelm Joseph Schelling	115
3.2.	Friedrich Hölderlin	118
3.3.	Auguste Böhmer	167
4.1.	Friedrich Schelling	199
5.1.	Frontispiece to the publication of Goethe's *Faust, Ein Fragment*	209
5.2.	Albrecht von Haller	214
5.3.	Johann Friedrich Blumenbach	217
5.4.	Johann Gottfried Herder	224
5.5.	Immanuel Kant	230
7.1.	Johann Christian Reil	256
7.2.	*Katzenclavier*, or cat-piano	272
7.3.	Christoph Wilhelm Hufeland	283
8.1.	Erasmus Darwin	300
A.1.	Plate from Luigi Galvani, *De viribus electricitatis*	318
A.2.	Plate from Alexander von Humboldt, *Versuche über die gereizte Muskel- und Nervenfaser*	320
10.1.	Johann Wolfgang von Goethe	324
10.2.	Carl August, duke of Saxe-Weimar-Eisenach	356
10.3.	Charlotte von Stein	358
10.4.	Apartments of Charlotte von Stein	360
10.5.	The river Ilm	364
10.6.	Justus Christian Loder	368
10.7.	*Anatomieturm* in Jena	370
10.8.	Illustration of the human intermaxillary bone	371
10.9.	Goethe in apartment in Rome	386
10.10.	Emma Lyon	388

10.11.	Anatomy sketch by Goethe	392
11.1.	Christiane Vulpius	412
11.2.	Goethe's house on Der Frauenplan	414
11.3.	Gardens at the rear of Goethe's house on Der Frauenplan	415
11.4.	Megatherium drawn by D'Alton	484
11.5.	Lorenz Oken	494
11.6.	Bust of Oken in Jena	502
12.1.	Johann Wolfgang von Goethe	505
12.2.	Dedication page to Goethe by Alexander von Humboldt	508
14.1.	Charles Darwin	527
14.2.	Richard Owen	529
14.3.	Richard Owen's illustration of the vertebrate archetype	530
14.4.	Charles Darwin	548
14.5.	Charles Darwin	553

Plates follow p. 148

PLATE 1. Johann Wolfgang von Goethe in the Roman Campagna
PLATE 2. Henriette Herz
PLATE 3. Johann Wolfgang von Goethe
PLATE 4. Friedrich Schiller
PLATE 5. Alexander von Humboldt

Acknowledgments

No book springs whole from the mind of a single author, a simple fact of which I am acutely aware. References to the large scholarship on the subjects I treat accumulate on my pages like stars against a dark sky, where they reveal a considerable debt. Without the generous aid of several institutions that provided access to this scholarship and the resources to pursue it, this book would still be in embryo. My gratitude goes, first of all, to my own university, the University of Chicago. The university used to enjoy the pungent atmosphere of the steel mills on the south side of the city. I have often thought that gritty, demanding, and dangerous environment served as a metaphor for the scholarship practiced in Hyde Park; the monuments of this scholarship stand like the marvels of the city's architecture and have been for me comparably inspiring. The Dibner Institute for the History of Science granted me a research fellowship for the academic year 1995–96, and that allowed me to begin my project. The Dibner's directors, Evelyn Simha and Jed Buchwald, showed gracious hospitality and furnished a stimulating forum for collegial discussion. The Max-Planck-Institut für Wissenschaftsgeschichte in Berlin, through a research fellowship for the year 2000, enabled me substantially to complete the work. The director, Lorraine Daston, and many friends there provided a critical and sufficiently skeptical environment that compelled me often to rethink arguments, if not always to change them. The following institutions made both printed material and manuscripts available: the Alexander-von-Humboldt-Forschungsstelle, Berlin; the Manuscript Room, Cambridge University Library; Haeckel-Haus, Jena; the Manuscript Room, London Natural History Museum; Special Collections, Regenstein Library, University of Chicago; the Stadtbibliothek, Berlin; and Widner Library, Harvard University.

Preliminary versions of sections of several chapters have previously appeared in the journals *Critical Inquiry* and *Studies in History and Philosophy of Biology and Biomedical Science*, and in the collections *Biology and the Foundation of Ethics*, ed. Jane Maienschein and Michael Ruse (Cambridge University Press), and *The Moral Authority of Nature*, ed. Lorraine Daston and Fernando Vidal (University of Chicago Press). These essays were initially composed, however, with this present volume in view. All of the translations in the text—except in a few noted instances—are my own.

Several individuals have read large portions of my manuscript. I gave them enough rough pages on which to sharpen their critical faculties, and none failed to employ their judgment to my great advantage. I am supremely in the debt of Karl Ameriks, Frederick Beiser, Thomas Broman, Frederick Gregory, Michael Hagner, David Hull, Nicholas Jardine, the late Lily Kay, Cheryce Kramer, Wolfgang Lefevre, Peter McLaughlin, Robert Proctor, Michael Ruse, and Joan Steigerwald. It is often a reflexive cliché for academics to say how much they've learned from their students. My own students will recognize ideas formed commonly in seminars that now have been appropriated in this book. Kristin Casady scrutinized the entire manuscript with an eye sensitive to infelicities of English and ragged translations from German. She made numerous suggestions for improvement, and I am deeply grateful for her efforts. Erin DeWitt's meticulous eye and steady hand rendered my textual references deceptively consistent; I could not have hoped for better. Susan Abrams, my editor at the University of Chicago Press, has served not only as a patient critic but also as a good friend. Her intelligent guidance has allowed my book to appear amid a glittering array of others under her purview. My wife, Barbara, never ceased to urge me on, and my dedication is but small recompense for what she continues to provide.

Prologue

My title, *The Romantic Conception of Life*, refers to two related aspects of this study of early German Romanticism. First of all, I mean life as experienced by the individuals whom I discuss. Theirs were romantic lives: the young poet whose inamorata dies at age fifteen and he himself before thirty; the German beauty who quickly becomes infatuated with a French soldier during the Revolution and then is thrown into prison while pregnant with his child; the philosopher-scientist who falls in love with the wife of a friend and the woman's daughter as well; the famous writer who escapes to Rome, where he celebrates sexual liberation in exquisite poetry. These poets, philosophers, and scientists defied bourgeois moral conventions, advanced the ideals of freedom, and left memorable portraits of their extraordinary careers in autobiographies and letters, and indirectly in poems, novels, and even in theological and scientific texts. They were individuals who self-consciously appropriated the name "Romantic," and their lives came to define the term.

I also mean to convey by my title that the individuals who led such lives attempted to understand animate nature in a Romantic mode through poetry, philosophy, and science. They were not uniformly opposed, as is usually assumed, to the Enlightenment emphasis on reason. Many of the Romantics might, rather, be classified as hyperbolically rational in that they believed the scientific mind could penetrate into all of the dark corners of the universe. They did argue, though, that aesthetic judgment offered another, complementary path into the deep structures of reality, a path overlooked by most Enlightenment thinkers. Indeed, they altered fundamental assumptions about the very character of philosophy and science, arguing that a poetical transformation of these disciplines might reveal features of nature not contemplated by their predecessors. Emboldened by this new understanding, the Romantics attacked a significant bulwark of earlier thought, one that advanced mechanism as the engine of progress in science. From Descartes and Newton to Hume and Kant, mechanism had been employed as the basic concept by which to understand not only the inanimate universe but the living world as well. The Romantics replaced the concept of mechanism with that of the organic, elevating it to the chief principle for interpreting nature.

During the couple of decades on either side of 1800, the age of Goethe, these individuals came together in various venues to *symphilosophize* and to *sympoetize*, as they put it. Even the scientific members of this loose confederation brought the conceptual concerns of the philosopher and the refined sensibilities of the poet to the experimental investigation of life. The Romantics formulated powerful ideas about the relationship between artistic modes of representation and scientific modes. Their conceptions, however, flowed not simply in abstract numbers from dispossessed minds but from embodied personalities, from individuals whose particular relationships, fired in love and in hate, shaped those conceptions as much as did the formal aspects of their inquiries. I have followed the philosophic and scientific ideas of these early Romantic thinkers as their ideas emerged from the intellectual legacy to which they were heir, from their immediate scientific experiences, and especially from their more intimate personal relationships.

Though the New Critics of the 1940s and 1950s attempted to coax the authorial genie out of the well-wrought urn—and the postmodernists of our day simply refuse to believe in him—I think it impossible to exorcise the daemon. Can it be seriously entertained that poets who dramatized the extremes of life or philosophers who attempted explicitly to reconstruct it—that they formulated conceptions unmarked by their own lived experience? That the mind of such individuals could create a work that would "float free, detached from the imperfect life that produced it"?[1] This book will offer a long argument against such presumption. My effort, though, will be not only to show how their lives affected their work, but also to show how their work affected their lives. They composed Romantic poetry and then became the passionate, ironic, and adventuresome individuals their verse described—thus did life imitate art. In these pages, then, more of the biographical will appear than might otherwise be expected in a volume devoted to the history of philosophy and science.

I have tried to sustain several themes across the four parts of this book. My intention has been to show how concepts of self, along with aesthetic and moral considerations—all tempered by personal relationships—gave complementary shape to biological representations of nature. These themes of German Romanticism, however, have been developed in the several parts of this volume in respect to different individuals and events. This means that each of the parts can be read as a coherent whole, and so the burden of the book need not be hoisted all at once.

1. In his review of a biography of Bertrand Russell, Nagel claims that the philosopher's productions did indeed escape the mess of his life; he suggests that this is the normal situation in philosophic creation. See Thomas Nagel, "How to Be Free and Happy," *New Republic* 224 (7 May 2001): 31.

The various themes of this book and the arguments on which they ride flow to a conclusion that will be most perspicuously realized in the epilogue, which describes the fundamental ways in which Romantic thought gave shape to Darwin's conceptions of nature and evolution. Most often anything called Romantic science has been thought at best a minor tributary of nineteenth-century scientific thought—really nothing but a backwater. My general conclusion is quite different. On the basis of the history recounted in the main part of this volume, I have become convinced that the central currents of nineteenth-century biology had their origins in the Romantic movement.

Chapter 1

INTRODUCTION:
A MOST HAPPY ENCOUNTER

> When sleep overcomes her, I lie by her side and think over many
> things. Often I have composed poetry while in her arms and have
> softly beat out the measure of hexameters, fingering along her spine.
>
> —Johann Wolfgang von Goethe, *Römische Elegien*

On 20 July 1794, Johann Wolfgang von Goethe (1749–1832) and Friedrich Schiller (1759–1805) happened to attend the same meeting of the Jena Natural Research Society. They knew each other, having met six years before, but their relationship remained distant and distrustful. Goethe had been instrumental in securing Schiller's appointment to the history faculty at Jena; initially, though, the position carried only a small stipend and provided neither the financial security the younger poet sought nor the freedom to continue his literary work. Goethe's aloof genius, which seemed so effortlessly to sail over the flood of poetry spilling from his pen, evoked in Schiller feelings, he confessed, not unlike those "Brutus and Cassius must have had for Caesar." "I could murder his spirit," he wrote a friend, "and then love him with all my heart."[1] For his part, Goethe disliked Schiller's sensational Sturm und Drang play *Die Räuber* (The robbers, 1781), which seemed to endorse the kind of anarchic attitudes that inspired talk of revolution; and he was vexed over what he believed Schiller had implied about him in the essay *Über Anmut und Würde* (On grace and dignity, 1793), namely, that his aesthetic talent lacked a deeper moral character. What particularly disconcerted Goethe was Schiller's Kantian subjectivism, which he thought dropped a veil between the artist and nature. So when they left the Jena meeting and could not avoid interchange, they began to discuss, with mutual wariness, the fragmented character of the lecture they had just heard. Contrary to the approach of the speaker, the botanist August Johann

---

1. Friedrich Schiller to Christian Körner (2 February 1789), in *Schillers Werke* (Nationalausgabe), ed. Julius Petersen et al., 55 vols. (Weimar: Böhlaus Nachfolger, 1943–), 25:193.

Georg Carl Batsch (1761–1802), Goethe suggested that the study of nature had to move from the whole to the parts rather than the reverse. Intrigued, Schiller invited this benign nemesis back to his house to continue the conversation. There Goethe turned to his botanical theories, and he sketched for Schiller the ideal plant, the morphological model for understanding all plants. Seven years earlier Goethe had roamed through southern Italy and Sicily attempting to discover this *Urpflanze*. But when Schiller looked at the sketch, he exclaimed: "Das ist keine Erfahrung, das ist eine Idee"— "That's no observation, that's an idea." Goethe, quite provoked, responded: "Well, I'm rather fortunate that I have ideas without knowing it and can even see them with my own eyes."[2]

Goethe's irritating confrontation with Schiller actually marked the beginning of their intimate friendship, which terminated only with Schiller's death a decade later. Goethe described this "happy encounter" in the first number of his zoological writings *Zur Morphologie* (1817–24). He meant the tale to be emblematic of the several features of the morphological doctrine that he formed before and after his encounter with Schiller. But this literary *Denkmal* also suggests, I believe, the way in which Goethe's theory, and the tradition it spawned, arose out of a decidedly Romantic sensibility: his ideal plant did stem from empirical observations, which, however, had been transformed by a creative imagination to reveal a deeper core of reality.

Goethe as a Romantic—a rather anomalous idea, perhaps. Certainly his *Die Leiden des jungen Werthers* (The sorrows of young Werther) is easily read as a Romantic novel; and much of Goethe's lyric poetry expresses that delicacy of feeling for nature with which the English Romantic poets resonated. But his dramas, such as *Iphigenie auf Tauris* or *Torquato Tasso*, are normally understood as classical in character. And his several other novels, such as *Wilhelm Meisters Lehrjahre* (Wilhelm Meister's apprenticeship), are usually thought to occupy the genre of Bildungsroman, rather than that of the Romantic portrait. Even his very living quarters exhibited his classical taste. The floors of his well-preserved (and postwar reconstructed) house on Der Frauenplan in Weimar sag with the weight of Greek and Roman statuary that he accumulated on his travels. Goethe might be more easily classifiable had he become reasonable, for then he would have been, as Lessing remarked, an ordinary man.[3] But his genius and energies far outstripped

2. Goethe recounts this "happy encounter" in the first number (1817) of his *Zur Morphologie*. See Johann Wolfgang von Goethe, *Goethe: Die Schriften zur Naturwissenschaft,* 21 vols. 1st division, vol. 9: *Morphologische Hefte*, ed. Dorothea Kuhn (Weimar: Böhlaus Nachfolger, 1954), pp. 79–83; quotations from pp. 81–82.

3. Gotthold Lessing is quoted by Friedrich Jacobi in a letter to Johann Heinse (20 October 1780), in *Goethe in vertraulichen Briefen seiner Zeitgenossen*, ed. Wilhelm Bode, 3 vols. (Berlin: Aufbau Verlag, 1999), 1:262.

those of his contemporaries. He appeared at the time (and now) a figure immensely larger than any small category that might capture individuals of more common clay. And particularly in this instance, he disdainfully discarded the very label that might allow his work to be easily tagged as Romantic. Late in his life, in conversations with a young writer on the make, Johann Peter Eckermann, Goethe mentioned that he had very simple definitions of "classical" and "Romantic." "I call the classical healthy," he explained, "and the Romantic sick."[4] Yet a year later, on 21 March 1830, he acknowledged to his young friend that Schiller had convinced him that "I myself, contrary to my own will, was a Romantic."[5]

Goethe's acknowledgment of the Romantic character of his own thought will serve as a leitmotiv for this book. Historians of nineteenth-century science, the really serious historians, usually dismiss anything sounding like Romantic science as an aberration and suspect that anyone following such a red thread will be traveling down a path that terminates in the higher nonsense. Historical investigations of Romanticism may be amusing enough for the moment but certainly not instructive about the authentic science of the period, the science that grounds our contemporary understanding. The biologist E. O. Wilson, for instance, indicts the Romantics precisely as those responsible for advancing irrational fantasy over scientific reason—the latter, fortunately, having escaped the specter that still haunts present culture.[6] Even a sophisticated historian like Timothy Lenoir, who has focused on German life sciences of the early nineteenth century, wishes to shield the "real" biologists of the period—such as Johann Christian Reil, Carl Friedrich Kielmeyer, Ignaz Döllinger, Karl Friedrich Burdach, and Karl Ernst von Baer—from this taint. Lenoir argues the certainly interesting thesis that these aforementioned scientists have to be regarded as materialists and mechanists and that their construction of living organisms as teleological can only be modeled after Kant's teleology *als ob*—a model that supposedly allowed good scientists to believe organisms were mere mechanisms while heuristically describing them *as if* they exhibited an intrinsi-

4. Johann Peter Eckermann, *Gespräche mit Goethe in den letzten Jahren seines Lebens*, 3rd ed. (Berlin: Aufbau-Verlag, 1987), p. 286. This conversation took place on 2 April 1829.

5. Ibid., p. 350 (21 March 1830). In the late nineteenth century, German literary historians began referring to the period of Goethe's association with Schiller as the classical period of German literature. But this did not mean that these poets simply emulated the works of the ancients, rather that their writings functioned to establish the modern German literary tradition—they were the "classical" sources. The confusion of these two meanings of classical (i.e., emulating the ancients and founding a tradition) has further served to make it quite heterodox to refer to Goethe as a Romantic. See René Wellek, "The Term and Concept of Classicism in Literary History," *Discriminations: Further Concepts of Criticism* (New Haven: Yale University Press, 1970), pp. 55–90.

6. Edward O. Wilson, *Consilience* (New York: Knopf, 1998), p. 40.

cally purposive structure.[7] I believe these historiographic attitudes have excised the heart of nineteenth-century biology, which pulsed to more fascinating rhythms than can be imagined when dissecting the dried corpus of
the discipline. When that biology has its lifelines secured by reattaching
them to the thought and culture that animated it, I believe we will discover
that many of its main themes have been played out in a Romantic mode, or
so is the central argument of this book.[8]

In the next three chapters, forming part 1 of this volume, I portray the
life and thought of the individuals who created the Romantic movement at
the end of the eighteenth century—Novalis (Friedrich von Hardenberg),
the brothers Wilhelm and Friedrich Schlegel, Friedrich Schleiermacher,
Friedrich Schelling, Goethe, and the incomparable Caroline Michaelis
Böhmer Schlegel Schelling, whose surname plots only part of her romantic
trajectory. I also consider associated figures who, while not members of the

7. See Timothy Lenoir, *The Strategy of Life: Teleology and Mechanics in Nineteenth-Century German Biology*, 2nd ed. (Chicago: University of Chicago Press, 1989).

8. Some may cavil at my use of the term "biology" to characterize botanical and zoological
thought in the late eighteenth and early nineteenth centuries, under the assumption that the word
has a more modern provenance. Actually, the term, in its Latin form, made an appearance in the
mid-eighteenth century. Michael Christoph Hanov, a disciple of Christian Wolff, used it in the title of his *Philosophiae naturalis sive physicae dogmaticae: tomus III, continens geologiam, biologiam, phytologiam generalis* (Halle: Renger, 1766). In this large work, biology encompassed the various parts
of zoology. (I am grateful to Peter McLaughlin for this reference.) In 1736 Linnaeus employed the
Latin term *biologi* in his *Bibliotheca botanica* to refer to botanists who wrote about the life cycle of
plants. See Carl von Linné, *Bibliotheca botanica* (Munich: Fritsch [1736], 1968), p. 148. In preparation for a translation of Linnaeus's works into German, Johann Planer used the German *Biologie* to
indicate the work of such botanists. See Johann Planer, *Versuch einer teutschen Nomenclatur der Linneischen Gattungen* (Erfurt: Müller, 1771), last page of the *Vorrede*. Kai Kanz describes Planer's construction in his "Zur Frühgeschichte des Begriffs 'Biologie,' die botanische Biologie von Johann
Jakob Planer," *Verhandlungen zur Geschichte und Theorie der Biologie* 5 (2000): 269–82. In 1797
Theodor Georg Roose employed the word once in a book on *Lebenskraft*. In the preface he referred
to his own tract as a "sketch of a biology" [Entwurf einer Biologie]. See Theodor Georg Roose,
Grundzüge der Lehre von der Lebenskraft (Braunschweig: Thomas, 1797), p. 1. Karl Friedrich Burdach, in 1800, also used the term, though to indicate the study of human beings from a morphological, physiological, and psychological perspective. See Karl Friedrich Burdach, *Propädeutik zum
Studium der gesammten Heilkunst* (Leipzig: Dyt'schen Buchhandlung. 1800), p. 62. In Germany, usage became solidified with the publication of Gottfried Reinhold Treviranus's six-volume treatise
Biologie, oder Philosophie der lebenden Natur (Göttingen: Johann Friedrich Röwer, 1802–22). Treviranus announced: "The objects of our research will be the different forms and manifestations of life,
the conditions and laws under which these phenomena occur, and the causes through which they
have been effected. The science that concerns itself with these objects we will indicate by the name
biology [*Biologie*] or the doctrine of life [*Lebenslehre*]" (1:4). With Treviranus the modern concept of
biology became firmly established. Instructive—or tedious—as this linguistic history might be, it is
a bit beside the point. The term "biology" quite efficiently and significantly captures the subject of
my study, and it is as applicable to the work of Aristotle as to that of those naturalists of the eighteenth and early nineteenth centuries who studied the phenomena of life.

charmed circle, nevertheless provided some of the ingredients necessary for the magic to work. In this latter category, I focus on Kant and Fichte in particular. My effort is first to observe the leading ideas of these individuals as they emerged from the interstices of personal interactions and then more carefully to explore those conceptions in order to reveal their inner logic and external relations. It is impossible, I believe, to understand either the more overt or the very subtle ways Romanticism shaped biology in the nineteenth century without coming to understand the Romantic mode of being and thought, and to feel its very complex and often heterogeneous constitution. In part 2 I consider philosophical and scientific formulations of the nature of life, especially in the works of Kant, Herder, Blumenbach, Kielmeyer, and Humboldt. These formulations merged with and were transformed by writers like Schelling and Reil. During this period the basic categories of investigation changed from mechanism to organicism, and with that change arose a consideration that would assume huge proportions later in the nineteenth century, namely, the theory of evolution. In part 3 I return to Goethe and his apparently contradictory evaluations of Romantic literature. In order to understand his attitude, it will be necessary, first, to assess the various ways nature exerted an erotic command over his thought and life. I then concentrate on what the Schiller-induced acknowledgment of Romanticism reveals about the structure of Goethe's aesthetics and his science, their relationship, and the sources of their construction. Goethe dominated scientific and even philosophical thought in the Romantic circle, and his fundamental ideas about biological structures would reverberate throughout the nineteenth century, becoming united with the evolutionary thought of Darwin and Haeckel. I will limn the impact of German Romanticism on Darwin's biology in the epilogue and undertake a more thorough analysis of its consequence for the science of the late nineteenth century in a subsequent volume.[9]

The biographical emphasis of this book stems from my conviction, which can only be partly acted upon in a work such as this, that we catch ideas in the making only when we understand rather intimately the character—the attitudes, the intellectual beliefs, the emotional reactions—of the thinkers in question. Without an initial plunge into personality, logical analysis of the connections of their ideas will be blind and social construction of their theories empty.

---

9. That volume is tentatively titled "The Tragic Sense of Life: Ernst Haeckel and the Struggle over Evolutionary Thought in Germany."

The Historical Meaning of *Naturphilosophie* and Romantic Biology

To demonstrate the accuracy of Goethe's own avowal of Romanticism requires a rather thick description of the Romantic movement at the turn of
the eighteenth century into the nineteenth. I undertake that description in
the next three chapters. But no matter how rounded this narrative, it will
seem hardly sufficient to prepare for the argument that Charles Darwin and
Ernst Haeckel were Romantic scientists. The historian of science David
Knight expresses a perfectly common view when he takes for granted that
"the *Origin of Species* was not rooted in Romanticism." The Romantics, he
conventionally presumes, were "not in the business of genealogy, constructing family trees, but searching for natural kinds."[10] So a scientist like
Haeckel, who planted more genealogical trees in the hard ground of his empirical studies than any of his English friends, likewise should not, upon this
representation, be counted among the Romantics.

Part of the difficulty of seeing all those evolutionary trees but not their
Romantic roots stems, I think, from the unstudied vagueness with which
the terms "Romanticism" and *Naturphilosophie* have been used. Cultural
historians have been studiously reluctant to venture a definition of Romanticism, regarding such effort a "trap," as Isaiah Berlin put it.[11] In one of the
best recent books devoted to the topic of Romanticism and the sciences,
none of the contributing authors attempts a definition of the subject of the
volume; rather, they rely on cultural assumptions and unanchored intuitions—a caricature of the Romantic approach itself.[12] Most of these authors undoubtedly assume, with decent precedent, that any attempted
definitions of Romantic science or nature philosophy must be flat, stale,
and unprofitable; that limp generalizations must fail to capture the rich di-

10. David Knight, "Romanticism and the Sciences," in *Romanticism and the Sciences,* ed. Andrew Cunningham and Nicholas Jardine (Cambridge: Cambridge University Press, 1990), pp. 22,
16.

11. Berlin could not avoid, however, some characterization of Romanticism as propaedeutic to
his study of its origins; and he concluded with the specification of two related features that seemed
ubiquitous. The first was that the Romantics created, "out of literally nothing," goals, values, and a
vision of the universe. The second was that there is no structure of things, "only, if not the flow, the
endless self-creativity of the universe." Both of these features do reflect the Romantic movement,
though only in a certain way and not without severe qualification, as I will try to show. See the recently published transcription of Isaiah Berlin's 1965 Mellon Lectures, *The Roots of Romanticism,*
ed. Henry Hardy (Princeton: Princeton University Press, 1999), pp. 1–20, 119.

12. See the previously noted *Romanticism and the Sciences.* In this volume, the editors do suggest
that Novalis's *Novices at Sais* can serve as an implicit paradigm for Romantic thought in general
(pp. 4–5). This incomplete novella, however, is a rather delicate web with which to capture an understanding of such a complex and fraught movement as German Romanticism. Moreover, not
even the silken threads of Novalis's imagination could hold the careening personalities that furnished the movement's life.

versity of a movement that involved so many extraordinary individuals, the luxury of their art, and the prodigality of their ideas.[13]

I do not think, however, that the effort at general characterization is as hopeless as it appears. I will attempt two methodologically different approaches to definition. The first utilizes an evolutionary construction, and it constitutes the principal task of this book: namely, to uncover the historical roots of the ideas captured by the terms "Romantic biology" and *Naturphilosophie*, and then to follow their evolution, as they develop and bend to the causal interactions radiating from their intellectual, psychological, and social environments. Thus at any one time in their evolution, the meanings constituting "Romantic biology" will be fixed by their particular causal interactions; and their entire development will constitute *the meaning of Romantic biology* in the eighteenth and nineteenth centuries.

Definitions of the more usual sort also have their uses. After all, just as in the case of biological systematics—when one attempts to trace the genealogy of, say, a species of Galapagos mockingbird—so in the genealogy of ideas: general notions must initially be used to discriminate the lineages of interest. We must, therefore, form some broad but definite conceptions about the composition of *Naturphilosophie* and Romantic biology, which will constitute their provisional definitions, in order to recognize their sources and trace their consequent developments. If such preliminary characterizations are to be strategically useful, though, they must be induced, partly at least, from the actual history of their constituent ideas. Definitions so constructed can then be employed to illuminate the further course of the participant ideas as they extend beyond their preliminary inductive foundations. On the basis of these extending historical analyses, we will be able to refine our understanding of the principal subjects of interest and, accordingly, readjust the definitions. From induction to definition to deductive inference and around again, in historical constructions we raise our understanding by our own semantic bootstraps. Let me then, as propaedeutic to my study, discriminate those ideas that have traveled together for a good part of their history; these will constitute, in a preliminary fashion, the meanings of "Romantic biology" and of *Naturphilosophie*. No system of thought in the eighteenth or nineteenth centuries will likely include all of these marks; but we will understand a system to be more or less Romantic or *naturphilosophisch* depending on how many of these notes they do incorporate.

13. In a wide-ranging and otherwise quite useful bibliographical essay on Romanticism and science, Levere also urges the futility of trying to define Romanticism, so protean, he maintains, were its forms. See Trevor Levere, "Romanticism, Natural Philosophy, and the Sciences: A Review and Bibliographic Essay," *Perspectives on Science* 4 (1996): 463–88.

Initially, I think, we must distinguish *Naturphilosophie* from Romantic science—or rather, Romantic biology as that discipline of science with which I am concerned. The terms *Naturphilosophie* and *romantisch* are not the historian's confections; introduced in the early eighteenth century, they were reoriented by authors at the end of the century for specific semantic purposes. They have different though closely related histories; so I think clarity demands that we define them differently, even if in recent historical literature they are often used interchangeably. Based on their respective histories and the constituent ideas with which they were associated, I wish to suggest that Romantic biology be taken as a species of the wider genus of German nature philosophy. Thus all Romantic biologists were *Naturphilosophen*, but not all *Naturphilosophen* were Romantics, at least in the way I propose to use the terms. Romantic thinkers added aesthetic and moral elements to the content of ideas traveling under the rubric of *Naturphilosophie*. Friedrich Schlegel, as will be discussed in the next chapter, employed the term *romantisch* precisely to indicate a specific kind of poetic and morally valued literature. I will use the term to distinguish a type of science that retains this aesthetic and moral heritage.

Naturphilosophie

The principal figures of this study adopted a conception, given currency by Kant, Schelling, and Goethe, that regarded living nature as exhibiting fundamental organic types, often called "archetypes" (*archetypi, Urtypen, Haupttypen, Urbilden,* and the like); the four most basic animal structures usually discriminated were the radiata (for instance, starfish and medusae), articulata (insects and crabs), mollusca (clams and octopuses), and vertebrata (fish and human beings). These archetypes themselves could be nested within more fundamental types—that of the general animal or the general plant—with the most basic of all being the archetype of the organic per se. Looking in the other direction, more proximate archetypes, as realized in particular organisms, were seen to exfoliate extraordinary subtype variations. Thus the articulata bulged with numerous classes of insects and species of beetles beyond reckoning. These types also exhibited a progressive hierarchy—so, for instance, within the vertebrata, horses were thought more progressively developed organisms than fish.

Kant had maintained that the archetypal structure of organisms suggested that they had been produced by the very ideal they embodied. Such an ideal might reside only in an *intellectus archetypus*, a mind whose conceptions would be productive—the Divine mind. Yet Kant also held that the proper scientific analysis of nature required the investigator to employ

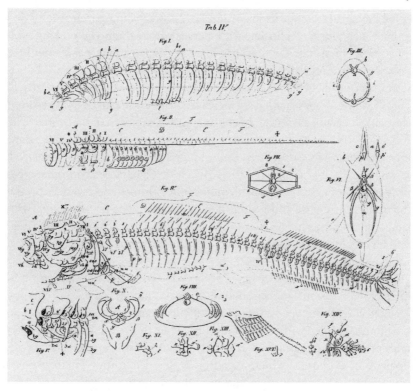

Figure 1.1 Illustrations of the vertebrate archetype (fig. 1) and of an ideal vertebra (fig. 3), from Carl Gustav Carus, *Von den Ur-Theilen des Knochen und Schalengerustes* (1828).

only the categories of Newtonian science. The necessary and universal laws of nature, including living nature, had to be parsed mechanistically—that is, so that an organism and its activities be understood as the determinate consequence of the operations of its parts. The Kantian biologist, then, should only deploy archetypal notions heuristically, *as if* organisms had been the products of an ideal plan, while yet searching for proper mechanistic causes. Schelling and Goethe—and those biologists following their lead—countered that if archetypes proved a necessary methodological assumption for the biologist, then there was no reason—especially on Kantian grounds—to argue that nature was not intrinsically archetypal, that is, essentially organic rather than mechanistic.

The *Naturphilosophen* usually invoked special causal forces to explain the instantiation of archetypes and their progressive variations. These scientists and philosophers, however, did not consider such forces incompatible with more mundane physical powers; rather, the forces were often conceived as special applications of physical powers (animal electricity and animal magnetism, for instance), emergent from them (*Lebenskraft, Bildungstrieb*, or natural selection) or even constituting them (e.g., Schelling's polar forces). Since the *Naturphilosophen* adopted the metaphysical position of monism, in which matter and *Geist* (understood indifferently as mind or spirit) were regarded as two features of the same underlying *Urstoff*, the causal activities of either had ultimately to express a unified force. The natural world—with its various organic types and their vertiginous varieties, its underlying substantial unity, and its particular forces—displayed, it was thought, higher-ordered patterns. Thus, for example, particular types of vegetation might be found at comparable latitudes and altitudes on the different continents; and similar animal forms might be associated with similar plant forms in the New and Old Worlds. Nature, to use Alexander von Humboldt's term, was a "cosmos"—a harmoniously unified network of integrally related parts.

The *naturphilosophisch* theory of archetypes spawned three different theories of their instantiation in nature. Initially, such theorists as Schelling characterized archetypes transcendentally, as features of *ideal reality* (not an oxymoronic phrase within this tradition). They then explained the appearance of archetypal variations in nature as a consequence of gradual development, or evolution, which instantiated the ideal forms. Theirs was not exactly Darwin's theory of evolution, rather a theory of *dynamische Evolution*, as Schelling termed it and Goethe adopted it. To the British mind of the early nineteenth century, the metaphysical position of Schelling and Goethe, grounded as it was in Spinoza's conception, exuded pantheism—a hardly less noxious form of atheism. Those in England who toed too closely the *Naturphilosophie* line, thinkers like Joseph Henry Green and Richard Owen, thus risked charges of irreligion. They consequently relocated archetypes in the Divine mind and regarded their appearance in nature as the result of God's creative activity. The final transformation of archetypal theory was undertaken by that superb English gardener Charles Darwin, whose conception of nature owed much to German Romantic sources. Utilizing his theory of evolution by natural selection, he rooted archetypal structures back in nature, not as abstract entities but as historical creatures. Darwin's evolutionary theory of archetypes was then replanted by Haeckel in native soil, where it flourished.

The *Naturphilosophen* commonly thought individual organisms and nature as a whole to be teleologically ordered. In the German context, this did not mean what it was taken to mean in the British context. Kant argued that we had to understand biological organisms *as if* they had been designed

them from one another—passions that transmogrified from lingering fasci-
nation, to erotic love, and finally to destructive hate. The Romantic "movement" or "school," then, is best conceived as constituted not by a group
displaying a unanimity of ideas but by sympathetically minded individuals, by thinkers whose mutually supportive considerations of philosophy,
literature, and science became enmeshed in the tangle of their personal and
professional relationships. Within their social and intellectual environments, particular experiences thus shaped their conceptual growth in di-
verging directions. Radical disjunction, however, was prevented by the
similarity of their intellectual heritages and by the co-adaptation of their
developing ideas to the views of one another and to those of their teachers,
friends, and, oppositionally, to their enemies. The implicit evolutionary
model that encourages my suggestion about how to understand the Romantic movement preserves the usual story—commonality of ideas, enough to
satisfy philosophers, but with an appreciation of particular differences,
which historians are anxious to retain.[4]

In conformity to our usual understanding of Romanticism, some of the
members of this early group turned decisively toward the night of cloudless
climes and starry skies, under which beauty revealed a more intuitive, emotionally marked, and even mystical path to reality's inner core. With the
poet Novalis, they desired a time

> When no more numbers and figures feature
> As the keys to unlock every creature,
> When those who join to sing and woo
> Know more than the deeply learned do.[5]

4. Recent historians have constituted the group of early German Romantics differently,
mostly by applying an ideational definition. Frank, for instance, represents the nucleus of the
group as consisting of the poets Friedrich von Hardenberg (Novalis) and Friedrich Hölderlin, and
the literary critic Friedrich Schlegel. Frank identifies members of the Romantic group by their essential commitment to two general ideas: anti-foundationalism and realism. This allows him to
include Hölderlin, who was never a part of the social-intellectual circle of Schlegel and Hardenberg, but to exclude Schelling, who was an intimate of the circle, sharing their common goals and
aspirations. Moreover, this ideational construction drops from view Caroline Böhmer and
Dorothea Veit, two women who exercised a huge intellectual and passional influence on the formation of the group and who ultimately destroyed it. Nonetheless, Frank's treatment is weighty
and extremely instructive. See Manfred Frank, Unendliche Annäherung (Frankfurt a. M.: Suhrkamp, 1997).

5. Novalis [Friedrich von Hardenberg], "Wenn Nicht Mehr Zahlen und Figuren," from studies
for his novel Heinrich von Ofterdingen, in Novalis Schriften, ed. Paul Kluckhohn and Richard
Samuel, 2nd ed., 5 vols. (Stuttgart: Kohlhammer, 1960–75), 1:344: "Wenn nicht mehr Zahlen und
Figuren / Sind Schlüssel aller Kreaturen, / Wenn die so singen, oder küssen / Mehr als die Tiefge-
lehrten wissen."

Two deeply learned brothers, the literary historians and critics August Wilhelm Schlegel (1767–1845) and Karl Friedrich Schlegel (1772–1829), nonetheless attempted to lower the consciousness of their friends, like Novalis, to a less ethereal plane, to the historical justification and philosophical explanation of the role of poetic literature in achieving this new kind of cognition. Their searching analyses and aesthetic criticisms, exemplifying scholarship of an exceptionally high order, also served as newly fashioned keys to unlock the secrets of man and nature. Their journal, the *Athenaeum*, which published the poetry and prose of their circle during the three years of the magazine's existence (1798–1800), created an institutional body that displayed this new perception of the Romantic.[6] Yet their intellectual accomplishments would not have had the purchase on the minds of their generation had they not also the ability to rekey and play out in their literary works those emotional chords that first resonated through their most intimate friendships, consuming loves, and despoiling enmities.

Karl Friedrich Schlegel, known simply as Friedrich to his friends, was the intellectual architect of the movement. In early essays he established the meaning *romantisch* would initially bear. He used the term to signify imaginative literature of a distinctively modern form, to be contrasted with the writings of ancient authors, especially Homer and the Greek lyric poets and dramatists. The word derives from the French *roman*, which referred to a story, usually a military tale of awful creatures, heroic knights, and chivalric love. When the word entered the German language, toward the end of the seventeenth century, it carried the meaning of *romanhaft*, novel-like, specifically the kind of story or attitude typical of the genre. It quickly came to indicate an action-filled and passionate adventure, as well as the wild, natural scenery that might be the setting for such a fanciful story.[7] By the

6. Margaret Stoljar provides an incisive thematic account of the literature published in the journal in her *Athenaeum: A Critical Commentary* (Bern: Verlag Herbert Lang, 1973). Ernst Behler focuses on the *Athenaeum*, but also extends his historical analysis to the various periodicals begun by the Schlegels. See his *Die Zeitschriften der Brüder Schlegel* (Darmstadt: Wissenschaftliche Buchgesellschaft, 1983).

7. Jacob and Wilhelm Grimm's *Deutsches Wörterbuch* cites this Latin definition as providing the original meaning of *romantisch*: "ex lectione quorundam romanticorum, i.e. librorum compositorum in gallico poeticorum de gestis militaribus in quibus maxima pars fabulosa est" [concerning the reading of certain novels, i.e., books poetically composed in French about military events in which the greatest part is fiction]. In Johnson's *Dictionary*, "romance" is defined as "a military fable of the middle ages; a tale of wild adventures in war and love." "Romantick" is comparably defined as "resembling the tales of romances; wild. . . . Fanciful; full of wild scenery." See Jacob Grimm and Wilhelm Grimm, *Deutsches Wörterbuch*, 16 vols. in 32 (Leipzig: Verlag von S. Hirzel, 1854–1952), 8:1155; and Samuel Johnson, *A Dictionary of the English Language*, abridged (1756) from H. J. Todd's corrected edition and reedited by Alexander Chalmers (Philadelphia: Kimber and Sharpless, 1842). See also Raymond Immerwahr, "The Word 'romantisch' and Its History," in *The Romantic Period in Germany*, ed. Prawer, pp. 34–63; and Pikulik, *Frühromantik*, pp. 73–79.

so that disparate parts functioned reciprocally as means and ends, and together contributed to the well-being of the entire individual. For example, the heart contracts for the purpose of circulating the blood, and the blood circulates for the purpose of supplying vital elements to the parts, including the heart, so that they might properly function to maintain the entire organism. After Kant, and especially because of the influence of Goethe and Schelling, biologists came to hold the teleological structure of nature not simply *as if* but as intrinsic: nature, whether in the individual or at large, really was purposively organized. But the *Naturphilosophen*, unlike the British natural theologians, did not appeal to a separate Creator who imposed final order on recalcitrant matter. Rather, they conceived nature in Spinozistic fashion—it was *Deus sive natura:* God and nature were one. This meant that the teleological structuring of biological organisms modeled the conceptual structuring of the ideas in terms of which nature was understood.

This organic conception of nature—given currency by Herder, Goethe, and Schelling—opposed the mechanical ideal stemming from Descartes and Newton. Within the tradition of *Naturphilosophie*, nature ceased to be mere product of the Creator's designs but itself became producer—of itself. Self-production and development revealed that nature moved from a simpler, less organized, earlier state to a more progressively developed later state. But this was only to say that nature became temporalized. Like a growing individual, it took on the form of a completely historical entity. The mechanistic conception of nature, by contrast, could not easily support intrinsic temporality: the clockwork mechanism that an intelligent Creator would produce was stable, coherent, and as perfect from the beginning as it would be at the end—save when it ran down a bit and needed a Divine rewinding.[14] The clockwork mechanism itself was, paradoxically enough, fundamentally atemporal and thus ahistorical. But nature as self-productive, as organic, could have a history. In the work of the *Naturphilosophen*, that history would be understood as inscribed in nature's very bones: they would argue, for instance, that individual organisms recapitulated the history of their species as they went through their own ontogenic development. The infusion of time into nature was not merely a necessary condition for the appearance of evolutionary theories; it constituted those very theories during the late eighteen and the nineteenth centuries.[15]

14. Laplace, it will be recalled, showed that Newton's fears about a destabilized nature came to nothing.

15. I am grateful to Robert Proctor for bringing me to see this important implication of organicism.

Romantic Biology

Those scientists to whom I refer as Romantic biologists generally accepted the metaphysical and epistemological propositions of *Naturphilosophie*. They took more to heart, however, Kant's analysis of the logical similarity between teleological judgment and aesthetic judgment, which he developed in the *Kritik der Urteilskraft* (Critique of judgment). Romantic biologists came to regard these two kinds of judgment as complementary approaches to nature, approaches that penetrated to the same underlying object. This meant that artistic experience and expression might operate in harmony with scientific experience and expression: the basic structures of nature might thus be apprehended and represented by the artist's sketch and the poet's metaphor, as well as by the scientist's experiment and the naturalist's observation. Further, Romantic biologists maintained, sometimes explicitly, often implicitly, that the aesthetic comprehension of the entire organism or of the whole interacting natural environment would be a necessary preliminary stage in the scientific analysis of respective parts: both in art and science, comprehension of the whole had to precede that of the parts—the theme Goethe played out for Schiller. Initially, for the biologist, then, ineffable aesthetic experience had to open the way to articulate scientific understanding. And for the reader, this meant that the sketches, drawings, figures, and metaphors that graced biological monographs could not be relegated to the status of dispensable pedagogic aids: images carried a scientific content often impossible to render precisely in words. Art became thus employed in the logic of scientific demonstration.

Romantic thinkers considered the activities of the scientist comparable to that of the artist, for both employed creative imagination. And when addressing nature, both found in their object a source of similar creativity. Nature's forms—various, unexpected, delightful, but exhibiting a deep unity—had to be regarded as creative expressions as well. Romantic biologists thus concluded that in nature "from so simple a beginning endless forms most beautiful and most wonderful have been, and are being, evolved."[16]

Kant also uncovered a deep logical similarity in the structures of aesthetic judgment and of moral judgment: in each we render a judgment on the basis of mere form (that of the structure of a beautiful object or of the structure of a moral maxim) and each such judgment produces a certain feeling claimed as valid for everyone. So, according to Kant, when evaluating a painting as beautiful or an act as moral, we simultaneously demand that others reach the same conclusion and experience the same regard to-

---

16. Charles Darwin, *On the Origin of Species* (London: Murray, 1859), p. 490.

ward the object or act. Because of these similarities, Schiller argued that the path to political freedom—a kind of moral freedom that respects the autonomy of a citizenry—had to be made through the fields of aesthetic experience. But for biology, given the aforementioned isomorphisms of judgment, this meant that the scientific and aesthetic comprehension of nature also involved a moral component. Romantic biologists thus understood nature to be the repository, not only of lawful regularities and aesthetic delights, but of moral values as well.

Along with nature, the individual self stood as a principal object of moral and aesthetic focus for the Romantics. Fichte's metaphysics laid the foundation and Schelling's transcendental philosophy constructed the framework for this concern. According to Fichte, the absolute self created and strove to develop, morally and aesthetically, both the empirical self and the nature that stood over against the individual. Schelling explored the various ways in which the empirical self and nature reflected each other and developed in intimate accord with each other. When a Goethe, a Humboldt, or a Darwin rambled along lovely littoral regions of the Mediterranean or ventured into the exotic jungles of South America, they discovered not only the sublime beauty of their surroundings but their own emerging selves as well. In this respect, nature became for Romantic adventurers the principal resource for the creation of the self—a self that ever hovered just on the horizon of their biological science. Given these preoccupations, the concept of development, of *Bildung*, helped channel biological research. As a result, most of the scientists I discuss showed an irresistible interest in theories of embryogenesis and species evolution, which they typically regarded as parallel and anchored in that deeper development of self and nature.

There are, of course, those thinkers in the nineteenth century—and today—whose intellectual skeletons display nary a Romantic bone. Arresting scenes of nature that would have moved Goethe, Humboldt, or Darwin to tears of febrile delight evoke only cool calculations or warm impatience. Such souls would be immune to the metaphysics and certainly the aesthetics that characterized the deeper structures of the Romantics' scientific thought. Fichte perceptively warned that the kind of philosophy adopted by individuals depended on their characters. Even highly ordered and rigorous systems of thought, he observed, were derivable from some first principle, which could not, of course, be demonstrated from within the systems themselves. The selection of a first principle, and thus the logical commitments it compelled, had to be a free choice of individuals.[17] From the per-

---

17. Johann Gottlieb Fichte, "Erste Einleitung in die Wissenschaftslehre" (1797), in *Fichtes*

spective of the other Jena philosophers and critics, such as Schelling, this
would be expected. Thought, they believed, does not dance naked in the
mind, in logically pure abstraction; rather, it must come imaginatively and
emotionally dressed. This suggests, correctly I believe, that only certain individuals would have been receptive to the metaphysical and aesthetic doctrines of Romanticism, since those doctrines would have been suited only
to a particular personal style. In its hyperbolic form, that style would be realized in the *genius* (a category given specific Romantic meaning by Kant
and Schelling); in its more attenuated form, though, it would reside in
other receptive individuals. After all, most intellectuals do, frequently
enough, respond aesthetically to given modes of thought, often before logical analysis.

This fusion of thought and feeling, along with the admonition of Fichte,
provides further reason for spending time on the biographies of the more
important individuals whom this history will treat. For these considerations
indicate vaguely, elusively, and, perhaps, maddeningly another criterion of
the Romantic: that to which the Romantic personality resonates. What is a
Romantic personality? One could answer by saying it is an individual who
pursues adventure, follows instincts, seeks to penetrate into the secret
depths of reality, responds to the beauties of nature, writes or reads poetry
and literature of a certain sort, and is often thrall to love. But it would be
better to say, it is Goethe in Rome, lying with his arms about his sleeping
mistress and composing a poem, while counting out its measures softly on
the vertebrae of her back.

These general definitions of *Naturphilosophie* and Romantic biology, of
course, lack specificity and a context that makes them more intelligible and
precise. They need to be rewoven within the mesh of the actual history that
gave rise to them. And that history begins with the Romantic movement in
Germany at the end of the eighteenth century.

Werke, ed. Immanuel Hermann Fichte, 11 vols. (Berlin: Walter de Gruyter, 1971 [1843–46]),
1:432–34: "There is no principle of decision possible for reason: for it is not a question of adding an
item in a series according to the rational principles governing the series; rather it is a question of the
beginning of the whole series which, as an absolute first act, depends only on freedom of
thought. . . . What kind of philosophy one chooses thus depends on what kind of man one is: for a
philosophical system is not a dead stick of furniture that one can lay aside or select; rather, it animates the very the soul of the man who has it."

Part One

The Early Romantic Movement in Literature, Philosophy, and Science

• ◆ • ◆ • ◆ •

Chapter 2

THE EARLY ROMANTIC
MOVEMENT

Perhaps a whole new epoch of science and art would be inaugurated
were symphilosophy and sympoetry to become so common and
deeply felt that there would be nothing odd were several people of
mutually complementary natures to create works in communion
with each other.

—Friedrich Schlegel, *Athenaeum Fragmente*, no. 125

The romantic mentality that ramified through Germany in the late eighteenth and early nineteenth centuries was nurtured in the friendships and
passions that held together the group that became known as "the early Romantics" (*die Frühromantiker*).[1] These individuals slipped away from the

---

1. Traditionally scholars have distinguished three phases of the Romantic movement: the early,
or Jena, Romantics, on whom I will focus; the middle, or Heidelberg, Romantics—including the
writers Achim von Arnim, Clemens Brentano, and the painter Caspar David Friedrich; and the
late Romantics, active in Vienna, Munich, and Berlin during the 1830s. This latter group retained
members of the first phase, such as the literary critic and historian Friedrich Schlegel and the
philosopher Friedrich Schelling, and added such writers as Johann Ludwig Uhland and E. T. A.
Hoffmann. Canonically, membership in the group referred to as the early Romantics is usually confined to the brothers Wilhelm and Friedrich Schlegel, their wives Caroline and Dorothea, the theologian Friedrich Schleiermacher, the poets and novelists Ludwig Tieck and Friedrich von
Hardenberg (Novalis), the theorist of the fine arts Wilhelm Wackenroder, and the philosopher
Friedrich Schelling. The list is not merely an artifact of literary historians—members of this group
had close personal ties and shared intellectual and artistic activities. Yet they certainly did not refer to themselves as "the early Romantics." That category serves certain historiographic purposes,
and its elements were chosen to meet those purposes. My own treatment will constitute the group
of early Romantics somewhat differently. For the orthodox account of the three phases of Romanticism, see Paul Kluckhohn, *Das Ideengut der deutschen Romantik*, 3rd ed. (Tübingen: Niemeyer
Verlag, 1953). The classic history of the early Romantic group is by Rudolf Haym, *Die romantische
Schule: Ein Beitrag zur Geschichte des deutschen Geistes* (Berlin: Rudolph Gaertner, 1870). Two recent, authoritative considerations of the early Romantic movement are Ernst Behler, *Frühromantik*
(Berlin: Walter de Gruyter, 1992); and Lothar Pikulik, *Frühromantik: Epoche, Werke, Wirkung*
(Munich: C. H. Beck, 1992). I have also found helpful: Oskar Walzel, *German Romanticism*, trans-
lation of 5th German ed., trans. Alma Lussky (New York: Putnam's Sons, 1932); Henri Brunschwig,
Enlightenment and Romanticism in Eighteenth-Century Prussia, trans. Frank Jellinek (Chicago: University of Chicago Press, 1974); Ernst Behler, *German Romantic Literary Theory* (Cambridge:

simpler rationalisms that dominated their native philosophical environment, especially as exemplified in the works of Christian Wolff, Gottfried
Wilhelm Leibniz, and the Immanuel Kant of the first *Critique*. They came
to disdain, with the sanguine intolerance of young revolutionaries, conventional and encrusted thought wherever they found it. The poet Joseph von
Eichendorff (1788–1857), then a student ready for the intellectual barricades, recalled in later years how they were regarded and what they strove
against:

> The common attitude, or rather enfeebled judgment, had been prosaic
> for so long that the Romantic approach was taken for a sacrilege against a
> debased human understanding and would be tolerated, at best, as a
> bizarre, youthful prank. The heavy wagon carrying the provisions of the
> meat and potatoes science moved slowly into the customary dock of a
> wooden schematism; religion had to assume reason and move with the
> rationalism of the schools; nature was dissected atomistically like a dead
> corpse; philology enjoyed itself like a childish old man slicing syllables
> and proposing endless variations on a theme that it had long forgotten;
> and visual art prided itself on a slavish imitation of so-called nature.[2]

The early Romantics were poets and painters, philosophers and historians, theologians and scientists, and mostly they were young. We usually
think of this group as forming a coherent movement, and readers have
certainly been seduced into adopting this view by critics and historians referring to the "Romantic school."[3] The Romantics, however, often appreciably diverged from one another in their conceptions of the operations of
sensation, imagination, and reason; and, indeed, they frequently assessed
the functions of these faculties differently at different stages of their own
intellectual developments. The powerful passions that held them together
refracted their philosophical commitments, but then eventually repelled

---

Cambridge University Press, 1993); Theodore Ziolkowski, *Das Wunderjahr in Jena, Geist und
Gesellschaft 1794/95* (Stuttgart: Cotta'sche Buchhandlung Nachfolger, 1998); and the collection *The Romantic Period in Germany*, ed. Siegbert Prawer (London: Weidenfeld and Nicolson,
1970).

　　2. Joseph von Eichendorff, *Erlebtes*, in *Joseph von Eichendorff Werke*, ed. Hartwig Schultz, 6 vols.
(Frankfurt a. M.: Deutscher Klassiker Verlag, 1993), 5:423.

　　3. Thus Heinrich Heine's *Die romantische Schule* (1835) (an unsympathetic treatment), in
Sämtliche Schriften, ed. Klaus Briegleb, 3rd ed., 6 vols. (Munich: Deutscher Taschenbuch Verlag,
1997), 3:357–504; and Rudolf Haym's *Die romantische Schule* (1870), a sympathetic treatment (see
note 1). Haym wrote during the latter part of the nineteenth century, when a negative reaction had
set in against the Heidelberg Romantics because of their religiously and politically conservative
character. Haym wished to recover the features of the early Romantic movement that he thought
formed "the conditions of development" and "the spiritual heritage" of German literature and philosophy of his own day. See ibid., pp. 4–5.

end of the eighteenth century, the term in common use bore many of the connotations with which we are familiar. Goethe captured several of those meanings in his early writing. His antihero, Werther, despairing over his beloved's rejection of him, tells her: "It is decided, Lotte, I want to die, and I write you that calmly, without Romantic expostulation" [*ohne romantische Überspannung gelassen*]. When Goethe returned from his journey to Italy in the late 1780s, he observed: "The so-called Romantic aspect of a region is a quiet feeling of sublimity under the form of the past, or, what is the same, a feeling of loneliness, absence, isolation."[8] Schlegel had in mind these various usages, but he wished to employ the term more specifically to describe a form of poetic literature developed in the modern period that expressed the subjective interests of the artist, that allowed conflicting elements to remain unresolved, that refused to restrain a freely playing and ironic imagination, that described the reactions and portrayed the sensibilities of individuals of common clay, usually the poet himself, and that focused on manners and times characteristic of a definite period in history. Ancient literature, by contrast, conformed to objective and formal principles of beauty, unified its various elements into a harmonious synthesis, described the activities of heroes and gods, transcended historically fixed times and places, and thus, indirectly, represented universal features of humanity.[9]

Schlegel's conception of the Romantic underwent an evolution during the late 1790s. In his early volume on Greek and Roman poetry (1795–97), he deemed modern literature to be inferior to classical literature, which latter, he believed, had achieved the most complete expression of beauty possible. But he quickly came to argue—after reading a monograph by Schiller—that a literature he would call *romantisch*, with its continual striving after the perfect realization of beauty, more conformed to the nature of man as a progressive being.[10] Ancient literature, the apotheosis of universal, formal beauty, thus belied human nature, the very essence of which melted away

---

8. Johann Wolfgang von Goethe, *Die Leiden des jungen Werthers* (1774), in *Sämtliche Werke nach Epochen seines Schaffens* (Münchner Ausgabe), ed. Karl Richter et al., 21 vols. (Munich: Carl Hanser Verlag, 1985–98): 1.2:280; and *Maximen und Reflexionen* (no. 181), in ibid., 17:749. Though the maxims were collected over a lifetime, this particular one was only published in 1821—hence it may already reflect coinages of the Romantic movement.

9. This is the theme of Friedrich Schlegel's *Über das Studium der griechischen Poesie* (1795–96), which was published as part of his monograph *Die Griechen und Römer: Historische und kritische Versuche über das Klassische Alterthum* (1797); see the *Kritische Friedrich-Schlegel-Ausgabe*, ed. Ernst Behler et al., 35 vols. to date (Paderborn: Verlag Ferdinand Schöningh, 1958–), 1:217–367.

10. The treatise he read was Schiller's *Über naive und sentimentalishe Dichtung* (On naive and sentimental poetry, 1795–96). Schlegel mentions the work in the preface to his *Die Griechen und Römer*; the body of the text was completed, however, before he read Schiller. I will discuss Schiller's treatise in chapter 11.

into that of an incomplete becoming, but with longings that drove the individual desperately toward the infinite. "Romantic poetry," Schlegel insisted in the famous fragment 116 of the *Athenaeum Fragmente*, "is progressive universal-poetry." "The Romantic mode of poetry," he proclaimed, "is still in the process of becoming; indeed, that is its very essence, that it eternally becomes and can never be completed." "Romantic poetry," he concluded, "is the only mode of poetry that is more than a mode, it is the poetic art itself; thus, in a certain sense, all poetry is or should be Romantic."[11] But Romantic poetry had a goal, an end point that literature of the time began to realize. This was the union of the distinctively modern mode of poetizing with the ancient, the merging of the greatest beauty possible consistent with human nature. Already in 1794, Friedrich wrote his brother: "The problem of our poetry seems to me to be the union of the essentially modern with the essentially ancient; if I indicate that Goethe, the first to have made a beginning for a wholly new period of art, has approached this goal, you will certainly understand what I mean."[12] Goethe's accomplishment thus became the model for the aspirations of Romantic literature.

The subtle shifts in Schlegel's attitude and the set of additional meanings *romantisch* would carry must be understood as the result not simply of more abstract definitions proffered by members of the Romantic circle, but of additional strands of personal feelings, political attitudes, aesthetic per-

---

11. Friedrich Schlegel, *Fragmente* (1798), in *Kritische Friedrich-Schlegel-Ausgabe*, 2:182–83 (fragment 116).

12. Friedrich Schlegel to Wilhelm Schlegel (27 February 1794), in *Kritische Friedrich-Schlegel-Ausgabe*, 23:185. In our century there has been a good deal of scholarly debate concerning Schlegel's shifting meaning for *romantisch*. Lovejoy, in a quite influential article, maintains that Friedrich Schlegel went through several stages in his evaluation of modern literature. Initially, prior to 1796, he was a classicist, deprecating modern poetry. Then he hit a transitional stage in 1796, after he had read Schiller's *Über naive und sentimentalische Dichtung*. Schiller, Lovejoy asserts, had shown Schlegel the potential in modern poetry and placed it in a more penetrating light. Then from 1797, so Lovejoy maintains, Schlegel endorsed modern or Romantic poetry as the highest kind of artistic endeavor, and so expressed that new view in the *Athenaeum Fragmente*. See Arthur Lovejoy, "On the Meaning of 'Romantic' in Early German Romanticism," *Modern Language Notes* 31 (1916): 385–96; 32 (1917): 65–77. Behler adds a further stage, holding that Schlegel finally felt that classical and Romantic modes of poetry had to be synthesized to achieve the highest perfection of beauty possible for progressive human nature. See Ernst Behler, "The Origins of Romantic Literary Theory," *Colloquia Germanica*, nos. 1–2 (1968): 109–26. Behler has subsequently rendered the matter even more complex in his *German Romantic Literary Theory*, pp. 72–180. Schlegel was undoubtedly encouraged to develop his position through the reading of Schiller's essay and through his experience with the poetry of that individual who embodied both classical and Romantic attitudes—Goethe. But we can see from the letter I have just quoted that Schlegel, even before reading Schiller, had foreseen that the highest goal of literature would be to join the classical with the modern.

ceptions, philosophical considerations, and scientific ideas that bound together the cognitive and emotional experiences of the individuals whose lives intersected at the end of the eighteenth century. Only by examining the development of their intellectual and passional lives—the ways in which individual experiences electrified their cognitive connections—can one understand the multifaceted meaning of the Romantic and how its elements became interlaced through early-nineteenth-century biology, giving it the particular structure it had. The intellectual development and fluctuating friendships of the Schlegel brothers offer entrance into the charmed sphere of the early Romantics.

Wilhelm and Friedrich Schlegel

The brothers were the youngest sons of Johann Adolf Schlegel (1721–1793) and Johanna Christiane Schlegel (née Hübsch, daughter of a mathematics professor).[13] Their father served as Lutheran pastor at the Marktkirche in Hannover and the Court Church of Neustadt (northwest of Hannover) and interested himself in matters literary and philosophical, as did several of his seven children. August Wilhelm—known to his family as Wilhelm—excelled as a gymnasium student and did brilliantly in literary studies at the university in Göttingen, where he matriculated in 1786. Friedrich, however, initially displayed little academic ability. He was sickly, somewhat recalcitrant, and aimlessly melancholic. He had the soul, we would say, of a Romantic, but a mind, his father rather hoped, that might be bent to business. In 1788 the young Friedrich was apprenticed to a banker in Leipzig, but within six months no one could doubt the failure of the effort. Friedrich was saved by his brother Wilhelm, who introduced him to Greek language and literature; and this seems to have completely altered his attitudes and kindled his ambitions. Under his brother's tutelage, Friedrich unveiled a considerable gift for tongues, quickly mastering Greek and Latin (as well as, later, Spanish, English, French—their modern and medieval versions—Sanskrit, and Persian). But Friedrich's desires pushed him beyond the goal of simple technical proficiency. He wanted to understand the minds that originally used these languages and the historical contexts that shaped their perceptions. In 1790 he joined his brother at Göttingen, where he ostensibly followed the curriculum in jurisprudence, the shadow of his father's hopes. He was usually discovered, however, in lectures on literature and language, especially those of his brother's mentor,

13. For further biographical details, see Klaus Peter, *Friedrich Schlegel* (Stuttgart: Metzlersche Verlagsbuchhandlung, 1978), as well as documents listed in the next several notes.

Figure 2.1

Friedrich Schlegel
(1772–1829) at age
twenty-two, sketched by
his love at the time,
Caroline Rehberg.
(Courtesy the Staatliche
Museem zu Berlin.)

the distinguished philologist Christian Gottlob Heyne (1729–1812). In summer of 1791, Wilhelm finished at university and became a tutor to a rich banking family in Amsterdam, while Friedrich entered the university at Leipzig, further to pursue, from a distance, his legal studies. He was initially earnest in his efforts, writing to his brother: "I regard the study of law much more seriously than you—it seems to me to accord more with middle-class [*bürgerliche*] destiny."[14] He quickly, however, fell into the arms of literature, philosophy, history, and fascinating women.

During their year together at Göttingen, both brothers suffered the pangs of slightly requited love. Friedrich pursued Caroline Rehberg, who drove him into a melancholic depression[15] and would later make an appearance as a sweet innocent in his "dissolutely Romantic" [*liederlich romantischen*] novel *Lucinde* (1799).[16] Wilhelm attempted to encourage the

14. Friedrich Schlegel to Wilhelm Schlegel (21 July 1791), in *Kritische Friedrich-Schlegel-Ausgabe*, 23:17.

15. Friedrich Schlegel to Wilhelm Schlegel (26 December 1792), in ibid., p. 75: "My first love was R. [Caroline Rehberg] . . . and she finally gave me a big shove into the deepest melancholy, into which, during the year in Göttingen, I sank ever deeper and out of which I only slowly and with some difficulty have climbed."

16. That was Heinrich Heine's description of the novel in his *Die romantische Schule* (in *Sämtliche Schriften*, 3:383). A convenient edition of Schlegel's novel, with added critical reaction

affections of a young widow, the most remarkable Caroline Michaelis Böhmer, who would play a dramatic role in the lives of both brothers. At Leipzig Friedrich continued the passionate pursuit of literature and love, detailing in letters to his brother his successes in the one and failures in the other. His movement along the trail of both undoubtedly intensified because of the new friend he made at the university, the poet Friedrich von Hardenberg (1772–1801), later known as Novalis.

Novalis: the Romantic Personality

The Friendship of Friedrich Schlegel and Friedrich von Hardenberg

Friedrich Schlegel met the nineteen-year-old Hardenberg, already a published poet, in January 1792, when they both were supposedly following the curriculum in law at Leipzig. Hardenberg had just come from a year at Jena, where he submitted to the spell of that greater master Friedrich Schiller— the reason his father pulled him back to Leipzig and serious study of law. The young poet, with dark golden hair flowing down to his shoulders and delicate features, left his imprint on the heart of his friend, as this description, which Friedrich supplied to his brother, attests:

> Fate has delivered to my hands a young man who could have come from the absolute. He pleases me greatly and I him. He has opened wide the inner sanctum of his heart to me, and I've taken up my place therein and search. A still very young man—a nice slim figure, a fine face with dark eyes, a magnificent expression when he reads something beautiful with fire—an indescribable fire—he reads three times more and three times as fast as we others—the quickest mind and receptivity. The study of philosophy has given him a luxurious lightness with which to form beautiful philosophical thoughts. He doesn't go after truth, but beauty—his favorite authors are Plato and Hemsterhuys.[17] On one of our first evenings,

---

during the last two centuries, is *Friedrich Schlegels Lucinde und Materialien zu einer Theories des Müssiggangs*, ed. Gisela Dischner (Hildesheim: Gerstenberg Verlag, 1980). The novel was judged during its time as by turns scandalous and boring—scandalous because of its suggestion of casual affairs and deprecation of bourgeois marriage. The novel, though, is a dithyramb to the ideal object of (male) love, a woman in whom one can find the quintessence of sexual and intellectual satisfaction. Schiller, who had a long-smoldering dislike for Friedrich, wrote Goethe that "the work, moreover, is not to be read straight through, for the hollow drivel will make you sick. . . . [T]his work is the height of the modern—unformed and unnatural." See Schiller to Goethe (19 July 1799), in *Der Briefwechsel zwischen Schiller und Goethe*, ed. Siegfried Seidel, 3 vols. (Munich: Verlag C. H. Beck, 1984), 2:241.

17. Franz Hemsterhuis (1721–1790) was a philosopher of aesthetics who had a great impact on the Schlegel brothers.

he lectured me with wild fire in his eyes on the idea that there was nothing evil in the world and that we were again approaching the golden age.
I've never seen such exuberance of youth. His sensitivity displays a certain chasteness but its foundation in his soul does not stem from inexpe-
rience. For he is very often to be found in society (he'll soon be
acquainted with everyone)—he had a year in Jena where he met the
beautiful spirits and philosophers, especially Schiller. Yet he was entirely
a student in Jena, and, as I hear, often in fights.[18]

The poet reciprocated these admiring sentiments. Hardenberg looked to
the "older" student—the difference in their ages (two months!) could
hardly be measured by the calendar—as a fount of deep learning in the clas-
sics and philosophy. He wrote Schlegel in summer of the next year: "I won't,
I think, ever see another man like you. You have become for me the high
priest of Eleusis. I have come to know heaven and hell through you—
through you I have eaten of the tree of knowledge."[19] Schlegel would discuss with him philosophical ideas from Kant to Schiller and the poetry of
the Greeks, Shakespeare, and Goethe, usually with Schlegel taking the tutorial upper hand, except on questions of Schiller's aesthetics and poetry.[20]
In their letters they hardly ever mentioned the law, and that subject soon
slipped entirely from Schlegel's formal course of studies.

Schlegel and Hardenberg also abetted each other in love. Indeed, they
had love affairs with two daughters of the same prominent Leipzig family,
Hardenberg with the seventeen-year-old Julie Eisenstuck and Schlegel
with her twenty-four-year-old sister Laura. Schlegel's affair was complicated by the fact that the lively, beautiful, and vain Laura had a husband—
a banker, dreary enough, however, not to be a great obstacle.[21] For both

18. Friedrich Schlegel to Wilhelm Schlegel (January 1792), in *Kritische Friedrich-Schlegel-Ausgabe*, 23: 40. Jena was known for the rowdiness of its students. See Theodore Ziolkowski, *German Romanticism and Its Institutions* (Princeton: Princeton University Press, 1990), pp. 228–36.

19. Friedrich von Hardenberg to Friedrich Schlegel (first half of August 1793), in *Kritische Friedrich-Schlegel-Ausgabe*, 23:115.

20. Hardenberg had been completely captivated by Schiller—the poet, the philosopher, and the personality. Just after Hardenberg had left Jena for Leipzig and learned of an illness Schiller contracted, he wrote his philosophy professor, the Kantian Karl Reinhold, about the virtues of Schiller, which the good colleague undoubtedly absorbed with sangfroid: "Oh, when I only men-tion Schiller's name, what an army of feelings arises in me; how various and rich are the features that assemble into that single, magnetic picture of Schiller and how they vie as magical spirits to perfect the magnificent portrait. . . . Schiller, whose soul nature seems to have formed *con amore*. . . . Schiller in whose nature so much art, and in whose art so much nature. . . ." The letter does go on. See Hardenberg to Karl Reinhold (5 October 1791), in *Novalis Schriften*, ed. Kluckhohn and Samuel, 4:93.

21. Richard Samuel's ingenious detective work uncovered the family identity of Laura and Julie; they were the daughters of Johann and Susanna Eisenstuck—he was a textile manufacturer

Figure 2.2

Friedrich von Hardenberg
(1772–1801), who wrote
under the name Novalis.
Copper etching by
E. Eichens after a portrait
by an unknown painter,
from Novalis, *Schriften*
(1837–46), edited by his
friends Friedrich Schlegel
and Ludwig Tieck.

friends their love relationships ended badly. The Laura of Schlegel's dreams threw him over and even ridiculed him within her circle of friends. "I sought revenge," Friedrich wrote his brother. "The paradise of which I had dreamed a few days before and which had been offered me—well, I would rather have had her blood."[22] It did not come to blood, either hers or his own—her sin was not quite sufficient, and honor, he explained to his brother, prevented him from the suicide he had contemplated.[23] Hardenberg's relationship foundered in a different, though no less ignominious way. His father, the Baron von Hardenberg, forced his son to break off with a woman of the petite bourgeoisie.[24]

Hardenberg left Leipzig for Wittenberg, where in summer of 1794 he completed his law degree with highest marks. Though he began the practice of law in one of its more mind-numbing aspects, taxes, his life took on

in Annaberg. Laura had married a banker in Leipzig, Christian Limburger. See Samuel's "Friedrich Schlegel's and Friedrich von Hardenberg's Love Affairs in Leipzig," in *Festschrift for Ralph Farrell*, ed. Anthony Stephens et al. (Bern: Peter Lang, 1977), pp. 47–56.

22. Friedrich Schlegel to Wilhelm Schlegel (19 and 27 February 1793), in *Kritische Friedrich-Schlegel-Ausgabe*, 23:83.

23. Ibid.

24. Friedrich Schlegel to Hardenberg (middle of May 1793), in ibid., 23:94. Schlegel reported to Hardenberg that the baron had gathered a group of cronies at Auerbach's tavern and proclaimed, "[Y]ou had wanted to marry a commoner [Bürgerliche], the sister of a businessman's wife. He spoke of you with great passion and continuous curses."

the pattern that would come to define the Romantic personality—a definition his friends would lovingly hone after his death. He continued to write
poetry and correspond with Schlegel over literature, philosophy, and love;
he engaged in many flirtations; and he dreamed of his destiny. That destiny
arrived in November 1794. While conducting some tax work, he was received by the family of the cavalry captain Johann von Rockenthien at
Schloss Grüningen (located a few miles from Erfurt). He met and "within a
quarter of an hour," he told his brother Erasmus, fell in love with the captain's twelve-year-old stepdaughter, Sophie von Kühn.[25] He was twenty-
two at the time. She was a beautiful and sweet child, who seems, from her
little-girl letters, to have had a genuine love for this older, brotherlike figure. She may have been as attracted, though, by his kind attention and long
hair; at least that is suggested by a letter she wrote thanking him for a small
jewelry case and a lock of his golden brown tresses:

> Dear Hardenberg,
> I thank you very much for your letter secondly for your hare and thirdly
> for the little kase whiches made me happy. You ask me whether you can
> write me? You can be sure that it allwaes is nice to get a letter from you to
> read. You know dear Hardenberg I cannot write long.
>
> Sophie von Kühn[26]

Hardenberg's relation to Sophie, comforted by the acceptance of her
family, quickly reached a climax: within six months, two days short of her
thirteenth birthday, she agreed to marry this ardent, playful, and decidedly
melancholic suitor. Through the next two years, Hardenberg pined after
Sophie, fantasized their life together, and waited for her to grow up. She
was, from all accounts, a charming, delicate girl, though with the odd habit
of smoking tobacco. Her large embracing family of the lesser nobility was
only a notch or two below the social rank of the Hardenbergs. And she was
sexually safe, yet undoubtedly erotically alluring for a fellow who seemed
constantly to be so aroused. Her unformed, innocent nature made of her a
young girl about whom poetry could be composed and meant; she was a tal-
isman of the poet's purer self. Hardenberg stayed often with the captain's

---

25. Hardenberg's brother Erasmus attempted to reason him out of this infatuation, without, fortunately, any success. See Erasmus von Hardenberg to Friedrich von Hardenberg (28 November
1794), in *Novalis Schriften*, ed. Kluckhohn and Samuel, 4:367.

26. Sophie von Kühn to Friedrich von Hardenberg (17–19 November 1795), in ibid., 4:408. In
the translation, I have tried to capture Sophie's misspellings and diction.

Figure 2.3

Sophie von Kühn
(1782–1797), beloved of
Friedrich von
Hardenberg.
Contemporary
watercolor. (Courtesy
Museum im Schloss
Weissenfels.)

family to be near Sophie and to coax her into thoughts of their married life together.[27]

In fall of 1796 Sophie fell quite ill. Likely she had contracted tuberculosis, and it migrated to secondary sites of infection. She underwent a series of painful operations at Jena, without, of course, any effective anesthetic or satisfactory outcome. Hardenberg made her plight known to all of his friends—even Goethe came to visit her in the hospital. Hardenberg venerated Goethe almost as much as he did Schiller; after the great poet's kindness, the young lover's devotion reached a new plane, as he wrote a friend in April 1797:

> Goethe's appreciation of Sophie's sublime form has made me happier than all of his excellent works. I now truly love him; he belongs to my heart. I won't hide from you that I could not regard him the Apostle of beauty if the mere sight of her had not affected him. It is certainly not

27. In a thorough and quite convincing analysis, William Arctander O'Brien portrays Hardenberg's relationship to Sophie as erotically charged but without sexual risk—perhaps the more charged because of the safety. O'Brien offers the most comprehensive account of Hardenberg's life and thought in English. See William Arctander O'Brien, *Novalis: Signs of Revolution* (Durham, N.C.: Duke University Press, 1995).

passion—I feel it too unremittingly, too coolly, too deeply in my soul, that she was one of the most noble, most ideal forms which has ever graced the earth or will.[28]

Sophie von Kühn died on 19 March 1797, two days after her fifteenth birthday. Hardenberg's grief, carried by his love, broke through all restraints, something he perhaps need not have explained to his friend Karl Woltmann, who also visited Sophie in the hospital:

> It is my tragic duty to report to you that Sofie is no more. After unspeakable suffering, which she bore in exemplary fashion, she died the 19th of March at 10 o'clock in the morning. . . . Three weeks ago I saw it threatening. Evening has surrounded me, even as I still gaze into morning dawn. My grief is boundless, as is my love. For three years she was my hourly thought. She alone bound me to life, the land, and my work. With her that is now all gone, since I cannot get hold of myself any more. It has become evening, and it seems to me that I should leave early, and since I would like to be at peace and see around me only well-meaning faces, I wish to live completely in her spirit, to be soft and content, as she was.[29]

For many months after Sophie's death, Hardenberg lived in sorrow and in love. He remained for several years transfixed by the tragedy, which itself became a kind of legend of Romanticism.[30] As truly awful as this event was, it yet stimulated one of the most sublimely beautiful poems to be associated with—indeed, becoming emblematic of—the early Romantic movement. The *Hymnen an die Nacht* was published under the name "Novalis" in the Schlegels' *Athenaeum*.[31]

Novalis's *Hymns to the Night*

Hymns to the Night arose, it seems, out of two kinds of experience: Sophie's death and transfiguration, and Novalis's reading of the philosophers Johann

28. Hardenberg to Karl Ludwig Woltmann (14 April 1797), in *Novalis Schriften*, ed. Kluckhohn and Samuel, 4:222.

29. Hardenberg to Woltmann (22 March 1797), in ibid., 4:206. A dissonant note plays through this event, however. In his letter to Woltmann, Hardenberg indicated that he could not bear to stay with Sophie during the final days: "Eight days before her death I left her; I had the strong conviction I would not see her again—It was beyond my strength to look on helplessly at the terrible struggle of this crushed flower of youth, the fearful anxiety of that heavenly creature."

30. O'Brien, in his *Novalis*, has many interesting things to say about the role of Sophie's death in providing a tone for the Romantic movement.

31. The poem appeared in the last number of the *Athenaeum* to be published, in August 1800. The poem, though, had a long gestation, beginning in spring 1797. It is reprinted (both in manuscript form and published form) in *Novalis Schriften*, ed. Kluckhohn and Samuel, 1:130–45.

Gottlieb Fichte (1762–1814) and Friedrich Wilhelm Joseph Schelling (1775–1854).[32] Beginning in 1795 and extending throughout the period of his relationship with his inamorata, Novalis undertook an extensive examination of Fichte's transcendental idealism; and in 1797, just after Sophie's death, he began a serious study of the work of Fichte's protégé Schelling. The supreme effort of the ego, according to Fichte, is to reach down into the depths of its own subjectivity, to uncover the absolute ego, that all-encompassing consciousness *in* which and *for* which the empirical ego and the natural world exist. In absolute subjectivity, all mundane reality, including the individual's consciousness and that of others, is taken up and united. Schelling, in early agreement with Fichte, emphasized the reciprocal relationship between the finite self and nature—both came into being simultaneously as the product of absolute subjectivity. (I will take up, in greater detail, the philosophical systems of Fichte and Schelling later in this chapter and in the next.)

Both Fichte and Schelling, in Novalis's estimation, allowed to consciousness an immediate awareness of its own foundations in absolute subjectivity, a knowledge of the condition of self-knowledge. The poet perceived, however, dangerously fractured struts in this bridge to absolute subjectivity. For the absolute ego to represent itself as identical from moment to moment, each representation would have to be compared to the nonrepresented ego, the ego supposedly doing the comparison. But how could such a comparison be made, a representation with the nonrepresented? Further, each representation must be different from the next, else you would not have several representations to compare. The idealists, Novalis concluded, have thus "abandoned the identical object in order to rep-

---

32. Hardenberg studied Fichte's philosophical works, especially the *Wissenschaftslehre*, intensely from 1795 to 1797. His notes on Fichte are collected in *Novalis Schriften*, ed. Kluckhohn and Samuel, 2:104–359, and conveniently in *Novalis Werke*, ed. Gerhard Schulz, 3rd ed. (Munich: Beck, 1989), pp. 293–318. Kathleen Komar develops in convincing argument the parallels between Fichte's philosophy of consciousness and Hardenberg's poem in "Fichte and the Structure of Novalis's 'Hymnen an die Nacht,'" *German Review* 54 (1979): 137–44. Her argument is, however, incomplete. She neglects the fact that after Sophie's death, Hardenberg turned to Fichte's follower Schelling as a further source of philosophic comfort. Schelling began as a disciple of Fichte, but in his *Ideen zu einer Philosophie der Natur* (1797) he began moving away from the concerns of his master. In that book the phenomenon of light becomes the *Urstoff* out of which the material world evolves; and in Novalis's poem, light serves that same function—it becomes consciousness estranged in nature. At this point that last idea must sit as a bright but opaque lump. For some clarification, see the discussion of Schelling's *Naturphilosophie* in the next chapter. Charles Larmore succinctly and elegantly portrays the gist of Novalis's study of Fichte in "Hölderlin and Novalis," in *The Cambridge Companion to German Idealism*, ed. Karl Ameriks (Cambridge: Cambridge University Press, 2000), pp. 141–60.

resent it." In truth, they can only provide "a philosophical parallelism," not an awareness of identity.[33]

Novalis did follow, however, a narrower passage to ultimate subjectivity. He detected a *feeling* of dependence on the absolute ego, if not a reflective awareness of it. Yet that feeling, which "proceeds from the infinite [*Unbeschränkte*]," could become a more definite presentiment. He believed the "godly beauty" of the absolute ego could be sighted in the "magic mirror" of poetry.[34]

These philosophical abstractions became the forms into which Novalis infused a lived epiphany, the feeling of transcendent union of two egos in the absolute, that he experienced during a visit to Sophie's grave a month after her death. The union, though, had its carnal side, an apotheosis of sexual desire. He recorded the event in his diary:

> I received a letter from [Friedrich] Schlegel with the first part of the new translation of Shakespeare [his brother's translation of *Romeo and Juliet*]. After dinner I went walking—then coffee. The weather was cloudy—first thunderclouds, then overcast and stormy. I felt very sexual [*sehr lüstern*]. I began to read the Shakespeare, and read myself into it. In the evening I went to Sophie. I was indescribably happy—a flashing moment of enthusiasm—I blew away the grave like dust—a century was a moment—her closeness was tangible—I believed she would always walk on.[35]

33. Hardenberg, [*Fragmente und Studien bis 1797*], in *Novalis Werke*, ed. Schulz, p. 293 (2).

34. Ibid., p. 293 (1). See Larmore, "Hölderlin and Novalis," pp. 153–56.

35. Hardenberg, *Journal* (entry for 13 May 1897), in *Novalis Schriften*, ed. Kluckhohn and Samuel, 4:35–36. Hardenberg's mention of feeling "very sexual" was one of a cascading series of similar remarks, usually in connection with thoughts of Sophie, recorded in his journal. Just the day before (12 May), he began his entry with "I was aroused with lasciviousness [*Lüsternheit*] from early morning to afternoon." He then indicates that in the evening of the thirteenth he visited Sophie's grave (ibid., 4:35). The passages in the journal from the time of Sophie's death to that of the grave scene track Hardenberg's fluctuating internal emotions: sometimes he simply cries; sometimes his head is clear and he is cheerful; he often says that he has eaten too much; he frequently feels sexual arousal and "manliness"; and always he thinks about Sophie. The importance of Hardenberg's experience at the grave site for stimulating the third section of his poem has long been observed, though sometimes disputed. See, for instance, the very instructive account of the poet's life and works, and the role of the grave episode, by Friedrich Hiebel, *Novalis: Deutscher Dichter, Europäischer Denker, Christlicher Seher*, 2nd ed. (Bern: Francke Verlag, 1972), p. 225. Hiebel also discusses the influence of Fichte on Hardenberg's thought; see pp. 125–48. Gerhard Schulz thinks the emotional impact of the grave experience could not be the progenitor of the poem, given its long gestation period. See Gerhard Schulz's commentary in his edition of *Novalis Werke*, pp. 620–21. Some of the phrasing from Hardenberg's journal entry is employed in the poem, and the emotional resonance seems exactly right, the long period of the poem's composition notwithstanding. Schulz has provided a highly nuanced analysis of Hardenberg's life and work in brief compass. See Gerhard Schulz, *Novalis, mit Selbstzeugnissen und Bilddokumenten*, 3rd ed. (Reinbek bei Hamburg: Rowohlt, 1993).

In Novalis's poem transcendental philosophy, the erotic remolding of Sophie in his thoughts, and the moving experience of the graveside—all became transmuted into an extraordinarily complex and delicate allegory of the relationship between day and night.[36] In the first two of the six hymns, the surface light of the sun, which seems to encompass the real world, evanesces when day is done and the heralds of evening—"the golden flow of the grape" and "the brown juice of the poppy"—make way for the night, whose grasp touches every "tender girl's breasts and makes a heaven of her womb." The night embraces the profound relationship of nature and other human creatures with the finite self of the poet. As the poet slips into the depths of night, empirical consciousness passes through into absolute subjectivity, that "vast heavenly expanse in which each shining star seems to us like the infinite eyes the night has opened in us."[37]

The third hymn sings of that transcendental night of the soul when the poet is reunited with his beloved:

The region reigned visibly above—over the region my newborn spirit glided free. The mound turned into a cloud of dust and through the cloud I saw the transfigured features of my beloved—in her eyes eternity rested—I grasped her hand and my tears became a glittering, unbreakable band. Centuries, like thunderstorms, flew into the distance—I shed on her neck enchanted tears for a new life.[38]

The union of the poet with his beloved occurs in the deeper reality, in the absolute subjectivity of the night, where all life and creative force dwell: "My secret heart remains true to the night and to her daughter, creative love."[39] The last part of the poem recounts the birth, death, and resurrection of Christ, who now becomes identified with the beloved. Sophie-Christ is entreated: "O, suckle me in, my beloved, so deeply that I can slip into that long sleep [of death] and into love."[40]

Novalis's poem expresses one of the characteristic features of Romantic thought—that beneath the dazzling surface of empirical nature, where all seems chaotic, variable, and mortal, lies a more enduring reality, ultimately

---

36. It is often thought a mistake to interpret literature biographically. After all, poetry and novels are created; they are fictions. But the imagination works over experience, and Novalis gives us cause.

37. Hardenberg, *Hymnen an die Nacht*, in *Novalis Schriften*, ed. Kluckhohn and Samuel, 1:133–34 (quotations are from the first and second hymns).

38. Ibid., third hymn, 1:135.

39. Ibid., fourth hymn, 1:137.

40. Ibid., 1:139: "O! sauge, Geliebter, / Gewaltig mich an, / Dass ich entschlummern / Und lieben kann."

infinite in essence, sublime in effect, and triumphant over death. The ecstasy of personal love would be the common means taken by members of
the Romantic group to penetrate the surfaces of life and reach the secret recesses of reality.[41] And among the several individuals, the geometry of their
love affairs might well suggest they had slipped into a new dimension. But
love unadorned, merely secreted through glands and adhering to the flesh,
lacked the cognitive grasp required to hold, if only for a moment, the infinite after which they longed. Love variously rendered through mind became the means. The Romantic poets strove to gain deeper and more
enduring experience of this underlying reality through the loving beauty of
metered language. With comparable intent, the philosophers tried to think
their way into the infinite and thus secure it by reason's cords; and the scientists synthesized the efforts of both by joining concrete experience with
systematic analysis in an attempt to uncover and establish the underlying
laws of infinitely creative nature. Goethe's biological archetypes (discussed
in chapters 10–12) express those creative forces that metamorphose into
the variegated phenomena of nature's shining surface. Darwin's *Origin of
Species* (chapter 14) also constructs nature as an infinitely creative, moral
force operating according to simple unifying laws that lie beneath the plane
of the great variability of life. Both Goethe and Darwin rejected the assumption that nature, like Locke's spinning jenny, creaked on in mechanical fashion.

The Poet as Mining Engineer

After Sophie von Kühn's death, Hardenberg's life took several odd turns
along a downward spiral. His father, who directed the salt mines of Saxony,
wanted his son to join its administration. He insisted that this poet-lawyer
now attend the famous mining academy at Freiberg. The great professorial
attraction of the institute was Abraham Gottlob Werner (1750–1817).
Werner taught a doctrine that became known as Neptunism, which held
that the various rock strata circling the globe had precipitated out of oceans

41. Friedrich Schlegel, for instance, thought of love as the human drive that arose from the mutuality of free men. He certainly believed it could move men cognitively beyond the bounds of conventional (i.e., Kantian) understanding. "Love is the enjoyment of free men," he proclaimed, "and
only man is its object. Since with one alone there can be no mutuality, there is no love without a responding love. It is, indeed, not madness to embrace all with love and to be in unity with nature.—
That human drive has an overflow of goodness, sprit, and abundance; human understanding has a
gap beyond the border of knowledge. Every overflow fills this gap and produces a representation of
a higher being and an inclination toward God." See Friedrich Schlegel, "Über die Grenzen des
Schönen," in *Kritische Friedrich-Schlegel-Ausgabe*, 1:40–41.

created by universal flooding. The doctrine, though not designed to accord with the biblical story of the Noachian tides, nonetheless acquired this association, which gave it something of an edge in competition with Vulcanism, the theory maintaining that the world was created by volcanic fire, a conception more congenial to French tastes. Hardenberg spent a year and a half under Werner's tutelage and also studied with Johann Wilhelm Ritter (1776–1810), who introduced him to the more outré aspects of natural science, particularly electrical physiology and alchemy. These earthly pursuits, as well as some persistent criticism rendered by Friedrich Schlegel, likely dampened Hardenberg's enthusiasm for Fichte's sublime subjectivism; in any case, his interest in the philosopher began to evaporate during this period. He found a new philosophical direction in the nature philosophy of Schelling, which he had begun reading in spring of 1797.[42] His interest in Schelling, though, also deadened rapidly, as Friedrich Schlegel poured poison into his ear about the boy-philosopher.

While at the Bergakademie, Hardenberg met another young woman of delicate health, Julie von Charpentier, the daughter of the inspector of mines in Saxony. She fell deeply in love with him, though his own deeper melancholy, disguised as poetic charm, masked other psychological states. Nonetheless, he became secretly engaged to her. He confided his poignant state to Friedrich Schlegel:

> I have much to tell you—the earth seems to want to hold me for a little while longer. That relationship, which I mentioned to you, has become more intimate and binding. I see myself more loved than I have ever been loved before. The fate of a young woman, *very worthy of love*, hangs on my decision—and my friends, parents, and siblings need me to make of it more than I do. A very interesting life seems to await me—but were I to tell the truth, I would rather be dead.[43]

In fourteen months he would be granted his wish.

After Hardenberg's death, his friends Friedrich Schlegel and Ludwig Tieck produced a small two-volume collection of his work.[44] Tieck prefaced the third edition with a memoir of the poet's life, which made the love and tragedy that beset it the very elixir that would intoxicate subsequent gener-

---

42. He wrote Friedrich Schlegel (3 May 1797) that "Schelling's philosophy of nature finds in me a very curious reader." See *Novalis Schriften*, ed. Kluckhohn and Samuel, 4:226.

43. Hardenberg to Friedrich Schlegel (20 January 1799), in ibid., 4:273; emphasis in original.

44. Friedrich von Hardenberg, *Novalis Schriften*, ed. Ludwig Tieck and Friedrich Schlegel, 5th ed., 3 vols. (Berlin: Reimer, 1837–46). The first edition (1820) was in two volumes and the third added Tieck's memoir of the poet.

ations of the Romantically inclined. The memoir, while not really falsifying Hardenberg's relationship to Sophie von Kühn, yet simplified it and concentrated it into the single creative force in his life. "The first sight of this beautiful and wonderfully lovely form," Tieck wrote, "determined his whole life; indeed, one can say that the feeling that shot through him and ensouled him became the content of his entire life."[45] Tieck believed that the pain Hardenberg experienced with Sophie's death produced a kind of elevated state, from which not only sublime poetry might flow but which would allow him to reveal another dimension of reality:

> At the same time [that he experienced this sorrow], his life became clarified and his whole being melted into a bright conscious dream of a higher existence. His nature and all his ideas can be understood as derived from the sanctity of this pain, from his intimate love, and his pious desire for death.[46]

For generations thereafter Hardenberg's life, as sublimed in Tieck's memoir, established the model for the Romantic personality and the poet of tragic genius.

Caroline Böhmer and the Mainz Revolution

After the Schlegels' father died, Wilhelm helped support his younger brother at Leipzig with what money he could, though Friedrich's debts constantly outstripped the financial capacities of them both. Wilhelm also offered support, emotional support, to his own early love, Caroline Michaelis Böhmer (1763–1809), who would become the femme fatale of the early Romantic circle. This fiery, beautiful, and omni-talented woman—she had mastered French, Italian, and English by age fifteen—was constrained by her father, Johann David Michaelis, the famous Göttingen biblical scholar, to marry an old childhood friend, Georg Böhmer, a country doctor. The match was deemed more financially stable than any that could have been arranged with one of the many students or young lecturers congregating at the Michaelis house—they gathered, of course, not simply for profound theological conversation with the famous professor. Böhmer dragged the twenty-year-old, thoroughly cultivated woman to Clausthal, a small village in the Harz Mountains, where she became unhappily domesticated. She bore three of Dr. Böhmer's children, though only one daughter, Auguste,

45. Ibid., 1:xiii.
46. Ibid., 1:xviii.

would survive infancy. Caroline herself almost died, it seems, of boredom. She wrote her sister, after two years in Clausthal, that

> I am no longer a girl, love offers me nothing to do other than light house-work—I expect nothing more from a rose-colored future—my die is cast. Nor am I an enthusiast for mystical religion—those are the two spheres to which the wives passionately turn. Since I thus have nothing here to keep me busy, the wide world remains open before me—and that, that makes me weep.[47]

Caroline's only solaces in that claustrophobic town were her children and the many books that she constantly solicited from friends.

Caroline Böhmer's life happily changed in 1788 when her husband died. She gratefully returned to Göttingen, where she met and toyed with the affections of Wilhelm Schlegel—toyed because, though fond of him, she had fallen deeply in love with another man, Georg Tatter, a government official with whom she would have an on-and-off relationship. When her finances became more difficult, she retreated to Marburg to live with her step-brother, a physician, all the while continuing to correspond with Schlegel and thereby nurturing his hopes. Fatefully in 1792 she moved with her daughter Auguste to Mainz to be near her childhood friend Therese Heyme (daughter of the Göttingen classicist), who lived there with her husband Georg Forster (1754–1794). Forster, as an eighteen-year-old, had sailed to the South Pacific with Captain Cook and later produced a celebrated tale that subsequently inspired Alexander von Humboldt (and thus indirectly Charles Darwin) to undertake his own great voyage of adventure.[48] Forster introduced the young widow into his social circle, including his dangerous republican friends, and encouraged her political education, giving her volumes of Condorcet to read and the letters of Mirabeau to translate. Caroline's reaction to Forster extended through a large emotional range; she later wrote a friend: "He is the most wonderful man; there is no one I have so loved or admired, or, again, thought so little of."[49]

---

47. Caroline Böhmer to Lotte Michaelis (28 May 1786), in *Caroline: Briefe aus der Frühromantik*, ed. Georg Waitz and expanded by Erich Schmidt, 2 vols. (Leipzig: Insel Verlag, 1913), 1:153.

48. Georg Forster, *A Voyage Round the World, in His Britannic Majesty's Sloop, Resolution, Commanded by Capt. James Cook, during the Years 1772, 3, 4, and 5*, 2 vols. (London: B. White, 1777). Alexander von Humboldt said it was reading Forster's account of Cook's voyage that brought him to plan his own trip to the Americas. Forster's book was also in Darwin's library. See *The Correspondence of Charles Darwin*, ed. Frederick Burkhardt et al., 8 vols. to date (Cambridge: Cambridge University Press, 1985–), 2:222.

49. Caroline Böhmer to Friedrich Meyer (17 December 1792), in *Caroline: Briefe aus der Frühromantik*, 1:279.

The Mainz Revolution

During the early 1790s, Mainz swung like a pendulum between the hot spirit of the French Revolution and the cold reaction of the German aristocracy. The city, lying on the Rhine close to the French border, became a refuge for French émigrés of the Old Regime; though many of the local intellectuals, Forster's group, for instance, yearned for the abolition of the rule of petty German nobles, like the prince archbishop of Mainz. The fever of revolution snaked across all national boundaries. The French republicans had been circulating broadsides in adjacent lands advocating the eradication of the ancient order, and several peasant communities in Saxony and the Rhineland had raised the pitchfork of freedom. The king of Prussia, Friedrich Wilhelm II, and the Austrian emperor, Leopold II, both feared spread of contagion in their own lands. The Germans, in addition to planning a preemptive strike, also thought warmly of the territorial possibilities of a nice little war with the disorganized French nation.[50] In anticipation, however, the French Assembly, on 20 April 1792, moved first, declaring war against the Germans. This act unleashed the hostilities. Friedrich Wilhelm met with the new Austrian emperor Francis II (his father Leopold having suddenly died) in Mainz to plan their strategy amidst much pomp and the rich cultural circumstances enjoyed in the city. They placed their combined regiments under the leadership of the competent but cautious duke of Brunswick. The armies of the duke totaled some eighty thousand men, including: his own King Friedrich Wilhelm; a cluster of Prussian nobles; and a troop of French émigrés, who had convinced the Germans that ordinary French soldiers would come over to the side of the ancient order when the imperial troops entered France. All assumed that the growing forces would be able to stroll into Paris.

One of the German regiments was led by Duke Carl August of Saxe-Weimar-Eisenach, who persuaded his confidant and civil administrator Johann Wolfgang von Goethe to join him in the triumphant march. For personal reasons, Goethe had been reluctant to enter the fray. But as a favor to the duke, he set out with a servant in his coach-and-four, doing what work he could on the manuscript of his *Farbenlehre* (Theory of colors), which he cautiously carried with him during the whole ordeal. He finally caught up with his sovereign in late August 1792, but along the way his foreboding had grown, as he recounted in his later memoir *Campagne in*

---

50. The story of the German alliance against France is told in James J. Sheehan, *German History, 1770–1866* (Oxford: Clarendon Press, 1989), pp. 220–23.

Frankreich (1822).[51] The Germans had randomly plundered their own countrymen's small border villages. And the elements were not condign. The weather that late summer had been extremely rainy and the roads became marshes, into which sank wagons attempting to supply the stretched-out armies. The miasma that hung over the exhausted troops helped spread dysentery, which in the end contributed the greatest number of casualties. The French towns of Longwy and Verdun were taken after a short siege, and the supplies found in those cities—some good wine and mutton for Goethe—did momentarily suggest a brighter outcome. Yet despite the early successes of the German forces, the French soldiers seemed unaccountably defiant, some after capture even committing heroic acts of suicide. The behavior of the German troops, as Goethe recounted, "alternated between order and disorder, between preserving and destroying, between robbing and paying, and this, indeed," he concluded, "is what it is that makes war really corrupting for one's character."[52] Morale in the German ranks continued to decline as foul weather and debilitating disease increased. Finally, on 20 September 1792, they engaged the main French forces at Valmy, a small town about a hundred miles from Paris. The French held their ground, all the while pounding the German armies with fierce cannonade. Goethe moved to the front and experienced, he later recalled, the kind of fever only caught when lethal shot rains all around. Afterward in camp, when it was obvious the French had the killing advantage, Goethe made an observation, the prescience of which was undoubtedly enhanced by the retrospective vantage of his memoir. He remarked to his friends: "Here and now a new epoch of world history is beginning, and you can say that you were there."[53] The actual battle had been small and the German retreat mostly embarrassing, if nonetheless quite deadly for the common soldiers.[54] Weather, disease, and poor rations made the experience for Goethe, if not of lethal consequence, at least a misery that bore deep into his soul.[55] But from Goethe's shrewd perspective, the victory of the French

51. Goethe, *Campagne in Frankreich, 1792*, in Johann Wolfgang von Goethe, *Werke* (Hamburger Ausgabe), ed. Erich Trunz et al., 14 vols. (Munich: C. H. Beck, 1988), 10:188–363.

52. Ibid. (3 September), 10:213–14.

53. Ibid. (night of 19 September), 10:235.

54. As Goethe recorded it, the loss of approximately one-third of the Germans' eighty thousand men was due mostly to illness, with only about two thousand having fallen from enemy fire. See ibid. (October), 10:302.

55. Goethe wrote his friend Karl Ludwig von Knebel seven days after the cannonade to describe the torture caused by the weather and shortage of food. He also wrote his mistress Christiane Vulpius to say simply that "the suffering we have gone through cannot be described." See Goethe

had a historical significance that outweighed the reality and transcended his own suffering. Two days later the National Convention, greatly encouraged by their successes over the once seemingly invincible Germans, abolished the monarchy and the new French republic began.

The French army, now with the victory of the Republic staining its bayonets, went on the offensive, invading German territory, and taking Speyer at the end of September, Worms at the beginning of October. Goethe and his company feared the worst: the small German principalities of the Rhineland and Bavaria clung to a desperate neutrality lest they be swallowed in war between the great powers, "while the masses, seized by revolutionary desires, became ever more active."[56] On 20 October the French General Adam-Philippe de Custine (1740–1793) walked into Mainz in the wake of the fleeing nobility and French émigrés.

Forster's wife Therese, with their two children, abandoned the city; but Forster himself remained to help establish a new democratic government. He thus dared treason, braced only with an enlightened faith in democracy and Caroline Böhmer by his side. Caroline moved into his house to help secure the new dispensation, and thus herself became, in the eyes of the opposing forces, a dangerous and degenerate traitor. Caroline's commitment to the revolutionary cause seems genuine, if initially based more on her personal ties to the men involved. Though she and Forster were rumored to be lovers, their relationship probably was not intimate.[57] For in January or February 1793, Caroline met and fell in love with a French lieutenant, Jean-Baptiste Dubois Crancé. Their affair flared brightly, if briefly; the German counterattack in the Rhineland was beginning. At the end of March, Forster traveled to Paris as representative of the new Rhenish Republic. But there the Terror had begun, and later in the year he watched as General Custine, and many other notables, climbed to the guillotine. Forster was never able to return to his country and died in Paris in July 1794 at age thirty-nine.

A few days after Forster had departed Mainz and as the German armies moved toward the city, Caroline fled with her daughter. Mainz became gradually encircled and the siege began in earnest in mid-June. Goethe and

---

to Karl Ludwig von Knebel (27 September 1792) and to Christiane Vulpius (15 October 1792), in *Goethes Briefe* (Hamburger Ausgabe), ed. Karl Robert Mandelkow, 4th ed., 4 vols. (Munich: C. H. Beck, 1988), 2:156, 158

56. Goethe, (Trier, 29 October), *Campagne in Frankreich*, in *Werke* 10:289.

57. In a letter to her friend Friedrich Gotter (16 May 1793), Caroline mentioned she had read in a French newspaper that her fate was known: "in the Moniteur it says qu'on a mené á la forteresse de K. la veuve Böh. amie du Citoyen Forster.—That's comforting, I am his friend, but not in the French sense of the word." *Caroline: Briefe aus der Frühromantik*, 1:290.

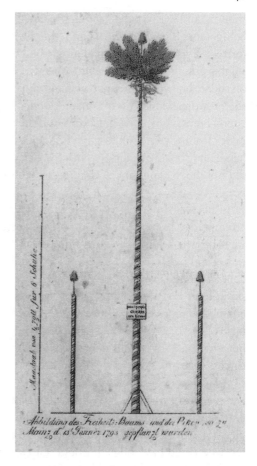

Figure 2.4

The symbol of the French Revolution, the freedom tree; frontispiece of the *Revolutionsalmanach von 1794*. The inscription reads: "Illustration of the freedom-tree and the pikes that have been planted in Mainz on the 13th of January."

Carl August enjoyed living conditions only slightly better than those they experienced during the French campaign: the rains poured down on their leaky tents for two weeks straight, and the temperature dropped to quite unseasonable levels. The weather did improve, only to make the killing easier.[58] With great hatred for the Mainz Jacobins, more than for the French troops, the German armies pounded the city with cannonade for almost a month. On 3 July, Goethe wrote his friend the privy councillor Christian Gottlob Voigt: "Half of this beautiful and well-laid-out city has probably burned. Only success can justify this cruel outcome. The situation of the Mainz émigrés is the saddest in the world."[59] After three more weeks of

58. In letters sent the same day (3 July 1793), Goethe described the miserable conditions to Christian Gottlob Voigt, member of the duke's privy council, and to Christiane Vulpius, his mistress. See *Goethes Briefe*, 2:167, 168.

59. Goethe to Voigt (3 July 1793), in *Goethes Briefe*, 2:167.

shelling and with food depleted, the city capitulated. On 23 July 1793 the belligerents signed a truce, and the French troops marched from the fortress. Typically, Goethe admired the courage of the French and their somber comportment on leaving the city. His approving gaze also fell easily on the several young German women who marched out with their lovers, as the crowd called for their tainted blood. He even intervened to protect a young French officer on horseback and the woman riding at his side, whose male clothing could not disguise from Goethe's discerning eye her splendid figure and striking beauty.[60] Thus the Rhenish Republic collapsed—but it did not die. Resuscitation would shortly be administered by a French corporal.

Just before the encirclement of the city, Caroline, with her daughter, had attempted to flee to Gotha, near Weimar, where she had friends; but on the way she was betrayed to the authorities. Both she and Auguste were arrested and jailed in the fortress of Königstein, near Mainz, where daily she could hear and even feel the bombardment of the city. Because of a confusion of her last name with that of her brother-in-law, a known Mainz Jacobin, but more especially because the authorities assumed her to be the mistress of Forster, they held her as both traitor and hostage for the return of her presumed lover. Her deprivations in the fortress were severe and the threats against her lethal—she witnessed the robbery, beating, and murder of many prisoners. This kind of experience, she wrote a friend, "creates dedicated sons of freedom."[61] When, however, she discovered she was pregnant with the French lieutenant's child, she became wildly desperate. Her pleas for salvation went out to all of her old friends. Her one-time lover Tatter, who could have helped, would not compromise his political position. But Wilhelm von Humboldt (1767–1835), a friend of her father, interceded with authorities to allow her to be brought to a more humane confinement. Wilhelm Schlegel, whom she informed of her pregnancy, flew to her side, bringing the poison with which she planned to blot out any further disgrace that might come to herself and her daughter. "My child would be better off," she wrote a friend, "as a complete orphan than to have a dishonored mother."[62] Schlegel was ready to stand with her in life or in death. Meanwhile, Caroline's youngest brother, the physician Philip Michaelis, made bold to petition the Prussian king, Friedrich Wilhelm II, offering as a bond of faith his services as surgeon to the army. Through her brother's ef-

---

60. Goethe, *Belagerung von Mainz*, in *Werke* (Hamburger Ausgabe), 10:389.

61. Caroline Böhmer to Friedrich Wilhelm Gotter (15 June 1793; the letter never made it through the censors at the prison), in *Caroline: Briefe aus der Frühromantik*, 1:292.

62. Caroline Böhmer to Friedrich Meyer (30 July 1793), in ibid., 1:303.

forts, Caroline was released on 5 July 1793. Because of her reputation—she had become known as the "whore of the occupation"[63]—she had no one to turn to, except the man who had remained faithful throughout, August Wilhelm Schlegel.[64]

Friedrich Schlegel and the Pregnant Radical

Schlegel brought Caroline to Leipzig, where she would be housed in anonymity with friends. Since he had to return to tutoring duties in Amsterdam lest his money dry up completely, he asked his younger brother to stay and care for her while she brought the child to term. On 4 November 1793, Friedrich wrote his brother: "I've just come from the baptism, where I stood as godfather. My godson is the little Citizen, Wilhelm Julius Cranz."[65] Since Caroline's life was becoming more stable and his own more economically precarious, Friedrich felt he had to leave her in the hands of local friends and see to his own finances. In January 1794 he traveled to Dresden, where his married sister would help provide for the necessities of his living. During his three and a half years in Dresden, he began producing those essays that would establish his reputation as a leading literary historian and critic.

The months Friedrich had spent with Caroline Böhmer profoundly affected him. He fell hopelessly in love with this beautiful, pregnant revolutionary, and undoubtedly would have pursued her had he not believed she was destined for his own brother. Just after he arrived in Dresden, he wrote Wilhelm: "My devotion to her [Caroline] is completely unconditioned. She is not only one of the few non-experts from whom one never ceases to learn, but she is good, the best, before whom I am ashamed because of my faults."[66] Later, in his novel *Lucinde*, Friedrich transformed his unrequited love into that of the hero Julius (who shared a name with Caroline's son). Julius meets the ersatz Caroline, who immediately "pierced his heart right in the middle. His passions heretofore played merely on the surface, or were

---

63. See Sigrid Damm's "Einleitung" to *Caroline Schlegel-Schelling in ihren Briefen* (Darmstadt: Luchterhand, 1980), p. 33.

64. Caroline wrote Friedrich at the end of August 1793 of her esteem for his brother Wilhelm. Friedrich quoted the letter in one to his brother (28 August 1793): "should a man be judged by his behavior toward a woman, it seems to me that W. [Wilhelm], in what he was to me, encompassed everything that one can call manly and, at the same time, childlike, disinterested, noble, and lovable." See the letter in *Kritische Friedrich-Schlegel-Ausgabe*, 23:125.

65. Friedrich Schlegel to Wilhelm Schlegel (4–5 November 1793), in ibid., 23:150. Caroline told Friedrich her whole sorrowful tale, including the true name of the father of her child, Crancé—hence the name of the "little Citizen."

66. Friedrich Schlegel to Wilhelm Schlegel (21 January 1794), in ibid., 23:175.

Figure 2.5

Caroline Böhmer
Schlegel Schelling (née
Michaelis) (1763–1809).
Steel-point engraving by
A. Weger, after a painting
(1798) by F. A. Tischbein.
(From G. Waitz, ed.,
Caroline, 1871.)

passing fancies without connection. But now he was struck by a new, un-recognizable feeling that she alone was the right one and that this feeling would be eternal." But because this woman was committed to his friend, "he kept all his love suppressed within, where he let his passion rage, burn, and devour him."[67] Schlegel's depiction of a love fevered with desire but ag-onizingly smothered in nobility reaches over two centuries to make even the heart of the modern reader beat with a reflective ache. In the novel this love transformed Julius, as in life it did Friedrich: "The worship of his sub-lime friend became for his soul the central point and foundation of a new world."[68] That world would circuit into view during his time in Dresden,

---

67. Friedrich Schlegel, *Lucinde*, p. 85. Caroline's magical, erotic power—the kind of power only a beautiful woman with a wonderfully creative intelligence can effect—has pulled writers into her embrace over the last two hundred years. In 1872 Kuno Fischer, the great neo-Kantian philoso-pher at Jena, published a large two-volume life and work of Friedrich Schelling, in which he spent over forty pages detailing the wonders of this woman. Fischer fell in love with her from a distance, and this historian, too, has succumbed. See Kuno Fischer, *Friedrich Wilhelm Joseph Schelling*, 2 vols.; vol. 6 of *Geschichte der neuern Philosophie* (Heidelberg: Carl Winter's Universitätsbuchhandlung, 1872–77), 1:74–115.

68. Friedrich Schlegel, *Lucinde*, p. 87.

while reconstructing the literary life of the Greeks and Romans. Hymns to his goddess would resound through these essays, wherein he played out the chords that would become identified as Romantic themes.

The Schlegels in Jena: The Break with Schiller and the Politics of Romanticism

In April 1795 Caroline, with her daughter, moved back to her mother's house near Braunschweig; and before she could send for her son, momentarily cared for by friends in Gotha, she received word of his death. Though Wilhelm Schlegel had been contemplating an escape to America, in July he returned to Germany and also settled in Braunschweig. Caroline wrote Friedrich that his brother now preferred to speak and write in French and that he "thinks differently of my friends, the republicans, and is certainly no longer an aristocrat. . . . And I will soon teach him passion—then will my instruction be complete."[69] So instructed, a year later, in July 1796, Wilhelm Schlegel married Caroline Böhmer and the couple immediately moved to Jena, where he had accepted a position at the university. Caroline's teaching, however, seems not to have had lasting effect, for passion drifted slowly away.

The city of Jena, unprepossessing in the Germany of today, was at their arrival a culturally lively town of about forty-five hundred citizens. It was part of the dukedom of Carl August, who resided in Weimar about twenty miles away. The city could boast of a university having eight hundred or so students; and for a decade and a half around the turn of the century, that university was ornamented with a faculty of extraordinary brilliance.[70] Karl

---

69. Caroline Böhmer to Friedrich Schlegel (August 1795), in *Caroline: Briefe aus der Frühromantik*, 1:36–67.

70. The university at Jena was founded in the sixteenth century. Through the eighteenth and nineteenth centuries, the population of students fluctuated wildly. By the middle of the eighteenth century, the number of students reached over thirteen hundred, but sank to less than six hundred by 1780. It climbed again to a high during the 1790s, but declined precipitously (as did the population of most German universities during the French occupation) in the first decade of the nineteenth century to less than two hundred students. Thereafter it began slowly to climb, having received impetus from the new *Burschenschaft* movement begun at the university in 1818. The movement slowly spread to other universities and, on its positive side, supported the ideals of constitutional government and German unification, but also left a legacy of vast quantities of beer consumed, dueling scars, and officers for the army. Only at century's end did the student population reach levels comparable to those seen at the end of the eighteenth century. In 1904 the student numbers hit one thousand. See *Geschichte der Universität Jena, 1548/58–1958*, 2 vols., composed by the collective of the Historical Institute of the Friedrich-Schiller University (Jena: Gustav Fischer, 1958), 1:309, 466. It is clear even from this East German production that the glories of the university are in its past.

Leonhard Reinhold (1758–1823) had established a redoubt in the philosophy faculty for the new approaches coming out of Königsberg; his lectures
on philosophy drew great numbers of auditors—in 1793, for instance, some
six hundred students attended his courses.[71] In promoting the thought of
Kant, Reinhold found an important ally in Friedrich Schiller, who served as
professor of history and a literary inspiration for his students. And when
Reinhold left for Kiel in 1794, he was replaced by Johann Gottlieb Fichte,
who had studied at Jena in the previous decade. Fichte, initially at least,
thought of himself as an orthodox Kantian, who simply wanted to clear up
a few residual problems in the critical philosophy. Fichte's transcendental
idealism became, via his influence on the Schlegels, a philosophic scaffold
for the construction of Romantic theory. In 1798, through the offices of
Goethe, Fichte welcomed to the philosophy faculty a new disciple, the very
young (age twenty-three) Friedrich Joseph Schelling, whose ideas about art
and nature would eventually depart from those of his mentor. Schelling's
Naturphilosophie would supply the important and enduring structural sup-
ports for Romantic theory both in Germany and England. In 1799 Carl August forced Fichte out of the university under the suspicion of atheism.
Shortly thereafter Schelling was joined at Jena by the roommate of his student years at Tübingen, Georg Wilhelm Friedrich Hegel (1770–1831),
who rose from Privatdozent to extraordinarius professor before leaving Jena
in 1807. Around this constellation of first-magnitude stars shone a halo of
other, somewhat lesser lights, but men whose names still produce a small
spark of recognition—including, in the medical faculty, Justus Christian
Loder (1753–1832), who instructed Goethe in anatomy; and Christoph
Wilhelm Hufeland (1762–1836), who lectured often to upward of five hundred students and acted as personal physician to Wieland, Herder, Schiller,
and Goethe.[72]

Jena and Weimar Society

During their first several months in Jena, Wilhelm and Caroline made their
way effortlessly through the thicket of society that extended from the city
to Weimar, about a two-hour coach ride away. Schiller and his wife entertained them frequently. Goethe visited them in Jena and invited them to

71. Ziolkowski, *Das Wunderjahr in Jena*, p. 24.

72. Through the nineteenth century, Jena retained an eminence far exceeding its size and location. And the Romantic traditions established at the century's beginning continued to invigorate
the philosophy and science emanating from the pens of those associated with the university during
the course of the century, such thinkers as Lorenz Oken, Jakob Fries, Karl Marx, Matthias Schleiden, Carl Gegenbaur, Kuno Fischer, Rudolf Eucken, and Ernst Haeckel.

dinner parties at his house on Der Frauenplan in Weimar. Caroline wrote her friend Luise Gotter about one of their trips to the capital of the duchy:

> Herder is the one who charmed me and made eyes at me. We had tea there, and Wieland was there all right—I found him in an extraordinarily good humor. . . . In the afternoon we were at Goethe's, as was Herder. . . . Goethe gave a wonderful dinner, really quite nice, without much fuss. He himself laid everything out, yet between this and that he still found time to paint us some beautiful word-pictures.[73]

During the dinner at Goethe's, Caroline caught brief sight of the great man's lower-class and not-exactly-attractive mistress, Christiane Vulpius, who was the mother of his son August. Caroline wondered aloud to Schiller why Goethe had not instead brought back from Rome some pretty Italian girl. Schiller guessed it would be more painful for him to send Vulpius away than to keep her.[74]

About a month after Caroline and Wilhelm had established themselves in Jena, Friedrich came from Dresden, poor in finances but wealthy in ideas and projects, to live near the couple. The Schlegel household vibrated with activity. Wilhelm wrote reviews and essays for Schiller's journal *Die Horen*, for which he was a coeditor.[75] He also began composing what would be an immensely influential translation of Shakespeare. Caroline advised her husband on the workings of those English plays.[76] And she helped create in

73. Caroline Böhmer to Luise Gotter (25 December 1796), *Caroline: Briefe aus der Frühromantik*, 1:410–11.

74. Ibid., 1:411. There may have been some dispute about how attractive Vulpius was; there was little, however, about her intellectual qualities. A contemporary thought that in her "the sylphe and the kitchen maid came together." See A. von Sternberg, *Erinnerungsblätter*, part 1 (Berlin: Schindler, 1855), p. 113.

75. It was Schiller who invited Schlegel to come to Jena, so that work on *Die Horen* could be facilitated. As Schiller put it: "Conversation would excite many more things than any written communication could." See Friedrich Schiller to Wilhelm Schlegel (10 December 1795), in *Schillers Werke* (Nationalausgabe), ed. Julius Petersen et al., 55 vols. (Weimar: Böhlaus Nachfolger, 1943–), 28:129.

76. Wilhelm Schlegel translated sixteen of Shakespeare's plays, with *Romeo und Julia* and *Ein Sommernachtstraum* appearing in May 1797. Caroline's notes to her husband on *Romeo and Juliet*, *Hamlet*, and *Macbeth* are quite shrewd. For instance, of *Romeo and Juliet* she wrote: "Juliet is nothing but love, and thus it would be impossible to take her for a glowing maiden who has been awakened for the first time and goes for any man she meets. These two seem to have been led to one another by a benign spirit—their eyes meet and every word follows from that first glance. One believes with them that no disappointment can occur. . . . And though we see nothing of Romeo and Juliet other than their passion, yet that very passion reveals in their souls a noble character. Don't blame Juliet that she is so easily won—she knows no other blameless way than to follow, without dissimulation, that mighty urge. . . . And whether *Romeo and Juliet* is really a tragedy, I leave to you and Friedrich to determine." See Caroline Böhmer to Wilhelm Schlegel (1797?), in *Caroline: Briefe aus der Frühromantik*, 1:427, 432.

Figure 2.6
August Wilhelm Schlegel
(1767–1845). Oil
painting (1793) by F. A.
Tischbein. (Courtesy
Archiv für Kunst und
Geschichte.)

the household a sustained enthusiasm for anything by Goethe, whose directness and firm command over emotional expression she greatly admired. Indeed, Caroline's constant reiteration of Goethe's poetic virtues came quickly to shape her husband's warm relationship with the Weimar genius and to focus Friedrich's concentration on him as the central figure in the new poetry of Romanticism. Her opinion of Schiller, though, had begun to curdle, passing from mild distaste for his poetic abstractions to derisive ridicule for his Swabian excesses. Friedrich completed the essays, begun in Dresden, constituting the first (and only) volume of his intended multivolume historical investigation of Greek and Roman poetry. He believed that such historical reconstruction would provide "a general natural history of the poetic art—a complete and legislative intuition [of the beautiful]." The historical analysis would reveal the Greek poetic achievement to have fully expressed universal human nature.[77] The publication of his historical

77. Friedrich Schlegel, *Die Griechen und Römer: Historische und kritische Versuche über das klassische Altertum* (1797), in *Kritische Friedrich-Schlegel-Ausgabe*, 1:276: "Already in the first stages of formation [*Bildung*] and while still under the guardianship of nature, Greek poetry comprehended in an even measured perfection, in the happiest balance and without one-sidedness or a forced deviation, the whole of human nature. Its strong growth soon developed into independence and

and aesthetic research marked Friedrich as a major critical thinker. But at this time as well, he also saw published those reviews that would finally cause a break in the relationship between the Schlegels and Schiller.

The Rupture with Schiller

The source of the rift, which sent waves of bitter antagonism through the social circles in Jena and Weimar, had literary, political, and personal causes. Schiller had briefly met Friedrich Schlegel in Leipzig, when the latter was a student. He initially judged the boy an "immodest, cold smart-ass [*Witz-ling*]."[78] Friedrich's fledgling essays hardly modified the older poet's attitude. After reading the young critic's piece "Ueber die Grenzen des Schönen" (On the borders of the beautiful, 1795), Schiller wrote a friend: "He [Schlegel] is knowledgeable and considers his object. But he does not clarify it and, consequently, lacks an ease in expression. I thus fear he has no talent as a writer."[79] This was a rather harsh and precipitous judgment on such a precocious twenty-three-year-old; yet as Schiller read further essays by Friedrich during 1795, his opinion remained little changed.[80] He took special exception to Friedrich's reevaluation, in the essay "Über die Diotima" (1795), of female character in ancient Greek literature. Friedrich— who had freshly impressed on his heart a striking example of a beautiful, intelligent, and strong woman—believed that the Greeks endowed their females with comparable, Caroline-like virtues.[81] Schiller never had an egalitarian view of women. In stated contrast to Schlegel, he did not think

---

reached that stage where mind, in its struggle with nature, achieved a decisive advantage; and its golden age reached the highest peak of ideality (complete self-determination of art) and beauty that could be possible for any natural formation [*Bildung*]. Its distinctive characteristic is the strongest, purest, the most determinate, simple and perfect expression of universal human nature. The history of the Greek poetic art is the general natural history of the poetic art—a complete and legislative intuition."

78. Quoted by Ernst Behler in his introduction to ibid., 1:xi.

79. Schiller to Christian Gottfried Körner (4 July 1795), in *Schillers Werke*, 28:2.

80. After he had read Friedrich's essay "Über die Diotima" (1795), Schiller did offer a tactful consideration of the young critic's abilities to the older brother. He wrote Wilhelm Schlegel: "The content still struggles in his work too much with the form, and it lacks easiness and clarity. But there is a lot that's real in him, and he succeeds in this battle, so one can expect he will become an excellent writer." See Schiller to Wilhelm Schlegel (29 October 1795), in *Schillers Werke*, 28:88. However, he offered a different judgment on the same essay to Körner (see the text associated with the previous note).

81. Schlegel would come to regard women—again, probably with Caroline as the model—as not merely men's equals but their superiors. Women, for instance, hardly needed the stimulation of poetry, as he wrote in his *Ideen* (1800): "Women require less of the poetry of poets because their very essence is poetry." See *Kritische Friedrich-Schlegel-Ausgabe*, 2:268 (fragment 127).

Homer's women, for instance, were anything special: Nausicaa is only "a naive country girl," "Penelope a shrewd and faithful Hausfrau," and Helen "merely a foolhardy woman."[82] When Schiller renewed his personal acquaintance with Friedrich Schlegel in Jena, however, the impressions gleaned from these early essays became tempered. The younger Schlegel, he mentioned to Goethe, "makes a right good impression and promises much."[83] It may be that this new attitude derived from Schlegel's use of Schiller's great monograph *Über naive und sentimentalische Dichtung* (1795–96)[84] in the preface to his *Die Griechen und Römer.*[85] In that preface Schlegel explained that he would have treated both modern and ancient poetry with greater insight had he read Schiller's treatise before his volume had gone to press. In the body of his treatise, he had disparaged modern poetry. But Schiller's analysis revealed to him a new understanding of the modern sentimental mode, which would become the central feature of the Romantic poetry he would champion.[86] Following the poet, he remarked: "The characteristic features of sentimental poetry are the interest in the reality of the ideal, the reflection on the relationship between the ideal and the real, and the focus of the idealizing imagination of the poetizing subject on an individual object. Only through the characteristic, that is, the representation of the individual, does poetry achieve a sentimental mood."[87] This kind of poetry was in sharp contrast to the ancient Greeks', which embraced the universal and objective. Through the next several years, Schlegel would take as his task the reconciliation of these two modes of artistic creativity. Likely he spoke to Schiller about these issues during those encounters when he seemed to promise much. But through fall and winter of 1796, Friedrich saw into print other essays whose acid criticisms quickly dissolved any links that momentarily joined him with the poet.

In late fall Friedrich published a review in which he unfavorably compared Schiller's poetry to Goethe's.[88] Then in another review early the next

82. Schiller to Wilhelm von Humboldt (17 December 1795), in *Schillers Werke*, 28:134.

83. Schiller to Goethe (8 August 1796), in ibid., 28:280.

84. The monograph was published in several parts in Schiller's *Horen* in 1795–96. See *Schillers Werke*, 20:413–503. I will discuss Schiller's treatise in chapter 11.

85. The volume finally appeared in 1797, bearing the preface that flattered Schiller's critical acumen. The preface itself was prepared the previous year.

86. Körner judged that "Schiller initially had made the classicist Friedrich Schlegel into a Romantic." See Josef Körner, *Romantiker und Klassiker: Die Brüder Schlegel in ihren Beziehungen zu Schiller und Goethe* (Berlin: Askanischen Verlag, 1924), p. 33.

87. Friedrich Schlegel, "Vorrede," *Die Griechen und Römer*, in *Kritische Friedrich-Schlegel-Ausgabe*, 1:212.

88. [Friedrich Schlegel], "Schillers Musen-Almanach betreffend," in *Kritische Friedrich-Schlegel-Ausgabe*, 2:8–9. Schlegel's reviews of Schiller's poetry and Schiller's journal (see the next note)

year, he panned several essays appearing in Schiller's journal *Die Horen* and showed one to be partly plagiarized.[89] Equally irritating to Schiller—and more irresponsible in his eyes—were Friedrich's political essays. The inspirational muse of these essays was perfectly obvious to the poet, that female revolutionary Caroline Böhmer Schlegel. Schiller believed she had simply led the Schlegels astray politically. In an epigram Schiller and Goethe wrote for the satirical series *Xenien*, they only thinly veiled the visage serving as target of their barb: "O! I am a fool. I am a raging fool! And anyone rages who, harkening to the advice of a woman, plants the freedom tree."[90] Just as Schiller believed women had no significant role in Greek literary and political life, he seemed just as sure that they should have no role in the German literary and political life of his own time.

The Politics of Romanticism

During the 1790s German governmental administrations frequently suffered from degenerative confusion, petty corruption, and revolutionary conspiracy. The exhausted remains of the Holy Roman Empire formed the German lands, which themselves were a loose confederation of ecclesiastical territories (like the archbishopric of Mainz), imperial cities (such as Frankfurt and Hamburg, and some forty-nine others), and dynastic states (led by Brandenburg Prussia and Austria, with the monarch of the latter state serving as emperor). Some of the more than three hundred political entities of the empire enjoyed stability and moderate government, for instance, the duchy of Saxe-Weimar-Eisenach under Carl August. Others, like the vast kingdom of Prussia, were frequently visited by turmoil and insecurity. From the reign of Friedrich the Great (1712–1786), patron of the arts and friend of Voltaire, to that of his nephew Friedrich Wilhelm II (1744–1797), who accepted the scepter at his uncle's death in 1786, the Prussian kingdom declined into bureaucratic and cultural sclerosis.

---

originally appeared in Johann Friedrich Reichardt's decidedly republican magazine *Deutschland*. In addition to journalism, Reichardt was a distinguished composer, who set many of Goethe's lyrics to music. Reichardt's republicanism, however, damaged his relationship to the poet. In chapter 7 I will discuss Reichardt's salon in Halle, which drew many Romantic philosophers, poets, and musicians. For a discussion of Reichardt's politics and his break with Goethe, see Walter Salmen, *Johann Friedrich Reichardt: Komponist, Schriftsteller, Kapellmeister und Verwaltungsbeamter der Goethezeit* (Freiburg i. Br.: Atlantis Verlag, 1963), pp. 166–88.

89. [Friedrich Schlegel], "Die Horen. Zwölftes Stück," in *Kritische Friedrich-Schlegel-Ausgabe*, 2:46–7. Schlegel had accused the historian Karl von Woltmann, Schiller's protégé, of plagiarizing Gibbon. The review caused Woltmann to give up his post at Jena and leave town.

90. Friedrich Schiller and Johann Wolfgang von Goethe, *Xenien*, in Goethe's *Sämtliche Werke*, 4.1:817.

Friedrich Wilhelm, in alliance with the Austrians and other contingents of the empire, had attempted to smother the infant French Revolution, but the child rose up to strangle its feeble aggressors. The victory of the French troops in the Rhineland forced Friedrich into a position of neutrality and financial collapse. Only higher taxes and greater repressive force could, he thought, keep the kingdom pacified. Friedrich's minister Johann Christoph von Wöllner, whom the king himself once referred to as a "swindling, scheming parson,"[91] instituted a special censorship commission, designed to combat unorthodox thought in religion and politics. Even Kant fell afoul of the commission because of his theological writings, and in 1794 the king issued a letter of censure to the old philosopher threatening "unpleasant consequences" should he continue to debase the principles of Scripture and Christianity.[92] While Friedrich and his deputy Wöllner bathed the land in pietistic cant, the court itself roiled with the traditional heterodoxies of petty corruption and hypocrisy. The king's private life reflected the life of the court: he twice committed bigamy and ensconced in his apartments mistresses too numerous to reckon. In these dispiriting times, the impetuous Friedrich Schlegel, in several essays, reminded the German public of how the French had reformed their government. He was emboldened by the experiences and convictions of his sister-in-law, and, of course, by the relative safety of dwelling in the more tolerant duchy of Saxe-Weimar-Eisenach, just beyond the borders of Prussia. He was still within conceptual shot, however, of Schiller.

Two of Friedrich's essays in particular seemed to display for Schiller the malign influence of "Madame Lucifer," as the poet took to calling Caroline.[93] The first, "Versuch über den Begriff der Republikanismus" (Essay on the concept of republicanism, 1796), had been stimulated by Kant's *Zum ewigen Frieden* (Toward perpetual peace, 1795), which had just appeared. Kant had argued that lasting peace could be procured only through a republican form of government. He thought the executive part of the government, however, had to be led by a king—thus a constitutional monarchy— since, as he argued, democracy united both the legislative and the executive functions in one body, which inevitably would lead to despotism.[94] Friedrich urged a different conclusion—derived, in the spirit of Fichte,

91. Quoted by Ernst Cassirer, in his *Kant's Life and Thought*, trans. James Haden (New Haven: Yale University Press, 1981), p. 376.

92. Ibid., pp. 393–94.

93. Cited by Gisela Dischner in her *Caroline und der Jenaer Kreis* (Berlin: Verlag Klaus Wagenback, 1979), p. 81.

94. Immanuel Kant, *Zum ewigen Frieden* (1795), in *Immanuel Kant Werke*, ed. Wilhelm Weischedel, 6 vols. (Wiesbaden: Insel Verlag, 1957), 6:206–7 (AB 25).

from the concept of a "social ego"—namely, that peace and international-
ism could only be accomplished through a democratic form of republicanism.[95] He urged that the deliberations and decisions of actual democratic
bodies—the empirical realities—constituted the only access we had to
that noumenal entity, the *general will* of the people, which was the foundation for both Kant and himself of any government that recognized the
moral dignity of humanity.[96] But where he differed strikingly from Kant was
in his conclusion that if the general will of the people should be thwarted by
its government, revolution would be justified.

Friedrich's analysis bore the marks not only of his conversations with
Caroline and his study of Fichte, but also of his recent investigations of
Greek literature. He believed the pinnacle of poetic beauty of any age had
been achieved by the Greeks during the flowering of Athenian democracy.
The literature of democratic Greece expressed, he thought, not merely the
pulse of a given poet but the very heart of the people—the poetry itself was
democratic and, in turn, fostered the spirit of freedom within the republics
of Greece.[97] Somewhat later Friedrich would discover in all true poetry this
democratic lifeblood. In his *Kritische Fragmente*, published in 1797, he
averred: "Poetry is republican speech—speech which is its own law and
purpose, where all the parts are free citizens and are able to come to common agreement."[98] In arguing for a return to the poetic standards of ancient
Greece, Friedrich thus implicitly and subtly urged political revolution. This
was a bold stand, taken at the height of the German reactions to the French
Terror and to the traitorous Rhenish Republic.

Friedrich's essay the next year, in 1797, must also have contributed to
Schiller's ire. It complemented the one on republicanism. The essay was an
admiring and emotionally driven sketch of the work and personality of

95. Friedrich Schlegel, "Versuch über den Begriff des Republikanismus," in *Kritische Friedrich-Schlegel-Ausgabe*, 8:14–17. The essay was originally published (1796) in the republican journal *Deutschland*.

96. Ibid., 8:17.

97. Jacob Minor, in the introduction to his edition of the early work of Friedrich Schlegel, credited him with infusing a social content into Romantic literature: "Initially it is through the Romantics that our literature has achieved a social character. It was Friedrich Schlegel who deduced
from the Fichtean foundational principle 'the ego should exist' the proposition 'the ego should
communicate itself.' Schlegel celebrated not only Forster as a social writer, he evaluated not only
Goethe's epigrams as poetic social pieces—he wished himself to shine as a social writer. To him the
Romantics owe (at least in a theoretical sense) their social character: he drove his friends to symphilosophize, to συμενθουσιάζειν, [to symenthuse]; he wanted even to symcry-out and symlaze-
about." See the "Vorrede" to *Friedrich Schlegel, 1794–1802: Seine prosaischen Jugenschriften*, ed. and
introduced by Jacob Minor, 2 vols. (Vienna: Carl Konegen, 1882), p. v.

98. Friedrich Schlegel, *Kritische Fragmente* (1797), in *Kritische Friedrich-Schlegel-Ausgabe*, 2:155
(fragment 65).

Georg Forster, the leader of the Jacobins in Mainz. Friedrich praised Forster
as one who "did not cling to the names of enlightenment and freedom . . .
[but] acknowledges and honors in his writing every glimmer of the true
spirit of lawful freedom wherever he meets it. . . . He doesn't hesitate to
proclaim: 'To be free is to be a man.'"[99] In the background of that proclamation, Schiller heard the rustling of petticoats.

The Schlegels, especially Friedrich, kept the ideals of the Revolution before their readers during that time when the German public was daily informed of its excesses. Yet even Schiller retained sympathy for democratic
ideals, if not for the practical politics of democracy. In his monograph *Über
die ästhetische Erziehung des Menschen* (On the aesthetic education of mankind, 1795), he focused on individual freedom—though of a Kantian moral
sort—as the goal of an enlightened society. He did not think this goal
achievable, however, without members of a state first having attained a certain kind of education—an education in beauty, which to his mind was also
an education in freedom.

Schiller argued, with no little impact on Friedrich Schlegel, that in the
Greek period, poetic sensibility and rational behavior, individual freedom
and enlightened government, all operated in harmony. In his own day, as a
condition of scientific, economic, and material advance, the harmony
within society and even within the souls of individual men had turned ca-
cophonous. Overspecialization had fragmented the life of the community
and deprived individuals of the autonomy that was their moral right—a
theme that Schiller heard enunciated a year before in Fichte's inaugural
lectures at Jena on the *Vocation of the Scholar*. But what might bring a healing to souls torn asunder? Enlightened governments might issue rational
commands, but their mandates could not touch the hearts of their citizens; compulsion thus became required. Ideally, the inner urgings of feeling
and imagination ought to harmonize with reason, and reason ought to accommodate to the particular needs of individuals for freedom and self-
realization. According to Schiller's Kantian aesthetics (discussed in more
detail in chapter 11), we actually have the experience of the requisite
harmony between reason and feeling, between law and freedom in the
appreciation of great art. Indeed, "the most perfect of all artworks is the
construction of true political freedom."[100] Hence, "the solution to each political problem in actual experience must proceed along the aesthetic way,

99. Friedrich Schlegel, "Georg Forster" (1797), in *Kritische Friedrich-Schlegel-Ausgabe*, 2:81.
Schlegel in his enthusiastic description moves from the past tense to the present, as if Forster were
still alive.

100. Schiller, *Über die ästhetische Erziehung des Menschen in einer Reihe von Briefen*, in *Schillers
Werke*, 20:311 (second letter).

for it is through beauty that one advances to freedom."[101] The goal of true political autonomy for the individual could not, therefore, be wrested from the state through force, but had to evolve through a more thorough, humane education of its citizens in the arts. Freedom and the capacity for appropriate political choice would percolate up from below, from those educated in the delights and liberating force of beauty.

Compared to the direct action of revolutionaries like Forster, Schiller's aesthetic approach to political freedom might seem anesthetic. The French Terror—the rule of the mob and their malign masters—during which such citizens as Condorcet died in his cell and the illustrious Lavoisier climbed to the guillotine, yet cautioned German intellectuals, including the Romantics, against precipitation. Moreover, though a German middle-class citizen might suffer restrictions under the petty tyranny of a prince, he could usually remove himself to another sovereign in a neighboring state— political competition in the Germanies kept the market in tyranny petty and low. The Schlegels had perhaps more keenly resonated to democratic sentiments, since they had a most effective teacher in Caroline Böhmer. But they, like Schiller, were content enough to live under the sovereign and benevolent eye of Carl August, duke of Saxe-Weimar-Eisenach.

Goethe, of course, was more than content to do so. He had a warm personal relationship with Carl August and served in the ducal court. His political attitudes during the French campaign were clear enough, and consistent with his initial feelings about the Revolution, which might be characterized as mostly irritation over the disorder fomented within the various societies of his concern.[102] While in Venice in 1790, he epigrammatically expressed his pique:

Every apostle of freedom is rather repulsive to me,
Each arbitrarily seeking his own end.
If you wish to free many, dare to serve many,
And if you want to know how dangerous that is, try it.[103]

---

101. Ibid., 20:312.

102. Wilhelm Mommsen explores Goethe's attitude toward the French Revolution in his *Die Politischen Anschauungen Goethes* (Stuttgart: Deutsche Verlags-Anstalt, 1948). Mommsen maintains that Goethe never considered the Revolution from the point of view of constitutional principle, only from that of the surface tension that mob rule and its threat introduced into peaceable, settled social arrangements (pp. 91–119).

103. Goethe, *Venezianische Epigramme*, in *Sämtliche Werke*, 3.2:94. Goethe's response was not simply that of an older intellectual. Henrik Steffens, the Norwegian geologist and Romantic enthusiast, recalled (in the tranquillity of age) that during his student years at Jena—1798–1801— he had complete contempt for the banner that carried the semblance of freedom but led to bloody excess: "I love freedom, I demand independence of every worthy man; yes, I seek to maintain with

Goethe undertook the adventure against the French forces not so much out of political conviction but out of curiosity and as a personal favor to his sovereign. It would be more correct to say, I believe, that Goethe, who certainly had to deal in local politics, felt himself above larger political matters. He marched on the German side against the French Republic, but he would later continue to admire the great Napoléon even after the Battle of Jena (1806), during which his own life was momentarily threatened.[104] Within the group of Romantic writers living in Jena and Weimar, Goethe was perhaps the most politically detached. He encapsulated his enduring political attitude at the conclusion of a one-act farce, *Der Bürgergeneral* (The citizen general, 1793), which he completed just after suffering through the French campaign:

> In a country where the prince denies no one his ear, where all stations of society think well of one another, where no one is prevented from behaving in his own fashion, where useful opinions and knowledge are generally broadcast—there no factions arise. What happens in the world will excite attention, but the seditious sentiments of a whole nation will have no influence.[105]

Goethe, whose ideas roamed completely unfettered through the realms of his imagination; Goethe, who was good-naturedly complicit in the frivolities of his own prince; Goethe, who as privy councillor could yet accomplish some practical reforms within the duchy of Saxe-Weimar-Eisenach; the poet, who would magnetically draw scores of young artists, intellectuals, and scientists to a small Thuringian town—this Goethe had no experience of the kind of spiritual and physical tyranny that might loosen the chaos of revolution. But the vividly imagined turmoil threatened him more immediately than any lack of abstract, political freedom.

We usually think Romanticism, of the sort that inspired Byron to fight for Greek independence, to be humane, democratic, anti-authoritarian, and golden with hope for the future, while nostalgic for a past in which that hope was believed to have been realized. Initially among some of the Germans, especially the Schlegels and Caroline Böhmer, as well as with the philosophers Fichte and Schelling, these heart-driven attitudes did color the surface of

---

all my power the very same for myself. But every abstract freedom is repugnant to my soul." See Henrik Steffens, *Was Ich Erlebte*, 10 vols. (Breslau: Josef Max, 1840–44), 3:55.

104. Goethe admired the genius of Napoléon as manifested in his military exploits. He also thought well of Napoléon's battlefield library, which contained a French version of his *Werther*, which Napoléon had read several times. See Johann Peter Eckermann, (7 April 1829), *Gespräche mit Goethe in den letzten Jahren seines Lebens*, 3rd ed. (Berlin: Aufbau-Verlag, 1987), pp. 299–301.

105. Goethe, *Der Bürgergeneral*, in *Sämtliche Werke*, 4.1:129.

their concerns, but beneath lay a deeper feeling for order. Most of the early Romantics—and the later ones as well, Haeckel, for example—endorsed, implicitly at least, a republican form of government, but one fixed by the bounds of reason, one that allowed a hierarchy of intellect, which thus recognized the authority of the well-educated. The Romantics' youthful ideal for government was that of an enlightened republic, one motivated, though, by high, philosophical, ethical, and aesthetic concerns, as they usually supposed the Athenian democracy to have embodied and the French Republic to have promised. Their political persuasions, therefore, did not differ appreciably from Enlightenment attitudes, but rather flowed directly out of them. Yet even among the very young, sentiments of a para-rational sort—of the kind that would later be identified with a harkening to tradition and a deep political and religious conservatism—were sounded: initially by one who had the sweetest voice of the early Romantics, Friedrich von Hardenberg.[106]

In 1798, shortly after the accession of the new king of Prussia the previous year, Hardenberg began work on a celebratory exhortation, really the declaration of a new political philosophy, directed at Friedrich Wilhelm III (1770–1840) and his consort Luisa. The young king—sober, dedicated, even expressing republican sentiments—was the very antithesis of his father. Hardenberg's *Rede* proposed a new beginning for the Prussian state, to be founded not on a contract or legal constitution but on the people's faith in the rectitude and wisdom of the royal couple, and on love for the ideal that these individuals embodied. But the proper state did require more. Government, according to Hardenberg, had to unite monarchy with republicanism—"no king without a republic, and no republic without a king can stand . . . both are as inseparable as body and soul."[107] The laws of the state would represent, though, not the will of the people, but rather the will of that higher man, the ideal of every person's striving. Like his mentor Schiller, Hardenberg believed a utilitarian contract could not generate a community, nor could a dispassionate Kantian rationality compel the entire person. The rational dictates of law had to appeal to the heart of each citizen. This could only occur when those laws issued from a beloved individual, the king, who would serve as a symbol of the higher man that each citizen longed to become. In a subtle amendment to Schiller's idea that aesthetic education had to prepare citizens for freedom, Hardenberg placed the artistic obligation in the care of the sovereign: "A true prince is the

---

106. My understanding of Hardenberg's political views has been abetted by Frederick Beiser's highly instructive *Enlightenment, Revolution, and Romanticism* (Cambridge: Harvard University Press, 1992); see especially pp. 264–78.

107. Hardenberg, *Glauben und Liebe*, in *Novalis Schriften*, ed. Kluckhohn and Samuel, 2:490.

artist of the artists; that is, the director of the artists. Each man should become an artist."[108] The king, in Hardenberg's conception, should use the artists of the realm as the material, which his own creative will must form, in order to procure the aesthetic education of the citizens, their *Bildung*.[109]

The shift in political view that Hardenberg enacted became more pronounced with the other Romantics as they aged; for with age—and the embittering experience of Napoléon's conquests—came a deeper political and religious conservatism that gave fuel to the obloquy of later critics, such as Heinrich Heine (1797-1856), that the Romantic movement embodied unthinking dogmatism and political despotism.[110] Hardenberg himself, in one of his last compositions, *Die Christenheit oder Europe* (1799), repositioned the responsibility for true character formation from king to church.[111] He discovered in medieval society that model of faith, community, and love that he thought necessary to promote the progressive development of the human spirit. Friedrich Schlegel drifted toward a harbor of similar security. He came to hold that monarchy had virtues he had previously overlooked. He also found a greater spiritual order in Catholicism, to which he finally converted in 1808.[112] By the end of the first decade of the new century, Friedrich foresaw the possibility of a new *Kaisertum*, comparable to that of Charles the Great, reuniting Europe against the ancient threats from the East. But this new political dispensation, grounded in a common Christian religion, would be possible, he thought, only under the wise leadership of the Austrian Germans, the preservers of the old *Reich*.[113]

108. Ibid., 2:497.

109. Ibid., 2:497–98. Hardenberg published his tract under the name of Novalis, and the question of the identity of the author was much discussed in Berlin. When Friedrich Wilhelm III read *Glauben und Liebe*, he protested: "More is demanded of a king than he can possibly perform. It is forgotten that he is only a human being. One has only to bring a man who lectures the king about his duties away from his desk and to the throne; then he will see the difficulties that it is not possible to resolve." Quoted by Beiser, *Enlightenment, Revolution, and Romanticism*, p. 273.

110. See footnote 3, above.

111. Hardenberg rose to this new position under the influence of Friedrich Schleiermacher's *Über die Religion*. See the discussion below.

112. Friedrich himself felt that his entrance into the Church was the culmination of a slow and gradual recognition of the spiritual authenticity of Catholicism, a recognition undoubtedly abetted by his intense study of the medieval period during the first decade of the nineteenth century. See Peter, *Friedrich Schlegel*, p. 52.

113. In 1802 Friedrich traveled to Paris, where he began the study of Sanskrit and Persian. Perhaps because of the failure of the Republic, his attitudes about the right kind of government to achieve universal peace underwent considerable alteration. At first he thought France might be the source, in confederation with the Germans, of a new political union in Europe. After the Franco-Prussian War and his own conversion to Catholicism in 1808, he elaborated a decidedly more Germanic ideal. A new Holy Roman Empire, guided by Austria, was the best hope for a reunited Europe. These themes are developed in Schlegel's *Über die neuere Geschichte* (1810–11), in *Kritische Friedrich-Schlegel-Ausgabe*, 7:125–407.

Schiller, who died in 1805, would undoubtedly have been even more despairing and disparaging of this last turn in the thought of his one-time adversary. Certainly Goethe believed Schlegel's capitulation to Catholicism a cowardly flight from reason and a depressing sign of the times.[114] It was precisely the reversion of Romantics like Hardenberg and Schlegel into the medieval mind-set that led later critics, such as Heine, to condemn Romanticism for its blinkered conservatism. In considering the religious and political character of Romanticism, however, what must be borne in mind is that its end differed significantly from its beginning.

Fichte, the Philosopher of Freedom

On the first day of his arrival in Jena, 6 August 1796, Friedrich Schlegel immediately visited the child-fiancée of his friend Hardenberg, Sophie von Kühn. She was confined to a clinic in the city, where her older sister looked after her. Friedrich was quite taken with Sophie, and the sisters with him—he had, it seems, a unique way with children—not at all patronizing, at times serious, at times light-hearted, but always with a scholar's mien. Then on the third day after his arrival, Friedrich ran into Johann Gottlieb Fichte, the thinker who would have a profound impact on him and his Romantic views. While in Dresden, he had kept up with the fashion in modern thought, the works of Kant and Fichte. He and Hardenberg had attempted to puzzle out the complexities of the new ideas—though often simply to rest delighted in the more obscure depths of the critical philosophy. Hardenberg quickly succumbed to Fichte's notions of transcendent unity founded in absolute subjectivity. Schlegel, initially, remained more ambivalent. He wrote a friend at the end of September:

> I see Fichte rather often, and find passing the time with him in conversation preferable to him in his flowing robes, whether that be on dry paper or in the lecture's chair. In that last, I find him wonderfully trivial. It is remarkable how he has absolutely no hint of everything that he lacks. The first time I had a conversation with him, he told me he would rather count peas than study history. Generally he is a complete stranger to any science [*Wissenschaft*] that has an object. . . . My brother has a kind of funny hatred for Fichte, a limitless worship for Goethe, and a certain ten-

---

114. Goethe to Karl Reinhard (22 June 1808), *Goethes Briefe*, 3:78: "It is worth the effort to follow Schlegel's conversion step by step, since it is a sign of the times. It is worthy also because you will not meet at another time so remarkable a case of an extraordinary and most highly cultivated talent being led astray in the full light of reason, of understanding, and of a cosmopolitan outlook, of one who disguised himself in order to play the puppet."

derness for Schiller, which I now, thank God, do not have any difficulty in overcoming.[115]

In his evaluation of Fichte, Schlegel did add, however, that perhaps nothing "great occurs without this sharp one-sidedness, without a certain limitation."[116]

Schlegel's feelings about Fichte, in all his guises, warmed during the course of the winter and spring as he continued to read the philosopher, meet with him, and attend his lectures. In May he told Hardenberg, "I am becoming an ever closer friend of Fichte. I love him a lot and I believe it is reciprocated."[117] In those gnomic *Fragmente*, published in the brothers' *Athenaeum* (1798), Schlegel would indicate the magnitude of Fichte's accomplishment: "The French Revolution, Fichte's *Wissenschaftslehre*, and Goethe's *Meister* are the great trends of the age."[118] While Schlegel's tether to these revolutionary tendencies in politics, philosophy, and literature became tangled and frayed as he grew older, during the 1790s they served as loose anchors for his impulsive, unpredictable, and elusive thought. The Fichtean anchor, though, often dragged bottom.

Fichte's Early Life

The Fichte whom Schlegel met was already a famous man. He had been hired, through Goethe's mediation, two years earlier, at age thirty-two, as professor of philosophy, and during those two years saw into print his principal theoretical work, the *Grundlage der gesammten Wissenschaftslehre* (Foundations of the entire science of knowledge, 1794–95). His arrival at Jena had been unlikely. He had come from a poor Saxon family; his father was a weaver and he himself, as a young child, toiled as a gooseherd. He

115. Friedrich Schlegel to Christian Gottfried Körner (30 September 1796), in *Kritische Friedrich-Schlegel-Ausgabe*, 23:333.

116. Ibid.

117. Friedrich Schlegel to Hardenberg (24 May 1797), in ibid., 23:367.

118. The *Fragmente* were fugitive thoughts honed by Friedrich, his brother Wilhelm, Novalis, and Friedrich Schleiermacher, and published in the *Athenaeum* in 1798. Friedrich was the author of the above-cited aphorism (no. 216). See *Kritische Friedrich-Schlegel-Ausgabe*, 2:198. Friedrich, despite his friendship with Fichte, was not uncritical of the *Wissenschaftslehre*. In private notebooks, kept during 1797 to 1798, he wrote: "The *Wissenschaftslehre* is a raw mix of systematics, polemics, mysticism, and logic" (no. 194). He was especially dubious of the abstract, unhistorical character of Fichte's thought. See "Geist der Fichtischen Wissenschaftslehre (1797–98)," in ibid., 18:36; see also below. The brothers got in a bit of a wrangle over the above-quoted epigram. Friedrich sent it, along with others to his brother, who became rather worried that it would insult Goethe. Friedrich did not believe Goethe would mind. See Friedrich Schlegel to Wilhelm Schlegel (25 March 1798), in *Friedrich Schlegels Briefe an seinen Bruder August Wilhelm*, ed. Oskar Walzel (Berlin: Speyer & Peters, 1890), p. 373.

later suggested that his rise from this lowly state to philosophy professor exemplified the great strengths of the common German people, as opposed to the nobility. Though without the kind help of the privileged class, Fichte could not even have dreamed of his eventual elevation. As a child, he had impressed the Baron Ernest Haubold von Miltitz by reciting verbatim a sermon the baron had missed. This nobleman and his family saw to Fichte's secondary education and financially secured his admission to the university at Jena, where in 1780 he undertook the study of theology. He then migrated to the universities of Wittenberg and Leipzig, finally obtaining a degree from the latter after years of financial distress. During his university studies and immediately thereafter, Fichte had to secure his own living by serving as a tutor to wealthy families. His arrogance and class sensitivity, however, made his various employments short-lived. His preoccupation with philosophical projects, as opposed to the mundane instruction of his wards, could not have helped. His initial philosophical disposition was likely that of a Leibnizian and Wolffian, with a dash of Spinoza—but this is uncertain. What is clear, though, is that after a study of Kant's three *Critiques* in 1790, Fichte became an immediate convert. Filled with the Kantian élan, he wrote a tract approaching theology from the critical standpoint. The anonymous publication of *Versuch einer Kritik aller Offenbarung* (An essay toward a critique of all revelation, 1792) persuaded many that it was by Kant himself. That and Kant's endorsement of the philosophical approach taken in the book guaranteed Fichte's future. For happily, just at this time, Goethe was searching for someone schooled in the Kantian philosophy who might replace Reinhold (called to Kiel). A young philosopher whose first major work had been mistaken for that of Kant himself seemed the perfect candidate.

Kant, though, remained a bit wary of Fichte, despite his initial approval of the new critique. Their first meeting in 1791, when the young disciple made a pilgrimage to Königsberg, had not been propitious. Perhaps it was because Fichte, though docile and worshipful toward the master, was not quite satisfied with the disparate ends of the critical philosophy. In his own early philosophical efforts, he would struggle to unite what Kant had initially torn asunder: the ego as knowing spectator and the ego as practical will. He would, in this pursuit, become, as Schlegel later described him, "Kant raised to the second power."[119] Had Schlegel taken a longer perspective, he might have added that the entire Romantic movement—not simply its philosophy, but its aesthetics, politics, and science—was Kant, if not squared, certainly rounded to a completion. The Kantian legacy must,

---

119. Friedrich Schlegel, *Fragmente*, in *Kritische Friedrich-Schlegel-Ausgabe*, 2:213 (no. 281).

therefore, be considered, if briefly, to appreciate the driving problematics of the movement.

An Excursus into Kant's Critical Philosophy

Critique of Pure Reason. In the first *Critique*, the *Kritik der reinen Vernunft* (Critique of pure reason, 1781), Kant had shown that Humean skepticism about science could be overcome by a radical examination of the presuppositions that grounded our conscious experience of the natural world. This examination demonstrated to many German intellectuals at the end of the eighteenth century that the phenomenal world of our experience arose from the interaction of necessary modes of thought and the mute impact on our senses of the reality standing beyond experience, the noumenal sphere (as Kant called it), of which nothing could be scientifically known. The proper object of scientific thought, according to Kant, is the phenomenal world. In our experience of this realm, we discover both unity and order, whose ultimate foundations could only be the mind. The world appears to one consciousness, the consciousness in which our ego (the transcendental ego in Kant's terms) is always peering around the corner, ready to claim ownership of its experience. This unity of consciousness structures the general unity of experience and thus the unified appearance of the world. The natural world further manifests temporal, spatial, and causal relationships, which ground the fundamental laws of nature. These natural laws, however, cannot be epistemologically anchored in any awareness of transcendent reality. For we have no access to such reality; and even if we did, inductive experience could not establish the *necessity* of its natural laws. Such laws, Kant argued, have to be logically isomorphic with the ways in which we mentally constitute the natural world—that is, through the ineluctable modes of sensibility (space and time) and the forms of judgment (which ground the categories of causality, substance, unity, necessity, and the rest). The subject-based forms of thought supply the universal and necessary character to relationships in the object-world, thus protecting Newtonian science from the corrosive considerations of Hume.

Reviewers reacted to Kant's first *Critique* with a mixture of objection to various details, complaint about its obscurity, and, usually, awareness that it was nonetheless an extraordinary book. One of the areas of deepest confusion concerned what looked like Kant's adoption of a kind of Berkeleyan idealism, one in which each individual created his own little world.[120]

---

120. An anonymous review appeared in the *Zugabe zu den Göttingische Gelehrte Anzeigen* (19 January 1782) that, while objecting to confusions and contradictions, charged Kant's *Critique* with purveying a thoroughgoing idealism. The review had been initially composed by Christian Garve,

Kant tried to counter that misunderstanding, as well as offer a more accessible introduction to his "Copernican revolution" in philosophy, with the *Prolegomena zu einer jeden künftigen Metaphysik* (Prolegomena to any future metaphysics, 1783). In that small tract, he maintained that mind did construct the features of the natural world of ordinary experience, but it did so jointly out of the necessary categories of thought, common to all human beings, and raw qualities delivered to sense from a real world beyond experience. As he baldly put it: "The existence of things, that which appears, is not destroyed as in real idealism; rather it is only shown that we cannot know anything about them, insofar as they are things in themselves, through the senses."[121] The noumenal world *appears* in consciousness as phenomena, but itself remains forever hidden from the gaze of reason.

Critique of Practical Reason. Kant's epistemological analysis in the first *Critique* had immediate moral consequences. It argued that all natural objects, including human beings, were caught up in a causal web that excluded any activities freely determined. A person thus could not, it seemed, make moral choices, could not form moral judgments, since by their very nature such judgments could not be effectively valid if coerced by physical causality. Moral judgment and behavior required freedom. But Kant was not inattentive to the need to preserve morality from his own theoretical threats. In the second *Critique*, the *Kritik der praktischen Vernunft* (Critique of practical reason, 1788), he laid out the presuppositions of an experience different from that of the objective scientist, the experience of moral choice. This experiential ground, no less valid than that fixing nature in mathematical formulae, required the supposition that the moral will was free. The moral will, that is, the conscious ego making practical decisions about behavior and acting on them—that moral will seemed ripe for identification with the transcendental ego of the first *Critique*. But in the first *Critique*, the transcendental ego served merely as logical subject; it was *the-looking-on* that might accompany every conscious perception and thought (thus unifying them in one consciousness) but itself had no content—it was always subject and never object. Nor was it understood to be active, as a moral will

---

a popularizer of philosophy. The editor of the journal, Johann Georg Feder, reduced the generally competent though critical review to one-third its original length and then added many criticisms of his own, including the likening of Kant's position to that of Berkeley. Kant reacted with particular vehemence. His response was the *Prolegomena*, which both offered an easier introduction to the argument of the first *Critique* and responded directly to the charge of idealism. Frederick Beiser provides a thorough and illuminating discussion of the Garve-Feder review in *The Fate of Reason: German Philosophy from Kant to Fichte* (Cambridge: Harvard University Press, 1987), pp. 172–77.

121. Kant, *Prolegomena zu einer jeden künftigen Metaphysik dies als Wissenschaft wird auttreten können*, in *Werke*, 6:153 (A 64).

would necessarily be.[122] Kant obviously intended the identification of the transcendental ego with the moral will, but the first two *Critiques* never offered a sustained argument for their union. The reconciliation of causal determination of nature with human freedom and the unification of the subject in theoretical and practical thought became driving problematics of Fichte's lasting concern.

Kant himself, of course, believed he had made the reconciliation between the transcendental ego, which grounded the causal determinations of phenomena (including the empirical self), and the moral ego, which acted autonomously. At one level of analysis, the resolution was simple: we think ourselves determined, since in the theoretical sphere this is the only consistent way to conceive of our behavior; but our deliberations and practical decisions occur under the assumption of freedom, which assumption is justified by the experience of practical action itself and has as much claim to validity as the necessary assumptions that make possible our experience of a causally determined world. Yet, most readers of Kant, and Kant himself, still needed some perspicuous way of grasping even the possibility of free action. A model was required. Kant suggested that model, albeit obscurely, obliquely, and, to Fichte's mind, unsatisfactorily in the third *Critique,* the *Kritik der Urteilskraft* (Critique of judgment, 1790), which ostensibly treated the logic of aesthetic and teleological judgments, the latter principally concerning biological nature. The suggestion that these purposive kinds of judgment furnished the wanted model had been tacked on to the analyses of the book by way of an opaque introduction to the volume.[123] To understand how purposive judgment might furnish a model for comprehending the possibility of free action in a determinate world, a few more words concerning Kant's *Critique of Judgment* are required. We need, though, to expand the consideration even a bit more, since the third *Critique* furnished the starting point for the Romantics' own theories of aesthetics and biological science.

Critique of Judgment. Aesthetic judgments and teleological judgments— the *Critique*'s main concern—are judgments about the purposive character

122. Kant was more than a little ambiguous on this point—for he at times spoke of the "spontaneity" of the ego, which term, in the normal Kantian parlance, meant "free activity." For a thorough discussion of this difficulty, see Robert Pippin, "Kant on the Spontaneity of Mind," in his *Idealism as Modernism* (Cambridge: Harvard University Press, 1997), pp. 29–55.

123. In his introduction to the third *Critique,* Kant made clear this much, namely, that we had to be able somehow to think the possibility of freedom in a determinate world: ". . . we must at least be able to think of nature such that the lawfulness of its form could agree with the achievement in it of effective purposes undertaken according to the laws of freedom." See "Einleitung," *Kritik der Urteilskraft,* in *Werke,* 5:247–48 (A xix–xx, B xix–xx).

of beautiful objects and of organic nature. Kant's use of the terms "purposive" and "purpose" vary somewhat in the third *Critique*, but the words retain core meanings. He defines them this way:

> ... a purpose [*Zweck*] is the object of a concept insofar as this latter is
> considered the cause of the former (the real ground of its possibility); and
> the causality of the concept in respect of its object is the purposiveness
> [*Zweckmässigkeit*] (the *forma finalis*).[124]

Thus a building is a purpose, since it is the product of a concept in the mind
of the builder; and the actions of the builder are purposive, since they bring
to reality the purpose, the building itself. By extension, one can speak of the
components of an artifact or of an organic being as purposive insofar as
they indicate design, and thus suggest the causal realization of a concept
whose object they are. Such components may also be understood to have a
purpose insofar as they are means to some end specified by the concept.

In the second part of the *Critique*, devoted to teleological judgment,
Kant argues that our understanding of organic structures and activity cannot proceed through a mechanically causal analysis alone. We observe the
various parts of animal and plant bodies to be reciprocally means and ends
for one another, and we see that the parts come into existence precisely in
order to realize these end products, these purposes. We thus have to understand the mammalian heart in terms of its function, its purpose, which is to
pump blood to the parts; simultaneously these parts—the liver, the brain,
and others—function to ensure the proper operation of the heart. Each organ of an animal body acts as a means by which the other parts operate and
exist, and each in turn serves as the end or purpose of the existence of the
parts functioning as means. Moreover, all of the parts are, according to
Kant, directed to a primary purpose, the welfare of the whole. The existence and relations of the parts, therefore, depend on the whole; and their
operations and existence cannot ultimately be understood without a concept of the whole.[125] Such holistic concepts, when applied generally to
understand large groups of animals—say, the vertebrates—will become
known as *archetypes*. In light of the archetype, a biologist is able to assess
the functioning of an organism or one of its organs and judge that it *ought* to
achieve certain ends (thus a heart ought to pump sufficient blood to the
parts, which could be said of even defective hearts).[126] By contrast, there is

---

124. Kant, *Kritik der Urteilskraft*, in *Werke*, 5:298–99 (A 32, B 32).

125. Ibid., 5:483–88 (A 285–91, B 289–95).

126. Hannah Ginsborg makes the normative character of Kant's analysis of teleological judg-
ments the central interpretative idea of her careful and clear study of the third *Critique*. See her
"Kant on Aesthetic and Biological Purposiveness," in *Reclaiming the History of Ethics: Essays for
John Rawls*, ed. Andrews Reath et al. (Cambridge: Cambridge University Press, 1997), pp. 329–60.

no particular end or goal that stones ought to achieve. The normative judgment allowed the biologist ultimately assumes that nature is designed to realize certain intentions. In short, organisms seem to be the product of a designing intelligence, since the arrangement of organic parts can only be understood as resulting from an antecedent *Bauplan* or archetype.

Machines, of course, also display design. Yet, according to Kant, an organism cannot be immediately comprehended even as a designed machine, say, a clock.

> In a clock, one part is the means by which another is moved; but a wheel is not the efficient cause of the production of another part; one part is for the operation of another, but it does not come to exist by means of the other. The productive cause of a part and its form does not dwell in nature (i.e., the nature of this matter) but outside of it, in another being, according to whose ideas of a possible whole, the part can be produced through the being's causality. Just as a wheel in the clock cannot produce another wheel, it is even less the case that one clock can produce another by using and organizing matter. . . . An organized being is thus not a mere machine: for the latter has only the power of motion, while the former has a formative power [*bildende Kraft*], of a kind that it imparts to materials not possessed of it (it organizes these materials). Hence such a propagative power of formation cannot be explained merely through the ability of motion that a machine has.[127]

Living creatures thus display an organization of parts that appear to exist for the sake of each other and that seem to realize an ideal of the whole of which they are parts. We simply cannot conceive of such organization as arising through contingent causal fluxes. Moreover, not only do living creatures exhibit parts that form mutual means-ends relationships, but those parts and the whole are propagative. In growth, parts form out of other parts; in maintenance, they sustain one another; and in generation, creatures reproduce themselves. Thus for objects to be constituted "organisms" or, as Kant also refers to them, "natural purposes," they have to meet the following criteria: their parts form reciprocal means-ends relationships; those parts come into existence and achieve a particular form for the sake of one another (through growth, maintenance, and reproduction); and the entire system has to be understood as resulting from an idea of the whole. No mere mechanism displays all of these features. And thus no natural purpose can be understood by us in purely mechanistic terms.

---

127. Kant, *Kritik der Urteilskraft*, in *Werke*, 5:486 (A 288–89, B 292–93).

But what is the source of organic functioning, and what sort of knowledge might we have of it? Kant argues that the assumption of formative force and design in organisms is strictly regulative, a heuristic assumption that allows us to work out, if only partially, the mechanical causes that we can suppose to be really operative in organic systems—the "really," of course, refers only to the phenomenal world of the biologist's experience. When biologists discover mechanically causal relationships in nature—for instance, that light striking the cornea of a mammalian eye is refracted toward the perpendicular of the corneal sphere—they bring objects or events under the category of efficient cause and thereby affirm a structural component of refined, scientific experience. Kant calls such judgments "determinative." They exhibit a universal and necessary character because they are ultimately rooted in the a priori categories of human cognition. As such, these judgments and their resulting propositions can form part of an authentic science—for natural science, strictly speaking, is a system of necessary and universal propositions (for instance, Newtonian mechanics). But when biologists attempt to discover the particular ways in which determinate structures might be related to one another, that is, to find the interactive patterns seemingly exhibited by organisms—for example, that the various refracting media of the mammalian eye are seemingly designed *for the purpose* of focusing an image on the retina—they will not have ready to hand any a priori categories needed to forge necessary laws of nature; rather they have to search out empirical regularities by reflectively considering how differing mechanical relationships might fit reciprocally together in a web of interactive causes forming means-ends arrangements—thus to discover how the various refracting media of the mammalian eye must be arranged for the purpose of focusing an image on the retina. Judgments of this latter sort Kant calls "reflective" to indicate two related features: (1) that a concept of the whole has to be empirically discovered by an initial examination of parts; and (2) that such a concept is ultimately grounded not in a necessary requirement of nature—that is, in a natural law ultimately based in the categories—but rather in a necessary requirement of our reflective capacities.[128] We might then allow, as a *regulative* measure, such subjective requirements to be attributed to nature herself, *as if* the very heart of nature beat out intrinsic patterns and pulsed with productive powers. Ineluctably we would then be driven to assume the existence of a divine intellect as the ultimate ground of purpose in nature; but this assumption

---

128. There remain some deep ambiguities about Kant's distinction between determinative and reflective judgments. But see his preliminary definitions in ibid., 5:251–53 (A xxiii–xxvi, B xxv–xxviii).

could only be regarded as a methodological aid to investigation. Such regulative postulations can never be brought into a complete scientific system of a priori laws, which, of course, means that biology can never become an authentic science—only a loose system of empirical regularities.

The system of empirical regularities, though, seems yet to conform to the requirements of our mental apparatus, to our needs as scientific researchers in biology. In this respect, as well, we have to think of nature in teleological terms, to consider that the larger web of natural complexities seems precisely coordinated to our cognitive powers. Einstein said that the most incomprehensible thing about nature was that it could be comprehended. This fact must strike every thoughtful naturalist. We might have found, for instance, that organisms could not be arranged in the Linnaean system or consistently in any other classificatory scheme. Yet nature does, as a matter of fact, display larger unities in terms of which its smaller parts can be understood. And these webs of complexity seem so designed as to be comprehended by us—at least, this is the assumption we must make. We would otherwise have no reason to pursue science, since the scientific endeavor presumes that patient investigation ultimately yields consistent results. This assumption constitutes, according to Kant, the a priori principle of the faculty of judgment: namely, that our cognitive powers are suitable for understanding nature.[129]

Kant deploys these several kinds of judgment of purpose in the introduction to the third *Critique* as a bridge between theoretical lawfulness (as determined by the categories) and the law of freedom (when the will determines itself to act in accord with the categorical imperative). The bridge is the apparent a priori purposiveness of nature, a nature that under this conception seems designed both for our understanding and for our practical action. It is *as if* nature were the product of a divine mind—an *intellectus archetypus*, Kant calls it—that provides for our practical needs.[130]

---

129. Kant regarded this a priori principle as ultimately an aesthetic one, since the suitability of empirical objects for our understanding was based on a kind of feeling, an aesthetic sense. He thus observed (ibid., 5:268 [A xlviii–xlix, B l–li]): "In a critique of judgment the part that the aesthetic judgment contains belongs to it essentially, since this part alone contains a principle upon which the judgment is grounded completely a priori to its reflection on nature, namely, that of a formal purposiveness of nature, in respect of its particular (empirical) laws, for our cognitive powers, a purposiveness without which our understanding could not find its way through nature."

130. Ibid., 5:526 (A 346–47, B 350–51). Kant first introduced the conception of an *intellectus archetypus* in a letter to Marcus Herz, his former student. Such an intellect would be actually creative of the object of its representation. Kant distinguished this kind of intellect from an *intellectus ectypus*, which had its representations causally produced by objects. Our intellect, Kant contended, was neither of these types. See Kant to Marcus Herz (21 February 1772), in *Briefwechsel von Imm. Kant in drei Bänden*, ed. H. E. Fischer, 3 vols. (Munich: Georg Müller, 1912), 1:119.

A nature so produced would be lawful yet accommodate our free action.[131] With this conception, Kant advanced a model by which to understand how human free acts might thread their way through a law-bound nature: that nature, it would seem, has been designed for our practical activity. We can thus comprehend the possibility of free human activity in a causally determined world.

Aesthetic judgment, which occupies Kant in the first part of the third *Critique*, also suggests a way of making intelligible the relation between determinate nature and free human behavior. When we judge, say, Novalis's *Hymnen an die Nacht* to be a beautiful poem, we engage, according to Kant, in a reflective consideration of the harmony of the parts and their contribution to the realization of their purpose, namely, the production of beauty. Kant maintains, however, that in our reflective judgment of the purposiveness of the elements of Novalis's poem, we can never uncover any concept called "beauty," the content of which could be specified. Had we such a concept—presumably a set of rules—then anyone might, by following the rules, produce a poem of comparable aesthetic value. Kant, of course, does not wish to deny that human artists may consciously follow quite well-understood rules for the production of artifacts—for instance, the laws of perspective, of color mixing, of proportion in the human form. What he denies is that the artist has any conscious knowledge of how to apply all such rules in producing a *beautiful* object. There are no explicit meta-rules of beauty.

In making the judgment that a piece of art—or nature—is beautiful, what happens, according to Kant, is this. Our understanding and imagination are set in "free play," virtually simulating the purposive harmony exhibited by the object. Aesthetic "taste," then becomes, in Kant's terms, "the ability to judge an object in relation to the free lawfulness of the imagination."[132] Such harmonic free-play of the cognitive powers produces an aesthetic feeling, which if strong enough allows us to call the object beautiful. The judgment of beauty is made, therefore, according to a feeling. But since the feeling derives from the universal structure of human cognition, we can reasonably expect any other person who considers the same object to have that same experience.[133] Beauty can thus be predicated of some piece of art or of nature with epistemological objectivity, even though it is subjectively grounded in a feeling. And since the judgment of beauty is anchored in a

---

131. Kant maintained that we had heuristically to assume, as a consequence of our thinking apparatus, design in nature; but, as he insisted, this permitted us neither to assert as constitutive any design operating in the phenomenal world nor, certainly, to vault beyond the phenomenal world to the reality of a designing creator.

132. Kant, *Kritik der Urteilskraft*, in *Werke*, 5:324 (A 67–68, B 68–69).

133. Ibid., 5:291–95 (A 21–26, B 21–26).

feeling rather than a rational formula spelling out the necessary criteria of beauty, Kant regards it as a judgment of purposiveness without any concept of the purpose—*Zweckmässigkeit ohne Zweck* (purposiveness without purpose).[134]

In creating a beautiful object, the artist is guided by the same kind of aesthetic feeling as the observer. The artist's productive imagination, according to Kant's analysis, constructs a beautiful object in respect of ineffable and subjectively felt laws of harmony. The individual who has the capacity to execute beauty in an exquisite fashion we call a genius. Kant had a particular theory of genius, which came to underlie many varying notions of genius produced by the Romantics.[135] He encapsulated the phenomenon in this definition:

> Genius is the talent (natural gift) that gives the rule to art. Since the talent, as an inborn productive ability of the artist, itself belongs to nature, we can also express it thus: genius is the inborn mental trait (*ingenium*) through which nature gives the rule to art.[136]

Kant's disciples took him to mean by this definition that the ineffable rules of beauty lay deep within the genius's nature. Talented individuals, in constructing a beautiful object, would act purposively, but only by feeling their way, as it were, to the final product. Artists would not be able to conceptualize or consciously determine these rules of beauty; the rules would express themselves only in feelings of aesthetic rightness. Artists would thus create objects according to laws that they would freely give to themselves, inarticulable laws of their own natures. Artistic production, then, suggested to Kant's followers another kind of reconciliation of physical determinism and human freedom.

We now might be able to see why Kant joined in the third *Critique* what seem like quite disparate kinds of judgment. Organisms and aesthetic objects both exemplify purposiveness in their constructions—their parts harmonize and stand in reciprocal relation to one another. Such purposiveness could not arise by accident, through sheer contingency—or at least our human reason balks at such notion. To understand purposive effects, we must appeal to a concept as a cause, according to which the parts come to be so integrated and according to which the whole comes to exist. Hence, the bi-

---

134. Ibid., 5:300 (A 34–35, B 34–35).

135. See the next chapter for Schelling's reworking of the concept of genius. For a brief but adroit discussion of the concept of genius during the Romantic period, see Simon Schaffer, "Genius in Romantic Natural Philosophy," in *Romanticism and the Sciences,* ed. Andrew Cunningham and Nicholas Jardine (Cambridge: Cambridge University Press, 1990), pp. 82–100.

136. Kant, *Kritik der Urteilskraft,* in *Werke,* 5:405–6 (A 178–79, B 181).

ologist, through reflecting on the parts of organisms and assessing their relationships, specifies the putative concept according to which the creature
is organized—its particular archetype or *Bauplan*. Yet the researcher of
Kantian acuity will not draw the theoretical conclusion that nature or nature's God has produced the organism according to this plan. He will,
rather, judge organisms *as if* they owed their existence to a set of rules governing the whole, ultimately to a designing mind. The art critic also judges
aesthetic objects as if the harmony of parts required definite rules, even
though these rules cannot be explicitly stated either by critic or artist. The
critic must assume, however, that the beautiful production comes from the
will of the artist. Yet a significant difference remains to distinguish these
two kinds of judgment: the biologist judges an organism to be purposive according to a specific plan of which he can become aware—even if he cannot determinatively claim that the plan was indeed the cause of the
organism; the art critic judges a painting to be purposive, but cannot specify the plan or rules by which beauty has been produced. This difference implies that only the artist and not the biologist (or other natural researcher)
can be a genius. For the artist realizes in his or her object ineffable rules of
beauty, while the naturalist is perfectly cognizant of the rules of organization and determination of the object of inquiry.[137] Any neophyte can adopt
the concepts of the biologist and make appropriate judgments about organisms; but no budding poet can simply imitate the rules that Pope followed in
creating his inspired verse (or at least Kant thought Pope an inspired genius).

If nature and her laws seem to have been constructed with us in mind,
then we might have, in this analysis, a way of understanding how the human agent can act freely in an apparently determined world—the world appears designed to accommodate our self-legislative acts. This, at least, is
what Kant suggested in the introduction to the third *Critique*. But his own
disciples took the *Critique* to offer a somewhat different, though related, solution to the problem of freedom. Fichte and Schelling, for instance, looked
to the aesthetic judgment per se as providing a model of self-legislation—
Kant's basic meaning of freedom—according to laws of the artist's own nature. Artistic production, then, might indicate how our moral actions could
have an outcome through a causally determinate human nature that, nonetheless, accommodated free action. With the Romantics, moral actions became modeled on aesthetic acts, and morality itself became aestheticized.

137. Ibid., 5:407–8 (A 181, B 183–84). Schelling would challenge this implication by showing how the scientist can also conform to Kant's notion of genius. See chapter 3.

Fichte's Transcendental Idealism

Despite reservations, Fichte adopted for his own ends this Kantian model of the free creativity of the artist. He used it to understand how the ego not only could act freely but could create a world.[138] He would maintain that Kant's transcendental ego was not an empty logical postulate; it was an activity that could be known through transcendental argument and intellectual intuition. Reflection, as engaged in by a philosopher of the right stripe, showed the empirical ego and its correlate the non-ego, which is to say, the natural world, to be constituted by the actions of absolute subjectivity. Fichte conceived this creative action of the absolute ego to be comparable to the productive activity of the artist's imagination.[139] The ego, like the artist, spontaneously and freely created structures that nonetheless exemplified determinate, law-governed features.[140]

Fichte began his analysis with what could not be doubted, the immediate representations of consciousness. Our immediate experience of the natural world, upon close inspection, shows itself to be a representation, which we assume quite naturally has been forced upon us by things existing out-

138. Among Fichte's early papers is a long treatise, apparently meant for publication, carrying the title *Versuch eines erklärenden Auszugs aus Kants* Kritik der Urteilskraft (Attempt at an explanation of an excerpt from Kant's *Critique of Judgment*). It dates from winter of 1790–91, just after the *Critique of Judgment* had been published.

139. Fichte's inaugural lectures at Jena (discussed below) occurred in two parts, of which only the first, *Einige Vorlesungen über die Bestimmung des Gelehrten* (Some lectures concerning the scholar's vocation, 1795), were published during Fichte's lifetime. A truncated version of the second half of the series was recovered and published in the early part of this century under the title *Ueber den Unterschied des Geistes und des Buchstabens in der Philosophie* (Concerning the difference between the spirit and the letter within philosophy). I have used the version of this latter that is included in J. G. Fichte—*Gesamtausgabe der Bayerischen Akademie der Wissenschaften*, ed. Reinhard Lauth and Hans Jacob (Stuttgart: Friedrich Frommann Verlag, 1964–), 2.3:307–42. In these latter lectures, Fichte identified spirit or *Geist* with productive imagination, *produktive Einbildungskraft*. Those thinkers and artists who have a greater productive imagination may then be said to possess more spirit. Like the artist, the philosopher who has greater spirit [i.e., genius] will be able with greater alacrity to take feelings and convert them into representations (ibid., 2.3:317). Out of feelings for beauty, the sublime, eternal truth, and divinity, philosophers have created significant representations, ultimately philosophical systems. This same productive imagination, according to Fichte, creates for us the picture (the reality) of a physical world of nature: "According to Kantianism properly understood, this image [of the physical world] is nothing other than a product of the absolutely creative imagination; since the time we have come to know ourselves [i.e., recognize our continuing identity], our imagination has projected this image with the greatest of ease and without any difficulty" (2.3:326–27).

140. Charles Larmore, in his elegant *Romantic Legacy* (New York: Columbia University Press, 1996), rightly emphasizes the premium placed by the Romantics on the creative powers of imagination (pp. 1–31). But in amalgamating English Romanticism with German, he underestimates the role of imagination in the theories of the latter. Schelling, for instance, followed Fichte in holding that imagination created the world; it did not merely, to paraphrase Larmore, enrich experience through expression.

Figure 2.7

Johann Gottlieb Fichte
(1762–1814). Aquatint
by Friedrich Jügel after a
portrait by H. A. Dähling.
(Courtesy Archiv für
Kunst und Geschichte.)

side the mind. But once caught within the flow of representations, no rational line can be cast outside to some putatively solid grounding. Any inferences via the causal principle from sensation to the supposed cause of sensation in external objects simply cannot be justified. Thus our knowledge and awareness of things "is not in things, and does not stream out of them. It streams out of [us], insofar as it exists, and it exists out of our own being."[141] The objectivity of the world, though, remains as it is. Indeed, objects of many colors and sizes, tastes and textures exist for consciousness in the entirety of their being; no hidden thing-in-itself lurks on the other side, making of the experienced world but a shadow of some supposed noumenal sphere.

In his *Grundlage der gesammten Wissenschaftslehre*, Fichte maintained that all individuals, as soon as they became conscious of the world around them, had as well implicitly to be conscious of several other aspects of that conscious experience itself: first, immediately of the object experienced, for instance, of a tree; second, implicitly that it is an *experiencing*, an activity; third, implicitly that it is I who am experiencing, and thus that over against the I is the object, apparently limiting the I's activity of experiencing.[142]

---

141. Johann Gottlieb Fichte, *Die Bestimmung des Menschen*, in *Fichtes Werke*, ed. Immanuel Hermann Fichte, 11 vols. (Berlin: Walter de Gruyter, 1971 [1843–46]), 2:229.

142. The self, according to Fichte, determines itself (i.e., actively construes itself) as determined by a not-self; but to determine oneself as determined by something else is only to say that the self determines itself. See Fichte, *Grundlage der gesammten Wissenschaftslehre*, in *Fichtes Werke*, 1:127.

But even more fundamentally, consciousness of objects requires that the representations of consciousness are connected in a continuous and identical activity of thought.[143] In this respect, the ego becomes the author of its own existence. By taking itself up at each moment through self-recognition—what Fichte called "self-positing"—the ego reproduces itself as an identical flow of consciousness.[144] The ego simply *is* this self-reflective, flowing activity, nothing more—no underlying Cartesian substance that thinks; only the activity itself. This understanding of the ego, he maintained, yields the first principle required to found a complete science of knowledge: "The ego posits itself [*setzt sich selbst*] and it *exists* by virtue of this simple positing of itself."[145] He called this self-positing ego the absolute ego, or absolute subjectivity, since its existence was a function only of its own activity.

Fichte's first principle had its prototype and inspiration in the defensive efforts of Reinhold, the Kantian philosopher whom he replaced at Jena. Reinhold had argued that the critical philosophy required an indubitable first principle whence its other claims might be deduced. Such a principle would insulate the system from the skeptical attacks it had been weathering since the publication of Kant's first *Critique*. Reinhold believed that he had discovered such a principle in the indisputable fact of consciousness itself. He observed that in every conscious representation—whether it be a perception, memory, idea, or the like—we can logically distinguish three elements: a representing subject, a represented object, and the representation itself.[146] Neither the subject itself nor the object itself, he urged, could actually exist in the representation; otherwise they would be identical with the representation and thus neither representing nor represented. The *form* of representation implies it to be something produced by a subject, and the

143. In reconstructing Fichte's theory of consciousness, I have been aided by Robert Pippin's "Fichte's Contribution," *Philosophical Forum*, 19 (1987–88): 74–96; Frederick Neuhouser's *Fichte's Theory of Subjectivity* (Cambridge: Cambridge University Press, 1990); and Frederick Beiser's *German Idealism* (Cambridge: Harvard University Press, forthcoming).

144. Fichte wished his readers to understand, however, that the unifying synthesis of many representations in one consciousness was not the essence of the ego. Rather, the timeless ground of any and every act of consciousness was reproductive self-consciousness. See Fichte's *Zweite Einleitung in die Wissenschaftslehre* (1797), in *Fichtes Werke*, 1:475–76. This second introduction was one of his several continuing efforts to make his doctrine clear.

145. Fichte, *Grundlage der gesammten Wissenschaftslehre*, in *Fichtes Werke*, 1:96: "Das Ich setzt sich selbst, und es *ist*, vermöge dieses blossen Setzens durch sich selbst."

146. Karl Leonhard Reinhold, *Versuch einer neuen Theorie des menschlichen Vorstellungsvermögens* (Jena: C. Widtmann and I. M. Mauke, 1789), p. 200: "Consciousness makes it necessarily clear that to every representation there belongs a representing subject and a represented object, both of which must be distinguished from the representation to which they belong."

content refers it to something existing extra-mentally. Reinhold had thus made clear what he thought Kant had only unsystematically articulated. Fichte agreed with Reinhold that the critical philosophy needed a first, unyielding principle to ground its analyses of the structure of intelligence and the features of the empirical world. He thought Reinhold's principle suggested the general strength of the critical view but also revealed a distinctive liability, at least in the conventional interpretation rendered by its supporters. The principle indicated that since the content of any representation lacked the form (i.e., the mode) of an externally existing object—but indeed had only the form of a representation—we could hardly represent that which existed as nonrepresented, namely, the thing-in-itself.[147] Yet Reinhold, like Kant, assumed that the content of irresistible representations was, nonetheless, furnished by an extra-mental world, the things-in-themselves.[148] Fichte found Reinhold's own principle to be an insurmountable barrier to such an assumption. From the isolation of consciousness one could not vault over its walls to compare representations to things-in-themselves, nor even to assert the bare existence of such things. The ego pole of the principle yet seemed more secure. It indicated that representations were necessarily related to a unifying subject. Fichte's genius was to see that a *fact* of consciousness might better be conceived as an *act* of consciousness—unification would be achieved not by a logical supposition but by an underlying action. His own postulation of a first principle thus enshrines this insight: the ego posits itself and exists only in this self-positing action. The self-positing ego would then be the source both for a unified consciousness and for the existence of the only world we could possibly know. Yet Fichte's Archimedean principle seemed even less durable than its predecessor. So how to justify it? He found the right way of steeling his principle with the aid of an instrumental model that Kant had casually introduced into the third *Critique*.

147. Ibid., p. 247: "The form by which the object comes to consciousness through the content is thus essentially different from the form that must be attributed to it outside the mind (in itself); and this latter form, through which the object is thought of as a thing-in-itself, is simply not otherwise conceivable [*vorstellbar*], *since we deny it to have the form of the representation.*" This appears, at first blush, to be a defective argument, since it seems plausible that the same content could have two different forms or modes of existence—that of a mental object and that of an extra-mental object. If, however, we consider a musical performance and its representation in the sheet music, we might be inclined to grant Reinhold his conclusion.

148. Ibid., p. 249: "The things-in-themselves can as little be denied as the representable objects themselves. They are those objects themselves in as far as they are not representable. They are those things that must lie outside of representation as the ground for the sheer content of representation."

In the first essay in which he formulated his fundamental principles of the *Wissenschaftslehre* (1794-95)—and later in the *Zweite Einleitung in die Wissenschaftslehre* (Second introduction to the Wissenschaftslehre, 1797) —Fichte referred to the productive (though implicit) awareness of the self in all acts of consciousness as an "intellectual intuition."[149] The phrase originated with Kant. The master had argued that human awareness of empirical objects could occur only through "sensuous intuition." This latter kind of intuition depended upon a manifold of sensation being *given* to human cognition. The faculty of intuition itself would weave spatial and temporal relations through the manifold and provide a conduit, as it were, for the understanding to further organize the manifold through the categories of cause, substance, and like formal structures. The result would be ordinary consciousness of the natural world. Kant held, however, that we could yet imagine another kind of understanding, one that cognized objects but did not require something to be passively received. This understanding would employ an intellectual intuition, an intuition that in its very cognizing would produce the object.[150] Moreover, since a mind so constituted would always be aware of itself, it would also be self-productive. Well, we could imagine such an understanding, with its intellectual intuition—God's understanding—but it was not the one with which we were endowed.[151]

Fichte slyly appropriated the notion of an intellectual intuition to endow the absolute ego with just those divine qualities Kant had denied to human cognition. The self, in the Fichtean mode, actively reproduced itself through self-awareness—indeed, the self essentially was the act of self-awareness. Yet the empirically produced self was constituted as well by its relationship to a constraining non-self, an external world. The productive act of self-awareness thus encompassed not only a finite self but a natural world. Both the finite self and the world of the non-self arose together in

149. Fichte, *Aenesidemus oder über die Fundamente der von dem Herrn Professor Reinhold in Jena gelieferten Elementar-Philosophie. Nebst einer Vertheidigung des Skepticismus gegen die Anmassungen der Vernunftkritik* (1794), in *Fichtes Werke*, 1:16. See also Fichte's *Zweite Einleitung in die Wissenschaftslehre*, in *Fichtes Werke*, 1:463.

150. Kant, *Kritik der reinen Vernunft*, in *Werke*, 2:144 (B 145).

151. Kant's notions about a creative, intellectual intuition were vintage. The medieval philosophers distinguished between our intellect, which understood objects by receiving something from them, and the divine intellect, which in understanding objects gave something to them, namely, their existence. Kant made this same distinction quite early on. In his *Inaugural Dissertation* (1770), he distinguished our mode of intuition of objects, which required something being given to sensibility, and divine intuition, which was productive, "an archetypal intellect and, accordingly, perfectly intellectual" [*est Archetypus et propterea perfecte intellectualis*]. See Kant, *De mundi sensibilis atque intelligibilis forma et principiis*, in *Werke*, 3:40–2 (A 12).

consciousness, and did so without any help from a completely external thing-in-itself. Self and its world, rather, issued from the creative activity of the absolute ego.

Intellectual intuition, whereby the self posited itself through self-awareness, stitched together acts of consciousness as *my* acts and posited them in relation to the empirical objects that they represented. When the natural world, therefore, appeared before consciousness, there too, silently, was the I. The I had to be mute, that is, implicitly present to consciousness, rather than objectively so. For if it were present to consciousness in an objective way, say, as a tree would be directly present in the perceptual awareness of an observer, that is, through a representation (*Vorstellung*), then there would be two representations in consciousness, one of the tree and one of the I. But then that representation of the I, as all representations in consciousness, would require the presence of another I to accompany it, which as represented, would require another I, and so on regressively into the infinite depths of dark confusion. Hence, according to Fichte, the intellectual intuition of the ever-present I, of the self-positing ego, could not come through a direct representation. Rather, he contended, we came to a knowledge of the intellectual intuition of the I through an inference.[152] From the fact that our consciousness had to be continuous and that we regarded it as "our" consciousness, we could conclude that synthetic consciousness was and had to be self-aware.[153] Indeed, to be self-aware formed for Fichte the very meaning of what it was to be a self, an I. Hence, within the structure of consciousness of objects, the self was *immediately* aware of an object of nature and *mediately* aware that it was the subject who perceived the object. The philosopher, through his inferential activity, would

152. Fichte, *Zweite Einleitung in die Wissenschaftslehre*, in *Fichtes Werke*, 1:464: "But if it must be admitted that there is no direct, isolated consciousness of the intellectual intuition, how does the philosopher come to the knowledge and the isolated representation of that intuition? I answer: without doubt, in the same way he comes to a knowledge and to the isolated representation of the sensible intuition, through an inference [*Schluss*] from the obvious facts of consciousness."

153. That each of us, with every moment of thought, commits an act of intellectual intuition seems well grounded. What would consciousness be like, *per impossibile*, without such an intuition? It might be like the disjointed frames of a movie film prior to being run through the projector. Each moment of consciousness would be flat, dead, disconnected, and obviously not a consciousness of the "real." Run through the projector and displayed on the screen, a real world appears. The action of the projector in synthesizing and bringing together the frames of the film would be something like the act of synthesizing consciousness. The crucial difference between this analogy and its analogate would be this: synthesizing consciousness must be implicitly aware that *it* is taking up each successive scene, implicitly aware that each frame of consciousness, as it were, is enacted by itself. The philosopher, then, attempts to bring this implicit awareness to assertable knowledge through the necessary conclusion he or she draws about its required existence.

make explicit these structural features of consciousness, and indeed might exercise with awareness what the self ordinarily accomplished below the limen of immediate reflection. In this latter instance, the philosopher's intuition would be both productive and *explicitly* aware of the act of production—an intellectual intuition properly so-called.

Fichte worked out many approaches to his position, and was never satisfied that he had achieved the convincing formulation. One of the most succinct presentations occurred in an article attacking a flabby Kantian, Christian Schmid. The gist of his analysis might be put thus: Consciousness is only a series of self-reproducing representations. The absolute ego is just this flow of representations. When considered from one perspective, these are representations of a natural world, and insofar as we become absorbed in the representations, all that appears to consciousness is that natural world of objects and their activities. If we step back and take another perspective on representations, we will be aware of them as subjective, as having an ego-pole. Fichte contended that Kant's thing-in-itself and ego-in-itself were just abstractions from representations. The thing-in-itself was only the abstract concept of thingness. But "it is simply dreaming, it is really madness when one takes the free productions of his imagination for reality."[154] (The skeptical realist might ask at this point, however, just who is taking the perspective, who is looking on the representations, regarding them now one way, now another.)

This absolute subjectivity, to which no antecedent category of causality was applicable—hence an absolutely free agent—became, in Fichte's analysis, the encompassing, productive activity that made superfluous any contradictory notions about things-in-themselves—that is, things supposedly not at all known, yet known enough for Kant to have posited them. Fichte, indeed, had caught Kant up in a dilemma: either reference to noumenal reality, as that standing behind our representations of the world, was a requirement of thought—but then it was only a feature of our cognitive apparatus, not of reality; or such reference to noumena was based on the belief that the manifold of sensations needed a cause—but then such reference became, in Kant's own terms, an illegitimate application of the causal category beyond the phenomenal realm.[155] The scientific conclusion to which we were driven was that absolute, free subjectivity created the world and everything in it. As Fichte baldly asserted to the philosopher Friedrich Jacobi:

154. Fichte, *Vergleichung des vom Herrn Prof. Schmid aufgestellten Systems mit der Wissenschafts-lehre* (1795), in *Fichtes Werke*, 2:450.

155. Fichte, *Zweite Einleitung in die Wissenschaftslehre*, in *Fichtes Werke*, 1:483.

> You are a well-known realist, and I a transcendental idealist, more un-
> compromising than Kant was; for he still had given a manifold of experi-
> ence—but God knows how and where it came from. I rather maintain—
> these are hard words—that even the manifold of sense has been produced
> by us out of our own creative faculty.[156]

The entire world of the not-I, everything that might have functioned as
limits to freedom, originated, according to Fichte, in the activities of the
absolute ego. Freedom, in this respect, existed as the soil from which sprang
all science and all human acts, from which grew a natural world.

Pragmatic and Metaphysical Interpretations of Fichte's Idealism

Fichte always claimed that his transcendental idealism was really "Kantianism properly understood,"[157] despite Kant's own protestations to the contrary. He had a compelling point. The world of nature and that of our
individual egos only come to us through consciousness. We are trapped
within the high, impenetrable walls of subjectivity. We may build up theoretical assumptions with the hope of leaping from them into an independent world not of our own making; but all known instruments for such
constructions have proven defective, allowing us only to climb up the rungs
of our own delusions. Kant made this much clear enough in the first *Critique*. But then he followed a more obscure passageway, attempting to
breach absolute subjectivity; he traced out the implications of natural resistance. We strive to move through phenomenal reality, but we are often
blocked. We wish for the moon, but it will not descend. For Kant the recal-
citrance of the apparent world suggested the reality of the nonapparent
world, that of things-in-themselves. These entities, he supposed, acted
upon human consciousness to produce sensations, which the Kantian mind
would organize into a coherent phenomenal realm. Fichte, though, regarded even sensation—something passively received according to Kant—
as actively constituted by intelligence, albeit constituted as limitations on
the self. Sensations, feelings yet have a givenness about them; they signal
resistance to and limitation of the self. What about these blind, stubborn
Anstossen, these checks and impulses directed against the self?

Fichte certainly did not deny them. They simply occurred within consciousness as ultimate facts. They appear as feelings of force—that is, as raw
feelings (the hots and colds, the reds and blues), which, at the highest level,

---

156. Fichte to Friedrich Jacobi (30 August 1795), in *Johann Gottlieb Fichte, Briefe*, ed. Manfred
Buhr, 2nd ed. (Leipzig: Verlag Philip Reclam jun., 1986), pp. 183–84.

157. Fichte, *Zweite Einleitung in die Wissenschaftslehre* (1797), in *Fichtes Werke*, 1:469.

become those resistances that we strive against in the world. It is these *Anstossen* that call absolute subjectivity to explain to itself their existence by positing a natural world whence they supposedly derive and a limited empirical ego into which they are supposedly received. These ultimate facts thus invoke absolute subjectivity to creative action.[158]

The ultimate contingencies, the resistances found within the very heart of subjectivity provoke the ego not only to create a natural world as countering the finite ego but also to strive against such contingency, to render the world of experience pliant to the demands of reason. Indeed, the very nature of absolute subjectivity reveals itself to be, as a condition of the possibility of any natural objects at all, an unending striving.[159] The Fichtean philosopher thus reaches below the limen of ordinary consciousness, and through a deductive conclusion brings to reflective awareness the very activity of ultimate subjectivity in its continuous, progressive striving against resistance. This means that the productive imagination, for Fichte the highest level of the ego's creative capacities, must have worked silently, subconsciously, to fabricate against a template of raw feelings of resistance both a world of external objects, as the barriers to overcome, and the finite ego, as that which receives the force of those presumed objects. The infinitely striving ego of the last part of Fichte's *Grundlage der gesammten Wissenschaftslehre* would become the foundation, in his more popular writings, for the moral imperative to act in the world to change it for the better. In that latter guise, it may seem to us less exotic, yet it had tremendous emotional impact on Fichte's students and friends, as I will describe in a moment.

But there is still the question of the *Anstoss*, that feeling of resistance that we meet in practical activity. According to some recent historians, Fichte's recognition of this irreducible constraint indicates that the philosopher was, as he claimed, in fundamental agreement with Kant—the *Anstoss*

---

158. Later in his *Erste Einleitung in die Wissenschaftslehre* (in *Fichtes Werke*, 1:441), Fichte suggested that the necessities and limitations experienced of objects ultimately were derived from limitations imposed on the self by the self: "Here then is simultaneously made comprehensible the feeling of necessity that the determined representation introduces: the intelligence thus feels not some impression from the outside, rather it feels in that action the limitations of its own nature." Fichte seems to have thoughts this explanation was required to eliminate any assumption of a force impinging from outside the mind. I suspect he borrowed this explanation from his young disciple Schelling. See the next chapter.

159. Fichte, *Grundlage der gesammten Wissenschaftslehre*, in *Fichtes Werke*, 1:261–62: "The result of our investigation up to now is, therefore, the following: the pure self-returning activity of the I is, in relation to a possible object, a striving [*Streben*]; and indeed, as demonstrated above, an unending striving. This unending striving, by reason of its endlessness, is the condition of the possibility of all objects: no striving, no object."

meant a mind-independent reality did exist.[160] Under this interpretation, Fichte's notion of a spontaneously self-positing ego, in which its activities would be completely free, was a regulative ideal: the self had to act in the real world *as if* it could make that world conform to the moral demands of reason. This would result in an infinite striving that could only rest in an ideal world fully compliant to the creative demands of the ego. The complete construction of a world by the ego would be, therefore, only a goal located at infinite temporal distance.

The difficulty with this interpretation lies in two fundamental considerations. The first is Fichte's frequent and unaltered decrying of Kant's notion of a thing-in-itself. This Kantian monstrosity never ceased to be, for Fichte, a risible idea. But the admission of a mind-independent source of the *Anstoss* simply reintroduces the derided conception. The second reason for hesitating to accept this interpretation is that if the freedom of the ego were to be merely regulative, really a demand to bring the world into compliance with the standards of moral reason, there would yet be no justification for this regulative ideal. Why would the Fichtean philosopher believe the causally determinate world would yield one inch to human striving? Kant himself thought the demands of freedom and the determinism of the natural world could only be reconciled by the regulative postulation of an *intellectus archetypus*, a divine intellect that would intervene, as it were, to adjust nature to the requirements of human freedom. Fichte had such an archetypal intellect, but it resided in the ego, not in a postulated God. And its actions had creative force, not merely methodological presumption. Only

---

160. Frederick Beiser, in his *German Idealism*, argues that Fichte regarded felt resistance as the sign of something outside the self. In this respect, Fichte was a realist much in the way Kant was. Beiser holds that Fichte's assertions about the ego creating a world were regulative: the individual has an obligation to alter the real world to meet moral ideals; the individual must act *as if* he or she were creating that ideal world. In like fashion, Wayne Martin—in his *Idealism and Objectivity: Understanding Fichte's Jena Project* (Stanford: Stanford University Press, 1997), pp. 75–77—maintains that Fichte did not abandon the thing-in-itself. But more fundamentally, Martin argues, Fichte was not interested in existence claims, either positively or negatively. Frederick Neuhouser, in *Fichte's Theory of Subjectivity*, pp. 53–65, maintains that Fichte, in his work of the late 1790s, turned his first principle of self-positing into a regulative principle. There is a sense, of course, in which it would be uncontroversial to say that Fichte retained the thing-in-itself. The ego might, after all, well fill that role—as Fichte himself suggested it might (*Erste Einleitung die Wissenschaftslehre*, in *Fichtes Werke*, 1:427–28). But in that case, he would not have denied, as Kant did, that we could have any knowledge of the thing-in-itself. Moreover, it is obvious that when Fichte explicitly denies the existence of the thing-in-itself, he was not denying the ego and its creative power. Aside from the admission of the ego as a thing-in-itself, however, the views of Beiser, Martin, and Neuhouser run across the grain of the account that I have put forward in the text. I believe my account represents the common understanding of Fichte during his period, an understanding, to be sure, that Fichte regarded often as a misunderstanding.

in the ego might a reconciliation between free self and determinate world be possible.

Fichte famously asserted that the kind of philosophy one engaged in was determined by the kind of person one was. The system of determinism, if one accepts its first principle of universal causality, has a coherency that is impregnable to theoretical objection. Fichte believed that only those philosophers whose education made them sensitive to the demands of moral reason, who felt the compulsion of moral imperatives, would be in a position to explore idealism as the system that explained the possibility of such recognition of moral constraint.[161] In this sense, practical reason, cultivated through the right kind of education, brought one to the position of reenacting the intellectual intuition that engendered the ego's creative self-realization. This practical propaedeutic does not, it seems to me, render the entire system of idealism a completely regulative affair, without theoretical implication. Were that the case, Fichte's idealism would have been defanged of metaphysical threat, which virtually all his opponents keenly felt.

Though Fichte never shifted his metaphysical position from what Schelling would dismissively call subjective idealism, in his later, more popular work he did emphasize the pragmatic foundation for that idealism. In *Die Bestimmung des Menschen* (The vocation of man, 1800), a friendly tract by which he hoped to salvage his finances after his expulsion from Jena in 1799, he made ever more clear that the choice of an idealistic system over a mechanistic one could not be made on epistemological grounds. The I would assert the idealistic system, not because it had to or because there were compelling antecedent reasons but simply because it willed to do so.[162] The decision was a free ethical choice, one, however, conforming to our "dignity and vocation" as human beings. This commitment to idealism amounted to the declaration of a metaphysical truth, namely, that the ego authored its own world. "In this way," Fichte concluded, "I always remain only *in myself* and in my own realm; everything that exists for me, develops purely and solely out of myself. Everywhere I intuit only myself and never any strange being outside of myself."[163]

---

161. Fichte, "Erste Einleitung in die Wissenschaftslehre" (1797), in *Fichtes Werke*, 1:432–34: "There is no principle of decision possible for reason: for it is not a question of adding an item in a series according to the rational principles governing the series; rather it is a question of the beginning of the whole series which, as an absolute first act, depends only on freedom of thought. . . . What kind of philosophy one chooses thus depends on what kind of man one is: for a philosophical system is not a dead stick of furniture that one can lay aside or select; rather it animates the very soul of the man who has it."

162. Fichte, *Die Bestimmung des Menschen*, in *Fichtes Werke*, 2:256.

163. Ibid., p. 300.

Though the warrant for the idealistic system was ethical, Fichte certainly did not regard its affirmation as merely regulative, an "as if" idealism. The truth of the system could not be made more secure, since all epistemological argument had to follow the acceptance of either the mechanistic or the idealist system. Idealism thus retained its metaphysical valence. It posited that the natural world was the self's double:

> The nature, in which I have commerce, is not an alien being that has been produced without regard for me, a being into which I could never penetrate. It is formed through my own laws of thought and must indeed conform to them. The natural world must be knowable and transparent even to its inmost depths.[164]

This kind of idealism offered the possibility that the world, through the procedures of natural science, could become fully open to reason. We might, within the domain of the idealistic system, treat physical objects as independent of the ego, but only in order to discover the laws governing their behavior, which laws would have their source in the ego. In this respect, Fichte allowed for a practical realism, though tucked safely within the embrace of a thoroughgoing idealism.

Schiller certainly understood Fichte's project in the more metaphysical way, as he made it sardonically explicit to Goethe: "According to Fichte's oral remarks, which do not appear in his book [*Grundlage der gesammten Wissenschaftslehre*], the I is creative through its representations, and all of reality is only in the I. The world is for him only a ball, which the I has thrown and which it again catches in reflection!! He ought, therefore, to have simply declared his divinity, something we expect any day now."[165]

Fichte's first disciples did not initially focus on the metaphysical ambiguities of his position. Rather they sought in Fichte's new philosophy a justification for their own projects. Novalis and Friedrich Schlegel, for instance, found in the Fichtean ego's longing to fly to a more perfect destiny the counterpart of the poet's desire to realize the beautiful object, to achieve infinite perfection in finite form, ultimately a desire that continually impelled

---

164. Ibid., p. 258.

165. Schiller to Goethe (28 October 1794), in *Der Briefwechsel zwischen Schiller und Goethe*, ed. Emil Staiger (Frankfurt a. M.: Insel Verlag, 1966), p. 62. Though Heine cautioned against this individualistic interpretation of Fichte's position, he could not resist describing the popular response to the philosopher: "Most people thought that the Fichtean I was the I of Johann Gottlieb Fichte, and that this individual I denied all other existences. What impertinence! exclaimed the good people. This man doesn't believe we exist, we who are fatter than he, and as mayor and administrator are really his superiors. The women asked: Doesn't he at least believe in the existence of his wife? No? And Madame Fichte puts up with this?" See Heine, *Religion und Philosophie in Deutschland*, in *Sämtliche Schriften*, 3:609–10.

but that could never be realized completely.[166] The scientific investigator, too, as Schelling would show, was driven by a passion to understand the natural world, to plunge to the very bottom of things—a quest that might succeed if the world sprang from the absolute ego of the researcher.[167] Such were the fruits plucked by Fichte's followers from the steady growth of essays, monographs, and revisions that constituted his *Wissenschaftslehre*. This doctrine of knowledge came to maturity during Fichte's time at Jena, and Friedrich Schlegel made it the seductive companion of his own aesthetic doctrines, though not without critical appraisal of this paramour's more tawdry traits.

Schlegel's Assessment of Fichte

Schlegel mused on the character of Fichte's improbable transcendentalism. In a notebook kept on Fichte's *Wissenschaftslehre*, he wrote:

> I have found no one who has believed Fichte. There are many who admire him, a few who know him, one or another who understands him. Fichte is really like a drunk who does not tire of climbing up on one side of a horse, transcending it, and falling off on the other side. He idealizes an opponent for the perfect representation of pure unphilosophy. Fichte's praxis goes more to the formation of men than of works (still less to the destruction of unworks).[168]

Schlegel became sensitive to two principal difficulties in Fichte's idealism. First, the starting point of philosophical understanding cannot be a single principle (such as the Fichtean principle that the ego posits its own existence for itself) whence all else is deduced. Rather, according to Schlegel, "philosophy must begin with infinitely many propositions"[169]—eclectic

166. Schlegel thought it the role of philosophy to quicken us to the desire for progressive development toward the infinite, just as political revolution (of the French sort) manifested the eternal becoming of man: "Philosophy tends toward the Absolute. From this proposition two corollaries for philosophy follow: 1) philosophy should develop in all men the desire for the infinite; 2) the appearance of finitude should be rejected; and in order to do that, all knowledge must be rendered revolutionary." This passage comes from lectures Schlegel delivered as *Privatdozent* at Jena in 1800. See *Kritische Friedrich-Schlegel-Ausgabe*, 12:11.

167. Friedrich Schelling, *Ideen zu einer Philosophie der Natur*, in *Schellings Werke*, ed. Manfred Schröter, 3rd ed., 12 vols. (Munich: C. H. Beck, 1927–59), 1:697: "As long as I myself am identical with nature, I understand what living nature is as well as I understand myself." Schelling's theory of the identification of self with nature will be explored in the next chapter.

168. This observation comes from Schlegel's manuscript *Geist der Fichtischen Wissenschaftslehre* (1797–98). See *Kritische Friedrich-Schlegel-Ausgabe*, 18:32.

169. This fragment comes from notes Schlegel kept during 1797. See *Kritische Friedrich-Schlegel-Ausgabe*, 18:26.

but historically structured observations that would contribute to the building up of philosophical doctrine. Second, Fichte's reduction of nature to a product of the absolute ego struck Schlegel as myopic. Employing lenses ground by Spinoza, Schlegel perceived nature and the absolute ego to be ultimately one. Neither had a privileged position. As Schlegel put it: "The minimum of the ego is the maximum of nature; and the minimum of nature is the same as the maximum of the ego."[170]

Though Schlegel harbored some decisive reservations about Fichtean idealism, he nonetheless would adopt the basic transcendental approach in his effort to construct a new kind of aesthetic criticism and in his understanding of the fit between human nature and the character of Romantic poetry. In the first instance, he conceived of the style of the fragmentary method itself—most perspicuous in the *Athenaeum*—in Fichtean terms. Critical remarks in the fragmentary mode were simultaneously "subjective," insofar as they came creatively from a particular consciousness, and "objective," insofar as they were structurally and morally valid. Moreover, they built up in tessellated fashion a progressive, critical project, one based on historical analyses of past literature. Fragments and the projects they gradually realized could thus be called "the transcendental component of the historical spirit."[171] The more important Fichtean impress on Schlegel, though, helped form the essential nature of the new critical approach to poetry, which I will discuss in more detail further on.

The complementary political side of Fichte's thought also greatly appealed to Schlegel. Just before he arrived at Jena, Fichte published, anonymously, a tract defending the French Revolution.[172] In that work he argued for a social-contract theory of democratically established monarchy, with a

170. This aphorism comes from Schlegel's lectures on "Transcendental Philosophy," which he delivered as *Privatdozent* in Jena in 1800. See *Kritische Friedrich-Schlegel-Ausgabe*, 12:6.

171. *Kritische Friedrich-Schlegel-Ausgabe*, 2:169. I have based this interpretation of the "transcendental fragmentary mode" on Schlegel's *Gespräch über die Poesie* (discussed below) and on fragment 22 of the *Athenaeum Fragmente* (in *Kritische Friedrich-Schlegel-Ausgabe*, 2:168–69). The latter reads: "A project is the subjective embryo of a developing object. A complete project would have to be simultaneously wholly subjective and wholly objective, an indivisible and living individual. According to its origin, it is subjective, original, and only possible in this consciousness; but according to its character, it is completely objective, structurally and morally necessary. The feeling for projects, which one could call fragments out of the future, is different from the feeling for fragments out of the past only by reason of the direction—progressive in the one case, regressive in the other. The essence is the ability to idealize and to realize directly and simultaneously, to complete and gradually to build up objects in oneself. Since the transcendental relates to connecting or separating the ideal and the real, one can just as well say that the feeling for fragments and for projects is the transcendental component of the historical spirit."

172. Fichte, *Beitrag zur Berichtigung der Urtheile des Pulbikums über die französische Revolution* (Contribution to the emendation of the judgment of the public on the French Revolution, first part in two sections, 1793 and 1794), in *J. G. Fichte—Gesamtausgabe*, 1.1:202–404.

defense of the right of a people to revolt should the contract be abrogated by the government. The authorship of this politically sensitive book could not remain secret, and a row occurred among officials at Weimar over the possibility of a revolutionary professor being hired for Jena. Gottlieb Hufeland, professor in the law faculty, wrote Fichte in December 1793 that he had to counter the opposition to the appointment by maintaining that the young philosopher "defended the democratic party only in respect of the law and completely in an abstract way."[173] The official invitation to join the university reached Fichte in January 1794, and in May he entered the philosophy faculty. His reputation as a revolutionary thinker and the particular nature of his theoretical considerations, though, continued to nourish the suspicion that the university had let slip into its midst not merely a philosophical Jacobin, but an activist ready to shatter the chains restraining student revolutionary inclinations.

The Professor at Jena

Fichte began his university career with a private evening seminar on his *Wissenschaftslehre*.[174] The course initially drew a quite modest number of students, a couple of professors, and the young, beautiful, and accomplished wife of the university librarian, Sophie Mereau (1770–1806).[175] Because of the severe difficulty of the philosophical inquiry, the private seminars never exceeded the capacity of the small seminar room in his home. His inaugural public lectures on *Die Bestimmung des Gelehrten* (The vocation of

173. Gottlieb Hufeland to Fichte (December 1793), quoted by Lauth and Jacob in the "Vorwort" of ibid., 1.1:198.

174. These private lectures became the source for Fichte's most famous work, the *Grundlage der gesammten Wissenschaftslehre* (1794-95). The oral delivery, however, must have had a rather different character than the printed version. For Fichte had the reputation of one who encouraged his students through friendly discussions at these evening gatherings at his home. He thus displayed an openness with students unusual for the time. He also, it seems, made an unstinting effort to clarify difficult problems in his doctrine. Henrik Steffens recalled his own experience as a student with Fichte in fall 1798: "I cannot deny that I was impressed the first time I saw this short, compact man with sharply defined features. His speech itself had a decisive sharpness; and recognizing the weakness of his audience, he sought every way possible to make himself understood. He made every possible effort to demonstrate what he said. But his lecture nonetheless seemed to be ordering us, as if he wanted by a single command, which we had unconditionally to obey, to dismiss every doubt." See Steffens, *Was Ich Erlebte*, 4:79.

175. The twenty-four-year old Sophie Mereau was a recently published novelist who attracted a number of admirers, as Johann Georg Rist, a young student at the time, recalled: she had "a charmingly small figure, delicate and refined, full of grace and feeling. . . . She was at the time greatly celebrated by everyone who had any taste and sensitivity; wherever she appeared, people pressed around her and a thick swarm of admirers, standing just around her, would cling to her every word or smile." Rist as quoted by Ziolkowski, in *Das Wunderjahr in Jena*, p. 72.

the scholar, 1794), by contrast, drew large and wildly enthusiastic audiences—often up to five hundred students, professors, and ordinary citizens crammed into the lecture hall. Friedrich Karl Forberg (1770–1848)—a biblical scholar at Jena who became implicated in Fichte's later troubles with the Weimar administration—recorded his initial reaction to hearing the philosopher lecture:

> He roars like a thunderstorm that unleashes its fire in a single strike. He doesn't becalm you like Reinhold, but elevates your soul. Reinhold regards you as if he would make you a good man; Fichte wishes to make you a great man. . . . By his spirit he wants to direct the spirit of this age. . . . He possesses more wit, more incisiveness, more penetration, more spirit, in short, more power of mind than Reinhold. He doesn't have a rich imagination but an energetic and powerful one. His images are not charming but bold and blunt. He plumbs the innermost depths of his subject and moves through the realm of concepts so naturally as to suggest he not only dwells in this invisible realm, he rules it.[176]

In his inaugural lecture, Fichte situated the obligation of the scholar within the context of the vocation of man in general. As human beings we were obliged to nothing less than perfection, both in life and in work, which—pace Rousseau—required us to discipline the irrational, rendering it transformed through intellectual culture. "The unending striving after perfection," he preached to his rapt audience, "is the vocation of man. He is the kind of being that exists to become always more virtuous, and thus to make everything sensible around him also better, and, insofar as he enters society, also to make it ethically better—and in this way he makes himself happier."[177] The goal of perfection is unreachable, but every human being as such—"rational, though finite, sensible, though free"—is compelled to pursue it. The scholar, by reason of his knowledge and abilities to communicate, has been specially anointed for leading others toward their human vocation. But he cannot, Fichte continued, perform his task from the armchair. Then in a peroration that must have sent a whiff of musket powder through the lecture hall, he exclaimed: "To stand around and complain about the corruption of mankind, without lifting a hand to alleviate it, is womanish. . . . Act! Act! That is why we are here."[178]

176. Friedrich Karl Forberg, *Fragmente aus meinen Papieren* (Jena: Voigt, 1796), pp. 71–72.

177. Fichte, *Einige Vorlesungen über die Bestimmung des Gelehrten*, in *Gesamtausgabe*, 1.3:32.

178. Ibid., 1.3:67. Friedrich Karl Forberg, a Kantian, remarked at the time: "Fichte intends to have an impact on the world through his philosophy. He carefully nourishes and cultivates—and thus attempts to bring to fruition—the tendency to disquieting activity that dwells in the breast of every noble youth. At every opportunity he announces the principle that the vocation of man should be action, action. But one has to fear that the majority of youth who take this to heart assume that the invitation to action is nothing more than an invitation to destruction. Moreover, the

Fichte's lectures hardly stanched fears that the university had indeed hired a subversive. Rumors about his performance shot back to Weimar, and Fichte himself felt compelled to assure Privy Councillor Goethe that he was not involved in any political movement and would only venture an opinion touching on politics insofar as it pertained to the very abstract courses he might offer in natural law and ethics. Nonetheless, as he further explained to Goethe, he would not repudiate his earlier defense of the French Revolution, something he had been advised to do. He could not abandon his book since it was hardly a secret whom the author was; and in any case, there were no propositions he wished to deny. Fichte concluded his defense with a mild threat, one familiar to academic administrators yet today: he knew that a position awaited him in the more tolerant Switzerland.[179] Goethe and Carl August assured their highly strung professor that they only held him in the highest regard. Kantian philosophers, they undoubtedly felt, were hardly the sort of souls that might rouse students to mayhem.

Friedrich Schlegel, a student ready for blood—of a more abstract kind—yet heard the call as he read the first five of Fichte's lectures while still in Dresden. He detected therein more than the Weimar administration did. He wrote his brother Wilhelm in August 1795: "Schiller and [Wilhelm von] Humboldt dabble much in metaphysics, but they have not digested Kant, and now suffer indigestion and colic. . . . Compare the spellbinding eloquence of this man [Fichte] in his *Lectures on the Vocation of the Scholar* with Schiller's stylized declamatory exercises. He is the kind of person after whom Hamlet sighed in vain. Every aspect of his public life seems to say: this is a *Man*."[180]

Fichte was perhaps more Hamlet's equal than Schlegel thought. He certainly had his Rosencrantz and Guildenstern, in the form of student fraternities. The Weimar administration had banned these societies of social drinkers and duelers, though they continued to meet and make life unpleasant around Jena. Fichte condemned them, and they in turn disrupted his classroom and harassed him at home. Likely the students received encouragement from some disgruntled faculty members, who had taken great exception to that vaunted ego of suspect political tendencies, especially when that ego had begun lecturing during the time of religious services on

---

proposition is false. Man is not made to act, but to act correctly, and if he cannot act without acting correctly, he must remain at rest." Quoted by Fritz Medicus, "Einleitung," in *J. G. Fichtes Werke*, 6 vols. (Leipzig: Fritz Eckardt Verlag, 1911), 1:lx–lxi.

179. Fichte to Goethe (24 June 1794), in *Johann Gottlieb Fichte, Briefe*, pp. 113–16.

180. Friedrich Schlegel to Wilhelm Schlegel (17 August 1795), in *Kritische Friedrich-Schlegel-Ausgabe*, 23:248.

Figure 2.8
Friedrich Schlegel, about
1800. Artist unknown.
(Courtesy Stiftung
Weimarer Klassik.)

Sunday mornings—he was even thought to have been attempting to establish a church of reason, a sect to promote the aspirations of the French Revolution. Goethe, receiving Fichte's numerous complaints about student rowdyism, wryly observed that having a stone thrown through your window was "the most unpleasant way to become convinced of the existence of the not-I."[181] But the contretemps with obstreperous fraternities proved less hazardous than, strangely, the more abstract considerations that ultimately proved his undoing in Jena.

From the beginning of his career in Jena, Fichte faced growing opposition from more orthodox Kantians, and then in 1799 from Kant himself, who rejected any effort to pass off Fichte's idealism as compatible with the critical philosophy.[182] The opposition came to final victory in Jena with the renewal of charges of atheism against Fichte. In 1798, in response to new stories about such tendencies in his philosophy, Fichte published an essay to clarify his position on the God-question.[183] He certainly agreed

181. Goethe, *Tag- und Jahres-Hefte* (1795), in *Sämtliche Werke*, 14:41.

182. Kant published an open letter declaring "that I hold Fichte's *Wissenschaftslehre* to be a completely indefensible system." See Kant, "Erklärung in Beziehung auf Fichtes Wissenschaftslehre" (7 August 1799) in *Briefwechsel von Imm. Kant* 3:267–69.

183. Fichte, *Ueber den Grund unseres Glaubens an eine göttliche Weltregierung* (1798), in *Fichtes Werke*, 5:177–89.

with Kant—usually a safe thing to do—that we could have no theoretical knowledge of God, whose existence must be beyond our experience. But we had another kind of access, as indeed Kant had also suggested, through moral experience. But whereas Kant had used such experience as a ground for postulating the practical necessity of belief in a transcendent God, Fichte simply identified God with the moral order of the universe—not an abstract, indefinite order, to be sure; rather the entire free activity of absolute subjectivity, through which individual acts occurred.[184] But this could not be disguised as anything other than idealistic pantheism, hardly more palatable than the similar, if harder, identification of God with nature made by Spinoza. Calls for Fichte's dismissal intensified among the Saxon duchies and at Weimar. The authorities considered offering a mild rebuke as a way of assuaging all parties. In the meantime, however, Fichte wrote an angry letter to Privy Councillor Voigt; he declared the charges of atheism ridiculous and threatened resignation should even a hint of censure be issued.[185] Fichte's arrogance, quarrelsomeness, and overripe sensitivity were more than Carl August could stomach. He sent a letter of censure, mild enough, to the university senate and included a note that the philosopher had tendered his resignation, which would, he said, be accepted.[186] Goethe, despite the difficulties, hated to lose so gifted and irreplaceable a thinker.[187] As Heine later remarked, Goethe, whose private religious views hardly differed from Fichte's, reproached the philosopher only because "he had said what he thought and had not spoken in the usual veiled terms."[188] Fichte made frantic efforts to retrieve his fortunes, but they proved unavailing. In June 1799 he left for Berlin.[189]

184. Ibid., 5:186: "That living and acting moral order is itself God; we require no other God and we can conceive of no other. . . . That world order is the absolute first [ground] of all objective cognition, just as your freedom and moral vocation constitute the absolute first ground of all subjective cognition."

185. Fichte to Christian Gottlob Voigt (22 March 1799), in *Johann Gottlieb Fichte, Briefe*, pp. 218–21.

186. Carl August to the University of Jena (29–31 March 1799), in *J. G. Fichte im Gespräch*, ed. Erich Fuchs, 6 vols. (Stuttgart-Bad Cannstatt: Frommann-Holzboog, 1980), 2:90–91.

187. Goethe to J. G. Schlösser (30 August 1799), in *Goethe-Briefe*, ed. Philipp Stein, 8 vols. (Berlin: Wertbuchhandel, 1924), 4:312–13.

188. Heinrich Heine, *Zur Geschichte der Religion und Philosophie in Deutschland*, in *Sämtliche Schriften*, 3:617.

189. W. Daniel Wilson depicts Fichte's dismissal, along with several other incidents of Goethe's early Weimar years, as another case of the great poet's conspiring to restrict freedom of thought and expression in the duchy. He suggests that initially Fichte rather cravenly accepted restrictions on his republican political sentiments in order to secure his professorship. He maintains that ultimately this self-censorship did little good and that the real reason for his dismissal was the determination of Carl August to extirpate any dissenting political sentiment. Wilson suggests, without any proof, that Goethe's expression of regret at losing Fichte was hypocritical. See W. Daniel Wilson,

The Salons of Berlin

While the Austrian empire remained at war with the French, Prussia enjoyed a period of relative calm. Friedrich Wilhelm II had signed a separate peace, the Treaty of Basel, with the Revolutionary government in 1795. The agreement required him to allow French occupation of the Rhineland, but it freed him for taking more than his share of the now completely dismembered Poland. During the next decade Prussia remained neutral, while the Austrian empire and its shifting alliances (usually including England and Russia) had to defend against the constant attacks of the French, led now by its supreme commander, Napoléon Bonaparte. The neutrality of Prussia allowed the arts and sciences to flourish.

During this period of Prussian peace, Henriette Herz (1764–1847) entertained in her Berlin salon some of the most interesting young writers and thinkers active at the turn of the century.[190] In her parlor could be found at various times the brothers Wilhelm and Alexander von Humboldt, the theologian Friedrich Schleiermacher, the novelists Carl Philipp Moritz (1757–1793) and Ludwig Tieck (1773–1853), and various high government officials. Henriette gathered around her the literarily inclined, while the taste of her husband Marcus Herz (1747–1803), physician and student of Kant, shaded toward those of a scientific bent. The Herz household thus became the location not only of Henriette's literary tea parties and dinners, but of informal lectures and scientific demonstrations, usually conducted by Marcus. This is where Alexander von Humboldt first encountered replications of Galvani's experiments, which would stimulate him to conduct his own research in electrophysiology. Since, however, Henriette was seventeen years younger than her husband, electrophysiology was not the only subject that attracted an admiring audience. Alexander von Humboldt said of Henriette that "she is the most beautiful and also the smartest—no! I should say the wisest of women."[191] Given Humboldt's own stunted heterosexual inclinations, the compliments must be regarded as understated. Other feminine interests also captured the attention of the salon-goers. Dorothea Veit (1763–1839), oldest daughter of the philosopher Moses Mendelssohn (1729–1786), frequented the discussions and parties held at

---

Das Goethe-Tabu: Protest und Menschenrechte im klassischen Weimar (Munich: Deutscher Taschenbuch Verlag, 1999), pp. 243–48.

190. For an engaging, if statistically spiked, portrait of Jewish intellectual life in Berlin at the turn of the century—especially the salons of Henriette Herz and Rahel Levin—see Deborah Hertz, *Jewish High Society in Old Regime Berlin* (New Haven: Yale University Press, 1988).

191. Alexander von Humboldt to Wilhelm Gabriel Wegener (8 May 1788), in *Die Jugendbriefe Alexander von Humboldts, 1787–1799*, ed. Ilse Jahn and Fritz Lange (Berlin: Akademie-Verlag, 1973), p. 7.

22 Neue Friedrichstrasse. The intellectual stimulation she received there compensated for her dreary existence at home with her banker husband—perhaps overcompensated. At the Herzes's she met and fell passionately in love with Friedrich Schlegel.

Dorothea Veit

Schlegel arrived in Berlin in June 1797. He had left Jena suddenly for reasons that are not clear. Certainly the difficulty with Schiller made life in the small city rather uncomfortable. Perhaps his still painful love for Caroline made the proximity unbearable. He wrote to Hardenberg just before his departure, "I now go out into the world so easily and so unselfconsciously as if I owned it."[192] He met Dorothea at the Herzes's home that summer and became immediately attracted to her. She was like Caroline in many ways, virtually the same age and with a young four-year-old son. She also had a gift for languages and considerable literary ability, which must have transformed her features before Schlegel's eyes—since in respect of physical beauty, she was not much like Caroline. By fall their relationship had intensified; and one can—with due caution about such exercises—recover Schlegel's feelings for her in the descriptions offered by his alter ego Julius in the novel *Lucinde*:

> Lucinde had a decided romantic penchant. He was struck by the similarities to himself he recognized, and he discovered yet more. . . . What was real for her was only what she loved and honored in her heart, nothing else mattered. . . . She confided to him, not without deep pain, that she had already given birth to a beautiful, strong boy, which death had torn from her. . . . He could not restrain himself, he pressed a modest kiss upon her fresh lips and fiery eyes. With infinite relish he felt that divine head of this noble form sink upon his shoulder, the black locks flowing over the snow of her full breasts and beautiful back. And he said softly "wonderful woman," just as the damned party unexpectedly came back into the room.[193]

The readers of these passages did not have their narrative hopes disappointed. In the next paragraph, the couple finds solitude and consummation:

---

192. Friedrich Schlegel to Hardenberg (14 June 1797), in *Kritische Friedrich-Schlegel-Ausgabe*, 23:371.

193. Schlegel, *Lucinde*, p. 91.

One doesn't seek from a goddess, he thought, what is deemed only transitional and a means to an end, but one makes known immediately, with
openness and confidence, the goal of all one's desires. So he asked of her,
with innocent frankness, for everything a lover can request, and described to her in flowing periods how his passion would destroy him were
she to become too coy.[194] . . . She couldn't make a decision, but would
leave everything to chance. . . . In only a few days, when they were alone,
she gave herself to him forever and opened to him her magnificent soul,
and all of power, nature, and blessedness that in it lay.[195]

When Dorothea left Berlin with Schlegel in 1799, she abandoned virtually
everything except her child. But she achieved freedom, freedom from what
she regarded as a "long slavery."[196] Because of her fear of losing her son in a
custody battle, she did not marry Schlegel until 1804. Until then, however,
she lived openly with Friedrich—and with Wilhelm and Caroline, in
something of tense ménage à quartre, more combustible than Friedrich had
anticipated.[197] The internal pressures within the house would pass beyond
containment when a fire was ignited in Caroline by the shy smile and
volatile intellect of the very young philosopher Friedrich Schelling. The
explosion would not occur, however, until the turn into the new century.

194. The twenty-first-century reader must fall into complete sympathy with Schlegel upon
coming across a line that seem to have been used by every male adolescent since time immemorial.

195. Ibid., p. 92. Joanna Dunn pointed out to me that what particularly attracted Julius to Lucinde, and presumably Friedrich to Dorothea, was her maternal character, as is suggested by such
passages as "Julius found his boyhood again in Lucinde's arms. The voluptuous figure of her beautiful ripeness was more stimulating to the passion of his love than the fresh charm of the breasts and
mirrorlike clarity of a virgin's body. The ravishing power and warmth of her embrace was more than
that of a young lady; she had an aura of feeling and depth that only a mother can possess" (p. 93).

196. Dorothea Veit to Karl Gustav von Brinckmann (2 February 1799), quoted in Franz Deibel,
Dorothea Schlegel als Schriftstellerin im Zusammenhang mit der romantischen Schule (Berlin: Mayer &
Müller, 1905), pp. 160–61: "Three weeks ago I separated from V., after many arguments and scenes,
and much hesitation and doubt; and I live alone. Out of this shipwreck—which has freed me from
a long slavery—I have saved only a very small income, from which I can live with only the greatest
economy, a good cheerful attitude, [and] my Philipp. . . ." Her son Philipp Veit (1793–1877) would
later achieve fame as a painter, especially of religious themes.

197. Friedrich saw the foursome engaged in something of a group trial marriage, or at least that
is what is suggested by one of his *Athenaeum Fragmente* (no. 34): "Almost all marriages are only con-
cubinages, clumsy unions, or, rather, provisional attempts, and distant approximations to real marriage. The proper essence of marriage, not according to the paradoxes of this or that system, but
according to every spiritual and earthly law, consists in this: that several people should become
one. . . . It is hard to see what fundamental objection one could have to a marriage *à quartre*." See
Kritische Friedrich-Schlegel-Ausgabe, 2:170. Hegel execrated such sentiments as these, as well as
Schlegel's celebration of free love and deprecation of bourgeois marriage in *Lucinde*. See Georg
Friedrich Hegel, *Hegel's Philosophy of Right*, trans. T. M. Knox (Oxford: Oxford University Press,
1967), p. 263 (addition to paragraph 164).

Friedrich Schleiermacher: The Poetics and Erotics of Religion

The other relationship that Schlegel formed in Berlin—as powerful in its own way as that with Dorothea—was with Friedrich Schleiermacher, whose views would profoundly influence his aesthetic and religious philosophy.[198] Indeed, Schleiermacher's ideas would reverberate through the nineteenth century, and many naturalists would find his way of reconciling science and religion quite congenial—certainly Ernst Haeckel did.

Friedrich Ernst Daniel Schleiermacher, the son of a Reformed Church pastor, was born in Breslau in 1768. After due consideration of proper academic and religious training, his father enrolled him in the school of the Moravian Brotherhood (Herrnhuter), where he received a pietistic education similar to that of Kant. In 1785 Schleiermacher transferred to the Seminary of the Brotherhood in Barby, where, in some secret, he began reading Goethe and Kant, who shifted the young student into an extreme heterodox mode. Schleiermacher finally had to confess to his father that he no longer believed in the divinity of Christ or in eternal punishment.[199] He then pleaded with his father to allow him to transfer to the university at Halle, where he might live with his uncle, who taught church history at the Reform gymnasium in the city. His father, after much soul searching—and money counting, since he had to pay the bills—acquiesced. Schleiermacher matriculated as a theology student at Halle (1787), but remained true to his first loves—Goethe and Kant. During his two years at the university, he broadened his philosophical fare to include Plato (whom he would later translate into German) and Spinoza, the philosopher whose mystical love of God transcended the Euclidean approach to metaphysics and morals that characterized his most famous book, the *Ethica*.[200] The theological faculty at Halle had retained its allegiance to the extreme rationalism of Christian Wolff (1679–1754), who had made the university famous during the early part of the century. The objections his teachers had to the new critical ideas of Kant merely convinced this recalcitrant student to engage in more profound study of the Königsberg sage. Through the writings of

---

198. For brief but illuminating treatments of Schleiermacher and his relation to science, see Friedrich Gregory, *Nature Lost? Natural Science and the German Theological Traditions of the Nineteenth Century* (Cambridge: Harvard University Press, 1992), pp. 34–38; and "Theology and the Sciences in the German Romantic Period," in *Romanticism and the Sciences*, ed. Andrew Cunningham and Nicholas Jardine (Cambridge: Cambridge University Press, 1990), pp. 69–81.

199. Friedrich Schleiermacher to J. G. A. Schleyermacher (21 January 1787), in *Friedrich Daniel Ernst Schleiermacher, Kritische Gesamtausgabe*, ed. Hans-Joachim Birkner et al., 11 vols. to date (Berlin: Walter de Gruyter, 1980–), 5.1:50.

200. Schleiermacher probably got most of his knowledge of Spinoza indirectly through F. H. Jacobi's *Über die Lehre des Spinoza in Briefen an Herrn Moses Mendelssohn* (Breslau, 1785).

Kant, remarked Schleiermacher's great biographer Wilhelm Dilthey, the young theology student "learned, as it were, to think."[201] While at Halle, Schleiermacher overcame the religious doubts he had harbored and, in 1790, took the series of exams in theology that permitted him to preach in the kingdom of Prussia.

Schleiermacher's father, in some despair of the future of his son's soul—and his son's finances—counseled him, while at Halle, to learn English well, so as to be able to teach it. He also wisely advised him to avoid acquiring an Irish or Scottish accent. After leaving the university, Schleiermacher made good use of his linguistic abilities—which included Latin, Greek, and Hebrew—in his capacity as tutor to a noble east Prussian family, that of the Count Dohna-Schlobitten. While attending to the education of the family's children, he awakened, as he confessed, to "a feeling for women," a delicate service provided him by the young Countess Friederike.[202] That feeling would quicken later in his relationships with Henriette Herz and Elenore Grunow, both, however, encumbered by husbands.

After a brief stint in Landesberg as a preacher, Schleiermacher moved in 1796 to a position as curate at Berlin's famous Charité hospital. In Berlin he renewed his acquaintance with Count Alexander von Dohna-Schlobitten (1771–1831), brother of his former intellectual wards, who introduced him to the society of Marcus and Henriette Herz. Despite warnings of friends about the dangers to his advancement in the church by consorting with even assimilated Jews, Schleiermacher become devoted to Henriette and she to him. She had married Marcus at age fifteen (engaged at twelve) and regarded him as more a companionable father figure. In the dense gatherings of her many friends and admirers, Schleiermacher yet filled a void. He would visit the Herz household frequently, usually in the afternoons when Marcus was away. He taught Henriette Greek—the ways to a woman's heart being mysterious; discussed with her the literature and philosophy of the day; and carried on a very large correspondence with "Dear Jette." So intense seemed their relationship that many gossiped that they were lovers.[203]

201. Wilhelm Dilthey, *Leben Schleiermachers*, vols. 13 and 14, *Gesammelte Schriften*, 3rd ed., (Göttingen: Vandenhoeck & Ruprecht, 1991), 13.1:88.

202. He later described this advance in his education to his beloved Elenore Grunow. See Schleiermacher to Elenore Grunow (19 August 1802), in *Schleiermacher/Briefe* (Jena: Eugen Diederichs, 1906), p. 240.

203. Friedrich Schlegel almost wished they had become lovers, since, as he believed, only in sensual love (with its complementary intellectual side) could the highest levels of poetic existence be realized. Schleiermacher's platonic relationship with this older woman—whom Schlegel would take to calling "the ancient"—would, Schlegel thought, corrupt his character and also their own friendship. Obviously Schlegel had in mind a conceptual age difference rather than a chronologi-

Schleiermacher made the casual acquaintance of Friedrich Schlegel at the meeting of a literary society, the "Wednesday Club," but came to know him on more intimate terms at the Herzes's. Schleiermacher, himself profoundly learned, reacted to Schlegel in the way many did, with amazement that so much knowledge could be crammed into a fellow so young. He described his new friend to his sister:

> He is a young man of twenty-five years, and of so extensive a knowledge that one cannot understand how it is possible that someone so young can know so much; he is an original mind that far surpasses all others, even here where mind and talent abound. In his comportment he displays a naturalness, an openness, and a childlike youthfulness, which added to everything else is by far the most amazing. Because of his wit, as well as because of his lack of self-consciousness, he is regarded the most pleasant companion wherever he goes. For me, though, he is more than that—he is a tremendously essential need.[204]

Their commonality of interests and mutual admiration, plus some undoubted convenience, suggested that the two new friends become roommates. They began sharing quarters near the Oranienburg Gate (not far from the Charité) in late December 1797, and their arrangement lasted almost two years, until the beginning of September 1799, when Schlegel returned to Jena. They worked out together a kind of domestic schedule of morning study, early afternoon dinner, late afternoon and early evening duties and visits; Schlegel did his serious work quite early in the morning and retired about 10:30 in the evening, while Schleiermacher would continue to read until 2:00 in the morning, arising at 8:30. Their friends joked that they had entered a kind of marriage, with the slight and more reserved Schleiermacher taking the wifely role and the energetic and intellectually aggressive Schlegel that of the husband.[205] This relationship produced an important intellectual offspring. Schlegel insisted that his companion had to write something significant, as Schleiermacher mentioned to his sister: "He raised a choir [with friends] . . . that I should get busy and write books. Twenty-nine years old, and nothing accomplished! He wouldn't lay off of that, so I had to give him my hand in promise that I would write something

---

cal, since Herz was only four years older than Schleiermacher; Dorothea was nine years older than Schlegel! See Friedrich Schlegel to Caroline Böhmer Schlegel (middle of December 1798), in *Kritische Friedrich-Schlegel-Ausgabe*, 24:211.

204. Friedrich Schleiermacher to Charlotte Schleiermacher (22 October 1797), in *Schleiermacher, Kritische Gesamtausgabe*, 5.2:177.

205. Friedrich Schleiermacher to Charlotte Schleiermacher (21 November to 31 December 1797), in ibid., 5.2:219.

of my own during this year."[206] Before the year 1798 was out, Schleiermacher had contributed some thirty or so fragments to the first volume of the Schlegels' *Athenaeum*, had published an essay on the foundations of sociality,[207] and had begun work on a document that would become both an electrified pole of religious thought in the nineteenth century—powerfully attracting some and repulsing others—and an animating stimulus to German Romanticism. The work was his *Über die Religion: Reden an die Gebildeten unter ihren Verächtern* (On religion: talks to the cultured people amongst its despisers, 1799).

Schleiermacher's *Über die Religion*

The cultured people that Schleiermacher's treatise addressed were the likes of his friends Friedrich Schlegel and Henriette Herz, and the circle that formed around them—those who have "no other household gods than the sayings of the wise and the songs of the poets, and who have completely taken up humanity and fatherland, art and science . . . so that nothing remains, no feeling for that eternal and holy being which lies far on the other side of the world."[208] Schleiermacher wanted to persuade his friends, to exhort them to look both inwardly into the depths of themselves and outwardly to the infinite universe, and to find in that commerce a source of authentic religion, a fount of pure rapture that hardly resembled the meandering streams of muddy principles trickling from contemporary churchmen and philosophers. He urged that immediate intuition and heightened sensitivity, both cultivated by art and love, could release oceanic feelings that rose up to the "holy being," that infinite, universal existence which sustains our finite, individual lives.

Schleiermacher built his conviction against two opposing forces: the church theologians then regnant in Berlin, those preachers of Enlightenment theology who attempted to reconcile Scripture with right reason, making religion safe for the prudent; and the philosophers, particularly the British empiricists and the Kantians, including Fichte. The British retained a polite acquiescence in religion, but their pedestrian empiricism allowed to belief only the function of social utility. The Germans, he thought, had a deeper insight into the nature of religion, but even they strayed from the path laid by Spinoza. Fichtean idealism, for instance, began from a prem-

---

206. Ibid., 5.2:213.

207. Schleiermacher, "Versuch einer Theorie des geselligen Betragens," in ibid., 1.2:165–84; first published in *Berlinisches Archiv der Zeit und ihres Geschmacks*, 5, part 1 (1799).

208. Schleiermacher, *Über die Religion: Reden an die Gebildeten unter ihren Verächtern*, in Schleiermacher, *Kritische Gesamtausgabe*, 1.2:189.

Figure 2.9
Friedrich Daniel
Schleiermacher
(1768–1834). Engraving
by A. Schultheitss, from a
portrait by L. Heine.
(From H. Meisner, ed.,
Schleiermacher als Mensch,
1922.)

ise—namely, that the individual was a mask for the absolute ego—that had
no purchase on our own intimate sense of individuality; and it reached a
conclusion—that the universe was merely a "degraded allegory, a vanishing
silhouette of our own finitude"—that denied a reality that the "holy, but rejected, Spinoza" had come to know through intuition, in an *amor Dei intellectualis*.[209] Kant, of course, foreclosed immediate *conceptual* access to the
divine. He sought a path through moral judgment, but that too Schleiermacher rejected: the Kantian mode subordinated authentic moral behavior
to an extrinsic force and it provided uncertain access to the infinite beyond.

Kant had argued that moral experience justified the postulation of freedom, immortality, and God—postulates that made such experience practi-
cally comprehensible. Moral experience occurred when the individual
acted from duty, which meant that he or she freely chose to act from maxims that conformed to the categorical imperative (which states: Act so that
the maxim of your action could be willed as a universal law). Hence, moral
experience, being under obligation, only made sense if the individual were
free. According to Kant, the highest good for which a good will might strive
would be consistently and constantly to act from duty, while recognizing
that such actions made one worthy of happiness. For human beings, how-

209. Ibid., 1.2:213. In the *Ethica*, Spinoza described the "intellectual love of God" as arising
from an intuitive knowledge of our dependence and that of all other things upon God as cause. See
propositions xxiv–xxxvii of book V, in *Opera*, ed. J. Van Vloten and J. Land, 3 vols. (Hague: Martinus Nijhoff, 1890), 1:255–62.

ever, achievement of the highest good could only come after an eternity of striving—hence the practical requirement for a belief in immortality. Further, the worthiness to be happy implied that there had to be a harmonizer, a moral God who would allow the agent to realize that worthiness. Schleiermacher objected to this concatenation, which Kant had forged in the second *Critique*. Even according to Kant, the theologian pointed out, we were morally obliged to act from duty, without regard to possible happiness; hence the highest good for which we strove should not dangle before us the carrot of happiness. And in any case, Kant's notion of happiness made no sense—it was an angelic happiness not dependent upon human desire and human satisfaction, the necessary conditions for the only kind of happiness we knew. Schleiermacher concluded that the impulse to religion and the imperative of duty were entirely distinct: one did not depend upon the other, nor did a Kantian happiness play a role in either.

Kant had assumed that the moral imperative to duty arose only in a purified intellect, one whose motives were not sullied by sensuous desire; the divine came trailing after as a practical inference from our need to make logical sense of moral judgment. Schleiermacher took the lower road. He rather thought the whole person, carnal nature and luminous intellect, strove after the infinite and achieved religious consummation in an intuition of the universe.

> Intuition of the *Universum*—I ask you to become acquainted with this concept. It is the hook of my entire speech, it is the most common and the highest mode of religion, through which you can find every aspect of religion, through which its essence and boundaries can be exactly determined.[210]

The intuition that formed the heart of religion produced in the individual a feeling of "the quiet disappearance of one's whole existence in the immeasurable."[211] In his tract Schleiermacher provided a kind of transcendental deduction of the conditions for such an intuition—which deduction echoed back ideas sounding through the Romantic circle. First, the intuition itself would immediately yield a feeling of submission to the whole (as suggested by Novalis and Spinoza). The proximate object of the intuition would be the immensity of nature, but nature as displaying regularity and law beneath the great chaos of her objects and actions. The intuition would be ultimately directed to the basic forces of nature into which all others could be resolved, namely, repulsion and attraction (as in Schelling); but

---

210. Schleiermacher, *Über die Religion*, in Schleiermacher, *Kritische Gesamtausgabe*, 1.2:213.
211. Ibid., 1.2:212.

these fundamental concepts of uniformity and polarity must have been forged in the interior of the mind (as in Kant, Fichte, and Schelling). "Hence, it is actually the mind into which religion looks and from which it takes the intuition of the world."[212] But the Adam in us could not traverse the path from the inner garden to the outer world without having come to know Eve. Carnal human love, Schleiermacher maintained, prepared us to be receptive to the intuition that grounded the religious. "To intuit the world and have religion," he maintained, "a human being must have found humanity, and he finds it only in love and through love."[213]

Schleiermacher found in finite love a means to transcend its own limitations. Schlegel thought it also the goal of human striving. In his novel *Lucinde*, he depicted a religion of love, while his friend, composing his tract in the next room, exhorted his readers to a love of religion. Yet Schleiermacher's religion, as he confessed, was a "religion of the heart"[214]—a religion whose founding intuition had roots deeply imbedded in the erotic impulse, certainly in his own case. As he described the religious intuition for readers of his volume, his language began to tremble with images as ancient as those of the *Song of Songs*, perhaps now given an urgency reflecting his recently ignited passion for Elenore Grunow:[215]

> Religious intuition is as fleeting and transparent as the first sweet breath of dew which the awakened blooms inspire, as modest and tender as a virgin's kiss, as holy and fruitful as the bride's embrace. Yet it is not simply *like* these, rather it *is* all of these. Quickly and magically it develops from an apparition to an event, then to the very image of the universe. Just as the beloved and desired figure takes form, so my soul flies to it. I embrace it, not as a shadow, but as the holy being itself. I lie on the breast of the infinite world: I become in this moment her soul, since I feel all her power

212. Ibid., 1.2:227. Schleiermacher's metaphysical position is that of idealism, as he expressly put it in his *Monologen* (1800), which he saw published a year after the *Religion:* "For me the spirit [*Geist*] is the first and only thing: since what I call the world is the spirit's most beautiful work, its self-created mirror." See Schleiermacher, *Monologen, Neujahrspredigt von 1792, und Über den Wert des Lebens*, ed. Friedrich Schiele (Hamburg: Felix Meiner Verlag, 1978), pp. 15–16.

213. Schleiermacher, *Über die Religion*, in *Schleiermacher, Kritische Gesamtausgabe*, 1.2:228.

214. Schleiermacher to Henriette Herz (3 March 1799): "My religion is so through and through religion of the heart that I have room for no other kind." See *Schleiermacher, Kritische Gesamtausgabe*, 5.3:30.

215. Schleiermacher had become passionately involved with Elenore Grunow by December 1798. In spring of 1800, he mentioned to his new friend August Hülsen that he had urged Frau Grunow to divorce her husband, a Lutheran preacher, so that they might marry. His friend counseled against the move. See August Hülsen to Schleiermacher (13 April 1800), in ibid., 5.3:481. Grunow left her husband several times, only to return, and thus greatly disappoint her lover. The relationship between Schleiermacher and Grunow broke off decisively in autumn of 1805.

and her infinite life as my own; she becomes in this moment my body, since I penetrate her muscles and her limbs as my own, and her intimate nerves move in harmony with my senses and anticipations, just as my own do. The faintest shudder and the holy embrace is released. And only now the intuition stands before me as a separate figure, I take her measure, and she is mirrored in my open soul, just as the image of the disentangled lover shines in the awakening eye of the youth. And now that feeling works its way from deep inside me toward the light and spreads itself over me, just as the blush of shame and lust spread upon his cheeks. This moment is the highest flowering of religion.[216]

The divinity of love and the erotic penetration of nature formed typical themes of the Romantic poets. These creative artists transformed and heightened emotional insight, driving it beyond the conceptual confines of finite consciousness. They touched the divine and returned with messages from Him who lives far away. Schleiermacher was quite explicit about the preparatory role that poets and artists performed for religion:

Such as these serve as true priests of the Most High, since they bring him closer to those who are wont to deal with only the finite and the small. They convey the heavenly and eternal as an object of pleasure and unity, as the one inexhaustible source of that toward which all of their poetry is directed. They strive to awaken the slumbering kernel of a better humanity, to inflame a love for higher things, to transform a common life into a higher one. . . . They are the higher priesthood who transmit the most inner spiritual secrets, and speak from the kingdom of God. They are the source of all visions and prophecies, of all holy works of art and stirring oratory, which are scattered on the chance that they might find a receptive mind and thus bring forth fruit.[217]

The poets and artists are thus the earth-tethered angels of the infinite.

---

216. Schleiermacher, *Über die Religion*, in *Schleiermacher, Kritische Gesamtausgabe*, 1.2:221–22. In subsequent editions, Schleiermacher dampened the heat of his prose in this passage.

217. Ibid., 1.2:193–94. Despite his admiration for the way poetry moved the soul toward the divine, Schleiermacher confessed an ignorance of how it performed its magic: "I would wish, were it not so iniquitous to wish to leap beyond oneself, that I could ever so clearly see how the artistic sense on its own passes over into religion, how despite the quietude into which the mind sinks through particular pleasures it nonetheless feels driven to make advances that can lead to the *Universum.* . . . I believe that more than anything else the sight of great and sublime works of art can achieve this wonder; I just can never grasp it" (ibid., 1.2:262). In the second edition of the work, in 1806, Schleiermacher no longer portrayed art as making the way clear for religion—perhaps because by that time his friend Friedrich Schlegel was advancing from sublime poetry toward the clutches of the tawdry harlot, the Church of Rome, who completely seduced him in 1808.

Physical love and creative inspiration revealed, Schleiermacher thought, the means by which human finitude could transcend the boundaries laid down by Kant; love and poetry opened a direct path to an independent, infinite *Universum* that lay on the other side of our subjective consciousness. Fichte flung back the veil of individual consciousness to reveal the hovering absolute ego. Schleiermacher pushed in the other direction; he penetrated through the confines of individual consciousness to reach the infinite beyond, and then, like a lover, he became absorbed into it. His intuition thus achieved something akin to Spinoza's third kind of knowledge, the *amor Dei intellectualis*. In Schleiermacher, as in Spinoza, this kind of intuition "widens the sharply delineated contours of our personality and gradually resolves them into the infinite."[218]

The object of religious intuition was, in Schleiermacher's own decidedly aesthetic account, an ineffable vastness, an infinity, something all encompassing—to use his own vague term, a *Universum*. Precise theological language, which enlightened theologians might have employed to convey this kind of intuition, could not, for Kantian reasons, be taken literally. In this respect, theology, at its best, could only be a kind of poetry that reflected this or that cultural age. The dogmas of organized religion might provoke an intuition in us but could not conceptually measure that intuitive feeling. In his later work *Der christliche Glaube* (1821–22; 1831), Schleiermacher would elevate this insight about church doctrine into a cardinal feature of his mature religious philosophy. In that powerful treatise, he would again argue that theological propositions only figuratively indicated a kind of feeling, a feeling of dependence in the presence of the infinite. This feeling of dependence, he maintained, constituted the essence of religion.[219] Religion construed in this way, of course, could not come into conflict with science. Indeed, religion and science might, in their respective ways, foster a feeling of reverence, of dependence in the face of a nature that had infinite depths.[220] Many scientists of the nineteenth century would embrace

218. Ibid., 1.2246.

219. In *Der christliche Glaube*, Schleiermacher held that "the essence of Christian piety is this, that we are conscious of ourselves as indeed being dependent, or what is the same thing, as in relationship to God." Christian dogmas, he concluded, were but expressions of this essential religious feeling: "Christian dogmas are expressions of a pious Christian attitude represented in speech." See *Der christliche Glaube*, 2nd ed., 2 vols. (Berlin: Walter De Gruyter [1830], 1960), 1.4:23, 1.15:105.

220. According to Schleiermacher, nature was not split into two kinds of events, natural occurrences susceptible of scientific explanation and miracles requiring religious explanation. All events, he thought, can be given both a scientific and religious account. See *Über die Religion*, in *Schleiermacher, Kritische Gesamtausgabe*, 1.2:240: "A miracle is only the religious name for an event, each of which, even the most natural and common, is a miracle, insofar as it lends itself to being interpreted predominately from the religious point of view. For me, everything is a miracle. . . ." Moreover, as he explained in *Der christliche Glaube* (vol. 1, sec. 47, p. 235), if a feeling of depen-

Schleiermacher's resolution of the seeming conflict of science and religion; the conflict would, however, be reinstated as one between empirical science and fundamentalist theology.

Everyone read Schleiermacher's *Religion* when it was published in early summer of 1799. The book bore no author's name, but, of course, his friends and their rippling circles knew the identity of the author. Schlegel immediately wrote a review of the book and he referred to it frequently in the set of fragments called *Ideen*, which were published in the third number of the *Athenaeum* (April 1800). He dolloped praise with artful swirls of his pen,[221] but the hard center of his review was critical. Schleiermacher had, he thought, so separated religion from philosophy, poetry, and morals as to suck out its core.[222] Nonetheless, Schleiermacher's intuitionism, with its easy glide into pantheism, fed Schlegel's own growing enthusiasm for religion. Hardenberg, after receiving a copy of the book from Schlegel, became "completely taken with it, thoroughly impressed, excited, and captivated by it."[223] In his study of Fichte, Hardenberg had arrived at a similar conclusion: we could only be aware of a feeling of dependence on the absolute (see above). His own composition *Die Christenheit oder Europe*, which looked to the Church as the principal source of character formation for individuals, bore the mark of his engagement with Schleiermacher's *Religion*. The young philosopher Schelling, however, did not respond with the same enthusiasm. He disdained Schleiermacher and Hardenberg's sort of religion, which seemed too sentimentally ethereal, too aesthetically precious. In a long poem that Schlegel thought wiser not to publish in the *Athenaeum*, Schelling expressed his sardonic attitude:

My only religion is this,
That I love a pretty leg,
Full breasts and slender hips,
As well as flowers with a fragrant scent,

---

dence is the essence of religion, then scientific analysis does not impugn the complementary religious interpretation of nature: "so we will attempt to interpret those facts in our purview, so long as it is possible, with respect to natural causal connections and without damage to our feeling of dependence."

221. Friedrich Schlegel, *Ideen* (no. 150), in *Kritische Friedrich-Schlegel-Ausgabe*, 2:271: "One can neither explain nor conceive the *Universum*, one can only intuit and reveal it. Give up calling the system of the empiricists a *Universum*, and learn the true religious idea itself, even if you have not quite understood Spinoza, through reading the *Talks on Religion*."

222. Ibid. (no. 149): "Without poetry, Religion becomes dreary, false, and evil; without philosophy it becomes obscene in its self-indulgence and lustful to the point of self-castration." See also Dilthey's evaluation of Schlegel's reaction in *Leben Schleiermachers*, 1:443–44.

223. Hardenberg, quoted by Dilthey in *Leben Schleiermachers*, 1:448.

The full satisfaction of every passion,
And the sweet surrender of every love.[224]

At the time of his versifying, Schelling had begun exercising his kind of worship on the divine form of Caroline Böhmer Schlegel.

Scholars have attempted to situate Schleiermacher's work in relationship to the Romantic movement, in which he was enmeshed. The two great authorities on Schleiermacher and the movement—Dilthey and Haym, respectively—have claimed that the Charité preacher was *in* the movement but not *of* the movement. Dilthey thought that Schleiermacher's religious insight outstripped that of Friedrich Schlegel, who did not place religion at the core of his thought and whose limp religious and moral convictions paraded in extreme mythological dress—hence, according to Dilthey, his easy seduction later by the popish Church.[225] Dilthey distinguished Schleiermacher's intuitive mode from that of Goethe and Schelling. While Goethe and Schelling also began with intuitions of the whole, the object of their intuitions was, respectively, nature and the ego; their common aim was scientific and aesthetic development, not religious formation.[226] Haym, as well, believed that though the similarities between Schleiermacher's various positions and those of the Romantics appeared great, the differences were greater still. "None of the young poets and aesthetes," he contended, "could stand measure with Schleiermacher according to the depth and moral earnestness of convictions. Only Hardenberg shared that particular religious trait, but he resolved that aspect of his personality in poetry."[227]

To cast Schleiermacher just outside the ring of Romantic writers supposes that Romanticism had a well-defined essence, a stable set of clear doctrines that would allow such precise discrimination. But the writers we have considered all differed from one another along a series of dimensions. It is rather their associations and interactions, their sympathetic responses and adaptations to each other's ideas, the congruence of basic positions and basic oppositions, and the harmony—which in several instances fell into dis-

224. Parts of Schelling's poem *Epikurisch Glaubensbekenntniss Heinz Widerporstens* appeared in a journal other than the *Athenaeum*. The full poem is in F. W. J. Schelling, *Briefe und Dokumente*, ed. Horst Fuhrmans, 3 vols. to date (Bonn: Bouvier Verlag, 1962–), 2:205–14. Schelling was probably reacting as much to Hardenberg's passionate endorsement of Schleiermacher's *Religion* as to the book itself. Later he wrote Wilhelm Schlegel that he himself was an "enthusiastic reader and admirer of the *Talks on Religion*." He thought Schleiermacher must have written it out of "the deepest philosophical study—or he has written it from blind, divine inspiration." See Schelling to Wilhelm Schlegel (3 July 1801), in ibid., 2:335. Schelling wrote the letter, though, after a profound shift in his philosophy of idealism.

225. Dilthey, *Leben Schleiermachers*, 1:451–53.
226. Ibid., 1:453–54.
227. Haym, *Die romantische Schule*, pp. 421–22.

sonance—of their personalities and ideas that constituted them, at least in the eyes of the historian, members of a common movement. Certainly they all took religion seriously, even if they created variously different articulations of that primal, sacred feeling. And whether that intuitive feeling got expressed in moving poetry, profound idealistic philosophy, or sublime religious language, it leapt the Kantian boundaries and pointed toward a reality independent of individual consciousness. That reality, according to Schlegel, Schelling, Goethe, and Schleiermacher, underlay the various appearances of nature and acted as a poetically creative agent in determining them. Hence, values, aesthetic and moral, they believed, could be found in nature, not merely extrinsically imposed from without. The source of such values, as well as of individual consciousness and nature tout court, was absolute mind. As Schleiermacher put it a bit later in his *Monologen* (1800): "What I call the world is the spirit's [*Geistes*] most beautiful work, its self-created mirror."[228] Communion with nature would thus reflect mind at its deepest level, and that communion would be fostered by religion, art, science, and most especially by a loving relationship with another. And for all of the Romantics, feelings of love, that creative energy fueling personal development, that spontaneous overflow of powerful emotion that reached out toward another individual—this all-too-human expression supplied the force that expanded into that higher love, the intuitive feeling encompassing infinite nature. But the energy of that force also fractured the foundation of friendships holding together the Romantic circle.

Friedrich Schlegel's Aesthetic Theory

Friedrich Schlegel's stay in Berlin did not mean that he lost contact with those remaining in Jena. He corresponded frequently with family and friends, especially with Caroline and her daughter, Auguste. The letters to Auguste show what great affection he had for the young girl—she was twelve years old when he left Jena. He told her of his friends in Berlin, he spun out little stories for her amusement, he teased her, he gently reprimanded her for neglecting studies, he examined her Greek translations—he drenched his missives in paternal love. A bit later, though, that innocent love encroached on its far boundary. When Auguste was thirteen, Friedrich wrote Hardenberg, someone who knew the experience of young love, a letter with a telling afterthought: "Lastly, you have observed Auguste. You do well understand, don't you, how one can so love such a child, as one can only love once, when one is so very happy and everything else by contrast

228. Schleiermacher, *Monologen*, p. 16 (15–16); *Geist* could, of course, be translated as "mind."

seems as nothing?"[229] Friedrich's letters to Caroline display a comparable solicitude, as well as undercurrents of muted desire. Both mother and daughter transfixed his heart, and each would, in time, reduce it to bitter ashes.

Friedrich also wrote frequently to his brother, particularly over the great enterprise they had begun while he was in Berlin—the *Athenaeum*, the journal that would be written principally by the brothers, their lovers, and a few selected friends—Schleiermacher, Novalis, A. F. Bernhardi, Sophie Tieck Bernhardi, Karl Brinckmann, and August Hülsen. The initial number, published in May 1798, began with Wilhelm Schlegel's essay on language and Novalis's epigrammatic fragments *Blütenstaub* (Pollen). The subsequent five issues would contain, among many other items: Novalis's *Hymnen an die Nacht*, the jointly produced *Fragmente* (principally by Friedrich, but with contributions by his brother, Novalis, and Schleiermacher), and, in the two numbers of the third and final volume (1800), the several essays that constituted Friedrich's more systematically developed theory of aesthetics, his *Gespräch über die Poesie*.

Schlegel's *Gespräch*—along with Kant's third *Critique*, Schiller's *Über naive und sentimentalische Dichtung*, and the last part of Schelling's *System des transscendentalen Idealismus*—established the aesthetic foundations of German Romanticism. Schlegel's treatise distilled the aesthetic ideas that had been brewing in his early essays and in the *Athenaeum Fragmente*, but now tinctured with the spirit of transcendental philosophy. Out of this mixture came a revaluation of previous values. In his monograph *Über das Studium der griechischen Poesie* (On the study of Greek poetry, 1795–96), part of his volume on Greek and Roman poetry, Schlegel had been thrall to the classical enthusiasms of Johann Joachim Winckelmann (1717–1768). The young critic had argued that in Greek poetry "the entire course of the organic development of art has been enclosed and perfected, and that this greatest period of art, wherein the possibility of beauty received its freest and most perfect expression, contains the complete measure of aesthetic taste."[230] The Greeks had manifested in their poetry the standard later set by Kant for the achievement of beauty, which was, in Schlegel's redaction: "the universally valid object of a disinterested pleasure [*Wohlgefallens*] that is equally independent of the pressure of requirement and law, free yet necessary, wholly without purpose and yet unconditionally purposive."[231]

229. Friedrich Schlegel to Hardenberg (26 September 1797), in *Novalis Schriften*, ed. Kluckhohn and Samuel, 4:491.

230. Friedrich Schlegel, *Über das Studium der griechischen Poesie*, in *Kritische Friedrich-Schlegel-Ausgabe*, 1:307.

231. Ibid., 1:253.

Against this measure, modern literature could not compare, with its pusil-lanimous emphasis on particular interests and subjective concerns. But af-ter he read Schiller's monograph *Über naive und sentimentalische Dichtung*, which dilated on the character of post-Greek poetry, Schlegel suffered the shock of the modern.[232] His habits of thought changed as a result—an ironic turn, given the enmity that quickly developed between himself and Schiller. Schlegel, at the time, though, generously credited Schiller with "broadening his insight into the character of the poetry of interest [i.e., modern poetry], and providing a new light on the limits of the sphere of classical poetry."[233] Though even before reading Schiller, as he was finish-ing his monograph on ancient poetry, he already mused that the goal of Ro-mantic poetry would be to join the ancient with the modern.[234] But there is little doubt of the impact of Schiller's monograph. Schlegel began an in-sistent reevaluation of modern Romantic poetry. He now recognized the importance of those qualities that characterized such poetry, namely: "the interest in the reality of the ideal, the reflection over the relationship of the ideal and real, and the connection to an individual object of the ideal-izing imagination of the poetizing subject."[235] What had changed in the two years between 1795 and 1797, when those last lines were published, was not only the reading of Schiller but Schlegel's engagement with Fichte and Schelling, thinkers who considerably sharpened his critical spirit and brought him into the field of transcendental thought, that space of greater aesthetic possibility wherein Schiller's essay could have a most potent ef-fect. Yet these more intellectual causes of Schlegel's engagement with the Romantic would not have had the purchase on his thought were it not for the stimulus of his old and new relationships: with Caroline and Auguste, the child who began to bloom in 1797; and with Dorothea and Schleierma-cher, both of whom he met in summer of that year, and with whom he formed a loving bond. These personal relationships touched Schlegel with the possibilities of creative love, and, in the case of Caroline and Auguste, with the feeling that loving consummation could occur only in the infinite beyond. These intellectual and personal forces brought an altered convic-

232. I discuss Schiller's *Über naive und sentimentalische Dichtung* in chapter 11.

233. Friedrich Schlegel, "Vorrede," *Die Griechen und Römer*, in *Kritische Friedrich-Schlegel-Ausgabe*, 1:209.

234. Friedrich wrote his brother Wilhelm on 27 February 1794: "The problem of our poetry seems to me to be the union of the essentially modern with the essentially ancient; if I indicate that Goethe, the first to have made a beginning for a wholly new period of art, has approached this goal, you will certainly understand what I mean." See *Kritische Friedrich-Schlegel-Ausgabe*, 23:185.

235. Schlegel, "Vorrede," *Die Griechen und Römer*, in *Kritische Friedrich-Schlegel-Ausgabe*, 1:212.

tion about the soul of poetry and its mode of expressing the deeper nature of human beings.

During the period in which he worked on the *Athenaeum Fragmente* (1798), Schlegel had become convinced that modern Romantic poetry spoke to individual needs and to the needs of the age, an age that had become more philosophically sophisticated. "Romantic poetry," he proclaimed in fragment 116 of the *Athenaeum Fragmente*, "is progressive universal-poetry."[236] It is that mode of experience that realizes the ontological longing of human beings, a mode that pushes them into that deeper sphere of becoming, which reveals their infinite destiny. In the *Gespräch*, Schlegel now brought these several hyper-blossoming concerns into a yet new arrangement, one that insisted that the greatest possible achievement of beauty was the orchestration of classical modes with the Romantic. With ear attuned, Schlegel recognized that Goethe's recently published *Wilhelm Meister* played the overture to this performance.

Schlegel had begun working on those essays making up the *Gespräch* during the last few months of his stay in Berlin. In May 1799 he wrote Caroline that he had given a lecture to a local group on the "differences of style in Goethe's earlier and later works,"[237] and this, with some rewriting, became part 4, the last part of the *Gespräch*. A previous letter to Caroline, in March, seems to contain the germ of part 3, the "Brief über den Roman" (Letter on the novel).[238] A few weeks after he had returned to Jena, in September, he completed the essays that constitute the treatise; and in January 1800 the first part appeared in the *Athenaeum*.

Schlegel rendered the *Gespräch über der Poesie*, or *Dialogue on Poetry*, in a mixed form: four essays ostensibly delivered by four male friends, each essay followed by a short argumentative dialogue among the friends. There are two women represented in the party. They have their chance to comment, but not to essay. In the prologue Schlegel indicates that the views expressed arose out of his own circle of companions, though no one of the characters in the dialogue is easily identifiable, except perhaps Amalia, who hosts the party and whose skeptical and argumentative responses seem distinctively Caroline's. Each of the essays certainly expresses Schlegel's strongly held views, elusive and fragmental though they be.

---

236. Friedrich Schlegel, *Fragmente* (1798), in *Kritische Friedrich-Schlegel-Ausgabe*, 2:182 (no. 116).

237. Friedrich Schlegel to Caroline Schlegel (May 1799), in *Kritische Friedrich-Schlegel-Ausgabe*, 24:287. As in this letter, the essay in the *Gespräch* argues with a woman correspondent about the nature of the novel, decries her lack of interest in English novels, and declares wit to be an essential ingredient in novels and all poetic creation.

238. Friedrich Schlegel to Caroline Schlegel (early March 1799), in ibid., 24:239–41.

The first essay, "Epochs of Literature," sketches the developmental history of the poetic art, beginning with the Greeks, whose plays and verses were "poetry itself."[239] The essay then limns the decline of poetry until the late medieval period, when the signal practitioners of modern Romantic poetry appeared: Dante, Petrarch, and Boccaccio. The heights of poetic achievement were reached with Cervantes's *Don Quixote* and Shakespeare's *Hamlet,* "the most artistic and insightful works of art in the entire realm of Romantic art."[240] In Schlegel's estimation, the only figure whose works could compare with these illustrious creations was Goethe. His poetry reflected that "of all nations and ages," and his magnificent inspiration, the novel *Wilhelm Meister,* accomplished what should be the goal of all literature—the harmony of the classical Greek style with the Romantic.

Two features of this initial essay must strike the modern reader. First is the medium of the essay itself, that of a historical sketch. Here Schlegel sounds a leitmotiv that will resonate through much of subsequent Romantic literature and science—namely, the historical or developmental approach. Schlegel argues—really asserts—that "art rests upon knowledge [*Wissen*] and the science [*Wissenschaft*] of art is its history."[241] By history, however, Schlegel really means the development of a work from its fetal essence. In this respect, a poetic creation is like an organic being: to understand such a creature scientifically, one must follow the evolution of the conceptus as it develops and becomes realized in its parts. Romantic writers and scientists will canonize this historical method: from Schelling and Goethe through Haeckel, development becomes the privileged way of understanding literature, science, and in the case of biology, also the objects of the science. For Schlegel, the developmental approach is *the* method of achieving a transcendental poetry. Just as Fichte would take for his subject the way in which thought creates a world of experience, so Schlegel, in the *Gespräch,* takes as his subject the way in which poetic thought creates its world. The corrective he would make to Fichtean transcendentalism, however, was precisely that his way into poetic thought would be historical and developmental; and in this insistence, he would help move his friend Schelling beyond Fichtean subjective idealism. Secondly, the modern reader is undoubtedly surprised to find Schlegel denominating Cervantes, Shakespeare, and Goethe as Romantic poets.[242] The surprise, however, is

239. Friedrich Schlegel, *Gespräch über die Poesie,* in *Kritische Friedrich-Schlegel-Ausgabe,* 2:293.
240. Ibid., 2:346.
241. Ibid., 2:290.
242. As already noted, Schlegel, following Schiller, perceived in Goethe the Romantic poet—the modern poet—whose style melted into that of the classical authors to produce the greatest beauty possible.

only a function of our usual understanding of Romanticism. Schlegel's analysis and the recommendation coming out of it provide, I think, correctives for our contemporary understanding of the historical meaning of the Romantic.

To use history for the proper conception of literature suggests that historical models might serve to rejuvenate the poetic spirit. And so, just as Greek poetry—its dramas and verses—grew out of and took as subject matter ancient mythology, so Schlegel calls for in his day a new mythology as a maieutic to the birth of a more thoroughly Romantic poetry. One needs to say "more thoroughly Romantic," since, according to Schlegel, all poetry has something of the Romantic about it—that is the essence of poetry—but not all poetry could properly be called Romantic, for not all poetry exemplifies the art in its highest form. Schlegel thinks the new mythology is already aborning, first in the shape of the new philosophy of idealism, which has moved beyond the old dichotomy of self and nature. The creative energy of idealistic philosophy has stimulated poetry to new accomplishment and itself to produce "out of its womb a new and comparably infinite realism," a realism not embodied in a philosophical system but in the highest kind of poetry. In this respect, then, idealist philosophy "will serve, in its mode of development, not only as an example for the new mythology, but as an indirect source of it."[243] What Schlegel seems to mean by this remark is that idealism has raised reason to a new height, though one still bound in absolute subjectivity and one that deprecated nature. But idealism, especially in the form of Kant's and Fichte's rehabilitation of aesthetic imagination, has, nonetheless, shown the way for the approach to the infinite, the infinitely real, through poetry.[244] For poetry unleashes imagination to pluck gently the ever-sensitive cords of love. It is via love, as Schlegel tried to portray in *Lucinde*, that human nature can contact the infinitely real. Poetry is the expression and vehicle for that love. Given what he thinks poetry capable of—through love to comprehend the infinite—it will come only as a small surprise that Schlegel names Spinoza as that author whose work can serve as an inspiration for Romantic poetry.[245] For Spinoza's *Ethica* preached that the highest form of knowledge was the *amor Dei intellectualis*—the intellectual love of God. This level of cognition, according to Spinoza, was a mode of understanding beyond mere reason; it

---

243. Ibid., 2:315.

244. Ibid., 2:319: "This, then, is the beginning of all poetry, to overcome [*aufzuheben*] the process and law of rationally thinking reason and to transport us again into the beautiful confusion of fantasy, into the original chaos of human nature, for which I know no more beautiful symbol than that colorful whirl of the old gods."

245. Ibid., 2:317.

was an intuition into the core of reality, an intuition in which the individual perceived himself part of that whole of nature, part, in other words, of God.[246] In the *Ethica*, Spinoza also represented infinite nature as both *natura naturata* and *natura naturans*—nature natured, that is, nature as product, and nature naturing, or nature as infinitely productive.[247] For the Romantics, for Schlegel, Schelling, Goethe, and later Darwin and Haeckel, nature herself would take on the form of the divine. She would be thought infinitely creative. This is why Schlegel can also refer to "the unconscious poetry" that "moves in the plant, that streams forth in light, that laughs out in the child, that shimmers in the bud of youth, that glows in the loving breasts of women."[248] Nature herself is a Romantic artist whose loving creations become our reality—a sentiment that emerges to explicit philosophical expression, as we will see, in Schelling's *System des transscendentalen Idealismus*.

In the third essay of his treatise, "Letter on the Novel," Schlegel specifies a bit more exactly what he means by the Romantic: "According to my view and my usage, that is Romantic which presents a sentimental content in a fantastic form."[249] The two terms "sentimental" and "fantastic" require some exegesis to avoid gross misunderstanding. By "sentimental" Schlegel, as he avers, does not mean anything like "maudlin." The term harkens back to Schiller's essay *On Naive and Sentimental Poetry*, in which "sentimental" refers to the poet's sensitivity to the disparity between the real and the ideal, between finite representations and the infinite to which they refer, between feelings entirely subjective and those grounded in the object of concern. Schlegel, however, goes beyond Schiller to suggest that the fundamental feeling resident in Romantic poetry ought to be that of spiritual love, a deeply felt emotion that reaches out to the infinite.[250] By "fantastic" Schlegel refers to that which is produced by the fantasy, by the imagination. Here he seems to have in mind a double-edged meaning for fantasy or imagination. First, imagination is that faculty described by Kant and Fichte—

246. Spinoza, *Ethica*, V, xxx–xxxvi, in *Opera*, 1:257–61.

247. Ibid., I, xxix, schol., 1:60.

248. Schlegel, *Gespräch über die Poesie*, in *Kritische Friedrich-Schlegel-Ausgabe*, 2:285. Schlegel's notion of the unconscious poetry of the plant seems an echo of a remark by Schelling in a series of articles he wrote for Fichte and Niethammer's *Philosophisches Journal*. In explaining how the mind created nature, he observed that "every plant is, so to say, the entwined aspect of the soul." See Schelling, *Abhandlungen zur Erläuterung des Idealismus der Wissenschaftslehre* (1796–97), in *Schellings Werke*, 1:310.

249. Schlegel, *Gespräch über die Poesie*, in *Kritische Friedrich-Schlegel-Ausgabe*, 2:333.

250. Ibid., 2:333–34: "What then is this 'sentimental'? That which speaks to us, where that feeling prevails, a feeling which is not sensible but spiritual. The source and soul of this emotion is love, and the spirit of love must defuse through Romantic poetry everywhere, visibly and invisibly."

the organ through which the artist reaches down into his or her own nature to touch that which is infinite, true, and real. Thus a "true history [*Geschichte*] is the foundation of all Romantic creation . . . an evocation, more or less, of the self-knowledge of the author, the burden of his experience, the quintessence of his personality." But secondly, what is imaginative, what is fanciful, what, in Schlegel's term, is "witty" [*witzig*], well therein lies the tale. It is the tale. The poet must display through irony, wit, and all the imaginative means at his or her disposal a variegated world, an "artfully ordered confusion" that is life.[251] Yet the poetry itself must be so ordered as to drive us through love to move below the surface of things to catch the infinite stream that carries us toward unity and perfection. The irony made famous by the Romantics—and Friedrich Schlegel in particular—was just this poetic signal, a linguistic turn that simultaneous pointed to the multiply refractive surface of becoming and to that which lay beyond, to the absolute—the contrast and poignancy of the tension constituting the wit of the trope.[252] The definition of the Romantic thus recapitulates the dominant theme of Romantic literature and science: the expression of a deep unity and perfection through the multiply variegated dance of life.

The Return to Jena and the Closing of the Circle

In September 1799 Friedrich Schlegel decided to return to Jena. Several forces operated to induce the move. His novel *Lucinde* was just appearing, and he undoubtedly sought the comfort of close friends and family to ward off what he suspected—correctly—would be a strong reaction.[253] And then, Dorothea and her son Philipp needed a new beginning after she had left her husband. Mother and son followed Friedrich to Jena in October, as did Ludwig Tieck and his wife, Amalie. An added pressure on Schlegel to leave Berlin was probably the slight tension that arose when he took some exception to Schleiermacher's view of religion. His friend's strong defense of *Lucinde* in *Vertraute Briefe über Friedrich Schlegels Lucinde* (Confidential

251. Ibid., 2:337, 318.

252. The range of meanings embodied in Schlegel's concept of Romantic irony is fluently discussed by Gary Handwerk in his "Romantic Irony," in *The Cambridge History of Literary Criticism*, vol. 5: *Romanticism*, ed. Marshall Brown (Cambridge: Cambridge University Press, 2000), pp. 203–25.

253. The reaction was indeed strong (see the next note). But Friedrich probably found only mild comfort in the bosom of his family. Wilhelm, in a diary entry that Caroline saw, versified: "Pedantry asked of fantasy a kiss, / But she warned him it was all amiss. / Saucily, though impotently, he yet came; / and she with dead child bore pain, / And christened it Lucinde by name." Or words to that effect. See Caroline Böhmer Schlegel to Schelling (February 1801), in *Caroline: Briefe aus der Frühromantik*, 2:41.

letters on Friedrich Schlegel's *Lucinde*, 1800) would, however, keep their relationship warm.[254] By October, then, the Romantic circle in Jena had its most complete closure: the brothers were together again, now complemented by Caroline and Dorothea, as well as by the budding Auguste; Tieck and his wife had also arrived; Novalis was a few miles away at Weissenfels and would frequently visit Caroline's salon; Schiller continued at the university (moving to Weimar at the end of 1799); Fichte had been cast out but yet visited from Berlin; and Goethe was at Weimar and would travel to the university town often, especially to confer with Wilhelm Schlegel. When Friedrich returned home, he found, however, a new element in the equation defining the circle: the very young, blond, intense follower of Fichte, Friedrich Wilhelm Joseph Schelling.

254. The immediate reaction of critics to *Lucinde* was harsh, and the judgment through the nineteenth century continued to be so. Heine prayed that the Blessed Virgin would forgive the author, since, as he judged, the Muses never would (*Die romantische Schule*, in *Sämtliche Schriften*, 3:408–9). In the last third of the century, Dilthey wrote that it had been well confirmed that the novel was "immoral as well as poetically formless and reprehensible" (*Leben Schleiermachers*, 1.1:496). Dilthey's own hero, Schleiermacher, thought differently. He gracefully and with deep consideration defended his friend's effort in his anonymously published *Vertraute Briefe über Friedrich Schlegels Lucinde*. The tract contains nine letters to and from an "Edward" and three women, Ernestine, Karoline, and Elenore. Henriette Herz suggested that the seventh letter, from Elenore, was actually written by Schleiermacher's beloved Elenore Grunow. See Henriette Herz, *Berliner Salon: Erinnerungen und Portraits* (Frankfurt: Verlag Ullstein, [1850] 1986), p. 62. In that seventh letter, Elenore writes to Edward: "It is rather unkind of you to press upon me so quickly your copy of *Lucinde* before I got my own. You probably don't know, you naughty, dear man, how close to me it has been, so that I place myself before this pure and beautiful mirror of my love as often as I can in my quiet loneliness. In that magical picture I soon catch a glimpse of your form and mine, and then all those other forms of unity and eternal love." See Schleiermacher, *Vertraute Briefe über Friedrich Schlegels Lucinde*, in Schleiermacher, *Kritische Gesamtausgabe*, 1.3:195.

Chapter 3

SCHELLING:
THE POETRY OF NATURE

Schelling, as he says, is going to wall himself in, but that certainly won't last long. He is rather a man who breaks through walls. Believe me, friend, he is a more interesting man than you give him credit for, a real primitive nature, something like mineral, real granite.

—Caroline Schlegel to Friedrich Schlegel

If any figure were to be crowned philosopher king of the Romantic circle, it would be the boyish, imperious, sullen, ingenious, and magnetic Friedrich Wilhelm Joseph Schelling. He championed and then reformulated Fichtean idealism, infusing it with the realism of Spinoza, to reinstate nature as the ego's coequal. He understood nature as fundamentally organic, virtually a living and generative being. In this respect, biology would become for him the paradigmatic science, the discipline that most conspicuously displayed the creativity and beauty of the one infinite reality. And at that deep level where the objective world of nature and the subjective world of the self merged, the forces that created nature and gave her form would be revealed, under the tutelage of the right philosophy, to be the same as those that created aesthetic objects and endowed them with beauty. Indeed, according to Schelling, "the objective world is only the original, though unconscious, poetry of the mind [Geist]."[1] Hence, the biologist's great aid in comprehending nature would be poetic, that is, aesthetic judgment. Schelling reworked Kant's aesthetic doctrine so that artistic genius and scientific genius would become a Janus-like individual, with one heart animating the two approaches to nature. And in this he realized the Romantic mandate,

1. Friedrich Schelling, *System des transscendentalen Idealismus* (1800), in *Schellings Werke*, ed. Manfred Schröter, 3rd ed., 12 vols. (Munich: C. H. Beck, 1927–59), 2:349 [III:349]. The volume and page reference in brackets are to *Sämtliche Werke*, ed. K. F. A. Schelling, 14 vols. (Stuttgart: Cotta'scher Verlag, 1857). The Munich edition is more accessible and contains unpublished material not included in the edition produced by Schelling's son Karl.

Figure 3.1

Friedrich Wilhelm Joseph
Schelling (1775–1854).
Portrait (1801) by C. F.
Tieck. (Courtesy Sciller-
Nationalmuseum,
Marbach.)

most powerfully expressed by Friedrich Schlegel, that science (*Wissenschaft*)
and poetry should become as one.[2]

The conjunction wrought by the Romantics between science and art is
easy to misinterpret. Most recent historians of science regard the Romantic
movement in general and Schelling in particular as having turned away
from cool reason to embrace emotional mysticism.[3] This historiographic
conviction, however, will not stand close scrutiny. Schelling's early work,
on the contrary, exemplifies a persistent effort to think rationally through
basic epistemological problems. He attempted to ground his philosophy
both in the certainty of self-reflection and in an experimental understand-

2. In the conception of Schlegel and Schelling, *Wissenschaft* would, of course, encompass a
range of disciplines, including natural science.

3. H. A. M. Snelders, for example, thinks Schelling's *Naturphilosophie* exemplified a "denial of
the value of experiment" and that he and his school "opposed empirical research as the basis for
studying natural sciences." See his "Romanticism and Naturphilosophie," *Studies in Romanticism* 9
(1970): 193–215; quote from pp. 196, 198. Snelders's view has remained the same in his *Weltenschap en Intuïtie: Het Duitse romantisch-speculatief natuuronderzoek rond 1800* (Baarn: Uitgeverij
Ambo, 1994); see especially p. 183. Walter Wetzels, though more positive in his evaluation, also
thinks that Schelling had a "mystical vision of what science is all about." See his "Aspects of Natural Science in German Romanticism," *Studies in Romanticism* 10 (1971): 44–59; quote from p. 46.
Ernst Mayr, in his *Growth of Biological Thought* (Cambridge: Harvard University Press, 1982), refers
to the work of Schelling and Oken as "fantastic if not ludicrous" (p. 388).

ing of nature. In those various works having the term *Naturphilosophie* in their titles, Schelling sought to demonstrate through extended argument and appeal to scientific practice that the basic principles given objectively in nature were isomorphic with those acts delineated subjectively in the self. He believed this symmetry could be most clearly modeled and understood through the productions of aesthetic genius. So when the biologist, the physician, or the naturalist of Schellingian disposition would explore nature—a scientist like Alexander von Humboldt, for instance—he would discover there the self's other kingdom, where the forces were familiar and the mind would meet its double, most beautifully displayed.

This boy-king was well conceived and adored. Caroline Schlegel exulted to him: Goethe "loves you like a father, and I love you like a mother—what wonderful parents you have!"[4] Goethe instructed the young idealist in the rights of nature, while Caroline nurtured him with a love that quickly escaped maternal constraint. Her passion embraced him but consumed everything else, including the bonds of affection that held together the Romantic circle and, finally, Schelling's own reputation. He has been judged responsible by recent historians for the ideal death of Romantic science and by his contemporaries for the real death of a young woman who sat engagingly in one corner of a romantic triangle.[5] The turmoil and tragedy of Schelling's youth realigned the trajectory of his thought and drove it into the science of the nineteenth century. In this chapter I will recount Schelling's intellectual development against the foreground of his role in the Romantic circle; and in chapter 8 I will detail his organic and evolutionary conceptions of nature, conceptions that had a marked impact on German biology after the turn of the century.

Schelling's Early Life

Seminary Studies

Schelling descended from preachers on both sides of his family.[6] He was born in 1775 in the small town of Leonberg (near Stuttgart), where his father, Joseph Friedrich Schelling (1732–1812), served as assistant pastor at

4. Caroline Böhmer Schlegel to Schelling (October 1800), in *Caroline: Briefe aus der Frühromantik*, ed. Georg Waitz and expanded by Erich Schmidt, 2 vols. (Leipzig: Insel Verlag, 1913), 2:6.

5. In her quite interesting and clearly wrought survey of Schelling's thought, Dale Snow describes the demise of Schelling's early idealism, the epistemological and metaphysical foundations of his Romantic science. See her *Schelling and the End of Idealism* (Albany: State University of New York Press, 1996). I will discuss the real death below.

6. In reconstructing Schelling's life, I have relied on Gustav Plitt, *Aus Schellings Leben. In Briefen*, 3 vols. (Leipzig: S. Hirzel, 1869–70), which contains most of Schelling's important correspondence; Kuno Fischer, *Friedrich Wilhelm Joseph Schelling*, 2 vols.; vol. 6, *Geschichte der neuern*

the Lutheran church. Because of well-known accomplishments in theology and oriental languages, his father was called, in the year of the son's birth, to the higher seminary in Bebenhausen (near Tübingen), and there the young Schelling grew up. Since at age ten the boy was still too young for the lower seminary (they only took pupils at age fourteen), his father maneuvered him into the higher seminary, where he spent the next four years with boys several years his senior. With a special exemption, Schelling's father next enrolled him in the Tübinger Stift, part of the university where young men took instruction in philosophy (two years) and theology (three years) to prepare them for the ministry. At age fifteen, in 1790, Schelling entered the institute and remained there for five years.

Schelling had to share a room at the Tübinger Stift with two other young men whose critical intensity would soon turn brightly stellar: Friedrich Hölderlin (1770–1843) and Georg Wilhelm Friedrich Hegel (1770–1831), both five years older than the new boy. These three remarkable philosophers and artists remained fast friends through the turn of the century, influencing and inspiring one another. While the sublime poetic talent of Hölderlin bloomed early and affected the aesthetic inclinations of his younger friend—only to wither into madness by the first few years of the new century—Hegel's genius was initially led by the philosophical prodigy. Indeed, Hegel's first book was a comparison of the thought of the great Fichte with that of the recently arrived competitor Schelling.[7] By the end of the first decade of the next century, however, the old friends would become new and bitter opponents.

Philosophie (Heidelberg: Winter's Universitätsbuchhandlung, 1872–77), a sympathetic treatment of the life and works by the great neo-Kantian at Jena and Heidelberg; and F. W. J. Schelling, *Briefe und Dokumente,* ed. Horst Fuhrmans, 3 vols. to date (Bonn: Bouvier Verlag, 1962–), which, along with splendid introductions and notes, contains letters and biographical manuscripts not hitherto published. Alan White has written a helpful general account of Schelling's thought, in his *Schelling: An Introduction to the System of Freedom* (New Haven: Yale University Press, 1983). A most interesting study of Schelling's philosophy flows firmly from the pen of Dale Snow, in her *Schelling and the End of Idealism.* A detailed introduction to various aspects of Schelling's philosophy is provided in Hans Baumgartner, ed., *Schelling: Einführung in seine Philosophie* (Munich: Karl Alber, 1975). A useful collection of more recent articles on Schelling's philosophy of nature is contained in Ludwig Hasler, ed., *Schelling, Seine Bedeutung für eine Philosophie der Natur und der Geschichte* (Stuttgart-Bad Cannstatt: Frommann-Holzboog, 1981). This latter collection includes several East German contributors, who wrote before the wall came down. They do often strain to make Schelling a dialectical materialist. Finally, Frederick Beiser has written an almost-too-clear account of Schelling's basic philosophy in his *German Idealism* (Cambridge: Harvard University Press, forthcoming).

7. See Georg Wilhelm Friedrich Hegel, *Differenz des Fichte'schen und Schelling'schen Systems der Philosophie* (Jena: Seidler, 1801). This work is translated with instructive introductions as *The Difference Between Fichte's and Schelling's System of Philosophy,* trans. H. S. Harris and Walter Cerf (Albany: State University of New York Press, 1977).

Figure 3.2

Friedrich Hölderlin
(1770–1843); he shared a
room with Hegel and
Schelling at the Tübinger
Stift. Pastel portrait
(1792) by Franz Karl
Hiemer. (Courtesy
Schiller-
Nationalmuseum,
Marbach.)

Though the Tübinger Stift attempted to sequester its young wards from the activities of the external world, the seminary authorities could not keep the casket hermetically sealed during those tumultuous times. Through the cracks seeped news of the Revolution in France, the very premises of which were as acid to the chains of the ancient political authority that the seminary draped around the necks of its future preachers. And the critical ideas of Kant penetrated the walls of the Stift, intoxicating the three friends. Their enthusiasm for the new philosophy seems to have been in direct proportion to their seminary professors' efforts to domesticate it, to render it compatible with traditional theological conceptions.[8] Schelling and his comrades discovered in the critical philosophy, however, not the traditional hymns to the Creator God but siege engines of theological destruction. The halls of the Tübinger Stift rang with philosophical and revolutionary discord, as one student of the period later recalled:

8. Dieter Henrich describes the ways in which the theologians at Tübingen attempted to use Kant's conception of the postulates of practical reason to argue for a literal interpretation of Scripture. See Dieter Henrich, "Dominant Philosophical-Theological Problems in the Tübingen *Stift* during the Student Years of Hegel, Hölderlin, and Schelling," in *The Course of Remembrance and Other Essays on Hölderlin*, ed. Eckart Förster (Stanford: Stanford University Press, 1997), pp. 31–54.

One admired in the French Revolution, the triumph of reason, and the decisive victory of philosophy. . . . There was no longer any discussion of theology. That was empty twaddle. . . . The overriding interest of reason lay in science [*Wissenschaft*], which taught men to become free and equal, and to chuck all intellectual or worldly despotism into the coals.[9]

Schelling translated "La Marseillaise" into German, and with that song on their lips, the young went blissfully down the slide into the new age.

The three friends moved rapidly in harmony with those quickened times, even if that was rather dangerous for their academic welfare. For they received stipends from the Catholic duke Karl Eugen (1737–1793), who would not brook any insubordination from those in his financial thrall at the Protestant Stift.[10] In 1793 Hegel and Hölderlin escaped the repressive atmosphere of their academic life through the expediency of graduation. By the time they left, neither had any intention of becoming a preacher; both took the usual route of graduates without traditional profession—they became tutors for well-to-do families. Like his friends, Schelling also remained through the required five years, graduating in 1795.

After finishing the two-year course in philosophy, Schelling, at age seventeen, stood his magister's exam and defended a dissertation—his own, rather than a thesis produced by one of his professors, as was usually the case. In his dissertation, Schelling argued that the Genesis stories in the Bible ought to be regarded as deeply symbolic myths rather than as historical occurrences since among ancient peoples, reason struggled to express itself in folk costume—an incendiary proposition that licked at the heels of the more literal minded in the theology faculty.[11] The dissertation caught the attention of H. E. G. Paulus (1761–1851), a family friend and professor of theology and oriental studies at Jena. Paulus published in his journal *Memorabilien* an essay of Schelling that extended the theme of his disserta-

9. J. G. Phahl, *Ulrich Höllriegel, Geschichte eines Magisters* (1802), quoted by Fuhrmans in Schelling, *Briefe und Dokumente*, 1:17. Schelling later, in his obituary of Kant, asserted that the same spirit underlay both the real Revolution in France and the ideal revolution in philosophy created by Kant. In both instances, a spirit of freedom expressed itself, first in the moral sphere and then in the political. Knowledge of the Kantian philosophy, he felt, made scholars in Germany initially more receptive to the French Revolution. See his "Immanuel Kant," in *Schellings Werke*, 3:588 [VI:4].

10. The resources of the Tübingen university, beyond the Stift, had been drained by Karl Eugen himself; he preferred to support his new academy, the Karlsschule in Stuttgart.

11. Schelling's thesis concerned the origins of human evil, to which he gave an account that borrowed notions of historical development from Herder and various Enlightenment philosophers. He argued that it was the growth of reason itself that forced humankind from the garden of natural simplicity. See Schelling, *Antiquissimi de prima malorum humanorum origine* (1792), in *Friedrich Wilhelm Joseph Schelling Historisch-Kritische Ausgabe*, ed. Hans Baumgartner et al., 5 vols. to date (Stuttgart: Frommann-Holzboog, 1976–), 1:59–100.

tion. This essay, "Über Mythen, historische Sagen und Philosopheme der ältesten Welt" (On myths, historical legends, and protophilosophy of the ancient world, 1793), discussed the general nature of myths, particularly origin stories found in the earliest oral traditions of peoples.[12] Borrowing considerations from Herder, Schelling maintained that such origin myths arose naturally in the youth of a nation, when philosophical thought could only be given sensuous expression—an idea that would reappear in his later work on the philosophy of nature and the philosophy of art.[13] His friend Hegel wrote to congratulate him on his effort "to clarify central theological concepts" and gradually "to clear away the dead wood."[14]

Impact of Fichte

In the winter semester of 1792–93, Schelling began his theological studies, but his interest therein had been compromised; he had already become, as he later recalled, theologically unorthodox.[15] His reading of the Enlightenment philosophers and the more recent works of Herder and Kant released in him a flood of rationalistic criticism that eroded the grounds of traditional theology. Yet he continued in the Tübingen theologate, though in a depressed frame of mind. In June 1793, however, his spirits took fire. Fichte visited Tübingen. The philosopher had already been recognized as the author of the *Kritik der Offenbarung* (The critique of revelation, 1792) and as the thinker who was boldly extending Kant's ideas. He visited again the following May, now the admitted author of a defense of the French Revolution, his *Beitrag zur Berichtigung der Urtheile des Publikums über die französische Revolution* (Contribution to the emendation of public opinion on the French Revolution, 1793). Schelling fell deeply under the spell of this bewitching thinker. So when Fichte's sketchy and preliminary effort at his *Wissenschaftslehre* appeared later in summer—the *Begriff der Wissenschaftslehre* (Concept of the science of knowledge, 1794)—Schelling quickly obtained a copy and plunged in. When he surfaced, he had embraced the letter

12. Schelling, "Über Mythen, historische Sagen und Philosopheme der ältesten Welt," in *Schellings Werke*, 1:1–44 [I:43–83].

13. In his *Philosophie der Kunst* (Philosophy of art;, lectures delivered in Jena in winter 1802–3 and repeated in Würzburg in 1803–4), Schelling maintained that the objects of philosophical inquiry—the quasi-Platonic ideas he would later endorse—initially appeared in the history of thought as mythical beings: "Ideas, insofar as they are intuited as real, are the stuff, the universal and absolute material of art, out of which all particular artworks arise as a ripened growth. These real, living, and existing ideas are the gods; the universal symbolism—or the universal representation of the ideas—is therefore given as real in mythology." See Schelling, *Philosophie der Kunst*, in *Schellings Werke*, 3:390 [V:370].

14. Hegel to Schelling (24 December 1794), in Schelling, *Briefe und Dokumente*, 2:53.

15. Schelling to his parents (4 September 1797), in ibid., 2:124.

and spirit of idealism. In a few weeks, he dashed off a complementary essay, *Über die Möglichkeit einer Form der Philosophie überhaupt* (On the possibility of a form of philosophy in general), which was published in September 1794, while he was still a teenager. Inspired by Fichte, the young philosopher observed that Kant had not provided a principle that would ground a formal structure for philosophy. And like Fichte, he attempted to fill this gap by establishing the principle, and thus the structure, in the absolute ego; then, with adolescent élan, he argued that all representations of the natural world could be deduced therefrom.

Schelling sent his little work to Fichte as an indication of intellectual debt, obvious enough to any reader. Fichte responded by providing his new philosophical companion the first fascicles of his more comprehensive *Grundlage der gesammten Wissenschaftslehre* (Foundations of the entire science of knowledge, 1794–95), which again quickly elicited from Schelling another tract, his *Vom Ich als Prinzip der Philosophie oder über das Unbedingte im menschlichen Wissen* (On the ego as a principle of philosophy or on the unconditioned in human knowledge), which came out at Easter 1795. Two months prior to this time, Schelling had sent a letter to Hegel that gave the gist of his considerations and indicated his commitment to the Fichtean point of view. The letter, however, also inserted two small wedges of interest that would gradually separate the young disciple from the master, namely, the problem of the non-ego's status—that is, nature's status—and the Spinozistic solution to the problem.

> I am immersed in the work of Spinoza—don't be astounded. Do you want to know why? For Spinoza, the world was everything (the absolute object, as opposed to the subject); for me, the ego is all. The proper distinction between critical philosophy and the dogmatic seems to me to lie in this, that the former derives everything from the absolute ego (determined by no object), the latter takes all from the absolute object, that is, from the non-ego. The latter, in its most developed form, leads to Spinoza's system, the former to the Kantian. Philosophy must derive from the unconditioned. Now it is a question of wherein the unconditioned lies, in the ego or in the non-ego. If this question is decided, then everything is decided. For me the highest principle of all philosophy is the pure, absolute ego, that is, the ego insofar as it is simply the ego and not determined by any object, rather is posited by freedom.[16]

Though he had made a start with considerations initially cultivated by Fichte, Schelling became uneasy, fearing that perhaps he had not really got-

16. Schelling to Hegel (4 February 1795), in ibid., 2:65.

ten to the root of the self and its flowering into a world. But just after Easter, he was comforted by his friend Hölderlin, who was visiting from Jena, where he had been attending Fichte's lectures. The poet told him: "Take it easy. You've gotten as far as Fichte. I've heard him."[17]

Hölderlin's consoling words were those of a friend. For, from his first encounter with Fichte to the end of the century, Schelling really remained a few steps behind his mentor, at least in the exploration of the grounding principles of subjective idealism. Quite generally, all of his philosophical work during this period conformed to the Fichtean point of view. But Schelling ventured far into that territory avoided by Fichte, the realm of the not-I—that is, of nature. Schelling's work during the period 1795–1800 would be divided almost equally between the two tasks that he believed the new critical philosophy had set: either to begin within the subjective sphere, the structure of self-consciousness, and show how an objective world of nature arose therefrom—and this would be the task of *transcendental philosophy;* or to begin with the objective world, and show how its various contents and laws derived exclusively from the activities of intelligence—and this would be the goal of *nature philosophy.*[18] Schelling's efforts at *Naturphilosophie* would yield a highly original product and would, at the turn of the century, lead to a break with Fichte and a rehabilitation of Spinoza.

In late summer of 1795, Schelling completed theological studies and in November took the final oral exam. And therewith he finished with orthodoxy as well. He had no inclination or intention of following in the footsteps of his father. The marker of this resolution appeared in print under the title *Philosophische Briefe über Dogmatismus und Kritizismus* (Philosophical letters on dogmatism and criticism, 1795).[19] In this tract Schelling waged a polemic against those pseudo-Kantians—namely, some of his theology professors—who used the master's moral argument in an attempt to prove God's existence as a thing-in-itself. With the righteous indignation of the newly converted, he insisted upon the futility of trying to conclude theoretically to the existence of an absolute being through a moral argument. Such a lamentable effort sinned multiply against Kant: God as thing-in-itself could not be legitimately conceived; practical reason does not yield theoretical knowledge; even if it could, a moral God would mean the absolute could do otherwise; and so on. Schelling additionally showed that

17. Friedrich Hölderlin, quoted by Plitt, *Aus Schellings Leben,* 1:71.

18. Schelling canonized this division of philosophical labor in his *System des transscendentalen Idealismus,* in *Schellings Werke,* 2:340–42 [III:340–42].

19. Schelling, *Philosophische Briefe über Dogmatismus und Kritizismus,* in ibid., 1:205–66 [I:284–341].

the philosophies of criticism and dogmatism, from their theoretical sides, faced the same problem, namely, the derivation of the finite (an empirical self and the natural world) from the infinite (either the absolute self or Spinoza's God). Serious philosophers from both camps recognized the theoretical impossibility of such a transition. A selection between the two systems could only be made, he thought, *pragmatically*: dogmatism led to a quietism in the face of causal determinism, while criticism led to action in order to overthrow what seemed determinate relationships in the world. It was this latter system, he suggested, that best conformed to our inmost self—an empirical individual that continually strives to realize in itself the productive acts of an infinite self. In short, the Gordian knot of theoretically deriving the finite from the infinite would be cut by repeating in consciousness the intellectual intuition through which a self and world came to be. Schelling thought his seminary professors, by contrast, had concocted a nauseating brew of flat criticism and benighted dogmatism. The young philosopher's rejection of the route for which his seminary days prepared him meant only one real possibility for his immediate future—academic servitude.

The Roving Tutor

In November 1795 Schelling became tutor to the family of the Baron von Riedesel in Stuttgart. He had the task of preparing the baron's two sons for the study of law, which meant that he had quickly to cram a legal education for himself. The effort did not expend itself merely in tutoring his wards. He began to think hard about the philosophy of law and ethics. And those thoughts came liquid to his pen, resulting in an article entitled "Neue Deduktion des Naturrechts" (New deduction of natural rights). It appeared a year later (1796), appropriately enough, in the *Philosophisches Journal*, for which Fichte was coeditor, along with the neo-Kantian Friedrich Niethammer (1766–1848). In the piece Schelling maintained that if free beings were to realize their moral intentions in the natural world, "the causality of freedom must be revealed through physical causality."[20] That is, the laws of nature and moral laws had ultimately to be the same. This conclusion, though initially startling to our eyes, would not have been viewed as completely outrageous by the readers of the *Journal*. Kant and Fichte both ar-

---

20. Schelling, "Neue Deduktion des Naturrechts," in ibid., 1:172–73 [I:248–49]. Kant had also asserted, though for a purely transcendentally intellectual world, the likely coincidence of physical law and moral law. See Immanuel Kant, *Kritik der Urteilskraft*, in *Werke*, ed. Wilhelm Weischedel, 6 vols. (Wiesbaden: Insel Verlag, 1957) 5:420–21 (A 338–39, B 342–43).

gued that the fundamental laws of nature derived from the judicial structure of rational consciousness. Since the ego, according to the critical point of view, acted freely, such legislation formally constituted moral law as well as physical law. A bit later, in 1798, Fichte himself would elaborate an ethical system—his *System der Sittenlehre* (System of ethics)—built on a similar identification of the natural and moral orders. It bespoke Schelling's genius that here, as in so much else, he drove to the philosophical heart of the critical matter. He was twenty years old at the time.

Schelling remained in Stuttgart and tutored the two young noblemen until March 1796, when he traveled with them to the university at Leipzig. During the first two months in his new position, he received visits from Hölderlin, who had much to report of his time at Jena, especially of the lectures of Fichte. Out of these discussions, and feeling his own way toward a more original footing, came a brief, unpublished tract that has been christened by historians in this century "Das älteste Systemprogramm des deutschen Idealismus" (The oldest systematic program of German idealism). The tract—variously attributed to Hegel, Hölderlin, and Schelling—set out the trajectory along which Schelling would move through the rest of his philosophical career.[21]

The "Program" begins with a confirmation of Fichte's starting point: "The first idea is naturally the representation of myself as an absolutely free being. With this free, self-conscious being an entire world arises simultaneously—out of nothing—the only true and conceivable creation out of

---

21. This small document was found among Hegel's papers and transcribed in his hand. It was first published in 1917. It seems to have been composed in late 1796 or early 1797. Early scholarly opinion ascribed the work to Schelling, though cases were also made for Hegel himself, Hölderlin, and the three friends in concert. The use of the first-person personal pronoun would seem to preclude this last possibility. Fuhrmans thinks it to be Schelling's. See his discussion of the question in Schelling, *Briefe und Dokumente*, 1:57 n. 3. Frank-Peter Hansen has examined the huge literature on the question (some 160 articles between 1965 and 1989 alone); and while the weight of opinion still rests on the side of Schelling, Hansen himself, after meticulous study, argues for Hegel's authorship. See Frank-Peter Hansen, *"Das älteste Systemprogram des deutschen Idealismus": Rezeptionsgeschichte und Interpretation* (Berlin: Walter de Gruyter, 1989). More recently, in a marvelously detailed and argued article, Eckart Förster maintains the essay to be Hölderlin's. He thinks the article hints at Goethe's view of science, which, he believes, Hölderlin at this time endorsed. See Eckart Förster, "'To Lend Wings to Physics Once Again': Hölderlin and the 'Oldest System-Programme of German Idealism,'" *European Journal of Philosophy* 3 (1995): 174–98. Förster seems to forget the very strong Fichtean idealism of the essay (see the text below), which Goethe at this time would have none of. Moreover, Hölderlin himself would have disavowed it. He had written Hegel in 1795 that he thought Fichte's *Wissenschaftslehre* "dogmatic," and he then brought serious objections to Fichte's conception of the absolute ego. See Hölderlin to Hegel (26 January 1795), in Friedrich Hölderlin, *Sämtliche Werke*, 3rd ed., 4 vols. (Berlin: Propyläen-Verlag, 1943), 2:315–16. Whoever the real author, the program of the tract does accurately forecast Schelling's intellectual development.

nothing." But then the tract suggests that this one-sided subjectivity needs to be balanced by a scientific investigation of nature, a new physics to "satisfy our creative spirit." During this period Schelling had undertaken natural scientific studies that would give substance to this programmatic demand. Those ambitious souls armed with the new philosophy of science would be able to transcend the mechanical conception of the state—which had individuals causally meshed with one another as cogs in a law-governed machine; and those newly enlightened thinkers would as well discover the moral world, including God and immortality, within themselves. The tract concludes that all of the ideas of self, nature, society, God, and the rest would find their union in the idea of beauty: "I am now convinced that the highest act of reason, that in which all other ideas are contained, is an aesthetic act and that truth and goodness become united like twins [*verschwistert*] in beauty—the philosopher must therefore possess as much aesthetic power as the poet."[22] During the next few years, Schelling would work out in detail the aesthetic vision of philosophy and science that is vaguely limned in the essay. Via Schelling's agency in particular, this conception would come to shape the course of biological science through the nineteenth century.

At Leipzig, Schelling oversaw the work of the brothers von Riedesel as they began their formal study of law. As a master tutor, he had the obligation to attend law lectures with his wards. But he also took the opportunity both to frequent lectures on various areas of natural science and to engage in considerable independent study of the newest developments in these disciplines—especially recent work in electricity, magnetism, chemistry, physiology, and medicine. The new work in those first three disciplines would be of a sort, in the words of the "Program," "to give wings again to our physics, which has under the weight of tedious experiments been progressing so slowly."[23] Schelling seriously engaged the disciplines of physiology and medicine as well. He wrote to his parents on 4 September 1797 to request some disputations that his father might have on physiology, since he was then working on "a theory of animal life." He also mentioned he had begun the study of medicine, which had made "great progress in short time," but was "simple enough" that one could "master it in a few years"[24]—a remark that reveals as much about the state of Schelling's con-

22. [Schelling], "Das alteste Systemprogramm des deutschen Idealismus," in Schelling, *Briefe und Dokumente*, 1:69–70.

23. Ibid., 1:69.

24. Schelling to his parents (4 September 1797), in ibid., 2:122.

fidence and genius as about the state of medicine in the late eighteenth century.

Out of this critical period of investigation in Leipzig came, with lightning speed, three monographs that would secure Schelling's reputation and bring him quickly to Jena. The first was a long article published over seven numbers of the *Philosophisches Journal* (1797–98). The article had been solicited by Niethammer, with whom Schelling had struck up a considerable correspondence. Ostensibly the article reviewed current literature in philosophy. But Schelling also used it, as he warned the editor, to develop his own philosophical position.[25]

In his literature review, Schelling attacked the obdurateness of the anti-Kantians but even more the presumptions of the slavish followers of Kant. These latter assumed that our representations of objects in the world stemmed, on the one hand, from certain formal structures (space, time, substance, cause, and the rest) contributed by the mind and, on the other, material components (the manifold of sensation) produced by objects existing outside the mind, the unconscionable things-in-themselves. But was it possible that a freely acting mind could impose its laws on a world presumably having necessities of its own and that this world would obey those laws? Such a system, the young philosopher chided, was not idealism, not even dogmatism; rather, "no system has existed that is so laughable or fantastic."[26] Knowledge, he insisted, was correspondence: to know something required that our representation of the object be identical to that object in

25. The article was originally published under the title "Allgemeine Uebersicht der neuesten philosophischen Literatur," but as *Abhandlungen zur Erläuterung des Idealismus der Wissenschaftslehre* in a second edition (1809). Niethammer, the editor in chief, published the review anonymously, but the identity of the author was well enough known—certainly to the contributing editor, Fichte. Friedrich Schlegel also knew who the author was. He wrote a review of the first four volumes of the *Journal* and mentioned to Novalis (5 May 1797): "I find the second number of the 'Journal' by Schelling, just between us, a bit trivial." See *Kritische Friedrich-Schlegel-Ausgabe*, ed. Ernst Behler et al., 35 vols. to date (Paderborn: Ferdinand Schöningh, 1958–), 23:362. Schlegel, however, had liked the first installment of Schelling's essay. See ibid., 23:350.

26. Schelling, *Abhandlungen zur Erläuterung des Idealismus der Wissenschaftslehre*, in *Schellings Werke*, 1:285 [I:361]. A central figure in Schelling's discussion of the recent literature in philosophy was Karl Leonhard Reinhold, the most visible and sophisticated of the Kantians writing at this time. In the *Abhandlungen*, Schelling tended to distinguish Reinhold from the other Kantians and suggested that it was a misunderstanding of Reinhold's position to think he postulated a thing outside the mind (ibid., 1:233–34 [I:409–10]). Reinhold had tried to provide a more systematic introduction to the critical philosophy, especially in his effort to establish the elementary principles whence the theory of cognition might be derived. In this respect, he served as a model for Fichte's similar attempt and he inspired Schelling as well. Despite Schelling's suggestions to the contrary, however, Reinhold clung to the thing-in-itself as necessary to furnish the sensible content for representations of the world. See the previous chapter and Frederick Beiser's lucid account of Reinhold's position in his *Fate of Reason: German Philosophy from Kant to Fichte* (Cambridge: Harvard University Press, 1987), pp. 226–65.

relevant respects.[27] This simply could not occur if representation and object existed as radically different and partitioned off from one another. If we were to have any hope of explaining the possibility of knowledge, we had simply to give up the Kantian assumption of a thing-in-itself.

The one case that conformed to the correspondence requirement of knowledge was that of self-knowledge: "For only in the self-intuition of a mind is there an identity of representation and object."[28] This appeal to the self as the only possible object of representation, once the thing-in-itself had been cast out, formed the ground in which transcendental idealism was rooted.

Schelling took the opportunity in his review to develop an alternative to unregenerate Kantianism. He sketched, in rather elliptical fashion, his (and Fichte's) conception of the operations of the ego in the creation of a world. In 1800 he would expand these considerations in his *System des transscendentalen Idealismus*, which I will describe below. The coeditor of the *Philosophisches Journal*, Fichte, was certainly pleased with the young philosopher's review—as it complemented his own position—and began working to secure Schelling's appointment at Jena.

The two other monographs written during this period also operated powerfully to ignite the fuel of this rising star. These works glowed with originality and introduced into the burgeoning idealism a new facet, that of nature philosophy. These works were Schelling's *Ideen zu einer Philosophie der Natur* (Ideas for a philosophy of nature, 1797) and his *Von der Weltseele, eine Hypothese der höheren Physik zur Erklärung des allgemeinen Organismus* (On the world soul, a hypothesis of the higher physics for the clarification of universal organicity, 1798). They were quickly followed by his *Erster Entwurf eines Systems der Naturphilosophie* (First sketch of a system of nature philosophy, 1799) and the introduction to this latter, his *Einleitung zu dem Entwurf* (1799). These tracts reflected the rapid evolution of Schelling's changing conceptions of nature and the proper philosophical approach to it. Each title suggests an effort still in the process of becoming (*Ideas for . . .* , *a hypothesis of . . .* , *First sketch of . . .*), a conception not having achieved completion, indeed, a Romantic adventure. Despite the tentativeness of their titles, these monographs introduced radical interpretations of nature that would reverberate through the sciences, and particularly the biology,

27. Frederick Gregory observes that Johannes Müller shared this fundamental notion of truth with Schelling. See his "Hat Müller die Naturphilosophie wirklich aufgegeben?" in *Johannes Müller und die Philosophie* (Berlin: Akademie Verlag, 1992), p. 153. See also Gregory's *Nature Lost? Natural Science and the Theological Traditions of the Nineteenth Century* (Cambridge: Harvard University Press, 1992), p. 286 n. 43.

28. Schelling, *Schellings Werke*, 1:290 [I:367].

of the next century. They developed the fundamental doctrines of *Natur-philosophie*.

Naturphilosophie

The Basic Program of *Naturphilosophie*

Schelling never thought of his *Naturphilosophie*, in its several incarnations between 1797 and the first few years of the new century, as a substitute for empirical natural science. His writings on *Naturphilosophie* groan with the weight of citations of the most recent, up-to-date experimental work in the sciences.[29] He considered *Naturphilosophie* as a framework for systematizing the laws of empirical discovery and grounding that system in higher, a priori principles. In this respect, his *Naturphilosophie* had the character of Kant's first *Critique* and *Prolegomena*, works that showed how the fundamental laws of physics, though empirically established, nonetheless derived their universality and necessity from the transcendental categories of mind. Schelling advanced upon Kant, however, in his attempt not simply to draw the basic laws of physics from a priori principles but also the laws of chemistry, biology, and eventually medicine, all of which latter Kant specifically excluded from the realm of demonstrable, authentic science.[30]

During the mid-nineteenth century, empirical scientists grew increasingly hostile to a degenerating strain of *Naturphilosophie* that did have its roots in Schelling's work. Matthias Jakob Schleiden (1804–1881), co-founder of cell theory and himself not a little touched by Goethe's Romantic construction of nature, issued a broadside from his redoubt in Jena against what he thought Schelling and Hegel had wrought, namely, a dogmatic and speculative effort to establish the facts and laws of nature through deduction alone.[31] For Schleiden, the inductive, experimental determination of facts and laws formed the ground of natural science, which

29. Snelders, along with others, holds a contrary view of Schelling's conception of empirical science. He believes that Schelling regarded empirical work trivial and opposed it—a belief, as far as I am aware, for which there are no grounds in Schelling's work. Nonetheless, see Snelders's "Romanticism and Naturphilosophie" and his *Weltenschap en Intuïtie*.

30. For a discussion of the ways in which physicians took up Schelling's *Naturphilosophie* as a means of scientific certification for their discipline, see Günter Risse, "Kant, Schelling, and the Early Search for a Philosophical 'Science' of Medicine in Germany," *Journal of the History of Medicine* 27 (1972): 145–58.

31. Matthias Jakob Schleiden, *Schellings und Hegels Verhältniss zur Naturwissenschaft. (Als Antwort auf de Angriffe des Herrn Nees von Esenbeck in der Neuen Jenaer Lit.-Zeitung, Mai 1843, insbesondere für die Leser dieser Zeitschrift.)* (Leipzig: Wilhelm Engelmann, 1844). Olaf Breidbach has reprinted this pamphlet with an extensive introduction, in his *Matthias Jakob Schleiden, Schellings und Hegels Verhältnis zur Naturwissenschaft* (Weinheim: VCH, Acta Humaniora, 1988).

could only be sullied by speculative philosophy. He certainly thought the review of his own book *Grundzüge der Wissenschaftlichen Botanik* (Foundations of scientific botany, 1845–46) by the Schelling disciple Christian Gottfried Nees von Esenbeck (1776–1858)—a review that occasioned his pamphlet—was clear indication of the befogged nature of *Naturphilosophie*.[32] Schleiden's polemic assumed that Schelling was attempting to replace experimental procedures with armchair musings. Alexander von Humboldt, in the wake of his return from his South American voyage, had initially been an enthusiast for Schelling's *Naturphilosophie*, thinking it quite compatible with his own conception of the interconnectedness of nature. In his later years, however, he became a bit wary of Schelling's disciples, like Nees von Esenbeck and Carl Gustav Carus (1789–1869), but remained quite good friends with Schelling himself.[33] In the early 1870s, Kuno Fischer directed part of his effort, in his two-volume study of Schelling's life and work, to stem the assumption that the philosopher's conceptions were inimical to empirical science.[34] Fischer obviously did not succeed, as that belief is still commonly held.

Ideen zu einer Philosophie der Natur

Schelling's first major work in *Naturphilosophie*, his *Ideen zu einer Philosophie der Natur*, displays immediately his respect for and employment of experimental research. In the first part of the volume, he examined the latest developments in various areas of the natural sciences—chemistry and physics, and their special subjects: combustion, light, gases, electricity, and magnetism. His inductive analyses of this research supported, he believed, his fundamental physical hypothesis: diverse natural phenomena might be regarded ultimately as transformations of two underlying, polar forces, those of attraction and repulsion. The particular phenomenal qualities of light or airs, for example, would derive from the different ways in which the

---

32. See Breidbach for a brief discussion of Nees von Esenbeck's views, in ibid., pp. 21–22.

33. Humboldt remained on good terms with Schelling to the end. A considerable number of letters passed between them in the 1850s, most of which remain unpublished in the Berlin-Brandenburgische Akademie der Wissenschaften, Alexander-von-Humboldt-Forschungsstelle. Friendship did not prevent Humboldt from poking fun at some of the philosopher's ideas. In a sporting mood, he wrote K. A. Varnhagen von Ense (28 April 1841) of the Schellingian kind of pronouncement: "The East is oxygen, West hydrogen; it rains when the eastern clouds mix with the western." This letter is quoted by Petra Werner, in her quite judicious and insightful assessment of the relationship between Humboldt and Schelling. See Petra Werner, *Übereinstimmung oder Gegensatz? Zum widersprüchlichen Verhältnis zwischen A. v. Humboldt und F. W. J. Schelling* (Berlin: Alexander-von-Humboldt-Forschungsstelle, 2000), p. 22.

34. Fischer, *Friedrich Wilhelm Joseph Schelling*, 2:433–39.

underlying forces constituting them operated on sense perception. In the second part of his book, Schelling had two general aims. First, he sought to demonstrate that theories postulating the small, indivisible, passive atoms of Newtonian matter, to which forces could only be extrinsically applied, constituted conceptions quite insufficient to account for the special phenomena of the natural sciences. Second, and this formed the heart of his project, he attempted to develop what he thought a more promising view than the Newtonian, one that would find its ultimate justification in transcendental idealism.

Following Kant, Schelling proposed—and tried to deduce transcendentally—a concept of matter that revealed it to be a dynamic equilibrium of the forces of attraction and repulsion.[35] Even according to the usual beliefs of dogmatic science, he observed, our experience of material objects and their qualities can occur only through the agency of forces that act on us. We can never experience even mediately material objects not expressive of force. Such objects without force would be quintessentially *Dinge an sich*. We thus have no rational grounds for assuming a completely passive matter to which forces would have to be extrinsically applied, as in the case of Newtonian physics. It is more reasonable, Schelling argued, to postulate as primary not some inaccessible matter but just those forces necessary to account for experience. If matter consists of an equilibrium of repulsive and attractive forces, then the various phenomena of matter—electricity, magnetism, chemical affinities, and the rest—ought also to be derivable from these forces alone; and this Schelling attempted in the several chapters of the *Ideen*.

The Newtonian mechanical physics and chemistry had the task of constructing the different kinds of matter from the extrinsic relations of homogeneous atomic bodies—a task, Schelling thought, doomed from the start. A dynamic chemistry of forces, by contrast, could explain our experience of diverse kinds of matter by assuming them to be products of specific disequilibria of attraction and repulsion.

> Dynamic chemistry does not postulate any original matter such that out of its synthesis all other phenomena can arise. Rather, since it considers all matter to be originally a product of opposing forces, the greatest diversity of matter is nothing else than differences in the relationship of these

---

35. Schelling mentioned that Kant had shown, in *Die metaphysichen Anfangsgründe der Naturwissenschaft* (Metaphysical foundations of natural science), "with evidence and completeness," that the analytical composition of the concept of matter was that of attractive and repulsive forces. See Schelling, *Ideen zu einer Philosophie der Natur als Einleitung in das Studium dieser Wissenschaft* (1797), in *Schellings Werke*, expanded vol., 1:238 [II:231].

forces. But forces are in themselves already infinite. Thus one can conceive for each possible force an infinite number of degrees . . . and since
all qualities rest only on degrees, from this assumption alone one can derive and conceive the infinite diversity of matter in respect of its qualities
(insofar as they are known to us in experience).[36]

From the assumption of the two underlying polar forces of attraction and
repulsion, one could thus generate the various phenomena that the sciences of physics and chemistry revealed. For instance, heat could be understood as a particular degree of repulsive force producing an expansion of
matter (i.e., a force producing a disequilibrium of initially constituting
forces of a particular kind of matter); and this force of heat might be communicated to other matter until equilibrium was again established. A sufficiently heated body would permit combination with oxygen to produce
combustion. Oxygen itself, when greatly expanded through the repulsive
force, would become the phenomenon of light. And so for the various other
kinds of matter and their properties—they could more easily be construed
as different ratios of the underlying polar forces of attraction and repulsion
than as different arrangements of homogeneous Newtonian matter with
added forces.[37]

While Schelling had employed Kant's concept of matter—a construction out of opposing forces—and had, on one track, followed Kant in deriving the concept from the possibility of experience, he quickly glided to
the Fichtean position that the possibility of experience of a material world
did not depend on any mysterious thing-in-itself. Rather, he insisted
(though in a quite elusive way) that the mind had an originary power, a positive power of thought that extended unrestrictedly outward; simultaneously, however, once a determinative, restricting force arose, then thought
became finite, shaped, of definite proportions. The Kantians assumed, quite
unjustifiably, that the restricting power on the activities of thought radiated
from things-in-themselves, lying beyond the borders of the mind. Fichte,
by contrast, recognized the constraining force simply as a fact of consciousness, which ordinary consciousness referred to the not-I. Schelling
advanced upon Fichte. He sought an explanation for this fact of consciousness. He argued that the passive, restricting power was a necessary product
of the mind itself: the limitation resulted from the absolute ego's coming to
know itself. Hence, according to Schelling's conception, the natural world
as well as the empirical ego arose through the interaction of two powers of
the mind: a principal, creative power, infinite in scope, which expanded

36. Ibid., 1:282–83 [II:275–76].
37. Ibid., 1:282–306 [II:275–99].

outward, as it were; and a constricting, formative force that gave determination to the original power. The limiting force was, however, only that original power in its effort to become self-aware. Schelling referred to this limiting force as the absolute ego, since it had no ultimate conditions on its actions. In an act of intellectual intuition, then, the absolute ego, hidden from our empirically conscious view, would impose limitations on the infinite, primary power of the ego; these limitations would, for empirical consciousness, become "felt and sensed."[38] They would be naturally interpreted as the limiting presence of the *not-I*, of the manifold of sensation. Hence, sensory intuition itself, as it were, created the matter intuited. Experience of the physical world was thus hermeneutically sealed within the infinite mind:

> In the mental being there is an original struggle of opposing activities, out of which there first comes—a creation from nothing—a real world. With the infinite mind there is also first created a world (the mirror of its infinity), and the entire reality is thus nothing other than an original struggle of infinite production and reproduction. No objective existence is possible without some mind that recognizes it; and, conversely, no mind is possible without a world existing for it.[39]

I will discuss Schelling's basic theory of the self-creating intellectual intuition more thoroughly below. Here I will schematically recapitulate the theory as it appeared nascently in the *Ideen*:

infinite ego (also called the real or objective ego) ... felt limitations → empirical world

↑↑↑ → struggle for self-consciousness → ↓ (assumption of the impact of the world on the self)

absolute ego (also called the ideal or subjective ego) ... feeling of limits → empirical ego

The power of infinite thought reflects (as the absolute ego) upon itself, recreating itself at every moment through the struggle for self-consciousness. The resulting limitations imposed by the absolute ego become interpreted as the impingements of an outside world on this individual's consciousness.

38. In a footnote (ibid., 1:227 [II:220]), Schelling emphasized that the restrictions that the ego imposed on itself were not realized as merely abstract considerations: "Here we are referring to opposing activities in us insofar as they are *felt* and *sensed*. And it is out of this felt and originally sensed struggle in ourselves that we want *reality* to proceed."

39. Ibid., 1:229 [II:222].

Though the language of Schelling's analysis makes his theory of mind sound a bit mystical, he merely recapitulated ideas about thinking that had been endorsed in the history of Western philosophy, especially in the works of Plato and Aristotle. First, he pointed to thought as potentially grasping all things and constituting them in consciousness—thought as infinitely capacious and ever striving to know all things. Second, he indicated that this infinite capacity became limited when a particular thought was actually entertained; a limited thought was the partial realization of the infinite capacity during the conscious experience of a particular object. Schelling added to this basic structural analysis a more contemporary (to him) insight that lent it considerable plausibility, namely, the recognition that mind cannot be tapped on the shoulder to get its attention, nor can external objects *enter* mind—they can only be, as it were, *performed* by mind. That is, the objects of experience, coated over with sense qualities and supported by an assumption of extrinsic forces, must be actively constituted in perceptions, constructed with the only resources at the mind's disposal, namely, its own activities. Thus "the system of nature is at one and the same time the system of our mind."[40] Assumptions about objects external to the mind— especially those inaccessible *Dinge an sich*—simply cannot be justified.

The task of *Naturphilosophie* (as distinguished from transcendental philosophy) is to begin with a refined understanding of nature, a nature articulated with the help of the latest empirical theories, and to show how its various phenomena and relationships can be regressively chased back into the ego as their only possible source. The absolute ego gives birth, however, not only to nature but also to the finite self. Indeed, nature and the empirical self develop, as it were, in mutual dependence upon one another:

> With the first consciousness of an external world, there also arises the consciousness of my self; and conversely, with the first moment of my self-consciousness, the real world appears before me. The belief in a reality outside of me is established and grows with the belief in the existence of my self; one is as necessary as the other. Both—not speculatively separated but in their complete and intimate interaction—are elements of my life and all my activities.[41]

Because, as Schelling held, one's individual self-consciousness and the world emerge together, the mental origin of the world lies beyond individual conscious awareness. That is, as the absolute ego attempts to grasp itself (in its infinite capacities), it imposes limitations on itself, which limiting

40. Ibid., 1:689 [II:39]. Note, the introduction to the *Ideen* is contained in the first volume of the Munich edition.
41. Ibid., expanded vol., 1:224–25 [II:217–18].

activity spawns both an empirical, sensing self and an empirical, sensed world. A natural world independent of the subject thus comes forth out of the mists of unconscious mentality, carrying with it the apparent marks of unintelligent, brute facticity. Yet, the closer inspection allowed by transcendental philosophy shows that even the givenness of nature has the ego as its ultimate ground.

This rather abstract philosophical theory of nature supported a lived conviction that had two reciprocal aspects: first, if nature were ultimately the product of the self, then the natural researcher could plumb the depths of nature, could eventually get to the very bottom of things. As Schelling put it with youthful élan: "As long as I myself am identical with nature, I understand what living nature is as well as I understand myself."[42] Second, Schelling's theory suggested that nature might furnish a path back to the self: one might go into nature, enter the lush forests of central Europe or travel to the tropics of foreign lands, and there in the wild, tangled growth of primitive nature discover the self. But not only might one discover oneself in nature; the exploration of nature might even be regarded as a necessary propaedeutic to the development of the self, of one's character and personality. This implication of Schelling's doctrine greatly appealed to Alexander von Humboldt, whose retrospective creation of his experiences in the jungles around the Orinoco—where he grew into that true self he became—those re-created experiences in his many descriptive volumes bear the mark of his engagement with Schelling.[43] By extension, one could

42. Ibid., 1:697 [II:47]. Dieter Sturma lucidly explores this aspect of Schelling's *Naturphilosophie* in his "The Nature of Subjectivity: The Critical and Systematic Function of Schelling's Philosophy of Nature," in *The Reception of Kant's Critical Philosophy: Fichte, Schelling, and Hegel*, ed. Sally Sedgwick (Cambridge: Cambridge University Press, 2000), pp. 216–31.

43. After his return to Europe from his travels in the Americas, Alexander von Humboldt wrote Schelling to say, "I regard the revolution that you have produced in natural science the most beautiful episode of these rash times." He agreed that "nature philosophy cannot harm the progress of the empirical sciences. On the contrary, it traces a discovery back to its principles and simultaneously provides the foundation for new discoveries." See Humboldt to Schelling (1 February 1805), in Schelling, *Briefe und Dokumente*, 3:181. Humboldt made these sentiments quite public in his *Ideen einer Geographie der Pflanzen* (Tübingen: F. G. Cotta, 1807), where he wrote: "Being not wholly unacquainted with the spirit of the system of Schelling, I am far from the view that the study of authentic *Naturphilosophie* should be damaging to empirical investigations and that empiricists and nature-philosophers should forever repel one another as opposite poles of a magnet" (p. 5). In his celebrated *Kosmos*, Humboldt unhesitatingly endorsed Schelling's aesthetics of nature. He urged that his own effort was not simply to portray the physical structure of the universe, but "to arrive at a higher point of view, from which all formations and forces reveal themselves as one, living, internally active whole of nature. Nature is not a dead aggregate. She is 'for the enthusiastic researcher,' as Schelling expressed it in his wonderful essay on the plastic arts, 'the holy, eternally creative, primary force of the world, who actively generates and produces all things out of her self.'" See Alexander von Humboldt, *Kosmos, Entwurf einer physischen Weltbeschreibung*, 5 vols. (Stuttgart: Cotta'scher Verlag, 1845–58), 1:39.

say the same of Darwin and his experiences in the rain forests of Brazil. And, of course, Goethe. Werther, Goethe's passionately romantic alter ego, searched for his creative self in his experiences of nature, "who alone forms the great artist."[44] Just this feature of Schelling's speculations—the forging of an organic bond between self and nature—encouraged Goethe to take, in Caroline Schlegel's terms, a paternal interest in the young philosopher. And that very close relationship with the great poet of nature, along with certain other events in Schelling's personal affairs, led him ultimately to shake off Fichte's subjective idealism.

Schelling often represented the various relationships among the infinite self, the absolute self's intuition, the empirical self, and nature as if they were formed with atemporal logical swiftness. He would portray the outcome of these creative logical-metaphysical relationships as the products of a self ever striving for reflexive comprehension, a self in the process of infinite becoming. But even these representational strivings appeared in Schelling's early writings to be the result of a critical mind of Fichtean disposition reflecting on its own activity. In these same early efforts, however, one could turn the page and find Schelling mentioning the historical dimensions of the relationships that had initially been depicted in abstract terms. He would suggest that the actions of the ego played themselves out, stage after stage, in a larger temporal sphere, through the developmental history of mankind, from the period before the Greeks introduced reflective thought to his own age of critical philosophy. These historical stages would recapitulate in time the reproductive activities of the critical mind in search of itself. This view implicitly guided Schelling in the introduction to the *Ideen*. He seems to have been influenced, in this regard, by the developmental and cultural analyses of Herder's *Ideen zur Philosophie der Geschichte der Menschheit* (Ideas for a philosophy of the history of humanity).[45] Kant's conception of the regulative teleology of historical reconstruction likely made an impact as well.[46] In a fashion similar to that of Herder and Kant,

---

44. Johann Wolfgang von Goethe, *Die Leiden des jungen Werthers* (Munich: Deutscher Taschenbuch Verlag, 1978), p. 15. The line suggests that nature both is an artist and helps create the artist. As Goethe gradually discovered that his own poetic intuitions found a like sentiment articulated in Schelling's system, he drew ever closer to the young philosopher. Schelling's mercurial, self-possessed, and dramatic personality also exerted a strong pull on the positive pole of Goethe's affections.

45. Herder's *Ideen zur Philosophie der Geschichte der Menschheit* was published in three parts, over the years 1784 to 1791.

46. See Immanuel Kant, *Idee zu einer allgemeinen Geschichte in weltbürgerlicher Absicht*, in *Werke*, 6:35 (A 389): "Since nature has set the life of individual men only to short compass (as, indeed, is the case), she requires what may be an incalculable series of generations, each of which transmits its enlightenment to the next, in order ultimately to drive that seed in our species to those stages of

Schelling offered, in the introduction to the *Ideen*, a sketch of the historical development of Western philosophical culture. He imagined a period in human history when individuals lived philosophically in a state of nature, when the minds of men had not really achieved the reflective capacities attained by the Greeks (and later, of course, by the Germans). In the infancy of human mind, the conscious separation between self and world had not yet arisen; initially, human intellect hewed out no distinction between ideas and objects, between perceptions and things. Neither Newtonian matter nor the thing-in-itself had yet appeared; rather, incipiently reflective individuals had to struggle in some philosophical terror to unleash those creatures from the depths of their own minds. "Nature," said Schelling, "releases no one willing from her tutelage; there are no naturally born sons of freedom."[47] Human spirit had, thus, to struggle for the self-realized freedom of reflection. Philosophy came aborning with the agonistic instruments whereby that freedom was achieved. And through the development of the Western tradition in philosophy—from the postulation of an external, passive, dead matter to the inference of an unknowable presence hovering behind the phenomenal world—these antinomian shades yet called forth an internal, active, free spirit. The bifurcation of mind and matter marked a phase in the cultural maturation of philosophy, a kind of willful adolescence difficult to pass through; when prolonged it became pathological, as witness the unregenerate Kantians. The rift within the absolute ego had to be healed. Schelling hoped that his approach would provide both a speculum in which the empirical self could view that original unity of mind and object, and a therapy by which it could become whole again, reunited with that nature it had created.

Kant's critical philosophy had the task of demonstrating how the general, categorical features of experience had to be products of rational mentality, as opposed to the deliverances of independent objects in the world. In the *Ideen*, Schelling attempted to follow the Fichtean program of reducing not simply the general features of experience to mentality but also the apparently contingent features as well, particularly those of a scientific character. He wished to demonstrate that specific causal sequences (such as a heart pumping blood to the organs), which the Kantians assumed had to be due to things-in-themselves, were also the free, if hidden, decisions of the absolute ego. He yet admitted that he had not mastered the deductions

---

development that are completely commensurate with her intention." Kant offered this perspective as a regulative idea to the historian; Schelling, I believe, considered such ideas constitutive of nature herself.

47. Schelling, *Ideen zu einer Philosophie der Natur*, in *Schellings Werke*, 1:662 [II:12].

to render these productions translucent. His introduction to the *Ideen* concluded with just this recognition:

> Nature should be visible mind [*Geist*], mind invisible nature. Here, there-
> fore, is the problem that must be solved: in the absolute identity of mind
> in us and nature outside of us, how is external nature possible? This goal
> of our further research is thus the idea of nature. If we succeed in reaching
> it, we can also be certain that the aforementioned problem will be re-
> solved.[48]

The obstacle to tracing out the mind-tethered features of nature was ultimate facticity. The ego may have imposed categories of substance, causality, plurality, and the rest on the manifold of the given in order to render
experience organized, but did it also dictate that flowers required insects for
pollination, that mammals needed livers to purify the blood? To show that
these empirically discoverable relationships also resulted from impositions
of the absolute ego was the daunting task that Schelling initially believed
to be required of a thoroughgoing idealism.

In the *Ideen*, Schelling approached the problem of a priori demonstration not from the perspective of nature—which a reader might have expected—but from that of the ego itself. He began with the kinds of
deductively derived ego structures that originated with Fichte and then
moved to show how these structures could be displayed in his surveys of inductive natural science. In the next two major tracts in *Naturphilosophie*,
his *Weltseele* and *Erster Entwurf*, he readjusted his conception of the proper
disposition of induction and deduction in natural science, matters I will discuss in a moment.

The Organic in Mind and Nature

One realm of nature that Schelling thought absolutely vital to consider if
he were to realize his goal of discovering invisible mind in visible nature
was that of life. In the *Ideen*, he had intended to discuss biology as well as
physics. But aside from preliminary remarks in the introduction, the book
came fitfully to a conclusion with a sketch of some first principles of chemistry. The movement to the planned goal of the *Ideen* seems to have been
stilled by the eruption of a new conception, the first hint of which appeared
in a review of philosophical literature done for the *Philosophisches Journal*.
Schelling began serializing his large review for Niethammer and Fichte's

---

48. Ibid., 1:706 [II:56].

journal in early spring of 1797, at the time the *Ideen* was published. In the third installment of the review, Schelling maintained, briefly and elliptically, that the concept of *organism* should be the foundational concept for a philosophy of nature.[49] This thesis was advanced in the introduction to the *Ideen* (written last) but completely absent from the body of the work—which suggests that he came to the notion while composing the review but after he had completed a large part of the *Ideen*. His formulation of this new consideration likely caused him to bring his book to an abrupt conclusion. After the appearance of the *Ideen*, Schelling immediately turned to recast his *Naturphilosophie* in a work to which he would give the exotic title of *Weltseele*. In this book the concept of organism dominated as the ruling principle.[50] I will discuss the *Weltseele* more thoroughly in a subsequent chapter, but its foundational considerations can be briefly indicated here.

Organisms, Kant proposed, can be understood only in terms of the purposes that their parts realize; and those parts will be reciprocally the cause and effect of each other (thus the heart provides, via the blood, nutrition to the liver, which, in turn, purifies the blood, thus acting for the welfare of all the parts of the body, including the heart). Further, as he argued, the functioning of the parts must be conceived in terms of the design of the whole, which the parts instantiate. In the *Weltseele*, Schelling sought to show, through extensive surveys of the latest scientific work, that teleological structures characterized all living creatures. He contended that Alexander von Humboldt, Johann Friedrich Blumenbach, Carl Friedrich Kielmeyer, and Johann Christian Reil all either explicitly or implicitly relied on the concept of teleological structure when describing living nature.[51] But fur-

---

49. In the *Abhandlungen*, Schelling said he did not wish to dwell on the concept of organicity, but he would venture a few words, the gist of which was "Insofar as it produces its own representations, the soul is mutually the cause and effect of itself. It will intuit itself thus as an object—that which is cause and effect of itself, or what is the same thing, as a self-organizing nature. . . . Since in our mind there is an unending striving to organize itself, so there must also be revealed in the external world a general tendency toward organization." The inchoate character of his remarks—more nebulous than usual—suggests he had not as yet thought through the implications and applications of the concept of organism. See Schelling, *Abhandlungen zur Erläuterung des Idealismus der Wissenschaftslehre*, in *Schellings Werke*, 1:310 [I:386]. I will return to these passages in chapter 8.

50. My thesis that Schelling came to the idea of the fundamental role of organicity after completing only a part of the *Ideen* means that the *Weltseele* should not be taken as a continuation of the former book, but rather as its complete recasting. This thesis makes sense of the otherwise enigmatic note in the preface to the *Weltseele*, in which Schelling simply stated that "this composition is not to be regarded as a continuation of my *Ideen zu einer Philosophie der Natur*." See Schelling, *Von der Weltseele, eine Hypothese der höheren Physik zur Erklärung des allgemeinen Organismus*, in *Schellings Werke*, 1:419 [II:351]. The title *Weltseele* refers to the ancients' belief that all of nature was organic, was alive.

51. The work of these scientists and their contributions to Schelling's evolving notions about the organic will be examined in subsequent chapters.

ther analysis of recent work in physics and chemistry—particularly that of Lavoisier and his followers—suggested that the concept of the organic was necessary for understanding the operations of the nonliving world as well. Kant had reduced matter to the polar forces of attraction and repulsion; but for these forces to structure a complete system of formative powers, they had to be thought of as reciprocally interactive, now in balance, now productive of new chemical or physical developments. Thus, in Schelling's estimation, all of nature had to be construed ultimately in telic, organic terms.[52] But as Kant had shown, such organic structures had to be considered the result of an intelligence, of mind. Kant supposed, as a heuristic, a divine mind was at play, while Schelling argued it was the absolute ego that gave rise to the natural world. The conclusion that Schelling drew from his inductive surveys and analyses stood clear: the ubiquity of organic structures in nature could be explained only if mind, whence the world derived, revealed itself to be organic.

The preparatory reading Schelling did for the *Weltseele*—especially the work of Carl Kielmeyer on organic powers, Alexander von Humboldt on electrophysiology, and the followers of Lavoisier on the role of oxygen—as well as his own efforts at experimentation, must have urged upon him the realization of the poignant facticity of natural relationships, that they could not be anticipated a priori even if one plunged headlong into the depths of the absolute ego. In the *Weltseele*, then, he recognized that the inductive explorations of *Naturphilosophie* could only show that certain patterns seemed to exist within and among the sciences, and that much experimental research was required to resolve still open questions. Throughout the *Weltseele*, this epistemological reserve is linguistically marked by such introductory qualifications as "thus we cannot do better than to maintain" and "it is not improbable that."[53] The empirical regularities that were revealed, nonetheless, did seem sufficient for an induction to the best explanation, which required a framework of interactive polar forces, that is, of fundamentally organic powers. Both the patterns and the framework that gave them their most cogent account could, then, be used to rule out certain other approaches to nature—that of the Newtonians, for instance.

---

52. Ibid., 1:449 [II:381]: "Both of these agonistic forces [of attraction and repulsion], represented simultaneously in unity and opposition, lead to the idea of an organizing and formative principle, the world as a system. The ancients wanted to indicate this by the notion of the world soul." The proposal that matter was essentially constituted by polar forces was Kant's; but Schelling may have made the connection between physical forces of attraction and repulsion and organic forces of contraction and expansion as the result of Goethe's influence. See chapter 11 for a discussion of Goethe's identification of these two sets of forces.

53. Ibid., 1:617 [II:549], 1:621 [II:553].

The a priori Character of Experimental Science

In the *Weltseele*, Schelling hesitatingly realized that the existence of the proposed polar forces could not be demonstrated a priori within *Naturphilosophie*; for that one had to move to transcendental philosophy. That is, one had to begin with the structure of the ego and its activities in order finally to deduce those forces that would constitute our concept of matter and its dispositions in the living and nonliving worlds. Schelling constructed this deduction in the last great work of this early period, that book by which he is best known even today, his *System des transscendentalen Idealismus*. By the time of its composition, his intellectual ties to Fichte had frayed to the delicate breaking point. In the year before, at Easter of 1799, he published his *Erster Entwurf eines System der Naturphilosophie* (First sketch of a system of nature philosophy) and a few months later his *Einleitung zu dem Entwurf eines Systems der Naturphilosophie* (Introduction to the sketch of a system of nature philosophy). And with these two works, he again, for the "first time," grounded *Naturphilosophie*; but this grounding, especially in the *Einleitung*, became hostile territory for Fichtean idealism.

The *Einleitung*, composed during the period when Schelling was also completing the *System des transscendentalen Idealismus*, shows a rapidly deviating conception of *Naturphilosophie*, even moving beyond the *Entwurf*, for which it was supposed to be the introduction. There appear to have been two precipitating causes for this new trajectory. The first was Goethe, who invited Schelling to read the *Einleitung* aloud to him. The young philosopher had to explain its various points, which the older man questioned in depth. Schelling confessed that the interchange produced a great "fluorescence of ideas."[54] The second stimulus for reexamining his basic framework seems to have been two belated reviews of the *Ideen zu einer Philosophie der Natur* in the *Allgemeine Literatur-Zeitung*. The ALZ was published daily out of Jena and prided itself on quite up-to-the-minute reviews. But the reviews of the *Ideen* were printed in October 1799, over two years after the book first appeared. One analysis was by a natural scientist, the other by a philosopher.[55] The scientist thought of himself as the hammer of

54. Schelling to F. A. Carus (9 November 1799), in *Goethes Gespräche*, ed. Flodoard von Biedermann, expanded by Wolfgang Herwig, 5 vols. (Munich: Deutscher Taschenbuch Verlag, 1998), 1:731: "A short while back he [Goethe] spent several weeks here. I was with him for a long time every day, and had to read aloud my work on *Naturphilosophie* and explain it to him. What a fluorescence of ideas these conversations have produced for me, you can well imagine."

55. The two reviews appeared anonymously in subsequent numbers of the *Allgemeine Literatur-Zeitung*, nos. 316 and 317 (3 October and 4 October 1799), pp. 15–30 and 33–38. The double review, as this historian has himself experienced, always signals something more afoot.

correct thinking. While he noted that Schelling discussed recent theoretical and experimental work in physics, he acidly suggested that the young man had utterly misunderstood some ideas (Euler's theory of light, for instance) or gave them a singular interpretation. These animadversions undoubtedly stung. The philosopher posted his objections from an almost precritical standpoint; at least the arguments seemed to Schelling the most degenerate kind of Kantianism.[56] In any case, Schelling obviously believed his project in *Naturphilosophie* had been incorrectly perceived. At the time he wrote the *Ideen*, though, perhaps he himself had not quite understood the direction of his own ideas. In a scathing letter to the editor, he dismissed the reviewers as being, respectively, either scientifically or philosophically naive. He put it to the editor—apparently quite seriously—that since he had changed his mind about several issues, he himself might have been the best candidate to review the *Ideen* for the *ALZ*.[57]

The *Erster Entwurf* and, even more, the *Einleitung* signaled a considerably altered course in Schelling's thought. In these works he expressly abandoned the Fichtean primacy given to transcendental philosophy, and he moved toward the "identity philosophy" that would cause the final rupture with his mentor. He wished now to demonstrate that *Naturphilosophie* could be established as an independent, systematic discipline. As propaedeutic to the goal of fixing the independence of *Naturphilosophie*, he had first to clear up some misconceptions concerning its character—a patient reader, though, might regard these clarifications more as further developments of his program. *Naturphilosophie*, he emphasized in the *Einleitung*, was not meant to replace empirical science. The heart of natural science (*Naturwissenschaft*), he maintained, was the experiment (*das Experiment*), in which "nature is forced to respond under determinate conditions, which usually do not exist in her or which must be arranged by others in order to exist."[58] Not just natural science, but all knowledge about the world must be ascertained initially through experience. Deduction from the armchair simply could not substitute for experiment in the laboratory. With a resolution that likely received warm embrace from Goethe, Schelling claimed that "originally we generally know nothing except through experience and by means of experience, and in this respect all of our knowledge consists of ex-

---

56. The philosopher (ibid., p. 35) argued, for instance, that temporal relationships might characterize the thing-in-itself—something, he urged, that Schelling had neglected to consider.

57. Schelling's letter to Christian Schütz, the publisher, was printed in the *Allgemeine Literatur-Zeitung* (2 November 1799).

58. Schelling, *Einleitung zu dem Entwurf eines Systems der Naturphilosophie*, in *Schellings Werke*, 2:276 [III:276].

perience."[59] The accumulation of such knowledge was the never-ending task of natural science.

Schelling believed, nonetheless, that one could cast empirically acquired knowledge into a deductive system. This could be done on the assumption that all of our factual knowledge reflected a world that displayed an organic and systematic wholeness—a *cosmos*, to use the term Alexander von Humboldt would later employ. What Schelling aimed to show, in short, was that natural science could be augmented by *Naturphilosophie* to become an objective, autonomous discipline, dependent neither on theology nor transcendental philosophy to establish its foundational principles. Now this was a conception that a thinking naturalist like Goethe could endorse with enthusiasm.

If *Naturphilosophie*—or "speculative physics" as he also called the enterprise—were to be autonomously established, then all occurrences in the world would have to be explicable by appeal to natural forces (*Natur-Kräfte*) alone.[60] This Schelling set out to demonstrate, though by transforming the principles governing the ego into those now to be applied to the fundamental structures of nature. Previously he had distinguished two initial phases in the activity of the ego: the ego as infinitely expanding and the ego as attempting to know itself and consequently as restraining its own infinitely expanding activity. He now transferred (without indication) these structuring principles to nature, thus animating nature with an intrinsic intelligence. He consequently distinguished two phases of the nonego: nature as *productivity* and nature as *product* (Spinoza's *natura naturans et natura naturata*). Productivity constituted the "subjective" aspect of nature, as he termed it, and product the "objective" aspect.[61] These philosophical

59. Ibid., 2:278 [III:278]. Goethe met with Schelling intensively for about a month, from mid-September to mid-October 1799, to talk over *Naturphilosophie*. He read Schelling's *Einleitung* on 23 September and talked with him about it; and then during four days, from the second to the fifth of October, they read through the work together. See Goethe's *Tagebücher*, in *Goethes Werke* (Weimar Ausgabe), 133 vols. (Weimar: Hermann Böhlau, 1888), III.2:261–63.

60. Schelling, *Einleitung zu dem Entwurf eines Systems der Naturphilosophie*, in *Schellings Werke*, 2:273 [III:273]. This goal of establishing the independence of *Naturphilosophie* is quite at odds with Schelling's remarks in the *System des transscendentalen Idealismus*. In that work, whose first part Schelling may have written before the *Einleitung*, he asserted: "The highest perfection of natural science [*Naturwissenschaft*] would be the complete mentalization of all natural laws into the laws of intuition and thought." See Schelling, *System des transscendentalen Idealismus*, in *Schellings Werke*, 2:340 [III:340].

61. Schelling, *Erster Entwurf*, in *Schellings Werke*, 2:17–18 [III:17–18]. In the *System des transscendentalen Idealismus*, Schelling retained the view that the original activity of the ego was infinitely expansive and only restrained by its own limiting effort to know itself. In the *System*, he referred to the original activity of the ego as the "objective" phase and the limiting activity as the "absolute" or "subjective" phase. These two principles of ego activity are not precisely the same as those attributed to nature in the *Einleitung*, but the parallel is obvious. The designations for the

distinctions conformed, he thought, to our ordinary understanding of nature as both creative and offspring of her own creative force—nature as mutually means and end of herself (the fundamental organic outlook).[62] So now, instead of deriving the fundamental forces of nature from oppositional movements of the ego, Schelling construed them as arising out of the dialectical processes of nature herself.[63]

Schelling postulated nature as an infinitely productive force that was continuously inhibited or limited by an opposing force. Every further determination or particular limitation on productive nature formed successive stages of products—from the inorganic powers of nature (magnetism, electricity, chemical processes) through their further delimitations in the organic powers (sensibility, irritability, and *Bildungstrieb*).[64] Various stages of matter (inanimate and animate) and their particular activities were but the meeting points of the several forces. In the *Erster Entwurf* and *Einleitung*, Schelling argued that nature could only be experienced as an unlimited or infinite becoming that went through stages of development. From this perspective, the limiting forces on the infinite productivity of nature arose from nature herself—at least this is what had to be assumed to explain our conscious awareness of her activity.

Here, even in his renewed approach to *Naturphilosophie*, Schelling could not quite establish the principles of nature without some appeal to a tran-

comparable phases of nature have been switched: nature as productive (expansive phase) is called the subjective aspect and nature as product (the limited), the objective.

62. Andrew Bowie, in his *Schelling and Modern European Philosophy* (London: Routledge, 1993), has many suggestive observations to make about Schelling's ideas, especially their relationship to later developments in philosophy. But Bowie's persistent effort to understand Schelling as a precursor to Heidegger and postmodern philosophy sometimes reflects back a distorting view. He argues concerning Schelling's concept of nature, for example, that "neither Schelling nor Heidegger sees scientific investigation in terms of a representation of the objective truth about nature. . . . A world of pure objectivity is and will remain inconceivable for Schelling, not least because it could never lead to that which can think of nature in even purely mechanistic terms" (pp. 39–40). On the contrary, for Schelling, the world of nature is completely objective—it is, after all, an object and product of mind, at least as the young philosopher maintained early on. Moreover, the mechanistic is not the only objective approach to nature. For Schelling, nature is objectively organic—that notion constituted one of his principal advancements upon Kant. Finally, if what is meant by "objective" is established as necessary and universal, and not dependent on the beliefs of particular individuals, then Schelling's *Einleitung* aimed to ground natural science as absolutely objective.

63. In the *Ideen*, Schelling referred to the original activity of the ego as expanding infinitely and only restrained by its own limiting effort to know itself (see above). In the *Einleitung*, these aspects of the ego's actions have obviously been transferred to nature in her productive and product phases, respectively. In the *System des transscendentalen Idealismus*, Schelling retained the same structural analysis of the ego, but he referred to the original activity of the ego as the "objective" phase and the limiting activity as the "absolute" or "subjective" phase (see above).

64. Schelling, *Einleitung zu dem Entwurf eines Systems der Naturphilosophie*, in *Schellings Werke*, 2:325 [III:325].

scendental argument: these natural principles were postulated as a condition of our conscious experience. Such direct and conscious intuition of an
unending, empirical becoming mirrored the unconscious intellectual intuition that provided the ultimate ground of our experience of nature.[65] In his
transcendental philosophy, Schelling would specify the limiting force as
that activity of the ego in its attempt at knowing itself, and thereby limiting
itself. Yet he could not employ such reference to the ego systematically in
this present effort, lest *Naturphilosophie* fail to achieve the independence he
sought for it—on the other hand, he could not help but allude to the ultimate ground that transcendental philosophy would furnish.[66] He was obviously struggling here with a concept of nature that was still in the process of
being formulated.

Schelling conceived the infinite productivity of nature as an unending
evolution (*unendliche Evolution*), with its products as momentary resting
places for the evolution, a slowing of the evolutionary process but not a cessation of it.[67] He employed the metaphor of a flowing river to capture what
he conceived as essential to this conception of nature:

> One thinks of a stream, which itself is pure identity; where it meets some
> resistance, it forms an eddy, which eddy is not an object at rest, but with
> each moment it disappears and then reestablishes itself.[68]

In referring to the productivity of nature as an evolution, or as a stream
in which eddying objects are held in dynamic tension, Schelling seems
to have had both a logical-metaphysical development and a temporal-
physical development in mind. I will discuss in more detail his ideas about
the organic development of nature—especially the question of species evolution—in chapter 8. I will simply point out here that Schelling borrowed
the term "evolution" from embryology, where in its original biological usage, it meant "preformation"—that is, the notion that the embryo was essentially the adult organism that simply had to unfold or evolve. When he

65. Schelling, *Erster Entwurf*, in *Schellings Werke*, 2:14 [III:14]: "Empirical infinity [*Das empirisch-
Unendliche*] is only the outer intuition of an absolute (intellectual) infinity, the intuition of which
is originally in us but would never come to consciousness without outer, empirical representation."
66. Ibid., 2:19–20 [III:19–20].
67. Schelling, *Einleitung zu dem Entwurf eines Systems der Naturphilosophie*, in *Schellings Werke*,
2:287 [III:287]: "Since nature must be thought as infinite productivity [*unendliche Produktivität*],
conceived really as occurring in an unending evolution [*in unendlicher Evolution*], so its fixity [*Beste-
hen*], the resting place constituting the natural product (of the organic, for example) must be represented not as an absolute rest, but only as an evolution continuing with an infinitely diminishing
rapidity or with an infinite retardation."
68. Ibid., 2:289 [III:289]. This model is first introduced in a footnote to the *Erster Entwurf*, in
Schellings Werke, 2:18 n. 2 [III:18 n. 2].

applied the term to the transformations of nature, he meant to suggest that the essential idea (or archetype) of a given organism—its species type—already ideally existed, but that its empirical realization required temporal development. In other words, Schelling was indeed proposing a real evolution occurring in nature and seems to have been the first thinker to apply the term to species alteration.[69] Schelling and Goethe, as will be seen in chapter 11, were of common mind on the question of evolution.

The Personal Factor in *Naturphilosophie*

Schelling's nature philosophy itself arose out of a matrix of opposing intellectual and personal forces. One can regressively trace its development from sources in Kant, Spinoza, and Fichte, as well as from his study of the most recent work in experimental natural science. As his ideas advanced, they traversed a path generally determined by the logical consequences of their intellectual origins; but they also gently swayed to the impingements of more personal powers generated during the fall of 1799. The new curve of his ideas can be detected in a revealing footnote to a passage in the *Einleitung*. The passage describes the kinds of oppositional struggle that give rise to nature's products. Schelling argued that nature's productive forces tended to arch toward full realization in overcoming their contrary limiting counterforces. But the struggle of contending powers must, he thought, be ever renewed. For example, in the animal realm, the perfection of the species can only come through oppositional forces expressed by the two sexes. In the footnote that elaborates this idea, Schelling wrote the following:

> Nothing is more indicative of the contradiction out of which life flows—and that life is only a more elevated condition of common natural powers—than the contradiction of nature in that which she attempts to achieve through the sexes. Nature hates sex, and where it occurs, it occurs against her will. The separation of the sexes is an unavoidable fate to which she, after she has once become organic, must resign herself and which she can never overcome. Through this very hatred itself of the separation of the sexes, she is involved in a contradiction; for what is distasteful to her, must lead to the most careful formation and to the apex of existence, as if she had intended it; for she always demands only the re-

---

69. I have discussed the history of the term "evolution" and its role in Schelling's thought in *The Meaning of Evolution: The Morphological Construction and Ideological Reconstruction of Darwin's Theory* (Chicago: University of Chicago Press, 1992).

turn to the identity of the species, which, however, is bound to the (never to be overcome) dualism of the sexes as an unavoidable condition.[70]

When Schelling composed this rambling note in fall of 1799, he was a raw youth of twenty-four, who had left the shelter of the seminary only a few years before. And during that fall, he entered the circle presided over by the Schlegels—a group whose ironic play and sexual freedom (*Lucinde* had just been published, and Dorothea had moved in with Friedrich Schlegel) must have excited a powerful eddy of conflicting feelings and attitudes in that previously placid pool of his spirit. Indeed, the *Einleitung* is rife with the struggle of the contending powers of nature, reflecting, one suspects, as in a glass darkly, the clashing forces of the Romantic circle.

After Schelling had been inducted into that ring of love, and then later spiraled out to other intense attachments, his interpretation of the natural relationship of the sexes changed dramatically. By 1810 he was arguing that love was the medium that joined the sexes in freedom. Indeed, love between the sexes revealed God's (i.e., infinite self-consciousness's) relationship to nature:

> The mystery of the separation of the sexes is nothing other than the instantiation of the original relationship of both principles [i.e., of the ideal and the real]. Each of these principles is actual for itself and to that degree independent of the other, and yet each is not and cannot be without the other. The medium of this duality, which does not exclude identity, and this identity, which does not exclude duality, is *love*. God is himself bound to nature through freely willed love; he does not require her and yet will not exist without her. For love is not the result of two beings requiring one another, but it occurs where each could exist for itself, . . . yet where neither can exist morally without the other.[71]

The mature Schelling had found a medium of connection that brought to unity what initially appeared to exist in antagonistic tension. The young Schelling, with his boyish and rather shy manner, regarded the sexual connection as rife with contradiction and turmoil. He may himself have initially hated sex, but then he had yet to try it. Caroline Schlegel introduced him to the subject about a year after he came to Jena as its new professor of philosophy.

---

70. Schelling, *Einleitung zu dem Entwurf eines Systems der Naturphilosophie*, in *Schellings Werke*, 2:324–25 [III:324–25].

71. Schelling, *Stuttgarter Privatvorlesungen* (from unpublished papers, 1810), in *Schellings Werke*, 6:345 [VII:453].

Schelling in Jena

From the time he left the Tübinger Stift through the period of his service as peripatetic schoolmaster, 1795 to 1797, Schelling had published several essays that marked him as a new star on the philosophical horizon. The most important of these articles came out in the *Philosophisches Journal*. At Easter of 1797, his *Ideen zu einer Philosophie der Natur* appeared, and he immediately sent off a copy to Niethammer, the editor of the *Journal* and his philosophical confidant. Niethammer quickly perceived that his young friend could not continue to act as schoolmaster and still fulfill the promise his many publications suggested and partly realized. Therefore, he, along with Fichte and Paulus, conspired to bring Schelling to Jena. But Schelling made it hard. When the possibility was broached, he politely refused the normal academic route—an extended habilitation and work as *Privatdozent*.[72] He undoubtedly thought his several hefty publications and demonstrated ability obviated the usual requirements to become a professor. He also knew that his father was attempting to bring him back to Tübingen as a professor, though he had little stomach for the constricted environment of his intellectual youth.[73] Niethammer's initial plan, however, fell through. The minister in charge, Christian Gottlob Voigt, insisted that the young philosopher had first to habilitate. But Niethammer persisted. He sought the help of Schiller, who quickly became an enthusiast for Schelling's critical philosophy, especially the idea that our experience of nature was conceptually constructed. Schiller in turn attempted to bring Goethe into the fold. But the poet balked. What Schiller liked, Goethe found distasteful:

> On reading Schelling's book [the *Ideen*], I've had other thoughts, which we must more thoroughly pursue. I gladly grant that it is not nature that we know, but that she is taken up by us according to certain forms and abilities of our mind. . . . [Yet] you know how closely I hold to the idea of the internal purposiveness [*Zweckmässigkeit*] of organic nature. . . . Let the idealist attack things-in-themselves as he wishes, he will yet stumble on things outside himself before he anticipates them; and, as it seems to me, they always cross him up at the first meeting, just as the Chinese is nonplused by the chaffing dish.[74]

---

72. Schelling, *Briefe und Dokumente*, 1:129.

73. The Tübingen Faculty Senate ultimately judged Schelling too young for a professorial position.

74. Goethe to Schiller (6 January 1798), in *Der Briefwechsel zwischen Schiller und Goethe*, ed. Emil Staiger (Frankfurt a. M.: Insel Verlag, 1966), pp. 537–38.

Goethe was not inclined to aid the cause of Schelling's appointment. He nonetheless kept dipping into the *Ideen,* not always happily emerging. The tide, though, turned at Easter. He met Schelling during a reception at Schiller's house, an encounter strategically planned. The great poet was unexpectedly charmed by the young philosopher. The Schelling he met had "a very clear, energetic, and, according to the latest fashion, a well-organized head on his shoulders." And below the shoulders, as he later assured Voigt, the young man gave "no hint of being a sansculotte" (unlike Fichte).[75] Goethe was surprised at his new friend's knowledge of optics, and the two spoke for hours on the topic. Likely Schelling let slip a few knowing references to Goethe's *Beiträge zur Optik,* which he had lately been reading—undoubtedly heartening to the older man, who felt his optical work had been unappreciated by other scholars. Shortly after their meeting, Goethe received an advance copy of the *Weltseele,* which had, as if at the poet's command, made intrinsic teleology central to the understanding of nature. Moreover, nature stood, as Goethe interpreted the book, objectively against the self and revealed her intimate parts only through direct, *anschauliche* experience.[76] This, to the older man's sensitive eye, suggested something rather different from Fichte's non-ego, as did the generous references in the book to his own *Beiträge zur Optik* and *Metamorphose der Pflanzen.* Now converted, Goethe quickly interceded on Schelling's behalf. As a result, on 30 June 1798 the twenty-three-year-old tutor received a call to Jena as extraordinary professor of philosophy.

Prior to traveling to Jena, however, Schelling spent six weeks in Dresden, beginning in the middle of August. He wanted to study the large art collections gracing the city, but that may have been a secondary reason. He appears to have been invited by Wilhelm Schlegel to join there the group of friends—his brother Friedrich, Novalis, and Caroline. Friedrich had been reading Schelling—probably prompted by Wilhelm—and even composed a semicritical fragment about the philosopher for the *Athenaeum Fragmente.*[77] And Novalis had met Schelling the previous winter. He wrote Friedrich Schlegel at the time, undoubtedly piquing his friend's interest:

75. Goethe to Christian Gottlob Voigt (29 May 1798), in *Goethes Briefe* (Hamburger Ausgabe), ed. Karl Mandelkow, 4th ed., 4 vols. (Munich: C. H. Beck, 1988), 2:349.

76. Goethe to Voigt (21 June 1798), in *Sämtliche Werke nach Epochen seines Schaffens* (Münchner Ausgabe), ed. Karl Richter et al., 21 vols. (Munich: Carl Hanser Verlag, 1985–98), 6.2:928: "I take the liberty to recommend his book *On the World Soul* to you. It contains some very beautiful insights and induces one to wish the author would continue to delve into the details of experience."

77. Friedrich Schlegel, *Fragmente* (from the *Athenaeum*), in *Kritische Friedrich-Schlegel-Ausgabe,* 2:216–17 (fragment 304). Schlegel remarked that significant philosophy must have a poetic component. He thought Schelling particularly good at uncovering the source of the fundamental forces of physics. He judged the philosopher's literary side, however, to be undeveloped, which prevented his philosophy of physics from attaining its professed end.

Johann Wolfgang von Goethe (1749–1832) in the Roman Campagna, oil portrait (1786–87) by J. H. W. Tischbein. (Courtesy Städel Museum, Frankfurt.)

Henriette Herz (1764–1847) as Hebe, the most beautiful of the goddesses. Oil portrait (1778) by Anna-Dorothea Therbusch. (Courtesy Preussischer Kulturbesitz, Berlin.)

Johann Wolfgang von Goethe. Portrait (1787) by Angelika Kauffmann. Goethe remarked of this painting: "He is indeed a handsome fellow, but no hint of me." (Courtesy Stiftung Weimarer Klassik.)

Friedrich Schiller (1759–1805). Portrait (1793) by Ludovike Simanowiz. (Courtesy Schiller-National-museum, Marbach.)

Alexander von Humboldt (1769–1859). Portrait (1806) by F. C. Weitsch; painted shortly after Humboldt's return from his five-year voyage to the Americas (1799–1804). (Courtesy Preussischer Kulturbesitz, Berlin.)

I have gotten to know Schelling. I candidly told him of our misgivings about his *Ideen*. He was quite understanding and he thought that in the second part he had begun to achieve a more elevated point of view.[78] We quickly became friends. He has invited me to an exchange of correspondence. Before the day is out I will write him. I like him a lot—a real universal tendency in him—true radiant force—from one point to infinity. He seems to have considerable poetic sensibility. Now he is into the ancients—he finds in the *Odyssey* Goethe's home ground.[79]

Schelling arrived in Dresden on 18 August and Novalis a week later. By 26 August the Schlegel brothers, Caroline, Novalis, and Schelling were all present and even Fichte showed up briefly. At the end of August, Friedrich Schlegel and Novalis left Dresden, but Wilhelm and Caroline stayed on with Schelling through September. By the end of his stay, Novalis's enthusiasm for his new friend cooled, likely because Schelling displayed some Swabian disdain for the religious delicacy of the poet's longings. When a bit later Novalis read the *Weltseele*, he found it, as he wrote Caroline, "immature,"[80] perhaps because the considerations of the book were highly empirical. Caroline reacted differently. She seems to have become immediately attracted to this penetrating intellect in a boy's body. She wrote Friedrich Schlegel in Berlin: "Schelling, as he says, is going to wall himself in, but that certainly won't last long. He is rather a man who breaks through walls. Believe me, friend, he is a more interesting man than you give him credit for, a real primitive nature [*Urnatur*], something like mineral, real granite."[81] The remarks abraded Schlegel's sensitive epidermis. He wrote back to inquire, "Where will Mr. Granite find a Miss Granite?"[82] What none of the party knew at the time, of course, was that he had already found her.

Schelling began his tenure at Jena with two lecture courses in the winter semester of 1798–99 on "The Elements of Transcendental Idealism" and "The Philosophy of Nature," the latter continuing through the summer semester, along with "The General System of Transcendental Philosophy." These two sets of courses yielded his *Erster Entwurf* and the *Einleitung* in

78. This "second part" of the *Ideen* never appeared; it was probably transformed into what became the *Weltseele*.

79. Hardenberg to Friedrich Schlegel (26 December 1797), in *Kritische Friedrich-Schlegel-Ausgabe*, 24:70.

80. Hardenberg to Caroline Böhmer Schlegel (9 September 1798), in Friedrich von Hardenberg [Novalis], *Novalis Schriften*, ed. Paul Kluckhohn and Richard Samuel, 2nd ed., 5 vols. (Stuttgart: Kohlhammer, 1960–75), 4:261.

81. Caroline Böhmer Schlegel to Friedrich Schlegel (14 October 1798), in *Kritische Friedrich-Schlegel-Ausgabe*, 24:179.

82. Friedrich Schlegel to Caroline Böhmer Schlegel (29 October 1798), in ibid., 24:190.

1799, and his most famous work, *System des transscendentalen Idealismus* in 1800. In the winter semester of 1799–1800, Schelling began to work on a topic that would round out the scheme earlier outlined in "The Oldest System of Idealism"; he lectured on "The Chief Principles of the Philosophy of Art." He also continued with a course on "The Organic Doctrine of Nature."[83]

Schelling's fame rapidly spread, and he became deluged with students. The English barrister Henry Crabb Robinson (1775–1867) came to Jena as a young man in 1802 and stayed three years—the happiest of his life, as he recalled. He attended Schelling's lectures and on occasion went to dinner parties at Schelling's house. As a fairly sober Englishman, he was often left in wonder after hearing one of Schelling's orations. In a letter written at the time, he observed:

> From Loder [professor of medicine] I shall proceed to Schelling, and hear him lecture for an hour on Aesthetics, or the Philosophy of Taste. In spite of the obscurity of a philosophy in which are combined profound abstraction and enthusiastic mysticism, I shall certainly be amused at particular remarks (however unable to comprehend the whole) in his development of Platonic ideas and explanation of the philosophy veiled in the Greek mythology. I may be, perhaps, a little touched now and then by his contemptuous treatment of our English writers, as last Wednesday I was by his abuse of Darwin and Locke. I may hear Johnson called thick-skinned, and Priestley shallow. I may hear it insinuated that science is not to be expected in a country where mathematics are valued only as they help to make spinning-jennies and machines for weaving stockings.[84]

More in tune with the tenor of German idealism was the Norwegian Henrik Steffens (1773–1845), the mineralogist and later philosopher, who also came to Jena as a foreigner. He arrived during the fall of 1798, just in time to attend Schelling's habilitation lecture (the formality was ultimately required) on the philosophy of nature.

> Professors and students had gathered in the great lecture hall. Schelling climbed the podium. He had a youthful appearance. He was two years younger than I, and now became the first of those important men whose acquaintance I greatly desired to make. He had something about him, as it seemed to me, determined, indeed, comforting—broad cheekbones,

---

83. The announcements for these lectures are reproduced in Schelling, *Briefe und Dokumente*, 1:163.

84. Henry Crabb Robinson, *Diary, Reminiscences, and Correspondence*, ed. Thomas Sadler, 3 vols. (London: Macmillan, 1869), 1:128.

temples wide apart, a high forehead, a face full of energy, a nose slightly turned up, and in large clear eyes an intellectually commanding power. As he began to speak, he seemed momentarily embarrassed. The object of his talk was what at the time consumed his soul. He spoke of the idea of the philosophy of nature, of the necessity to grasp nature in her unity, of the light that would be thrown over all objects when one dared to consider them from the standpoint of the unity of reason. He completely captured me, and the day after I hurried to visit him.[85]

Steffens became close friends with Schelling—the man, as he later reminisced, who "transformed my life" and gave "an inner meaning to my whole past from my earliest childhood on."[86]

Transcendental Idealism and Poetic Construction

The work that contributed most to Schelling's fame, both at Jena and beyond, a work that arose under the spiritual guidance of Fichte but that portended the break that would come between the two, was his *System des transscendentalen Idealismus,* which he finished in spring of 1800. The patches of relative clarity, the streams of repetition, and the large tracks of ill-fitting fragments bespeak its origin as a series of lectures. (Friedrich Schlegel jibed regarding the *System:* "In the dark all cats are gray"—a witticism later borrowed by Hegel.)[87] I will sketch the fundamental position Schelling takes in this book and mark those features that most distinctively indicate the future development of his thought. More difficult to convey, however, is the passion, the excitement, and the sheer exuberance that the work evoked in those young men and women (like Caroline) who felt themselves at the barricades of thought and of life. For us, the idealism of Fichte and Schelling might remain desiccated on the page; for them, it flowed with the blood of freedom, with the promise that they could make the world anew and merge in intimate congress with the other.

Construction of the Self and Nature

Like Fichte, Schelling argues in the *System* that self-consciousness, the awareness of the self, comes in an intellectual intuition. And like Fichte, he preserves something of Kant's original meaning for such an intuition. The intellectual intuition constituting self-knowledge, like God's presumed in-

---

85. Henrik Steffens, *Was Ich Erlebte,* 10 vols. (Breslau: Josef Max, 1840–44), 4:75–76.
86. Ibid., 4:2.
87. Ibid., 4:312.

tuition, creates its object in the very act of knowing. That is to say, the act of consciousness by which I come to know myself as thinking constitutes the self, and hence creates the self: "For since the I (as object) is nothing other than the very knowledge of itself, the I arises only through this, that it knows itself; the I itself is thus a knowing, which at the same time produces itself (as object)."[88]

The idealist philosopher first becomes aware of this productive intellectual intuition by an inference, a transcendental argument that concludes to the conditions that must obtain for any conscious awareness at all. But then, he freely repeats the very intellectual intuition, this time trying to become aware of it, fighting to reveal what for most individuals is an unconscious process. In this fashion, the intellectual intuition itself becomes the object of an intuition, that is, of an immediate awareness. But let me explore these relationships further.

Schelling has a rather vivid, virtually geometrical way of understanding this process of self-knowledge, which is also self-creation. Insofar as he focuses on that pure activity of consciousness, he regards it as undetermined, but infinitely determinable. It expands to the possibility of becoming all things, not unlike Aristotle's passive intellect.[89] Schelling refers to this as the "objective" self. The consciousness that attempts to know or intuit the objective self he refers to as the "ideal" or "absolute self"—absolute because the self in this phase is conditioned or constrained by no other activity. The act of the absolute self, from this perspective, is absolutely free; but like God, also absolutely necessary, since the act follows "from the inner necessity of the nature of the self."[90] The absolute self continuously attempts to capture itself, to know itself (in the objective phase) through intellectual intuition, the act of reflective awareness that, in Schelling's terms (following Fichte), is a kind of returning or reverting activity (*zurückgehende Thätigkeit*). It is through this reverting action that the absolute self strives "to intuit itself in that infinitude."[91] The persistent striving after self-cognition is also, for the idealists, a productive striving, since it creates the very object of its awareness—and, in this respect, a bit like Aristotle's active intellect. If this intuitive awareness did not, as it were, stitch together in delimited form the momentary existences of the outward flowing self, there would not be a con-

88. Schelling, *System des transscendentalen Idealismus*, in *Schellings Werke*, 2:369 [III:369].

89. A principal difference between Aristotle's passive intellect and Schelling's infinite self is that the former is thought of as indeed receptive, passive, while the infinite self actively reaches out. See, for comparison, Aristotle's *Physics*, bk. 1, chap. 5.

90. Schelling, *System des transscendentalen Idealismus*, in *Schellings Werke*, 2:395 [III:395].

91. Ibid., 2:391–92 [III:391–92].

tinuously existing empirical self as owner of its representations. To claim, via the intellectual intuition, representations as mine is the same as establishing and maintaining the empirical self that owns them.

The next step in Schelling's analysis becomes the hinge opening the door to all else—this is the introduction of finitude. Every philosophical or theological position that begins with an infinite substance or act—Parmenides' one Being or Spinoza's *Deus sive natura*, for instance—must explain why all is not simply engulfed by an oceanic tide of undifferentiated being. Here the idealists invoke the expediency provided by their Greek antecedents: thought itself instantiates limits. Both Fichte and Schelling assume, without any real argument, that thought of an object must be determinate and therefore limited. Hence in the intellectual intuition of the self, that is, in the positing (*Setzen*) of the self, to use Schelling's Fichtean vocabulary, the self limits itself, makes finite an originally infinite activity or producing. This delimiting has two aspects: as activity, it is the finite, empirical self; as product of that activity, it is the natural world of objects. But let me elaborate a bit further.

If, for the moment and *per impossibile*, we regard the self as infinite, but not as a self-reflecting activity—hence the self not "existing for itself"—then, according to Schelling, we would have an infinite activity but not a self. For the very nature of the self, whereby it comes into existence, is through the intellectual intuition, its own taking up of itself, reproducing itself, though in limited form, at every moment. In short, unlike objects that exist only "in themselves," the self exists "for itself." But now we have a paradox (at least one, it can safely be said). An infinite self becomes finite through its own activity, and can be infinite only insofar as finite and vice versa. This paradox can be resolved, according to Schelling, if the self is really not immediately infinite in act but rather an infinite *becoming*, an infinite striving that continuously reproduces a finite ego.[92]

Limitation, even if part of a process of infinite becoming, implies something producing a barrier, a countervailing force. Indeed, according to Schelling, "the I cannot limit its producing without opposing something to itself."[93] Hence, in the striving after self-awareness, the self-reverting ego introduces, in its very act of limiting, in its knowing of itself, something seemingly opposed to itself, a not-I. One has to add "seemingly," because what appears to be something other than the I can really only be the self in another guise. This self-deception, as it were, arises from the fact that the intuition of the self, and hence its limiting activity, cannot be conscious—

92. Ibid., 2:383 [III:383].
93. Ibid., 2:381 [III:381].

we are simply not immediately aware of the character of the intuitive, reverting act itself while engaged in the very act. The act of self-limiting, though not the consequences of the act, thus remains hidden to immediate scrutiny.[94] The transcendental philosopher learns of the true structure of consciousness—that its limitations are self-imposed—by way of a series of inferences, not, initially, through a direct, conscious awareness; but then he repeats in freedom and with attention what ordinary individuals do unconsciously.[95] So what initially began as a transcendental argument, whereby the underlying structure of consciousness is inferred, becomes through persistent attention a direct, intuitive awareness of that structure—or at least this is the goal of the transcendental philosopher.

In ordinary consciousness (not that of the philosopher)—because its intuitive acting stays unconscious (*unbewusst*)—the constraints imposed by the intellectual intuition appear to come from a non-self, that is, from the outside. Schelling concluded then that "self-intuition in its limiting activity [*in der Begrenztheit*] . . . is nothing other than what in common language is called sensation."[96] And quite normally we (who have achieved a certain level of philosophical maturity, short of the transcendental philosopher's, however) believe that this sensation comes from outside the self, from the thing-in-itself.

According to Schelling, then, the self's effort to know itself through intuition yields several distinct outcomes. Let me summarize their main features, as Schelling himself does, in three stages. First, the original intuition of the objective self by the absolute, or ideal, self imposes a limit on the objective self, restraining it. We interpret this constraint as something *sensed*—a feeling of passivity is induced, though we remain unaware of the

94. This unconscious source of limitation, produced by the self, furnishes the feeling of restriction (the *Anstoss*), which Fichte suggested might be a primitive given. This is a significant opening wedge beginning to separate Schelling from his mentor.

95. Ibid., 2:395–96 [III:395–96]. Initially, the philosopher, according to Schelling, becomes aware of the activities of the absolute self through a kind of transcendental argument—that is, from certain facts of conscious experience to the inference that particular conditions are necessary to produce these facts. As he put it in the *System*: "Now the question arises of how the philosopher assures himself of that original act [of the absolute ego], that is, how he knows it. Obviously not directly, but only through inferences [*Schlüsse*]. That is, I discover through philosophy that I arise for myself at every moment only through such an act; I thus conclude [*schliesse*] that I can have arisen originally only through such an act. I discover the consciousness of an objective world, at every moment, to be intertwined with my consciousness; and so I conclude that something objective must be involved originally in the synthesis of self-consciousness and must again issue from an evolved self-consciousness" (ibid.). But then the philosopher, with an intense act of freedom, can repeat the productive intellectual intuition, but this time with reflective attention: "Philosophy generally is thus nothing other than the free imitation, the free repetition of the original series of acts, into which the one act of self-consciousness evolves" (ibid., 2:397 [III:397]).

96. Ibid., 2:403–4 [III:403–4].

true source of this feeling. Second, the limitation imposed by the absolute (also called the "ideal") self can become the object of another intuition— hence, an intuition of an intuition.[97] Insofar as the limit is of a particular sort, it will come to consciousness as a particular sensation. The self, in a third phase of self-cognition, will recognize in the ideal self the active limiting and will regard it as a sensing.[98] And hovering back and forth (*Schweben*) between these two poles of its activity—awareness of the limited aspect of the objective self and the limiting activity of the ideal self—there arises the limited empirical self (as both sensing and receiving sensations) and the assumption of an external world as the source of those sensations. The interaction of self with the assumed world of external objects generates, as a consequence, an empirical intellect outfitted with those derived categories of unity, substance, cause, necessity, et cetera of the sort that Kant had elaborated. The incautious Kantian philosopher, however, might interpret the work of productive intuition as necessarily assuming a substrate causing sense qualities. The result of this incaution, as Schelling depicts it, induces the critical error of Kantianism: "The ideal activity has transformed itself into the thing-in-itself, and so the real [objective self], through the same act, will transform itself into the opposite of the thing-in-itself, that is, into the self-in-itself."[99] But this noumenal world of thing-in-itself and self-in-itself is a necessary creation of the productive intuition of the self. It does not exist beyond the self, and thus is really self-delusion.

What Schelling initially designates as a series of fundamental acts (yielding various consequences) stands for an infinite number of acts performed in the self's constant effort at self-cognition. Out of that infinite striving comes the multitude of objects in the world—the complex web of nature— and the particular individual reactions of the empirical self. Hence, the external world of nature and the internal empirical consciousness derive from the unconscious activity of the self in opposing its own tendency toward infinite expansion. Thus, according to Schelling: "We see nothing present outside of us in the objective world other than the internal limitation of our own free activity."[100] A world of nature has been generated simply out of the infinite longing for self-awareness.

The *System* makes its arguments in a hyperbolically abstract way, so that Schelling's cat often becomes Cheshire, but with nary a smile to guide the reader. Yet the cogency of his theory can, I think, be understood more easily with a concrete example (of the sort he deigns not to supply). If I per-

97. Ibid., 2:426 [III:426].
98. Ibid., 2:422 [III:422].
99. Ibid., 2:423 [III:423].
100. Ibid., 2:379 [III:379].

ceive a dog on the couch, this experience, upon some reflection, will be discovered to be a representation in consciousness, even if I initially assume an external object impressing its appearance on me. My own existence as a consciousness exhibits two poles, the awareness of the representation of the dog and the awareness that I am doing the representing. Fundamentally I will be aware that my dog representation is a conscious state, which potentially can represent all things, an infinite possibility—thus emerges the infinite, objective self. I am simultaneously aware that the representation of the dog is limited, determinate. Such feeling of limitation I automatically ascribe to something outside of consciousness as its source. Yet the limitation I feel is self-imposed (which is the second level of Schelling's analysis of the intellectual intuition). An awareness of limitation requires simultaneously that I am aware of the determinate features of the limitation—namely, the sensations (black and white colors, sound of growling, etc.) that I seem to receive from without. Finally, I can revert still further to my own previous reversion: that is, I can become aware that it is I who am both sensing and having this determinate sensation—thus the full emergence of an empirical self (Schelling's third level of analysis).

Only the philosopher, of course, can construct arguments of the transcendental type that lead to the conclusion that the self is an infinite becoming, a becoming that in the act of intuition creates itself, creates a finite self, and creates the world of nature. The task of the transcendental philosopher is, as it were, to repeat the acts of self-consciousness in a perspicuous way, to reveal the basic structures of consciousness, and to trace their consequences in forming the external world. But since the self, both in its initial real and ideal phases, is infinite in act, the task itself would be infinite—and thus impossible to carry out in time. For Schelling, this means that the transcendental philosopher can only reveal those very general processes that evolve from the self and constitute the basic structural relationships of nature—a large-enough task one would imagine.[101] The *System*, then, attempts to generate, from the fundamental acts of the self's imposing limits on itself, the basic structures of intelligence (i.e., the basic Kantian categories and forms of intuition) as well as the general material features of the world (thus sensation, matter, magnetism, electricity, and life). The more particular events of the world, however, cannot, in their individuality, be deduced from the structure of self-consciousness, though they are generated from that structure. Here one must simply rely on experience. Indeed, the self's initial awareness of itself comes indirectly through the experience of nature, whose features Western man has slowly retraced to mind. The tran-

101. Ibid., 2:398–99 [III:398–99].

scendental philosopher completes this task, showing how a world evolves from the self. As Schelling pointedly phrased it:

> As a physicist, Descartes said: Give me matter and motion, and I will fashion out of them a universe. The transcendental philosopher says: Give me a nature of opposed activities, of which one proceeds into infinity and the other strives to intuit itself in that infinity, and out of this I will let intelligence and the whole system of its representations arise for you.[102]

Life and Death

One ubiquitous feature of consciousness, which it imports into all its creations, requires special consideration. This is organicism. The idea that consciousness and, consequently, nature have a fundamentally organic structure seems to have come to Schelling as he was writing his review of philosophical literature for Niethammer in 1797. The conception had no presence in the body of the *Ideen*, published that year, but came to prominence the next year in the *Weltseele*. In the *System*, Schelling attempted to demonstrate that organicity forms the basic structure of consciousness. This organic structure, then, imprints itself surreptitiously on the fundamental relationships found in nature. The transcendental argument for universal organicity, typically subtle and worrisome, amounts to this: by the organic we mean that which is reciprocally cause and effect of itself; but in the continuous striving of the self to know itself, it is both the creator and the created, mutually the cause and effect of itself; hence, the self is organic.[103] The effort at self-knowledge, though, continuously produces representations, which, because of the unconscious activity of the ideal self, yields a complex interactive field of natural objects, which themselves are reciprocally related.[104] From top to bottom nature is organic.

Several important consequences for the conduct of empirical science in general and biology in particular follow from Schelling's demonstration of the basic organic structure of intelligence and, thus, of nature. First, it means that all of nature both reflects intelligence and has all of its structural principles and laws isomorphic with those of intelligence. As he sugges-

---

102. Ibid., 2:427 [III:427].
103. Ibid., 2:490 [III:490].
104. Ibid., 2:481–82 [III:481–82]: "If the I is itself originally limited, then a universe—that is, a general mutuality of substances—is necessary. By reason of this original limitation, or what is the same, by reason of the original struggle of self-consciousness, there arises for the I the universe, not gradually, but through one absolute synthesis."

tively phrased it: "Every plant is a symbol of intelligence."[105] Thus, for Schelling, even the principles of biology will express the kind of constitutive necessity and universality that Kant reserved for basic physics alone. Second, the internal succession of the self's representations, whereby a world comes to be, provides that world with an "internal principle of motion" (the mirror of the self's acts), which is what we commonly mean by life.[106] This consequence also immediately implies that everything, including the apparently inorganic, is alive and that, therefore, a strictly mechanistic approach will fail to capture nature's inner operations. Kant, of course, thought only mechanistic laws could scientifically explain organisms. As an heuristic expedient, we could regard plants and animals as purposive; but teleological accounts, he maintained, had epistemic value only as suggestions leading to efficiently causal explanations. Schelling, by contrast, argued that organicity, that is, purposiveness, was lodged in the very structure of plants and animals (and other matter as well). Transcendental idealism best preserved, he thought, the attitude of the real scientist. One does not have to accept the Aristotelian thesis, which would have matter itself, quite independent of mind, express purpose—an unintelligible proposition, according to both Kant and Schelling, since purpose could only be a function of mind. Nor does one's science have to be equal parts theology, as the Newtonians must practice it; for they believe organic matter intrinsically devoid of purpose, which must then be extrinsically imposed by the Deity. And finally, one need not accept the position of the Kantians, who would deny intrinsic purpose to organisms—while returning it as a pale suggestion about how to understand them. Only transcendental idealism reveals how nature can be intrinsically both blindly necessary (reflecting the unconscious activity of self) and purposive (reflecting the free acts of intelligence). Only transcendental idealism, Schelling concluded, remains true to the autonomous and rationally scientific approach to nature.[107]

Schelling's organicism would have another important implication, important in interconnected ways for both his personal and professional life. This concerns illness and death. Since representations of a world are produced by the absolute self unconsciously, they first come to empirical consciousness as if they were objects existing independently in nature. Within these limits, according to Schelling, lies the truth of empiricism.[108] Our own body is the result of an unconscious, productive intuition, that must

105. Ibid., 2:490 [III:490].
106. Ibid., 2:494 [III:494].
107. Ibid., 2:607–11 [III:607–11].
108. Ibid., 2:497 [III:497].

find its place within the interactive nexus of other natural objects; like a Leibnizian monad, our organism reflects all of those other objects. When, therefore, our body is "no longer a perfect reflex of our universe . . . it is sick."[109] Should this disjunction between the body and the rest of the world become radical, death naturally ensues. From the transcendental perspective, however, that disjunction must also be the free, albeit unconscious, decision of the self. As Schelling somewhat confusedly expressed it: "The absolute dissolution of the identity between the organism and intelligence, which in sickness is only partial, is death—a natural occurrence that itself falls within the original series of the intelligence's representations."[110]

Even a cautious reading of Schelling's explanation of sickness and death suggests that somehow the self wills these maladies, either for the body represented as its own or for that of another which it represents. Life and death, our own or that of others, almost seem, then, to be the arbitrary decisions of the will. When the adolescent philosopher in the privacy of his garret ruminates on the self's creations and destructions of natural bodies, the philosophical implications remain quite theoretical and emotionally innocuous. When a bit older, he enters the bedroom of a lover or presides over the death of a young woman, these implications cast a more terrible light.

The Poetry of Nature

In his *Aesthetics*, Hegel paid tribute to his one-time friend. "With Schelling," he said, "science (*Wissenschaft*) achieved its absolute standpoint, and although art had already begun to assert its special nature and dignity in relation to the highest interests of humanity, yet it was now that the conception of art and its place in scientific theory were discovered."[111] The discovery appears to have occurred while Schelling was composing his *System*. In none of his previous publications had he worked out the role of artistic experience in relation to transcendental philosophy or philosophy of nature.[112] But in the introduction to the *System*, he unexpectedly an-

109. Ibid., 2:498 [III:498]. Schelling discusses illness and death from a complementary perspective in his *Erster Entwurf*, in *Schellings Werke*, 2:220–40 [III:220–40].

110. Schelling, *System des transscendentalen Idealismus*, in *Schellings Werke*, 2:498–99 [III:498–99].

111. Hegel, *Vorlesungen über die Aesthetik*, vol. 1, in *Sämtliche Werke*, 5th ed. of the Jubiläumsausgabe (Stuttgart-Bad Cannstatt: Friedrich Frommann Verlag, 1977), 12:98. When Hegel speaks of "science" (*Wissenschaft*) here, he means conceptual disciplines broadly understood, including philosophy and natural science.

112. In the tract "The Oldest Systematic Program of German Idealism," there is just a hint of the importance that the philosophy of art would have in his considerations. See my description of

nounced that in order to demonstrate the identity between the subjective world of the self and the objective world of nature, he would have to turn to aesthetics, where such identity became revealed. The repair to art was required, he maintained, since "the objective world is only the original, yet unconscious poetry of the mind [Geist]; the common organ of philosophy— and the keystone of the whole arch—is the philosophy of art."[113]

The abrupt elevation of the philosophy of art to such prominence at the beginning of the *System* and the relatively brief discussion at the end—with no mention of the aesthetic through the bulk of the treatise—suggest that he rediscovered the philosophical and scientific importance of aesthetics during his first sustained encounter with the Romantic circle, which occurred while he was working on the treatise through fall and spring of 1799–1800.[114] During these months the group often came together in Caroline Schlegel's parlor to converse and to listen as one of their members would read from a recently composed manuscript. Deep into the night they would, to use their own jargon, "sympoetize" and "symphilosophize"—aided, to be sure, by sufficient quantities of decent wine and cold beef. During some of these gatherings, Friedrich Schlegel, who had just returned from Berlin that fall, likely read from the manuscript he was about to complete, his *Gespräch über die Poesie*, which would be serialized in the *Athenaeum* beginning in January 1800. In that dialogic essay (see the previous chapter), Schlegel called for philosophy to become poetic. Philosophy's poetization was ultimately the only way, he argued, that the infinite might be captured in finite form and thus beauty be brought to earth. Schelling, feeling the urgency and justness of this Romantic conception of philosophy, began lecturing on the philosophy of art in the winter term of 1799–1800. He also undertook, together with Schlegel and Caroline, the reading of the *Divine Comedy*. He thought Dante would provide him the model for recasting his own philosophical work as a great poem, something of pleasing prospect to his new friends. Though urged on by Schlegel, Schelling never got very far in this poetic effort.[115]

---

this tract above. Also in the *Einleitung*, in a fleeting passage that may have been written about time of the composition of the *System*, Schelling remarks that "we see in the animal kingdom, those products of blind natural forces, activities that are comparable in regularity to those which occur with consciousness, or indeed which arise externally in artwork that has achieved a kind of perfection." See Schelling, *Einleitung zu dem Entwurf eines Systems der Naturphilosophie*, in *Schellings Werke*, 2:272 [III:272].

113. Schelling, *System des transscendentalen Idealismus*, in *Schellings Werke*, 2:349 [III:349].

114. Schelling undoubtedly discussed the importance of aesthetics with the friends of his youth, Hölderlin and Hegel. During his time in Leipzig and first year in Jena, however, he focused on natural science and medicine.

115. During the winter of 1799–1800, Schelling began to study poetry carefully to prepare for

Standing just outside the Romantic circle and balancing its passionate activity was the monumental poet of Weimar, Goethe. He and Schelling met frequently during the fall of 1799 to discuss the relations of empirical science to transcendental philosophy (see chapter 11). Goethe had, though, another keen interest, really a compulsion. He desired some way of uniting his two loves, science and poetry, which, at least intellectually, he thought necessary to keep chastely separate. Given the general sway that Goethe exercised over all the Romantics, but especially Schelling, discussions with the poet might also have helped stimulate the seductive new proposals that made art the means to reveal the secret union of the poetically creative self and the scientifically perceived world. In the concluding sections of the *System*, Schelling attempted to lay bare this hidden relationship.

The ultimate aim of transcendental philosophy was, Schelling argued, to bring to intuitive identity the conscious and unconscious activity that constituted the unity of the self. Such an intuition would unite in one vision both nature, which was the unconscious product of self, and the conscious self—the objective realm with the subjective, necessity with freedom. But this sort of clarifying intuition, according to Schelling, had to be directed to the aesthetic act. For out of the unconscious of the artistic genius sprang ineffable laws that were executed with conscious intention.

Schelling's theory of genius depended upon ideas drawn from Kant, Schiller, and Friedrich Schlegel. Kant had defined genius as "the inborn mental trait (*ingenium*) through which nature gives the rule to art."[116] By this he seems to have meant—certainly Schelling took him as having meant—that the artist's unconscious nature determines those principles of beauty that express themselves in nonconceptual feelings, which, in turn, drive his conscious actions. The artist applies paint to canvas, therefore, not according to a fully conscious set of rules but by relying on aesthetic feeling—the conscious surface of underlying unconscious rules—to guide the brush. In Schelling's view, the inarticulable rules governing an aesthetic production surge forth with irresistible determinacy from the unconscious nature of the genius. These rules have to be followed. Yet every necessary blow of hammer on chisel, every required brush stroke, every per-

---

his great poetic composition, about which Friedrich Schlegel reported to Friedrich Schleiermacher (6 January 1800): "Schelling is quite full of his poem, and I believe it will be something great. Until now he has only made a study and tries to learn stanzas and terza rima. He will probably choose the latter for the whole thing. I am reading Dante with him and Caroline, and we are already halfway through." See Plitt, *Aus Schellings Leben*, 1:289.

116. Kant, *Kritik der Urteilskraft*, in *Werke*, 5:405–6 (A 178–79, B 181). See the preceding chapter for a discussion of Kant's aesthetic theory.

fect poetic metaphor, also stem from the free will of the genius. Insistent forces thus well up from the unconscious nature of the artist and rush in turbulent cascades through the narrows of consciousness. This creates, according to Schelling, violent eddies of contradiction that "set in motion the artistic urge."[117] Such contradictions can only be calmed in the execution of the work of art. As the artist comes to rest in the finished, objective product, he or she will sense the union of nature and self, of necessity and freedom, of—finally—the unconscious and the conscious self. In this respect, the goal of transcendental philosophy will be reached: what was originally an identical self has—through a kind of dialectical development in which self-reflection issues in the subjective structures of intelligence and the objective structures of nature—the one self will have returned to its original identity. The intelligence "will feel surprised and very happy by this union, that is, it will see this union as a generous gift of a higher nature, which through this connection has made the impossible possible."[118]

In his analysis of artistic genius, Schelling attempted to portray, in another key, the creative essence of the self as it constructed both itself and nature. He summarized this analysis by contending that "the aesthetic intuition is simply the intellectual intuition become objective." In this way, art became the model for nature, not the reverse—and hence his Romantic conceit that "nature is a poem that lies enclosed in a secret, wonderful script."[119]

This aesthetic grounding of nature has two immediate and crucial implications for the natural scientist. First, it indicates how nature in her particular productions—from animals to stars—and in the mutual relationships that constitute the grand scheme of nature can be understood as intrinsically both mechanical and purposive. Nature is blindly mechanistic and deterministic in that her products flow ultimately from universal and necessary laws; but nature is also purposive in that her entities are the creative products of intelligence, the unconscious manifestation of the self's striving to know itself. This is why a natural creature appears to arise, as Schelling put it, "as a product that is purposive without a purpose"—a designation that echoes Kant's formula for aesthetic objects.[120] Second, this grounding suggests that the reflective scientist must have the sensibilities of the artist.

If nature is a poem that the self has composed, we may then ask: Is it the scientist who interprets the poem through aesthetic intuition? Put another

---

117. Schelling, *System des transscendentalen Idealismus*, in *Schellings Werke*, 2:616–17 [III:616–17].

118. Ibid., 2:615 [III:615].

119. Ibid., 2:625 [III:625], 2:628 [III:628].

120. Ibid., 2:607 [III:607].

way, can the natural scientist penetrate nature in an act of genius? Kant had expressly denied the possibility of scientific genius, since scientific laws could be formulated in conscious, *conceptual* terms; but genius, by its very nature, operated according to *nonconceptual* feelings. One can read Newton's *Principia* and learn how to derive Kepler's area law and calculate the paths of planets. Reading Homer or Goethe with a comparable intention will not, however, enable one to formulate the law of beauty and use it to compose a wonderful poem.

Kant's sequestering of genius supposed that feeling and rational ability remained quite distinct faculties. Friedrich Schiller, in his great tract *Über die ästhetische Erziehung des Menschen* (On the aesthetic education of mankind, 1795), suggested that these aspects of mind could yet be synthesized "through aesthetic culture." Such cultivation would produce, he thought, an alteration in humanity: "Through beauty the sensuous man is led to form and thought; through beauty the intellectual man is returned to matter and again given to the sensible world."[121] Only when human beings have become able to *feel* the rational necessities of moral law, he contended, will true human freedom be achieved. But when it is achieved in the new man, reason would mingle with feeling so that desires might be rational and reason might be compelled by passion. Friedrich Schlegel most enthusiastically endorsed this anthropological consummation and promoted it as a requirement for the progressive, poetical science that would pulse life into the activities of his Romantic vanguard. Schelling resonated to these advancements in Kantian aesthetics and crucially expanded the Kantian notion of genius.

He certainly agreed that it took no scientific genius to read Newton's *Principia* with understanding. But what of Newton himself, in those acts of discovery that brought to existence a system of science that radically altered our conception of the world? Schelling allowed that this could, indeed, have been an act of genius. Two general conditions, he thought, would be necessary to postulate genius as the source of such scientific discovery.[122] First, we could infer the workings of genius if the creative scientific act initially comprehended a whole and only subsequently determined the operations of the parts—for example, if Newton perceived from the beginning, in an intellectual-aesthetic intuition, an entire system (however dimly) in which the laws of motion and gravity would play central roles. Such a vi-

---

121. Friedrich Schiller, *Über die ästhetische Erziehung des Menschen in einer Reihe von Briefen*, in *Schillers Werke* (Nationalausgabe), ed. Julius Petersen et al., 55 vols. to date (Weimar: Böhlaus Nachfolger, 1943–), 20:388 (twenty-third letter), 20:365 (eighteenth letter).

122. Schelling, *System des transscendentalen Idealismus*, in *Schellings Werke*, 2:623–24 [III:623–24].

sionary creation would be in contrast to the mechanical building up of a scientific theory through the tessellation of parts without initial presentment of the entire scheme. A second situation might also allow us to infer the presence of genius: when an individual formulated ideas that he or she could not have understood fully, thus suggesting a welling up from the unconscious of a conception whose significance could only be later consciously articulated. No doubt Schelling believed he himself met these conditions, being lured by a muse singing from beyond the harbors of consciousness.

In Schelling's theory, the external world of nature and the internal world of artistic creation proceeded from the same kind of productive intuition. The principal difference between the two worlds—one real, the other ideal—was that the world of nature only imitated the infinitely richer inner world that true art disclosed for us. Both the scientist and the philosopher attempted to retrace the intuitive acts of production by which these worlds spun into view.[123] If by his re-creations a natural scientist or philosopher intuitively illuminated the structures of nature or of art with felicitous comprehension and suggestive prevision—revealing that natural talent of which Kant spoke—then genius could not be denied them. Schelling certainly held that science and philosophy done at the margins of creative possibility were aesthetic-intellectual activities of the highest sort, of a piece with artistic creation. He prophesied that in the future a new poetics of nature would arise that would explicitly unite the scientific and aesthetic.[124] Through the nineteenth century, that conception of the possibilities of scientific creativity would lie at the heart of one science in particular, biology.

Evaluation of the *System*

Schelling's initial formulations of transcendental idealism followed directly from the work of Fichte and formed a part of the legacy of Kant. Schelling's analyses of the ego and its works, while seemingly a bit precious, can only be regarded as a persistent effort to live without the Kantian thing-in-itself. The sympathetic reader, whether today or at the beginning of the nineteenth century, must appreciate the force of the epistemological-metaphysical arguments of Fichte and Schelling: if we are going to explain experience, especially knowledge in science, but must do so without the aid of the thing-in-itself, then we can have as the objects of our consciousness only

123. One must again recall that when Schelling refers to *Wissenschaft*, he means to encompass all intellectual activities that articulate conceptual principles that can be systematized. In the comparison of art and science, by the latter he has natural science and philosophy chiefly in mind.

124. Ibid., 2:629 [III:629].

the very structures of subjectivity.[125] In one sense, this metaphysical position best accords with our native instincts, in conformity with which we regard objects of the natural world as completely real and not the shadow of hidden entities whose names, like those of the Lord, we may never pronounce. "If indeed the most perfect realism consists in recognizing things-in-themselves and doing so immediately," Schelling observed, "then such realism is possible only in a nature that perceives in things only its own reality through its own activity."[126] Idealism, then, is the soundest realism. Well, Schelling's claim was a bit disingenuous, but he had a point—though nature remains the self's complement, it is as real as the self. With the *System des transscendentalen Idealismus*, Schelling stood precariously at the edge of a decidedly more realistic metaphysics. He would shortly plunge into it. And one of the hands offering a gentle shove would be Goethe's.

After receipt of a copy of the *System*, Goethe wrote Schelling in September 1800 to say that "since I have shaken off the usual sort of natural research and have withdrawn into myself like a monad and must hover over the mental regions of science, I have only occasionally felt a tug this way or that; but I am decisively inclined toward your doctrine. I wish for complete harmony, which I hope to have effected sooner or later through the study of your writings, or preferably from your personal presence."[127] By the time of this letter, Goethe's admiration and affection for the star-crossed philosopher had grown. His message might be interpreted thus as more a sign of personal rather than philosophical harmony. Yet the shrewd poet undoubtedly detected in Schelling's work those considerations that would push the young philosopher beyond Fichte's subjective idealism. The *System*, and certainly the *Einleitung*, demonstrated a deep appreciation of empirical science and the autonomy of its methods, which separated him from Fichte and brought him within Goethe's intellectual sphere.[128] Schelling's organic conception of nature also appealed to the poet, especially its anti-

---

125. Schelling seems to have recognized that the deductive analyses he offered in the *System* had their force only because the thing-in-itself could be eliminated: "The act through which the I limits itself is none other than the act of self-consciousness, with which we must remain in our effort at explaining the ground of all limited entities, since the notion of an affection from outside of us transformed into a representation or knowledge is totally inconceivable" (ibid., 2:406 [III:406]).

126. Ibid., 2:428 [III:428].

127. Goethe to Schelling (27 September 1800), in *Goethes Briefe*, 2:408. In a previous letter that also remarked on the *System* (19 April 1800), Goethe suggested that the book offered much for "those who practice art and those who observe nature." See ibid., 2:405.

128. Entries in his journal for 19 September to 5 October 1799 indicate that Goethe assiduously read Schelling's *Entwurf* and *Einleitung*, and discussed these works with the young philosopher. See Goethe, *Goethes Tagebücher*, in *Goethes Werke*, II.3:259–63.

mechanistic character. At this time Goethe began speaking quite publicly and frequently of his admiration, of his "particular affection" for his young friend's *Naturphilosophie*.[129] Finally, Schelling's argument for the symmetry and virtual identification of artistic and scientific genius strikingly marked his ideas as disjoint from Fichte's but as golden links in the chain of Romantic thought, which Goethe himself would certainly not cast off. These theoretical considerations would bring Schelling intellectually closer to Goethe, but more personal needs drove him to the comforting embrace of the poet.

Schelling's Affair with Caroline and the Tragedy of Auguste

Schelling's intellectual and personal relationships in Jena abruptly changed in July 1799, when Fichte was forced to leave the city because of charges of atheism. Simultaneously Schelling lost a good friend and philosophical mentor, but gained intellectual breathing room and a new center of social interaction. From the time of his arrival in Jena, the young philosopher associated with Fichte, Schiller, and Goethe. Indeed, he came to regard Goethe as not only a close friend but really a kind of spiritual father; and the great man reciprocated with an intellectual and emotional solicitude only to be compared with that he showed for Schiller. With Fichte's departure and Friedrich Schlegel's return to Jena in September 1799, Schelling gradually moved toward the circle of friends that often met at Caroline and Wilhelm Schlegel's home.

During the fall and early winter of 1799–1800, the Schlegel's parlor became the regular gathering place of poets, philosophers, and lovers. They would read poetry and plays to one another, discuss the activities at court, and enjoy the bantering barbs that close friends can hurl at one another. During late summer and fall, Caroline's daughter Auguste, approaching age fifteen, was studying music in Dessau. Caroline's constant flow of letters to her beloved daughter, now regarded really as a sister, vibrated with a magnetic desire for her return. The letters attempted to keep Auguste apprised of the activities at home and thus of the circle of friends. For instance, at the end of September 1799, she wrote to her daughter:

> This evening Friedrich and I supped at Schelling's, to dedicate his new nest. He was glad that you had christened him Bacchus, since you named him the giver of wine, and soon he will be able to be called the giver of

129. This was observed by Friedrich Schlegel: "He [Goethe] talks constantly about Schelling's *Naturphilosophie* with particular affection [*Liebe*]." See Friedrich Schlegel to Wilhelm Schlegel (26 July 1800), in *Friedrich Schlegels Briefe an seinen Bruder August Wilhelm*, ed. Oskar Walzel (Berlin: Speyer & Peters, 1890), p. 431.

Figure 3.3
Auguste Böhmer
(1785–1800), daughter of
Caroline Böhmer
Schlegel Schelling.
Steel-point engraving by
A. Weger after a portrait
(1798) by F. A. Tischbein.
(From G. Waitz, ed.,
Caroline, 1871.)

joy, since he is soft and lovable and witty, and says that when you return you ought not act like a half-mademoiselle.[130]

The radiant sun that shone during this period on the Jena Romantics gradually passed behind more somber clouds of incipient and shifting loves, and of growing jealousies and hatreds. From the beginning Caroline found the young Schelling—twelve years her junior—a fascinating intellect and more. Initially they waged the typical sexual-intellectual wars that bespoke an underlying deeper attraction. She gave an account of their preliminary skirmishes to Hardenberg:

> Concerning Schelling, no one ever dropped so impenetrable a veil. And though I cannot be together with him more than six minutes without a fight breaking out, he is far and away the most interesting person I know. I wish we would see him more often and more intimately. Then there really would be a wrangle. He is constantly wary of me and the irony of the Schlegel family. . . . He is always rather tense, and I have not yet found the secret to loosen him up. Recently we celebrated his twenty-fourth birthday. He has time to become more relaxed.[131]

130. Caroline Böhmer Schlegel to Auguste Böhmer (30 September 1799), in *Caroline: Briefe aus der Frühromantik*, 1:557.

131. Caroline Böhmer Schlegel to Hardenberg (4 February 1799), in ibid., 1:497.

The bantering battles between Caroline and Schelling gradually faded into a deep love. She would later expressly pour out her feelings—"I am yours, I love you, I revere you, no hour passes that I do not think of you."[132] Schelling was, she confessed to her husband during the period of their "understanding," "the first and only love of my life."[133] Even in the fall of 1799—while Auguste was still in Dessau—their comportment toward each other became so obvious that Fichte, now banished to Berlin, wrote to his wife still in Jena that she should be wary of getting involved:

> Concerning Schelling and the Schlegel woman, be careful! I ask you for our love's sake. I am already opposed from one side, so I must ask extreme discretion of you. Schelling is making a bad name for himself, and I am sorry for that. Were I personally in Jena, I would warn him. The problem is that the actors do not think anyone notices that sort of behavior, while actually no one says anything until a completely public scandal occurs. Won't the husband make an end of this?[134]

Wilhelm did not. And the indiscreet behavior continued.

How did Caroline and Schelling understand their growing love for one another? From Caroline's side, the emotional pitch arched to the heavens. She fell fanatically in love with the young philosopher—and she was a woman practiced in love. Her own romantic élan and the Romantic circle's cultivated images of love freely given and received propelled her toward him with ever-greater acceleration. Schelling, completely naive in the ways of passion, was more tentative. There was the age difference; indeed, Caroline's daughter Auguste was closer in age to Schelling than she herself. Moreover, she was married to an estimable man. The moral and legal barriers, not to mention those of sheer hospitality, set up obstacles, enough to inhibit most normal men. But Schelling was decidedly not most men, and in a pre-Nietzschean way, he may well have thought himself beyond good and evil. In his *System des transscendentalen Idealismus*, which he completed in the first months of 1800, as his relationship with Caroline slipped all bonds of propriety, he observed that some few superior individuals manifested "not only genius for the arts or sciences, but also genius for action," a genius that might "act with freedom and a transcendence of spirit even above the law."[135]

---

132. Caroline Böhmer Schlegel to Schelling (February 1801), in ibid., 2:42.

133. Caroline Böhmer Schlegel to Wilhelm Schlegel (6 March 1801), in ibid., 2:65.

134. Johann Gottlieb Fichte to his wife (23–26 October 1799), in *Briefe*, ed. Manfred Buhr, 2nd ed. (Leipzig: Verlag Philip Reclam jun., 1986), p. 273.

135. Schelling, *System des transscendentalen Idealismus*, in *Schellings Werke*, 2:549 [III:549]. See above for a discussion of Schelling's conception of genius.

As the relationship between Caroline and Schelling gradually took form, that between Caroline and Dorothea, who followed Friedrich from Berlin in October 1799, slowly came undone. Caroline was obviously the star of their little society—beautiful, vivacious, demanding, creative, smart—a mélange of the traits that make intellectual men succumb and careful women distrustful. Dorothea certainly had a keen intelligence and a passionate intensity. She authored a successful novel, *Florentin* (1801), under Friedrich's name, that clearly manifests her creativity and knowledge of the heart. But she was not a beautiful woman, and her charms did not dazzle as did Caroline's. From a different motive than Fichte's, Dorothea began to grow more irritated with Schelling and spiteful of Caroline, whom she thought spent too much time playing "the beautiful woman." In a long letter to Schleiermacher (28 October 1799), Dorothea coolly dissected the characters of Schelling and Caroline:

> Schelling? I do not know much of him, he speaks little. His external appearance is what you would expect; through and through forceful, stubborn, raw, and noble. He should really be a French general; being a professor doesn't really suit him, still less, I believe, a position in the literary world. . . . Caroline is lovable and remains so despite the close acquaintance. It is really too bad that, with her unusual talent for art, she doesn't practice and with effort refine it. Being the beautiful woman, rather, occupies all her thought and time. Basically, I believe that she doesn't have any love for life as it is led here. She probably wishes to travel along a glittering path. She is friendly and lively, but never satisfied, never cheerful.[136]

The tensions that threatened the stability of the Romantic circle increased to the breaking point by March 1800, when Caroline, who had contracted some mild but lingering illness, grew worse. Schelling had plans to travel to Berlin in order to continue his studies of medicine at the Charité—a radical break for it meant departing from Caroline and from Goethe. But Caroline's illness, which seemed to accelerate as Schelling's planned departure grew nearer, constrained him to put off his Berlin plans. Instead, he offered to take her to Bamberg, where she might consult with doctors there and take the baths. Schelling also thought he might begin his medical

---

136. Dorothea Veit to Friedrich Schleiermacher (28 October 1799), in *Friedrich Daniel Ernst Schleiermacher, Kritische Gesamtausgabe*, ed. Hans-Joachim Birkner et al., 11 vols. to date (Berlin: Walter de Gruyter, 1980–), 5.3:223, 225. Later Dorothea wrote Schleiermacher (4 April 1800) of the deterioration of relationships in Jena. While she indicated her gratitude to Caroline for graciously making her feel at home, she nonetheless described Caroline as "completely imprudent, and extremely egotistical, though like an immature child who only thinks for the present; she is simply incapable of plans beyond that" (ibid., 5.2:452).

studies in Bamberg, which was famous for its orientation toward the ideas of
the English physician John Brown (1735?-1788). Neither Friedrich, nor
Dorothea, nor certainly Wilhelm regarded this offer as pure altruism. By
this time Friedrich had turned against Caroline, undoubtedly because of his
own dying love and for his brother's sake. Dorothea became reserved and
proper, as she indicated to Schleiermacher:

> My entire manner with Caroline lies precisely within the bounds of com-
> mon civility. Each day I make one or two short visits, but turn aside any
> closer relationship, since she is Friedrich's enemy, so why should I be con-
> cerned?—She takes daily walks with Auguste and Schelling, but that
> does little good, she says, so that a complete change of place will be nec-
> essary for her fully to recover. She will, therefore, travel this week with
> Schelling to Bamberg and there take the required baths. . . . They will
> leave shortly, and we'll be able to breathe again. I doubt that she'll
> quickly return, perhaps never! But she indicated to Wilhelm that she
> would soon come back—just so that she wouldn't completely leave him,
> or he her.[137]

With the departure of Caroline, Auguste, and Schelling in the first week
of May 1800, the Jena circle of Romantics had spiritually dissolved. Friedrich was beside himself in anger and frustration; Wilhelm was resigned, if
harboring some faint hope.[138] Dorothea poured out her resentment to
Schleiermacher:

> Caroline has left. . . . Her character is coquettish and low, while her spirit
> is prudish and superficial—that she has learned in her life not a bit of tact
> is obvious. . . . She won't have any greater happiness with Schelling than
> with Wilhelm, since she won him for herself with greater craft; he is not
> inclined toward her, or toward any mentally significant woman, so how
> can she believe that such a nature predestined to rawness can take on an-
> other nature that will love her (for she knows his dislike of educated
> women)? How can she take this recent passion for love. . . . Wilhelm still
> really loves her, and probably will continue to love her until some other
> one of her disposition is able to tie him down. Caroline never really loved
> him![139]

137. Dorothea Veit to Schleiermacher (28 April 1800), in ibid., 5.4:9.
138. Wilhelm Schlegel appears to have been rather saintly throughout the time of his wife's
love affair with Schelling. Indeed, Wilhelm and Schelling had planned, along with Fichte, to produce a journal of criticism in literature and philosophy. That plan, with its many twists, is discussed
below.
139. Dorothea Veit to Schleiermacher (15 May 1800), in *Schleiermacher, Kritische Gesamtausgabe*, 5.4:39–40.

Caroline consulted doctors in Bamberg and then traveled to Boklet (close by) with Auguste to take the baths. They stayed for some time. In June Schelling went to visit his parents. Caroline ached in her longing for him and would only be consoled when Auguste assured her that Schelling loved her "more than anything."[140] At the beginning of July, he returned to find Caroline much improved but Auguste now ill with dysentery. The local doctor promised an easy recovery in a few days. But on 12 July, Auguste—Caroline's most beloved daughter, now barely fifteen, a young women of infinite grace, refined education, and lively charm—suddenly died.[141] Caroline was devastated. She wrote her friend Luise Gotter: "I am only half alive and wander like a shadow over the earth."[142] Those who knew Auguste and loved her reacted with comparable feeling. Wilhelm wrote to Ludwig Tieck that "it was as if I had stored all my tears for this, and at times I have had the feeling that I should completely dissolve into tears."[143] Henrik Steffens, who had quite fallen for her, immediately sent an anguished letter to Schelling:

> I cannot bear to say—what for me, yes, for me, what Auguste's loss means. That beautiful girl—I cannot grasp her death.—So full of life, so much promise—and now dead. I can't speak about it—Oh! She was more dear to me than one knows, more than I want to confess. . . . When I am able to work in peace, when healthy and in a good frame of mind, I consider everything that Jena was to me, the source of my higher life, that child stands before me like a bright angel. When I was last in Jena, she became even closer to me—and now. Never—never, after so many years, has death come so close to me—I've seen accidents and people die, but I

140. Auguste wrote Schelling (4–5 June 1800) telling him of her manner of consoling her mother: "I tell her 'how much he loves you' and she gets all soft; the first time I told her, she wanted to know how much you loved her. Since that was out of my ken, I quickly responded: more than anything. She was satisfied, and I hope you will be as well" (*Caroline: Briefe aus der Frühromantik*, 1:599).

141. The doctor who attended Auguste blamed Schelling, in part, for the death, since he had slightly altered dosages of drugs ordered. Schelling apparently only removed from the prescription some drugs that he thought would weaken Auguste too much, a tincture of rhubarb among them. The drugs prescribed did contain opium, a standard enough ingredient. Schelling also reduced the prescribed amount. See Schelling's account to Wilhelm Schlegel (3 September 1802), in Schelling, *Briefe und Dokumente*, 2:431–32. Schelling made his explanation to Schlegel sometime after the event because of a later article in the *Allgemeine Literatur-Zeitung* that virtually accused Schelling of killing Auguste. See the discussion of this event below.

142. Caroline Böhmer Schlegel to Luise Gotter (18 September 1800), in *Caroline: Briefe aus der Frühromantik*, 1:610.

143. Wilhelm Schlegel to Ludwig Tieck (14 September 1800), quoted in Gisela Dischner, *Caroline und der Jenaer Kreis* (Berlin: Verlag Klaus Wagenbach, 1979), p. 154.

saw only change. I didn't see death—and now—well, I shouldn't renew the pain. Greet the unhappy mother for me.[144]

Dorothea's response did not quite touch the same emotional depths. She characterized Auguste's death as a "sacrificial offering for sin."[145] And perhaps Caroline may have even believed something like that. But now she clung to Schelling more desperately than ever: "My heart, my life, I love you with my whole being. Do not doubt that for a minute." She would test, she told him, "whether I can bring out of the combination of death and pain with that of bliss and love yet life and peace."[146] Faithful Wilhelm hurried to Bamberg to comfort Caroline, and amazingly he, Caroline, and Schelling remained together in the city until the following October. Wilhelm then traveled with Caroline, who was still psychically paralyzed, to Braunschweig, where she could be with her mother and sister. In February 1801 Wilhelm left for Berlin, never to return to Jena. Wilhelm Schlegel, who had been tempered by love and death, proved to be a man of great forbearance, concern, and character. Friedrich simply wanted Caroline's blood.

Auguste's death deeply affected Schelling. He too went into emotional collapse and reached such depths that Caroline thought he might commit suicide. She wrote Goethe from her mother's home in Braunschweig to plead with him to care for the young philosopher:

> He [Schelling] has been through a series of terrible experiences that have reduced him to a such a state that he is completely prostrate. . . . It cannot have escaped your notice how much he has suffered in body and soul, and he is now in such a sad and delicate emotional state that soon some guiding light must be shown him. I myself am tired and sick, and am not in a position to reinstate that powerful view of life which is his destiny. You can do it. . . . You have influence over him, of a kind which nature herself would have if she could talk to him as a voice from heaven. Extend your hand to him.[147]

Goethe did extend his hand and invited Schelling to spend the Christmas holidays with him, a grace for which the despairing philosopher was so very grateful.[148]

144. Henrik Steffens to Schelling (20 August 1800), in Plitt, *Aus Schellings Leben*, 1:305. Steffens's reaction to Auguste's death, as his letter to Schelling indicates, was obviously profound. He said he sent it to his friend just as it came from his pen, without correction.

145. Dorothea Veit to Schleiermacher (28 July 1800), in *Schleiermacher, Kritische Gesamtausgabe*, 5.4:175.

146. Caroline Böhmer Schlegel to Schelling (October 1800), in *Caroline: Briefe aus der Frühromantik*, 2:4, 6.

147. Caroline Böhmer Schlegel to Goethe (26 November 1800), in ibid., 2:19.

148. Goethe was in Jena from 12 to 26 December and traveled back to Weimar with Schelling on the twenty-sixth. Among other diversions of the season, Goethe went with Schiller and

It is hard to believe that Schelling was not in love with Auguste as much as with Caroline, if on a different plane.[149] The conclusion is easy to draw, since everyone seemed to be in love with her. The pitch of Schelling's response, however, had a sharper curve, since he was blamed for her death— and he did suffer with aching guilt. Immediately Dorothea gossiped that Caroline and her lover had not called in a proper doctor when the child became ill and that Schelling had "meddled" in the treatment.[150] The rumor of his culpability steadily grew over the months after the tragedy—and has hardly abated even today, at least among the cognoscenti. The circumstances are unusual and perhaps help explain a great shift in Schelling's philosophical position.

When Schelling returned from visiting his parents at the beginning of July, he found Auguste ill. The local doctor, who followed Brownian medical practice, prescribed opium mixed with gum arabic and tincture of

---

Schelling to a masked ball at court (2 January 1801), though the three finally escaped to talk about aesthetics. Schelling stayed until 4 January, when Goethe had to take to his bed because of a severe catarrh. See *Goethes Tagebücher*, part 3, in *Goethes Werke*, 2:315–16, 3:1. Schelling wrote Goethe at the end of January to give thanks that the poet was feeling better. He then expressed his gratitude for the invitation to spend time in Weimar: "The recollection of the healing and happy stay in your house and under your gaze has not left me for an instant, and was, for me at this time, of infinite value." See Schelling to Goethe (26 January 1801), in *Schelling als Persönlichkeit: Briefe, Reden, Aufsätze*, ed. Otto Braun (Leipzig: Fritz Eckardt Verlag, 1908), p. 88. Goethe later advised the family on the kind of memorial they might erect for Auguste—nothing elaborate at the grave site, rather they should have two urns finely wrought for their home, one depicting scenes from Auguste's life, the other their own reactions to her. See Goethe to Wilhelm Schlegel (28 February 1801) in *Schriften der Goethe-Gesellschaft* 13 (1898): 99–101.

149. There has always been a persistent belief that Schelling had really loved Auguste and that Caroline arranged an engagement between her daughter and her own beloved in order to keep him near. Dorothea certainly believed that "the mother had earnestly striven for Schelling to have married her [Auguste], since she would thereby be able to keep within the true nature of the relationship." There is no evidence that Schelling in fact loved Auguste in other than a brotherly way. Dorothea believed that Caroline had raised her daughter as a coquette and that her husband, Wilhelm, had been blinded to that fact. Wilhelm simply thought of the girl, according to Dorothea, as having a sweet nature and a marvelous "head for Goethe, for Shakespeare, for ancient poetry, for his poetry, and God knows what else." See Dorothea Veit to Schleiermacher (15 May 1800), in *Schleiermacher, Kritische Gesamtausgabe*, 5.4:42–43.

150. Dorothea Veit to Schleiermacher (22 August 1800), in ibid., 5.4:222. Dorothea maintained that Caroline treated Auguste as an adult much too soon, which, along with the affair with Schelling, had a debilitating effect. She went on to say: "The Brownian technique, in this case, is not to blame. They had no physician with her other than a completely unknown man from the region of Boklet, who was no less than a Brownian. To top the whole thing off, Schelling meddled in it [*hinein gepfuscht*]. They sent for a physician from Bamberg only as she grew cold from the waist up. Röschlaub came and found her already dead. He maintained that her sickness was lethal right from the beginning; all the more unforgivable, then, is the confidence they showed in not sending for a doctor right from the beginning. Shortly—And now the ostentation of the sorrow!—We are going to remain completely silent about all those people. I won't write you anything more on this, since I am simply too indignant."

rhubarb (this last ingredient presumably to moderate the constipation that opium would induce). Schelling removed the gum and rhubarb—he thought them too much an emetic for her condition—and saw to it that the opium was in a very small dose. Schelling's actions were confirmed by Wilhelm Schlegel, who saw the receipt Schelling obtained for the medicine.[151] Since dysentery causes severe dehydration and opium by itself constricts fluid loss—and Schelling fully recognized these features of the situation[152]—his intervention was probably a wise move. Andreas Röschlaub (1768–1835) and Adalbert Friedrich Marcus (1753–1816), both highly respected Bamberg physicians, explained to Wilhelm Schlegel that Auguste had contracted a fatal disease and that really nothing could have been done to save her.[153] Schelling did confess to Schlegel that he felt terribly guilty because he had trusted the local physician—he undoubtedly himself feared that had Röschlaub been called sooner, Auguste might have survived.[154] The circumstances were nonetheless sufficient—and the moral offense at his relations with Caroline so heated—that rumors of Schelling's culpability took wing.

The sinister shadows of the rumors grew, accelerated by a disdain for Schelling's idealism—which was seen as a fog that hid a lurking atheism—until they even darkened the public forum in summer of 1802. The precipitating event was the appearance of a small pamphlet, anonymously broadcast (by Franz Berg, a theologian at Würzburg), that made sport with the "new philosophy." The broadside reported that a medical candidate, a one Joseph Reubeln (1779–1852), had produced a Schellingian thesis that showed how death could be overcome. To the sardonic description of Reubeln's views, the author added: "Heaven protect him [Reubeln] that he does not meet a patient whom he idealistically cures but really kills; a misfortune that befell Schelling at Boklet in Franken in the case of M.B.

---

151. Schelling explained his procedure to Wilhelm Schlegel, and this was confirmed both by Johann Andreas Röschlaub, a well-known physician who was called too late, and by the receipt describing the medicine administered, which Schlegel saw. See Schelling to Wilhelm Schlegel (3 September 1802) and Wilhelm Schlegel to Friedrich Schlegel (27 August 1802), in Schelling, *Briefe und Dokumente*, 2:429–37, 425–26. The tenets of Brownian medicine and the career of Röschlaub are depicted with care and authority by Nelly Tsouyopoulos, in her *Andreas Röschlaub und die romantische Medizin* (Stuttgart: Gustav Fischer Verlag, 1982).

152. Schelling specifically mentioned to Wilhelm Schlegel that the tincture of rhubarb caused the bowels to loosen. See Schelling to Wilhelm Schlegel (3 September 1802), in Schelling, *Briefe und Dokumente*, 2:431.

153. The letters are included in Wilhelm Schlegel's defense of Schelling, in *An das Publicum. Rüge einer in der Jenaischen Allg. Literatur-Zeitung begangen Ehrenschändung* (Tübingen: J. G. Cotta'schen Buchhandlung, 1802). See below for a discussion of this pamphlet.

154. Schelling to Wilhelm Schlegel (3 September 1802), in Schelling, *Briefe und Dokumente*, 2:432.

[Mademoiselle Böhmer], as some naughty people say." This bit of scurrility might have died peacefully of neglect, except the *Allgemeine Literatur-Zeitung* in Jena, a widely circulated and influential journal, ran a review of the pamphlet and quoted that damaging sentence.[155] Schelling was benumbed with fury at the remark. His first thoughts were to seek a judicial action against the *ALZ* or to go directly to the ducal court for redress.[156] Such measures, however, were sidetracked by the actions of Wilhelm Schlegel, with whom Schelling remained on astonishingly good terms.[157]

Schlegel took up the defense of his wife's lover with a pamphlet that recounted the medical treatment advised by Schelling and then added letters from Marcus and Röschlaub testifying to the hopelessness of Auguste's condition. Both Schlegel and Schelling believed the editor of the *ALZ*, Christian Gottfried Schütz (1747–1832), with whom both men had been at odds for some time, was behind the review and complicit in the libel. Schlegel meted out his rage with characteristic eloquence:

> It must deeply pain me that he [Schelling], who was animated with desire to heal the sick, should be made a criminal. It must enrage me more than any other person, since it provokes my saddest and holiest memory, to see that terrible fate—which has so untimely snatched away the most pure grace and loveliness of a beautiful life—that this fate should be made an instrument of low revenge and contemptible passion by scandalous knaves. Yes, it must wound me deeply that a beloved name in this disgraceful instance should be publicly mentioned or even only suggested.[158]

Schlegel's defense of Schelling, while justified in fact, also bespoke a nobility that astounded his friends, especially as it came during the negotiations for his divorce from Caroline, which, with the help of Goethe, was

155. Anonymous review of "Lob der allerneusten Philosophie," *Allgemeine Literatur-Zeitung*, no. 225 (10 August 1802): 329. The screed against Schelling was more than a veiled protest against the untimely death of a young girl—indeed, it was hardly that. Franz Berg, the author of the *Lob der allerneusten Philosophie*, was a religiously conservative theologian, who connected Schelling's ideas with those of the atheist Fichte. And Berg and Schütz, the editor of the *ALZ*, reacted as well against the Brownian medical theories and the Romantic attitudes that supported those views. Urban Wiesing discusses these intertwining reasons for the charge against Schelling in his "Der Tod der Auguste Böhmer: Chronik eines medizinischen Skandals, seine Hintergründe und seine historische Bedeutung," *History and Philosophy of the Life Sciences* 11 (1989): 275–95.

156. He also thought of having Goethe attempt to exact a retraction from the journal. See his letter to Wilhelm Schlegel (3 September 1802), in Schelling, *Briefe und Dokumente*, 2:429–30.

157. Schlegel's continued friendship with Schelling, which was actually quite deep, was probably abetted by his growing interest in Sophie Tieck Bernhardi, the sister of Ludwig Tieck.

158. Wilhelm Schlegel, *An das Publicum*, p. 14.

officially granted some eight months later, on 17 May 1803.[159] The outrage
of both Schelling and Schlegel, though, brought scant retraction from the
ALZ—merely the feigning reply that no reasonable person would assume
that the reviewer claimed Schelling had actually killed the girl, only that
"some naughty people" had said this.[160]

Schelling's Identity Philosophy

The Break with Fichte

During the time that Schelling bore the considerable weight of his affair
with Caroline and then the crushing feeling of guilt, however unwarranted,
for the death of Auguste, he was moving toward a break with his mentor
Fichte. Though Schelling initially endorsed Fichte's fundamental epistemological-metaphysical conception that everything exists for and in the
ego, his scientific work and his developing ideas about the independence of
natural phenomena moved him slowly away from that original starting
point. Schelling interpreted Fichte's ego as an individual subject (one of
many individual consciousnesses), and as such it too needed to be explained,
not simply assumed. Schelling believed that explanation had to invoke an
absolute, a state that was neither subject nor object, but both indifferently—
something akin to Spinoza's *Deus sive natura*—whence individual egos and
their world would emanate. Methodologically, this meant that Fichte's *subjective* idealism had to be subordinated to the one-time disciple's new *objective* idealism. As Schelling expressed it in January 1801, casting a mirror
image of his earlier position: "There is an idealism of nature and an idealism
of the ego. The former is for me the original, the latter derived."[161]

The precipitating causes of the split between Schelling and Fichte were
manifold. They ranged from bitter feelings over intellectual politics to the
solace of reconciling love, from a desire for professional independence to

159. Goethe helped to make the proceedings as smooth as possible. A main difficulty was that
the divorcing couple was supposed to appear before a consistory together, to see if reconciliation
was possible. Schlegel was occupied in Berlin and seems not to have wished to go to too much trouble in the affair. Schelling pleaded for Goethe's help to annul the requirement, which, with some
effort, the great poet accomplished. See the letters from Goethe to Schelling (2 October 1802 and
7 January 1803), and the account given them by Karl-Heinz Hahn, "Zwei ungedruckte Briefe
Goethes an Schelling," *Goethe, Neue Folge des Jahrbuchs* 19 (1957): 219–25.

160. Schelling, of course, reacted violently to this "correction," which appeared in the *Intelli-
genzblatt der Allgemeine Literatur-Zeitung*, no. 173 (25 September 1802): 1399. See also Schelling's
letter to Wilhelm Schlegel (4 October 1802), in Schelling, *Briefe und Dokumente*, 2:445–49.

161. Schelling, *Ueber den wahren Begriff der Naturphilosophie und die richtige Art, ihre Probleme
aufzulösen*, in *Schellings Werke*, 2:718 [IV 84]. See his assertion to Hegel (4 February 1795), quoted
in the first part of this chapter, for the converse sentiment.

the recognition of subtle intellectual differences, and from the experience of the autonomy of science to the remorse over the death of a beloved. The growing intellectual differences felt by Schelling and his appreciation of the logic separating his theory from that of Fichte have been standardly noted by historians,[162] but I do not think these intellectual factors would have had purchase on the mind of the young philosopher had they not been caught in the matrix of the other causes. What might seem trivially mundane and personal to a more purified ideal of philosophical comportment nonetheless, I believe, had epistemological and metaphysical consequences.

One sharp unpleasantness that frayed the relationship between Fichte and Schelling had to do with an attempt to start a journal.[163] From the time Schelling had arrived in Jena through the time of Fichte's departure, the two had spoken of launching a critical review that would be the principal organ for idealism. During this same period, the Schlegels also conceived of starting a magazine devoted to philosophy and literature.[164] The brothers and especially Schleiermacher, who was in on the plan, hesitated to invite Fichte to become a member of the editorial panel, lest he simply take over. Finally, as Schelling traveled from Bamberg to visit his parents (just before Auguste's death), he met the publisher Cotta, who was willing to put out a journal edited by the young philosopher alone. The Schlegels urged Schelling nonetheless to come in with them, and, at his insistence, they did invite Fichte to join the group. Fichte, however, became quickly incensed over what he thought to be shabby treatment; and he wrote Schelling angrily about the self-aggrandizing and pecuniary actions of the Schlegels.[165]

162. See, for example, Reinhard Lauth's excellent study *Die Enstehung von Schellings Identitätsphilosophie in der Auseinandersetzung mit Fichtes Wissenschaftslehre (1795–1801)* (Freiburg: Verlag Karl Alber, 1975). Lauth traces subtle but, he believes, significant differences between Schelling and Fichte to Schelling's *Philosophische Briefe über Dogmatismus und Kritizismus* of 1795. Lauth thinks this signaled the beginning of their dispute. Kuno Fischer finds the division located in Schelling's *System des transscendentalen Idealismus*. Fischer marks out the parallelism between *Naturphilosophie* and transcendental philosophy, which the *System* postulated, as the beginning of Schelling's identity philosophy. See Fischer's *Friedrich Wilhelm Joseph Schelling*, 2:707. Fischer's view became largely the orthodox one. See, for instance, White, *Schelling*, pp. 70–73. None of these historical analyses suggests other than purely philosophical causes for the split between Fichte and Schelling.

163. Fuhrmans discusses in some detail the various plans for journals with which Schelling was involved. See Schelling, *Briefe und Dokumente*, 1:201–8.

164. A bit later, in 1803, after Schütz moved the *Allgemeine Literatur-Zeitung* from Jena, Goethe, Wilhelm Schlegel, Ludwig Tieck, and Friedrich Schleiermacher began work to found a competitor, the *Jenaische Allgemeine Literatur-Zeitung*. Goethe told Henrik Steffens that the philosophical foundation for the journal would be Schelling's and he invited Steffens to contribute an account of Schelling's philosophy for the first number. See Steffens, *Was Ich Erlebte*, 5:11.

165. Fichte to Schelling (13 September 1800), in Schelling, *Briefe und Dokumente*, 2:251–54.

He must also have been piqued by a hostile review in the Schlegels' *Athenaeum* that his recently published *Bestimmung des Menschen* (Vocation of man, 1800) received from Schleiermacher,[166] whom he called a "pure liar" about the journal plan.[167] In the meantime, the friendly relationship between Schelling and Friedrich Schlegel cooled to the brittle breaking point as quickly as the philosopher's love affair with Caroline heated to incandescence. It became quite clear that the two could not work together on the same journal. It also became clear that Schelling would not really be able to work harmoniously with Fichte either. Fichte seemed to regard his younger colleague as still a disciple, even though Schelling's philosophical fortunes had shot to the celestial regions. Indeed, despite Schelling's frequent urgings, Fichte had read very little of his colleague's work, and what he did look at, he did not like.[168] In spring of 1801, after briefly examining Schelling's new presentation of his system (his *Darstellung meines Systems der Philosophie*), Fichte wrote his friend a cordial but firm letter, expressing confidence that if Schelling, certainly a "clear, deep, and perceptive" fellow, were to consider more carefully the *Wissenschaftslehre*, "you would in time enough correct your mistakes"—a schoolmaster's condescension that could only irk a pupil who felt his own intellectual power.[169] Schelling, albeit naively, harbored the same belief—if Fichte studied the *Darstellung*, he would grant priority to the absolute. Thus each of the founders of German idealism waited, staring across a widening gulf, for the other to make his way back. Patient waiting came to an end, however, when Schelling negotiated with Cotta in late fall to establish a new journal. The *Kritisches Journal der Philosophie* appeared the next year but with Hegel as coeditor.

166. Schleiermacher's review is included in *Schleiermacher, Kritische Gesamtausgabe*, 1.3:235–48.

167. Fichte to Schelling (13 September 1800), in Schelling, *Briefe und Dokumente*, 2:251.

168. Schelling had sent Fichte a copy of his *System des transscendentalen Idealismus* in spring of 1800. On 31 October 1800, he wrote Fichte asking whether he had received the work since he had heard nothing from his friend. Fichte did finally read the book, which he thought was everything "to be expected from your ingenious representation." However, he did not think that they were of one mind about the opposition between *Naturphilosophie* and transcendental philosophy. He yet attempted a formulation that he thought caught an agreement: "The object does not come to me through consciousness, nor does consciousness come to the object; rather both are directly united in the ego, the ideal-real and real-ideal." Schelling replied (19 November 1800) that the ideal activity of consciousness and the real activity of things do not lie in the same ego, in his view; rather, they both stem from an objective nature: "Certainly you should not suggest to me that I regard the matter even so [as you have implied]; you wish to consider it superfluous, in regard to my system, that I have the ideal and the real activities simultaneously produced (in the theory of productive intuition). And so you find that I, just as you, place these two activities in one and the same ego." See the exchange of letters in ibid., 2:289–92, 294–300.

169. Fichte to Schelling (31 May 1801), in ibid., 2:339.

As this tangle of misperceptions and injured feelings became more knotted, Schelling continued to immerse himself in scientific study, performing his own electrical and physiological experiments, and becoming fully apprised of the best work in those areas (such as that of Alexander von Humboldt, Johann Ritter, and Christoph Pfaff).[170] As a result, he felt convinced that nature more readily gave up her secrets to probings with the electrician's voltaic pile rather than to the philosophical investigations of the hidden corners of the ego. The infinite resources of the self could only be understood through an experimental examination of its transformed existence in nature. Concomitantly, his initial embrace by the Schlegels—particularly by Friedrich, whose critical objections to Fichte mounted in step with his poetic requirements for philosophy—pulled the young thinker further away from the foundational assumptions of subjective idealism. And Goethe, in his many discussions of *Naturphilosophie* with Schelling, undoubtedly continued to press against the subjective interpretation of nature, which the poet himself simply could not abide (see part 3). That interpretation made occurrences in the natural world—including the unreasonable demands of love and the more unreasonable death of a young girl of promise—somehow the ultimate responsibility of the ego. And from the emotional perspective (despite the logic of the situation), the responsible ego had to be Schelling's. A subjective idealism made other individuals the productive responsibility of the self. A young philosopher, closed off in his study and communicating to his students only from the high chair of the German professorate, might imagine a world only of his own making and with an imperious gesture take responsibility for it. But that same philosopher, now pulled down by the grappling hooks of love, then dashed against the rocks of his own conscience because of the death of a beautiful spirit, must have his isolated self torn asunder. Only the abandonment of Fichtean egoism and the adoption of an austere and deterministic absolutism might mitigate the responsibility for love and for death—or so, I think, the emotional dialectic would have proceeded.

All of these sensible pressures and intellectual considerations expanded until the break had to occur. The final push came from Caroline. She wrote

---

170. For a sketch of Ritter's (1776–1810) early work in electrical experimentation, his collaboration with Goethe in this area, and his subsequent interactions with Schelling in Munich, see Stuart Strickland's interesting, if vexing, "Circumscribing Science: Johann Wilhelm Ritter and the Physics of Sidereal Man" (Ph.D. diss., Harvard University, 1992). When Christoph Pfaff (1773–1852) stopped in Leipzig on his way to accept a position at Kiel as professor of medicine, he met Schelling and for eight days they discussed and performed galvanic experiments. See Schelling to Christoph Pfaff (6 March 1798), in Schelling, *Briefe und Dokumente*, 1:119–21.

Schelling in March 1801, as their desperate love hardened to an impervious shield against the world:

> It occurs to me that for all his [Fichte's] incomparable power of thought, his powerful mode of drawing conclusions, his clarity, exactness, his direct intuition of the I and the inspiration of the discoverer, that he is yet limited. . . . When you have broken through a barrier that he has not yet overcome, then I have to believe that you have accomplished this, not so much as a philosopher—if I'm using this term incorrectly, don't scold me—but rather because you have poetry and he has none. It leads you directly to production, while the sharpness of his perception leads him to consciousness. He has light in its most bright brightness, but you also have warmth; the former can only enlighten, while the latter is productive. . . . In my opinion, Spinoza must have had far more poetry than Fichte—if thought isn't tinctured with it, doesn't something lifeless remain therein?[171]

Schelling's ideas could hardly have found a more loving efflation to lift them from the reflective plane of his speculations.[172]

Establishment of the Identity Philosophy

Several treatises published by Schelling in the years 1801 to 1803 widened the philosophical break that Caroline encouraged. The principal among them were *Ueber den wahren Begriff der Naturphilosophie und die richtige Art, ihre Probeme aufzulösen* (On the true concept of nature philosophy and the correct way to solve its problems, 1801); *Darstellung meines Systems der Philosophie* (Presentation of my system of philosophy, 1801); *Bruno oder über das göttliche und natürliche Prinzip der Dinge. Ein Gespräch* (Bruno or on the divine and natural principle of things. A dialogue, 1802); and a second edition of the *Ideen* (1802), with considerable amplifications. Each of these

---

171. Caroline Böhmer Schlegel to Schelling (1 March 1801), in *Caroline: Briefe aus der Frühromantik*, 2:58.

172. Caroline's judgment would find echo, negative echo, in Heine's evaluation: "Poetry is Mr. Schelling's force and weakness. It is that by which he is distinguished from Fichte, both to his advantage and disadvantage. Fichte is only a philosopher and his power lies in dialectic and his strength in demonstration. But these are the weak aspects of Mr. Schelling. He lives more in intuitions. He does not feel at home in the cold heights of logic; rather he likes to wander in the flower-strewn valleys of symbolism, and his philosophical strength lies in construction." See Heinrich Heine, *Religion und Philosophie in Deutschland*, in *Sämtliche Schriften*, ed. Klaus Briegleb, 3rd ed., 6 vols. (Munich: Deutscher Taschenbuch Verlag, 1997), 3:629.

works announced and expanded upon Schelling's theory of absolute identity, though each did so in a different form.

The *Begriff der Naturphilosophie*, which appeared in January 1801, was a direct response to a critical but not unsympathetic analysis of *Naturphilosophie* by Karl Eschenmayer (1768–1852), a medical doctor who would soon become a supporter of the transcendental position.[173] Schelling published Eschenmayer's article in his own *Zeitschrift für Speculative Physik*. The essay, which discussed some of the apparent paradoxes of idealism, prompted Schelling himself to face such questions as "How does it happen that a tree, which someone planted fifty years ago for posterity, should be produced by me now, just as it is, through productive imagination?" Or, "Is not the idealist fortunate that he is able to consider as his own the divine works of Plato, Sophocles, and all other great minds?"[174] Schelling had to allow such objections might have some basis, even if ultimately mistaken, in his *System des transscendentalen Idealismus*. He contended, however, that they really had no justification in light of his *Naturphilosophie*, which was the balancing side of his thought. He wished to use his response to Eschenmayer as a propaedeutic to a more comprehensive and systematic presentation of his unified view. His *Darstellung* was to be that presentation.

In the *Darstellung*, Schelling claimed he was not changing his philosophical position, only bringing it to a more complete, public expression, to unite in one system the transcendental philosophy and the *Naturphilosophie* that he had previously developed separately.[175] The novelty of the *Darstellung* lay not so much in its basic metaphysical position—though that changed more than Schelling would explicitly admit—but in its mode of presentation. He adopted the quasi-geometrical method of Spinoza's *Ethica*, casting his theses into the forms of definitions, demonstrations, and remarks. The very mode of presentation itself was revelatory, as Schelling himself acknowledged, of a new affinity for that revered Jewish philosopher's attempt to establish an objective and realistic system grounded in an absolute substance.[176]

The austere mode of the *Darstellung* left many readers nonplussed, as it still does today. Fichte detailed his complaints to Schelling in a letter of May 1801 (though he only sent it in August due to "lethargy"). Essentially

---

173. Karl Eschenmayer, "Spontaneität = Weltseele oder über das höchste Princip der Naturphilosophie," *Zeitschrift für Speculative Physik* 2 (1801): 1–68.

174. Ibid.

175. Schelling, *Darstellung meines Systems der Philosophie*, in *Schellings Werke*, 3:3–4 [IV: 107–8].

176. Ibid., 3:9 [IV:113].

he protested that Schelling was simply engaged in a renewed "Spinozism," a charge that, when relayed to Schelling, stung—since he knew Fichte took that description as synonymous with benighted dogmatism.[177] Because of such reaction and perhaps also because of the model he had in Fichte's popular work *Bestimmung des Menschen*, Schelling decided to remold what he came to call his identity system in the form of a dialogue, in which one of the interlocutors would represent Fichte (the character Lucien) and another himself (Bruno). The *Bruno* appeared early in 1802. The dialogue form and the doctrine of ideas the tract contains would conjure up the shade of Plato for many readers; but Fichte's initial judgment was not far wrong, at least if "Spinozism" is given a particular twist.

To understand Schelling's system, three things must be kept in mind. First, he was basically correct about the continuity with his older views. Much of the metaphysical structure of the *System des transscendentalen Idealismus* remained intact, though he positioned that structure differently— as I will explain in a moment. Second, the motivation for the system also held constant: Schelling simply wanted to get to the bottom of it all. His urge to construct such a comprehensive metaphysical framework is hardly different from that impelling most other major philosophers—and not a few scientists, who as well have had ambitions to formulate a "grand unified theory" (GUT) of everything. In Schelling's case, he, along with Fichte, desired to provided a system of knowledge that eliminated Kant's surd thesis about the thing-in-itself. Fichte attempted to found all of knowledge through an analysis of the structure of self-consciousness, and then to construct nature out of the resources of the ego. Schelling came to recognize a need to explain the origins of self-consciousness itself and to set the natural world as the ego's coequal. And in the latter, he understood his position to differ from Fichte's, who, he maintained, began with the subjective self-construction of the ego and never really got behind that construction to reach the absolute, objective condition for subjectivity itself.[178] Schelling would attempt to show that the *absolute ego* and Spinoza's *infinite substance* were identical, thus constituting an ideal realism. According to Schelling's formulation, Fichte's was an idealism of the ego, his an idealism of nature.[179]

---

177. Schelling to Wilhelm Schlegel (10 December 1801), in Schelling, *Briefe und Dokumente*, 2: 363: "Fichte is really hard to understand when he says of me that I am in pursuit of a renewed Spinozism; you know what he thinks of Spinozism and the kind of things he so names." See also Fuhrmans's discussion of the relation between Fichte and Schelling at this time, in ibid., 1:228–29.

178. Schelling, *Ueber den wahren Begriff der Naturphilosophie*, in *Schellings Werke*, 2:719, 722 [IV:85, 88]; *Darstellung meines System der Philosophie*, in *Schellings Werke*, 3:5 [IV:109].

179. Schelling, *Ueber den wahren Begriff der Naturphilosophie*, in *Schellings Werke*, 2:718 [IV:84].

The final aspect of Schelling's system that must be borne in mind—something Schelling himself did not highlight—is that the real method of justification for the postulation of an absolute was a transcendental argument. Spinoza began his *Ethica* with a statement of definitions and axioms à la Euclid, and proceeded to deduce everything therefrom. The absolute, or God, was demonstrated through an ontological argument—that is, from the definition of *substance,* as that which cannot be produced by another substance (thus a being whose essence is existence), and the definition of God (a substance of infinite attributes), one immediately deduces that God must exist and is the only substance.[180] Schelling, though he aped the *ordo geometricus* of Spinoza, employed a different method of demonstrating the necessary existence of the absolute. He used, at least initially, a transcendental argument; he recognized that without the thing-in-itself, the postulation of an absolute (with the properties he described) was the only way of explaining the possibility of an intellectual intuition and thus, ultimately, our experience of the natural world. As Schelling developed his identity theory, however, he came explicitly to assert that we could have a kind of direct, intuitive knowledge of the absolute and its properties. In response to Fichte's criticism that postulation of the absolute and knowledge of its features could only come from individual, subjective consciousness, Schelling argued that knowledge presupposed an identity between the knower and the known—a proposition he had earlier maintained; so that in the intellectual intuition, it must be that the absolute was knowing itself and that we were simply its instrument.[181]

Before commenting on the salience for biology of the newly oriented system, let me briefly sketch its major features. The system would become the weighted keel for the doctrine of archetypes, which flowed through

180. Baruch de Spinoza, *Ethica,* I, xi, in *Opera,* ed. J. Van Vloten and J. Land, 3 vols. (Hague: Martinus Nijhoff, 1890), 1:43.

181. Fichte protested that his one-time colleague had no way of postulating the absolute, except from within the boundaries of self-consciousness, and so could not vault beyond subjectivity to arrive at its starting point. Thus there was and could be no evidence for the identity system. See Fichte to Schelling (31 May 1801, sent 7 August 1801), in Schelling, *Briefe und Dokumente,* 2:341–42. Schelling's formal answer to this complaint came explicitly in his lectures in Würzburg in 1804. The lectures, though, were not published. See Schelling, *Das System der gesammten Philosophie und der Naturphilosophie insbesondere,* in *Schellings Werke,* expanded vol., 2:73 [VI, 143]: "In reason, that eternal identity is itself both the knower and the known. It is not I who recognizes this identity, but the eternal recognizes itself, and I am merely its instrument. This is why reason is reason—simply because the knowing agent is not subjective, since what is the same in it comes to recognize itself as the same and the opposition between subjectivity and objectivity is in the ultimate instance neutralized." Schelling's position here was quite similar to that of Spinoza. See, Spinoza, *Ethica,* II, xlv, in *Opera,* 1:108–9.

nineteenth-century biological science. By 1801 Schelling was ready to argue explicitly that, Fichte notwithstanding, the philosopher must step behind self-consciousness to its unconditioned condition, the *absolute*. He modeled the absolute on Spinoza's *Deus sive natura* and Plato's Good. The absolute constituted the fullness of being, yet without expressed differences. It formed the essence of every entity, so that every being yet subsisted with every other in radical identity. Indeed, the *essence* of the absolute was what Schelling meant by this state of self-identity. Though the absolute was essentially indifferent to differences, it did contain differences virtually and organically. More precisely, within the absolute, *ideas* subsisted in intimate relationship with each other. Like Plato's Forms, Schelling's ideas were spheres of individual content, each with an *essence* (by which it was organically related in Leibnizian fashion to all other ideas and identical with them) and *form* (which implicitly provided its individuality). Concerning the ideas, one could not say that any one was empirically real or merely ideal, existing or merely possible—the ideas stood indifferent to these designations. Simultaneously, therefore, they were real, finite beings as well as the universal concepts of those beings. As such, the ideas served as the eternal archetypes for individuals existing within the finite world of experience.

Even after this epitome of Schelling's theory of the absolute, the reader will undoubtedly plead for more light. The description can perhaps be made clearer (even if conundra lie just below the surface) by considering the relationship between concepts and their real referents. The universal *concept* (*Begriff*) of dog refers indifferently to Fido, Rex, and Max, individuals that have distinguishing traits (e.g., Max is black-and-white spotted; Fido is a golden tan; Rex, alas, died several years ago). The concept includes indifferently, as it were, the differences separating individuals—that is, the concept of dog is applicable to the black-and-white animal just as to the tan and to the dead one; in this sense, the concept includes the differences, but without noting them. This same relationship will obtain if we speak of a concept of this particular individual. The concept of Max is empirically realized throughout a period of time; and during that time, it exhibits many distinguishing and sometimes opposite traits. The concept of Max, then, will include indifferently its many actual instantiations over time. Now the *idea* (*Idee*) of dog, Schelling would say, encompasses or refers to both the concept of dog and individual dogs indifferently. It thus includes the infinite concept and the finite individuals, but without distinction, only as pure identity—namely, dog. Now if one ascends to consider the idea of all such ideas (of dog, cat, sea, mountain, etc.), it too must form a pure identity that includes all of its differences indifferently. The idea of all ideas is what Schelling means by the absolute, whose essence is sheer identity.[182]

In the realm of empirical time, where individual objects of nature appear to individual egos—though, for Schelling, such objects arrive in individual consciousness only as sensible intuitions—these objects are caught up in a causal web of other finite objects. This dog Max was begotten by its parents; it breathed oxygen; it ate, assimilating other objects; it bit the postman, altering another object; and it reproduced its kind. Max became empirically real through the antecedent causal situations of past time, and his existence will alter causal relationships down to the end of time. Max thus was actually caused by past events and is the potential cause of future events. All of these causal relations, actual and potential, will be mirrored in the full concept of that animal—comparable to the way in which Leibniz held apparently external causal relations to exist only within the monad, which had no windows open to other monads.[183] One might extend the Leibnizian model further—as Schelling does—and refer to the concept of an individual entity as its "soul."[184] Now both the actual dog Max and his universal concept (or soul), his finite existence and his infinite ideal representation, will consequently be implicated, respectively, with all other real entities causally and with all other concepts of those entities ideally. Thus "every single thing," according to this view, "represents the universe after its own fashion."[185]

In Schelling's theory, the ideas that constitute the absolute are essentially identified with each other. Yet their differences (e.g., the idea of a dog as compared to that of a human being) are not simply dissolved into complete unity. It is a hard saying, but the ideas subsist within the absolute, such that differences among them are preserved virtually. Thus, in some sense, the dog Max has four inclusive modes of existence: as an empirical creature within time; as an individual within its concept; as a concept within its idea; and as an idea within the absolute. But how to conceive the finite as part of the infinite? Schelling employs the model of the organic in an attempt to convey how the finite can exist within the absolute, not as a separate entity but virtually. "As in the case of an organic part [of an entire organism], when it is considered as real, it is not understood as a separate in-

182. Schelling, *Bruno oder über das göttliche und natürliche Prinzip der Dinge*, in *Schellings Werke*, 3:139 [IV:243]: ". . . the idea of all ideas is the only object of all philosophy."

183. Schelling draws the explicit parallel with Leibniz's concept of the monad in ibid., 3:212–17 [IV:316–22].

184. Ibid., 3:159 [IV:263]: "the relative unity of this finite thing, that is, the concept which is related directly only to this thing, serves as the soul of this thing."

185. Ibid., 3:163 [IV:267].

dividual within the organism, but when considered ideally, or for itself, then we do regard it an individual—so also for the finite individual as it exists in the absolute."[186]

Even if the internal relationships that Schelling posited in the absolute stood completely clear, there would still be the problem of how finite individuals come to exist in or to emanate out of the absolute. There is a definite sense, in Schelling's conception, in which the finite world appears only as an illusion. The true unity and integrity of the absolute become, as it were, spread out through finite perception into independent objects existing in space and time.[187] The concept of a thing, its soul, and its finite body thus do not really exit in themselves, but only "in the all-blessed nature, wherein possibility is not separated from reality, nor thought from being, in the archetype [*Urbild*], which is uncreated and truly immortal."[188]

Schelling's conception of the absolute stands intimately with Spinoza's idea of God, the being which, in the latter's system, constitutes the only substance: "Whatever is, is in God, and without God nothing is able to be conceived."[189] The truth of this proposition for both philosophers, though, could only be arrived at through a kind of knowledge beyond ordinary understanding. Thus, for both, therefore, our perception of the natural world, with its finite, temporal individuals, was entirely misleading; it was based, in Spinoza's terms, on "confused ideas" [*ideae confusae*].[190] It still remains unclear, however, even if we accept their respective accounts, why illusion and confusion should arise out of the absolute—thus disguising its true colors. This kind of objection, of course, has been mounted against every philosophical postulation of absolute being from Parmenides on, and does not fail to tell, as well, against more orthodox conceptions of God as an infinite and all perfect being. Perhaps, then, it was not incumbent on Spinoza or Schelling to resolve the difficulty.

The Beauty of Ideas

Schelling's *System des transscendentalen Idealismus* concludes with an argument that makes art the organon of philosophy. In producing the work of

186. Ibid., 3:146 [IV:250].

187. Ibid., 3:147 [IV:251]: "Since the true universe of an infinite fullness has nothing in it external to anything else, nothing separated, but everything absolutely one and in one another, that universe must spread itself out in the image [*Abbild*, i.e., the finite world] into a limitless time, so that the unity of possibility and reality, which in the organic body is timeless, must be drawn asunder in reflection so as to develop in a time that can have neither beginning nor end."

188. Ibid., 3:179 [IV:283].

189. Spinoza, *Ethica*, I, xv, in *Opera*, 1:47.

190. Ibid., II, xxviii, 1:96–97.

art, he contended, the unconscious, absolute ego, which functioned through artistic genius, melted into the conscious strivings of the artistic self; and thus an ineffable and determinate nature expressed itself consciously and freely. It was in this sense that he could offer the aesthetic intuition as a kind of substitute for the intellectual intuition, which latter had to remain shrouded in the unconscious operations of the absolute self. The relationship of art to philosophy and science, however, subtly changed as Schelling made the turn to identity theory. Three features of the new relationship stood out. First, the art object and the artist's intuition—insofar as they *stemmed from unconscious nature*—would no longer be regarded as the principal instruments of his system; rather the conscious recapitulation by the philosopher of the intellectual intuition would perform that service.[191] Schelling now thought the philosopher capable of this, since a systematic thinker no longer had to be considered a subjective consciousness trying to penetrate an absolute self through transcendental argument alone; rather, as he expressly put it in his Würzburg lectures in 1804, in the philosopher's intellectual intuition, it was really the absolute becoming aware of itself through the medium of a finite individual.[192] A fully conscious intellectual intuition, identical with what the absolute worked on itself, would be the prize of the new identity philosophy. This readjustment may reflect not only Schelling's enthusiasm for the orientation provided by Spinoza but also his competition with Friedrich Schlegel. Schlegel began lecturing in Jena as *Privatdozent* in fall of 1800 and professed a transcendentalism that indeed made the deepest philosophy a prerational kind of poetic thought that resisted subsumption into an absolute being.[193] Schelling re-

---

191. Schelling, *Ueber den wahren Begriff der Naturphilosophie*, in *Schellings Werke*, 2:223 [IV:89]: "the system of art will be subordinated" to the real-idealism.

192. Schelling, *System der gesammten Philosophie und der Naturphilosophie insbesondere*, in *Schellings Werke*, expanded vol., 2:73 [VI:143]: "In reason, that eternal identity is itself both the knower and the known. It is not I who recognizes this identity, but the eternal recognizes itself, and I am merely its instrument." Later on, in his *Philosophische Untersuchungen über das Wesen der menschlichen Freiheit* (Philosophical investigations on the essence of human freedom, 1809), in *Schellings Werke*, 4:295 [VIII:403], he came to understand the absolute, or God, as also organically developing. This metaphysical conception gives greater account of the possibility of finitude as a function of the absolute. He argued: "If the creation has an end-point, then why is this not directly achieved, why is everything not perfect from the very beginning? There is no other answer than what has been given: since God is 'Life,' not merely a being. Every life, however, has a destiny and is subject to suffering and becoming. . . . Being can only perceive itself in becoming." (I am indebted to Frederick Gregory for pointing out this passage to me.)

193. Schlegel habilitated in philosophy at Jena and began lecturing as *Privatdozent* in the winter term of 1800–1. In his unpublished lectures (in the transcription of a student), he, like Schelling, attempted to unite the idealism of Fichte with the realism of Spinoza. From a distance, their respective constructions appear quite similar, and there seems much that Schlegel may have

garded Schlegel's efforts as the merest "poetic and philosophic dilettantism."[194]

The second adjustment among the disciplines of art, philosophy, and science occurred by way of an explicit specification of the objects of these cognitions. According to the identity theory, ideas subsisting in the absolute were to serve as the exclusive standards or archetypes for both beauty and truth.[195] Truth would be achieved insofar as the images in our understanding came to adequate "things as they are preformed in the archetypal understanding [urbildlichen Verstande]." And beauty would be predicated of archetypes particularly insofar as they united "the universal and the particular, the species [Gattung] and the individual."[196] This meant that, in light of the objects of their cognitions, philosophy and art were ultimately the same.[197] Archetypes in themselves were adorned with truth as well. In their temporal incarnations, however, they might appear less than that,

---

borrowed from Schelling. There are, though, significant differences. One is that Schelling emphasized the rational and conscious character of philosophical thinking, and he conceived of the absolute as fully rational in character. Schlegel, more like the Schelling of the System des transscendentalen Idealismus, argued that philosophy must be guided by a prerational, aesthetic intuition into the nature of the absolute, which itself had attributes of thought and extension, but whose essence was neither. Further, and most importantly, for Schlegel the absolute (or Divinity, as he called it) was in the process of becoming; it was not complete. See Friedrich Schlegel, Transcendentalphilosophie, ed. Michael Elsässer (Hamburg: Felix Meiner Verlag, 1991), p. 79. Leonard Wessel discusses the metaphysical foundations of Schlegel's Romantic aesthetics, with special reference to the Transcendentalphilosophie, in "The Antinomic Structure of Friedrich Schlegel's 'Romanticism,'" Studies in Romanticism 12 (1973): 648–69.

194. When Schelling returned to Jena from Bamberg with Caroline, he found Friedrich Schlegel in the midst of his lecture series on Transcendentalphilosophie, with some sixty to eighty students. The lectures were more poetic and enthusiastic than analytic and penetrating. The students were disappointed and Schlegel ceased lecturing in the spring term of 1801. Schelling wrote Fichte (31 October 1800), "I find it impossible to watch as the well-laid foundation should be destroyed in such a manner, and stand by as the poetic and philosophic dilettantism of the Schlegel circle spreads among the students instead of an authentic scientific spirit, the foundations of which still remain here. . . . After four hours of lectures which I have held, he was already dead in the water and now is buried." See Schelling to Fichte (31 October 1800), in Schelling, Briefe und Dokumente, 2:283.

195. In lectures on Philosophie der Kunst—given in Jena in winter term 1802–3 and repeated in Würzburg in 1803–4—Schelling expressed it this way: "Just as in philosophy, the absolute is the archetype [Urbild] of truth, so in art, the absolute is the archetype of beauty. . . . The absolute is indeed one, but this one is absolutely intuited in particular forms, so that the absolute is not relinquished therein—this is the idea. The same is true for art. Thus art intuits primal beauty [Urschöne] only in ideas as particular forms, of which each is for itself divine and absolute. The difference is this, that in philosophy the ideas are intuited as they are in themselves, while in art they are intuited as real." See Schelling, Philosophie der Kunst, in Schellings Werke, 3:390 [V:370].

196. Schelling, Bruno, in Schellings Werke, 3:116 [IV:220], 3:139 [IV:243].

197. Ibid., 3:123 [IV:227].

since they would often be inadequately represented by finite minds of dimmer discrimination and understanding.

To come to the third adjustment made through the identity theory—this concerns precisely the nature of genius in art and philosophy (science, being included in the latter). The object of both artistic expression and philosophical or scientific analysis is the idea, which incorporates both the universal concept and the individual. Further, since every idea is organically related to every other in the absolute, that poet or scientist would achieve a greater measure of perfection (i.e., be a greater genius) to the degree he or she could capture the infinite ways one idea or archetype would be related to all others. Philosophic and scientific genius would, however, yet differ from aesthetic genius. The artistic genius would be seized by an idea or archetype that would act through the individual, while the individual, like Plato's poet, would remain unconscious of exactly the nature of his inspiration. The philosopher, on the other hand, and the philosophically minded scientist would "exercise the same divine activities, though in an inward fashion, that the artist exercised externally and without knowing it."[198] The highest artistic endeavor and the highest philosophical and scientific effort would, thus, be directed toward the same objects, the archetypes, the eternal standards of beauty and truth. The philosopher and scientist, though, would be consciously guided by the very ideas that the artist could only unconsciously follow.

In his lectures on *Philosophie der Kunst* (Philosophy of art) for the winter term 1802–3, his last in Jena, Schelling developed these aesthetic ideas in reference to specific kinds of art. Reaching back to notions that he first developed in his master's thesis at Tübingen, he argued that the first and most fundamental expression of true poetry occurred through the medium of mythology. In Homer and the Greek dramatists, the absolute comes particularized in the form of various myths, which embody the ideas resident in the absolute and differently expressed in the natural world. In Greek poetry, these ideas come forth from Olympus wielding trident and bow. "Indeed, the gods of every mythology are nothing else than the ideas of philosophy intuited as objective and real."[199] The mythology of the Greeks, according to Schelling, came up from the society as a whole, even if spilling out of the mouth of a particular poet. Having passed out of that age of wholeness, the modern artist stands as an individual, who must, if of the requisite genius, formulate a mythology for his or her own time: Dante and Shakespeare had

198. Ibid., 3:127 [IV:231].
199. Schelling, *Philosophie der Kunst*, in *Schellings Werke*, 3:390 [V:370].

to create myths out of materials of their societies that captured the absolute in a particular form, making its beauty and sublimity accessible to the period. In Schelling's own age, the achievement nonpareil came from the pen of Goethe:

> Insofar as one can judge Goethe's *Faust* from its [published] fragment, which is all we presently have, this poem is nothing else than the most intrinsic, purest essence of our age: matter and form are produced from that which the entire period contains and, indeed, from that with which the times were pregnant and still are. Thus we may call it a truthful mythological poem.[200]

Since "originality" marks the productions of modern poetic genius, it ought to be possible to develop a philosophy (or "speculative physics") that wears the guise of mythology.[201] And perhaps, insofar as the poetic life cannot be divorced from the philosophical endeavor, philosophy ought to become mythological. This, at least, was the rationale for the drift of Schelling's later philosophy into the murky currents of religious mythology.

The Impact of Schelling's Identity Philosophy

Schelling's theory of absolute identity left many knots still tightly constricting the arteries of the system. There was, as mentioned, the problem of the derivation of the finite from the absolute. And Fichte protested that his one-time colleague had no way of postulating the absolute, except from within the boundaries of self-consciousness, and so could not vault beyond subjectivity to arrive at its starting point. To Fichte's mind, therefore, there was and could be no evidence for the identity system.[202] Despite, however, formidable problems such as this, Schelling's new departure did garner admirers. Friedrich Schlegel, hardly an unbiased critic, grudgingly admitted that the *Bruno* could not be denied praise, though he did prefer the real Bruno to the "weak shadow" that appeared in Schelling's work.[203] Hegel, who had moved to Jena in 1800 and lived next door to Schelling, not surprisingly perhaps, celebrated the identity system in his book *Differenz des*

---

200. Ibid., 3:466 [V:446].
201. Ibid.
202. Fichte to Schelling (31 May 1801, sent 7 August 1801), in Schelling, *Briefe und Dokumente*, 2:341–42.
203. Friedrich Schlegel to Wilhelm Schlegel (16 September 1802), in Xavier Tilliette, ed., *Schelling im Spiegel seiner Zeitgenossen*, 6 vols. (Torino: Bottega D'Erasmo, 1974–81), 1:99. He wrote the day before to Schleiermacher (15 September 1802) saying that after Schelling's *Darstellung*, the *Bruno* had to please him. See ibid., 1:94.

Fichte'schen und Schelling'schen Systems der Philosophie (1801). Hegel seems genuinely to have thought very highly of Schelling's philosophical qualities (though he objected to the geometrical method of demonstration); indeed, Schelling's identity theory provided the grounding for the development of Hegel's own philosophical system. And then there was Goethe. Goethe himself had become convinced, through the reading of Kant and his discussions with Schiller, that rational ideas stood guard at the gates of consciousness and that any direct messages from nature could not legally pass by these sentries. He yet felt in his heart that the scientist could come to know the deep structures of nature, that objective science could have a purchase on reality. Schelling's identity philosophy showed him that consciousness might have ideas objectively representing the deep structures of nature and that those ideas were actually productive of the real features of organisms (see part 3).

The further scientific impact of this aspect of Schelling's thought was considerable. I will mention here only two crucial areas: his organic monism, in which the mental and physical aspects of natural beings were understood to be identical; and his conception of absolute ideas. The first notion drew biologists and physicians to his side, even deflecting the great Johann Christian Reil from his initial commitment to materialism. For biologists and physicians such as Reil, Ignaz Döllinger, and Karl Friedrich Burdach, Schelling's organicism secured a position from which biologists could reject materialism but without requiring them to embrace a dualistic vitalism.[204] Moreover, from this vantage they could perceive purpose in nature without, however, being forced to deliver it up immediately to the Most High. And for this latter benefice Goethe was especially grateful; and from his hand it was passed, later in the century, to Darwin and Haeckel.

Schelling's metaphysics of absolute ideas, that second important area, launched a signal theory that sailed through the nineteenth century—the theory of the archetype, which provided explanatory power for morphology during the period. A few important connections might briefly be mentioned here. Goethe's conception of a fundamental form—or archetype—of plants and animals received both support from Schelling's construction of absolute ideas and inspiration to a fuller expression of that theory. The theory of the archetype was elaborated by Döllinger, Lorenz Oken, Carl Gustav Carus, and Karl Ernst von Baer, and was transmitted to Joseph Henry Green through Karl Wilhelm Solger, a fervent disciple of Schelling.

---

204. For a brief but lucid discussion of Schelling's impact on the medical community, see Thomas Broman, "University Reform in Medical Thought at the End of the Eighteenth Century," *Osiris* 2nd ser., 5 (1989): 36–53.

Green, in turn, brought Richard Owen in England to the notion of an archetype, and he developed it into a grand theory, which, after suitable transmutation, Darwin and Haeckel adopted. This is but a bare sketch of the bloodlines that flowed through morphological theory in the nineteenth century. It will be fleshed out in subsequent chapters and in the epilogue. I wish here simply to indicate that Schelling gave the theory of the archetype the pulse that sent it coursing through the scientific and intellectual channels of the nineteenth century.

Can one seriously be an idealist in the Schellingian mode today? In philosophical circles one can seriously be anything, of course. But idealism does seem to have some rather spiky objections to overcome. To ask Eschenmayer's questions: Am I responsible for the tree in the garden that was planted fifty years before my birth; and, Should I be able to claim authorship of the complete works of Plato? To put the questions seems immediately to dismiss the idealist response. But, of course, Schelling's contention was not that this particular ego, this historically bound individual secretly penned the *Dialogues*. It was, rather, that both the individual ego and nature had a deeper, common source—a banal claim that a realistic metaphysics might well allow. Perhaps from this deeper source—call it the absolute ego or just the absolute—might flow both necessary laws of nature and the complementary structures of the empirical ego. Contingent facts appear to be the final obstacle. But is there any significant difference between holding that the tree in the garden arose from causal factors external to mind, residing in nature, and holding that it arose out of the relations of ideas lying beyond the individual ego, residing in something called the absolute? It hardly seems easier to believe the world is really a ball of mathematical strings that reveals itself to our consciousness as natural objects of ordinary experience than to believe it is an organic structure of ideas that reveals itself in comparable fashion. Idealism cannot be defeated, only forgotten. And intellectuals of the nineteenth century had not yet forgotten it, at least not all of it.

Chapter 4

DENOUEMENT:
FAREWELL TO JENA

There the magic veil of poetry's youth / Lovingly wound itself around the truth.

—Schiller, "Die Götter Griechenlandes"

At the turn of the eighteenth century, with the catalysts of the French and Kantian revolutions, everything seemed to change. A group gathered in Jena—poets, novelists, philosophers, historians, literary critics—individuals who self-consciously attempted to create things anew. Their very style of life announced that ancient and corroded chains of philosophic, moral, and scientific assumptions had been broken. They criticized everything and everyone—everyone, that is, except Goethe. Their ideas flashed with originality, cracking encrusted modes of thought like stale bread. They were very young, and therefore much in love, much in despair, but filled with élan for the possibilities of life. An older Henrik Steffens, in his quiet reveries back in Denmark, left a poignant depiction of those years that so transformed his life:

What this happy time in Jena principally meant was unceasing effort and seriousness that prevailed in everyone. The conviction vibrated through the group that one had to combat opponents on their own ground, that empty abstractions or spirited turns of phrase would not suffice, that a fight would be insignificant if it were not carried on with insight and knowledge. Those who reached the pinnacle of this period had distinguished themselves with their adroit works. Like Lessing, who strove against the received literature of his time, they established a capital and a significant redoubt in the literature of the world. These were men who knew what they wanted, who had their own fixed goal, which they incessantly pursued. . . . It was a time of productivity, of youth, that detected in all directions traces of the universal, cohesive spirit. It was a bubbling, yes joyous life, not the knotted spasms of death. Some faulted this confederation, especially the brothers Schlegel, because they hunted after paradoxes. But must not everything that arises out of greatness seem strange, incomprehensible, and paradoxical to those who are satisfied with details in the fractured shards of life? . . . When I lose myself in those times, I recognize an unusual similarity between them and my quiet life in Roskilde. What possessed me and governed me at that time, I hope I now have succeeded in making my own possession. So it might be claimed as the ultimate goal of all reflection: that one recognizes those traits at their beginning point, and in the quiet region of an originally sound mind one rediscovers in oneself their deeper significance.[1]

But those magical times of the hot joy of youth, of the new spirit of poetry, philosophy, and science, those times did end. And in the usual ways.

At the very beginning of spring 1801, at age twenty-nine, Friedrich Hardenberg quietly took the step he had been poetically anticipating for some time. He died from a lingering lung infection, likely tuberculosis. Friedrich Schlegel informed his brother of the tragic event:

> Yesterday I returned from Weissenfels, where the day before, on the afternoon of the 25th, I saw Hardenberg die. . . . It is certain that he had no suspicion of his death, and generally one would hardly believe it possible to die so gently and beautifully. He was, as long as I saw him, of an indescribable cheerfulness, and although the great weakness hindered him considerably from talking during the last days, he yet took a praiseworthy part in everything, and I treasure it over everything to have seen him.— Share this news with Tieck and Schleiermacher. The particulars will remain reserved for oral communication.[2]

Wilhelm Schlegel received this letter in Berlin, whither he had just moved. He presumably did tell Schleiermacher, but Friedrich Schlegel did not leave it to his brother and chance. He also wrote his friend with the sad news.[3] The report of the poet's death came to the preacher just as his own fame had come to life, in the aftermath of his *Über die Religion*. In April 1804 Schleiermacher received a call from Halle to become ordinary professor in the theology faculty and preacher to the university community. Henrik Steffens, late of Jena and new professor of natural philosophy and minerals, reported that Schleiermacher's parishioners would leave their church "with the conviction of the nothingness of all earthly relations,

1. Henrik Steffens, *Was Ich Erlebte*, 10 vols. (Breslau: Josef Max, 1840–44), 4:136–40.

2. Friedrich Schlegel to Wilhelm Schlegel (27 March 1801), in Friedrich von Hardenberg, *Novalis Schriften*, ed. Paul Kluckhohn and Richard Samuel, 2nd ed., 5 vols. (Stuttgart: Kohlhammer, 1960–75), 4:680–81.

3. Friedrich Schlegel to Schleiermacher (30? March 1801), in ibid., 4:682.

even the greatest, as his divine voice moved against them."[4] Schleiermacher's stay in Halle, however, was short-lived, for in fall of 1806 the university closed as Prussia sank back into war with Napoléon. He returned to Berlin and in 1810 received an appointment as professor in the theology faculty of the newly founded university. Steffens thought there was no one who so "elevated and ruled the mind of the city's inhabitants, and through all ranks spread a national, a religious, and a deeply spiritual attitude."[5] In Berlin, Schleiermacher wrote his famous *Christliche Glaube* (1821–22), and lectured in competition with Hegel, who taught in the philosophy faculty. His influence on later theology was as profound as Hegel's in philosophy. He died at age sixty-six in 1834.

Wilhelm Schlegel stayed in Berlin for a few years, lecturing privately on aesthetics and beginning his studies of Sanskrit. After his divorce from Caroline was officially declared, he considered other venues. In 1804 he moved to Switzerland and became companion to Madame de Staël and tutor to her children. He had a hand in her classic study of German culture, her *De l'Allemagne* (1813),[6] to which Heinrich Heine's *Die romantische Schule* (1835) was a strongly reactive antidote.[7] After her death in 1817, Schlegel moved to the new university in Bonn and there lectured on aesthetics and linguistics. He had Heine among his students, which did his later reputation no good at all. The young poet's festering anti-Catholicism infected his critical judgment of the Schlegel brothers. Only because his wicked remarks are so delicious can they yet be stomached—as for instance, when he compared Wilhelm's later marriage to the daughter of Heinrich Paulus with that of Osiris to Isis. After the god Typhon had torn Osiris apart, his wife had to piece him together again. Alas, a part was missing, and she had to substitute a wooden shaft for his most vital member. Heine suggested that Schlegel's prosthetic intellect had a comparable utility. Wilhelm Schlegel died on 12 May 1845 at age seventy-eight.

In April 1801 Friedrich Schlegel followed his brother to Berlin, and then in summer of the next year he relocated to Paris, where he began an intensive study of Persian and Sanskrit, lectured to private groups, and, on

4. Steffens, *Was Ich Erlebte*, 5:149.
5. Ibid., 6:271.
6. Madame de Staël Holstein, *De l'Allemagne*, 3 vols. (Paris: H. Nicolle, 1810). The book ran into trouble with the French censors and was published first by John Murray in London (1813).
7. Heinrich Heine's *Die romantische Schule*, in *Sämtliche Schriften*, ed. Klaus Briegleb, 3rd ed., 6 vols. (Munich: Deutscher Taschenbuch Verlag, 1997), 3:357–504. Heine wrote *Die romantische Schule* as a response to de Staël's *De l'Allemagne*. He complained particularly about the ultramontanism and enthusiasm for things medieval that he thought Schlegel encouraged in de Staël's representation.

6 April 1804, married his Lucinde, Dorothea Veit. Four years later, after having moved to Cologne, both were received into the Catholic Church—a "sign of the times," according to Goethe, one that flagged the conservative turn taken by the older Romantics.[8] In April 1808, on the eve of Napoléon's invasion, they resettled in Vienna. During the conflict Schlegel worked for the general staff, putting out an army newspaper. After the peace of Schönbrunn was signed in October 1809, he returned to his studies of Austrian history and world literature—which even Heine grudgingly admired. In 1813 he contributed—along with Count von Stein, Wilhelm von Humboldt, and Varnhagen von Ense—to the working out of the terms of the German Confederation during the Congress of Vienna. While on a lecture tour in Dresden, Schlegel suffered a stroke and died on 12 January 1829, age fifty-six. Dorothea moved to Frankfurt to be with her son Philipp, a well-known religious painter, and his family. She lived to be seventy-four, dying in 1839.

On 26 June 1803, a month after her divorce, Caroline changed her name for the third time (actually the fourth, if you count the "Madame Cranz" on the baptismal certificate for her son of the brief liaison), becoming Frau Schelling. After all the departures from Jena and the lingering rancor about their relationship, little remained for her and her husband in the once-beloved city (except, perhaps, Hegel and Goethe). Though social darkness prevailed in the region, Schelling's reputation as the leading light of German philosophy could not have been brighter. When he received a call to the reestablished university in Würzburg that fall—and with the knowledge that many acquaintances and enthusiasts for his ideas would be there—the decision was easily made to accept. The physicians Marcus (in Bamberg) and Röschlaub (in Landshut) were close by, and his new disciple Ignaz Döllinger (1770–1841) was at Würzburg. Moreover, his longtime friend, the theologian Paulus, had been called from Jena as well, and initially provided housing for the Schellings. The following year Niethammer also arrived in the theology faculty, and thus a bit of the old Jena intellectual life reconstituted itself in Würzburg. At this time Schelling also began his acquaintance with Lorenz Oken, who would become a powerful influence on the spread into biology of a certain kind of *Naturphilosophie* (see chapter 11). During the years in this Bavarian city, 1803 to 1806, Schelling further developed his identity theory, lecturing on *Das System der gesammten Philosophie und Naturphilosophie inbesondere* (The System of all of philosophy and nature philosophy in particular), the philosophy of art, and

8. Goethe to Karl Reinhard (22 June 1808), *Goethes Briefe* (Hamburger Ausgabe), ed. Karl Robert Mandelkow, 4th ed., 4 vols. (Munich: C. H. Beck, 1988), 3:78.

elementary philosophy. He also embarked on studies of religion, which would mark the beginning of what is usually called his "later philosophy"— a departure that soon would fly toward the mystical empyrean. Schelling's courses were attended even by the professors of the university, and Caroline reported to her friend Luise Gotter that her husband's lectures were "the talk of the day."[9] Schelling himself, while very pleased with the initial response to his presence, voiced a complaint to Hegel, familiar enough to many professors. "The spirit of study," he remarked, "is still far from that which obtains in Jena, and they find philosophy still completely unintelligible."[10] Undoubtedly his own philosophy proved a special challenge to students and faculty alike.

The environment at Würzburg, despite all its promise, turned quickly hostile. Part of the strife began with the women. The wives of Paulus, Niethammer, and others of the faculty took a strong dislike to Caroline. Karoline Paulus wrote Charlotte Schiller, who could not abide Caroline, that "Schelling has on this occasion shown that he is a docile husband and that the evil influence of this Madame Lucifer has had a powerful effect on him."[11] The troubles, which spread to the husbands, became compounded by the energetic battles fought over Schelling's identity philosophy and his turn to religion. Schelling also found himself in constant wrangles with the Catholic administration of the university, especially over questions of theology.

An offer of appointment to the Academy of Sciences in Munich in 1806 came as a way out for the renowned philosopher. The next year Wilhelm Schlegel, in the company of Madame de Staël, visited the Schellings in Munich, obviously demonstrating a lingering affection for both his former wife and his friend—an affection Friedrich Schlegel thought besmirched his brother's noble nature.[12] In 1808 the Bavarian king named Schelling general secretary of the Academy of Fine Arts in the city. The relative tranquility of his time in Munich, though, did not last. At the beginning of September 1809, Caroline contracted a case of dysentery from which she could not recover. Schelling wrote her friend Luise Gotter to describe the final moments:

9. Caroline Schelling to Luise Gotter (4 January 1804), in *Caroline: Briefe aus der Frühromantik*, ed. Georg Waitz and Erich Schmidt, 2 vols. (Leipzig: Insel Verlag, 1913), 2:380.

10. Schelling to Hegel (3 March 1804), in F. W. J. Schelling, *Briefe und Dokumente*, ed. Horst Fuhrmans, 3 vols. to date (Bonn: Bouvier Verlag, 1962–), 3:56.

11. Karoline Paulus to Charlotte Schiller (11 March 1804), quoted by Gisela Dischner, in her *Caroline und der Jenaer Kreis* (Berlin: Verlag Klaus Wagenbach, 1979), p. 159.

12. Friedrich Schlegel to Wilhelm Schlegel (6 January 1808), in August Wilhelm Schlegel, *Briefe von und an August Wilhelm Schlegel*, ed. Josef Körner, 2 vols. (Zurich: Amalthea, 1930), 1:493.

That last evening she felt light and joyous. The whole beauty of her lovely soul opened one more time. The always beautiful sound of her voice became as music. Her soul seemed already free from her body and only hovered over the shell, which it would soon leave. She departed on the morning of 7 September, gently and without struggle. Even in death she did not relinquish grace. She made a lovely turn of her head, and, with an expression of cheerfulness and the most wonderfully peaceful look on her face, she died.[13]

With Caroline's death, the Jena period of Romantic philosophy evanesced.

Schelling turned decidedly in that year of 1809 to a Christian Gnosticism of a kind to convince Heine that the Romantic school represented a mystical retreat from reason to dwell in the incense-clouded bastions of Rome. Schelling's personal life, though, preserved ties to a happier time. During the next few years, he found consolation in Pauline Gotter, the daughter of Caroline's friend Luise Gotter. Pauline was fourteen years Schelling's junior and also a young friend of Goethe.[14] They married in 1812. In 1821 Schelling and his family retired to Erlangen, where he wrote (without publishing), cultivated a growing circle of friends, and was invited to lecture at the university. He returned to Munich in 1827 to take the chair of philosophy at the university and to marshal forces against his one-time colleague Hegel. Hegel's influence at the Prussian university of Berlin continued to grow even after his death in 1831. To stem his impact, university officials in 1841 called Schelling, whom they deemed a more conservative force, to the chair once held by his friend. His lectures reached the eager ears of the likes of Friedrich Engels, Jakob Burckhardt, and Søren Kierkegaard. But he soon disappointed his auditors; the youthful figure of fire who had led his followers to touch the core of reason, that figure existed no more. Schelling retired from the university in 1846. On 20 August 1854, he died while attempting to recover his health in Switzerland.

13. Friedrich Schelling to Luise Gotter (24 September 1809), in *Caroline: Briefe aus der Frühromantik*, 2:569–70.

14. Schelling described Pauline to his brother Karl (10 July 1812): "Beginning from the outside, it is difficult to describe Pauline. Perhaps you still have a picture of her, since you saw her once as a child with me in Weimar. She is twenty-three years old, tall, slim, and seems almost more a work of fantasy than of nature. Without being a beauty, she has a loveliness in her demeanor, a loving creature, that wins all hearts for herself. She is tender and of delicate health, but completely free from all female illnesses. She has healthy juices, good color, and an inextinguishable and not easily disturbed cheerfulness." See Gustav Plitt, *Aus Schellings Leben. In Briefen*, 3 vols. (Leipzig: S. Hirzel, 1869–70), 2:322.

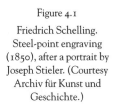

Figure 4.1
Friedrich Schelling.
Steel-point engraving
(1850), after a portrait by
Joseph Stieler. (Courtesy
Archiv für Kunst und
Geschichte.)

The Meaning of Romanticism

Friedrich Schlegel initially employed the term *romantisch* to designate an organic phenomenon that came to birth through the sympoetizing and symphilosophizing—to use only two of his neologisms—of friends in Berlin and Jena during the last years of the eighteenth century. If we gaze back, say from about 1809—marked as it is by the death of Caroline and by Schelling's turn from the philosophic and scientific occupations of his youth—we can take the measure of the several features of this developing conception of Romanticism. That conception prominently displayed two faces connected to one body, thus making it appear a rather Romantic creature from the beginning. To be a Romantic was to think and act in a certain aesthetic and philosophical mode. The aesthetic considerations stemmed from Kant's *Critique of Judgment,* Schiller's essays, the novels and poetry of Novalis and Goethe, and the critical studies of Greek literature and modern poetry undertaken by the brothers Schlegel—just to mention the more prominent sources. The philosophical origins of Romanticism also reached directly back to Kant and became formed through the transcendental structures developed by his erstwhile disciple Fichte. Neither Kant nor Fichte, of course, could properly be considered a Romantic thinker; but without

them, that particular organism would never have been born. Indeed, the
first use of the term *romantisch* by a member of the movement radiated
waves of Fichtean transcendentalism. In a fragment from an unpublished
study (1798), Friedrich Hardenberg wrote:

> The world must be romanticized. In this way one finds again its original
> meaning. Romanticizing is nothing other than a qualitative potentializ-
> ing. The lower self becomes identified with a better self in this operation.
> Thus we ourselves are such a qualitative potentialized series. This opera-
> tion is yet completely unknown. Insofar as I give the common an ele-
> vated meaning, the usual a secret perspective, the known the value of the
> unknown, the finite an infinite appearance—I thus romanticize.[15]

This kind of transcendental romanticizing elevated the ordinary events of
life and thought. It occurred, to be sure, in Hardenberg's poetry, as well as in
Schleiermacher's religious writing, Friedrich Schlegel's critical and philosophical work, and, of course, Schelling's idealism.

Yet, as I have argued, it would be a mistake to perceive Romanticism as
arising simply from the stitching together of literary and philosophical
parts. This creature had a heart that pulsed to the changing friendships and
loves within the Romantic circle—that magical circle of friends. Their
very personal experiences with each other made the being come alive.
Their initial enthusiasm for the French Revolution and the freedom it exemplified gave their often very abstract philosophy a tremendous urgency.
The longings and desires, the loves and passions, the sorrows and hatreds
that swirled around Caroline alone would have given life to the most anemic of ideas. Certainly many of the concepts that received ghostly formulation and then apotheosized—concepts like freedom, love, even the
absolute—drew their driving energy from those electrified personal relationships. Surely this must be the secret of how something as initially forbidding as Schelling's *System des transscendentalen Idealismus* magnetically
drew a generation to its pages.

Out of these intellectual and personal interactions came a mode of
thought that emphasized creative becoming, development, and self-realization. Romantic poetry was conceived to be modern, progressive poetry, an
as-yet-unfinished aesthetic effort to touch the deeper sources of reality, to
reveal them in witty, ironic, and striking numbers. Schelling, most perspicuously, regarded the aesthetic conception of reality as the complement to
the philosophical and the scientific. The artist of genius revealed in beauty

15. Friedrich von Hardenberg, "Poeticismen," in *Novalis Schriften*, 2:545 (no. 105).

the same deep structures the analytic thinker of genius experientially and rationally determined. In the poet's imagination, where Schiller's "Gods of Greece" did dwell,

> There the magic veil of poetry's youth
> Lovingly wound itself around the truth.[16]

Romantic philosophers, scientists, and poets—given the idealistic-realistic metaphysics that grounded their conceptions—found in nature the self's other kingdom. Novalis, for instance, after having read Schelling's *Ideen*, began a novel—*Die Lehrlinge zu Sais* (The novices at Sais)—in which the young quested to remove the veil of nature and to become one with her. He dramatically captured Schelling's abstractions with the distich: "One succeeded—he lifted the veil of the goddess of Sais—/ But what saw he? He saw—wonder of wonders—he saw himself."[17] Nature and the self were doubles, each welling up from a common source. That common source was not a personal God who might be lurking in the dark, ready to condemn any breach of moral convention—and with the Romantics, that would have taken a truly divine effort—or to unveil all to be a conjuror's trick, a bit of thing-in-itself here and an arbitrary fiat there. Rather the laws of morality and the laws of nature arose from the self; they were our laws freely imposed and, at the same time, the laws of that greater reality of which we were all a part. That reality simultaneously was both the creator—*natura naturans*—and the created—*natura naturata*. Its creations, whether of natural beings or of literary life, did not drop from the heavens as the unintelligible commands of a hidden divinity; rather they grew as arabesques of willful reason and archetypal structures.

Isaiah Berlin, in his many essays, has emphasized the heterogeneous character of Romantic idealism.[18] The Romantics, he argues, encouraged the notion that moral and aesthetic values, the goals of life, would likely be multiple and irreconcilable, and that reason itself would struggle through infinite time, without, however, finally ordering individual lives according to one general principle of evaluation. The strife of becoming would be the

---

16. Friedrich Schiller, "Die Götter Griechenlandes," in *Friedrich Schiller: Sämtliche Gedichte* (Frankfurt a. M.: Insel Verlag, 1994), p. 190: "Da der Dichtung zauberische Hülle / Sich noch lieblich um die Wahrheit wand."

17. These lines were written in May 1798, as part of Novalis's study for the uncompleted *Die Lehrlinge zu Sais*. See "Paralipomena zu 'Die Lehrlinge zu Sais,'" in *Novalis Schriften*, 1:110.

18. See for example, Isaiah Berlin, "European Unity and Its Vicissitudes" and "The Apotheosis of the Romantic Will," in *The Crooked Timber of Humanity*, ed. Henry Hardy (New York: Knopf, 1991); also Berlin's "The Romantic Revolution," in *The Sense of Reality*, ed. Henry Hardy (New York: Farrar, Straus and Giroux, 1997).

eternal lot of human nature. Friedrich Schlegel quite clearly embodied in his activity and thought a dramatic disorder of values and designs: he endorsed classical aesthetics, only quicksilver to advance poetic ideals of the new Romanticism, but then to urge a synthesis of both; love of Caroline scorched his soul, but he gave her up to his brother and then erupted in volcanic hatred when she turned to Schelling; and in simmering consequence his politics went from liberal to conservative, as did his religion. But even deeper than the contradictions that beset the iconic life of Schlegel (and to a less dramatic extent, of course, that of most of the others of the circle), Romantic theory itself held heterogeneous and conflicting values in warm embrace: human beings, as part of nature, were confined by the iron strictures of natural law; yet nature herself fell as handmaiden under the command of an absolutely free ego. Freedom, for the idealists, formed the very essence of human reality, and so unconstrained intelligence might from moment to moment pursue conflicting goals. Certainly the dialectic of freedom and necessity would sow a garden of mixed delights, aims, and values that would grow in antagonistic and perpetual struggle, perhaps never achieving final harmony. But this was, after all, the message of a philosophy of *becoming*. Berlin's assessment thus finds confirmation in the lives of the Romantic thinkers we have here surveyed.

Berlin's evaluation, however, has to be severely qualified. The Romantics strove to unify, at least as an ideal goal, what they found separate and fragmented: the unconscious with the conscious, the self with nature, freedom with necessity, the individual with the community, poetry with science, the classical Greek aesthetic with the Romantic modern aesthetic. Friedrich Schlegel envisioned poetry as the medium of this ideal consummation:

> Romantic poetry is a progressive universal poetry. Its vocation is not merely to reunite all the separate genres of poetry and to put poetry in touch with philosophy and rhetoric. It also seeks, rightly, sometimes to mix and sometimes to melt together poetry and prose, genius and criticism, the poetry of art and the poetry of nature; it rightly seeks to make poetry lively and sociable, to make life and society poetical.[19]

Schlegel's poetic aspiration was not essentially different from Schelling's philosophical endeavor to unite all being in the identity of an absolute ideal-realism or Goethe's desire, as Faust expressed it, to rest in a moment of which one could say, "Tarry awhile, you are so beautiful" [*Verweile doch, du*

---

19. Friedrich Schlegel, *Fragmente*, in *Kritische Friedrich-Schlegel-Ausgabe*, ed. Ernst Behler et al., 35 vols. to date (Paderborn: Verlag Ferdinand Schöningh, 1958–), 2:182–83 (fragment 116).

bist so schön]. Human beings had to face the fragmented and contending elements of life, yet as progressive creatures, they continued to strive toward the ideal of unity, perhaps never achieving it. Yet the goal of unity lay before them as that paradise they occasionally glimpsed in their most intimate personal relationships.

In the previous two chapters, I have examined Romanticism's implications for biology only in a general and somewhat desultory way. In the next part, I will focus more specifically on the ways in which Romantic thought transformed biological ideas during the late eighteenth and early nineteenth centuries. In the epilogue, I will sketch the central roles those transformed notions came to play in the development of Darwin's evolutionary theory.

Part Two

Scientific Foundations of the Romantic Conception of Life

• ◆ • • ◆ • • ◆ •

Chapter 5

EARLY THEORIES OF DEVELOPMENT:
BLUMENBACH AND KANT

Ideas so monstrous that reason shudders before them.

—Kant on Herder's *Ideen*

Faust and Mephistopheles enter the lab of Wagner, who has just concocted in a glass retort a most unusual precipitate. Mephistopheles asks what it is, and the researcher quickly responds: "A man is being made." The devil assumes it is in the usual fashion, done with "an amorous pair" tucked away somewhere. But this modern man of science dismisses that suggestion:

> God forbid! Though that was once the fashion,
> We think it now only a silly passion.
> The tender point that gave rise to kin,
> The charming power that surged from within,
> Which got and begot with self-reflection,
> With near and far made close connection,
> All that's now unworthy of our protection.[1]

After its chemical conception, a homunculus, still encased in its vial, uses a magic cape to transport Faust and Mephistopheles back to ancient Greece, where they anticipate a Walpurgis night, when naked witches will dance and the divine spirit Helen preside. Once there, the manikin mounts the back of Proteus, the morphic god of infinite forms, while the philosopher Thales encourages him on:

> Give in to that laudable desire,
> Which all creation does require,
> Prepare for fast work without abate

---

1. Johann Wolfgang von Goethe, *Faust: Der Tragödie, zweiter Teil*, in *Sämtliche Werke nach Epochen seines Schaffens* (Münchner Ausgabe), ed. Karl Richter et al., 21 vols. (Munich: Carl Hanser Verlag, 1985–98), 18.1:177 (6838–44). The jangle of the German verse has a comic ring, which I have tried to preserve.

And use the eternal norms
To move though many thousands of forms,
Until you've reached the human state.[2]

The homunculus, as if heeding the advice, becomes transfixed by the power of Eros and immolates himself against the chariot of the goddess of the sea. The verse only suggests this, but perhaps by the power of love, the progeny of the homunculus will be reborn from the sea, to rise through the multitude of forms to the fully human condition. Even in advanced age, when he composed the tale of the homunculus, Goethe's powers of fantasy and romantic irony remained supple.

Goethe worked on the second part of *Faust* during his last years. It is a romantic arabesque, an elusive mélange of songs and reveries, an epic of ironic and melancholic proportions, by turns witty and serene. It ends in that beautiful and enigmatic lyric of the Chorus Mysticus, which echoes back the dedicatory poem of the first part. The lines seem to express an undying romantic sentiment, at least in the heart of an aging male poet:

Alles Vergängliche
Ist nur Gleichnis;
Das Unzulägliche,
Hier wird's Ereignis;
Das Unbeschreibliche,
Hier ist's getan;
Das Ewig-Weibliche
Zieht uns hinan.[3]

Goethe's experiences, his ideas, even his scientific theories can be found transmuted, emerging under protean forms, in his poetry and plays. Sometimes the original seed-leaf, as it were, can be clearly seen in the mature growth; at other times, as in *Faust*, the final morphos only hints at the original shape.[4] There is, though, in the story of the homunculus, some definite suggestions of a theory of life that he shared with his young protégé Friedrich Schelling. It is a theory of a *naturphilosophisch* metamorphosis of

2. Ibid., 18.1227 (8321–26).

3. Ibid., 18.1351 (12104–11): "Everything fleeting / Is only an image; / What exceeds our grasp, / Is here performed; / The indescribable, / Is here put down; / The eternal-feminine / Draws us on."

4. For rather straightforward translations of scientific and philosophical ideas into lyric form, see, for example, his poems "Die Metamorphose der Pflanzen" (The metamorphosis of plants, 1798) and "Metamorphose der Tiere" (Metamorphosis of animals, likely 1798), in ibid., 6.1:14–19. Poems such as these must temper our reading of Goethe's remark to Eckermann that "it was generally not my way as a poet to attempt to embody any abstract ideas." See Johann Peter Eckermann, *Gespräche mit Goethe in den letzten Jahren seines Lebens*, 3rd. ed. (Weimar: Aufbau-Verlag, 1987), p. 547 (6 May 1827).

Figure 5.1 Frontispiece to Goethe's *Faust, Ein Fragment* (1790). Designed by Johann Lips, after an engraving by Rembrandt.

creatures out of the "living dust" of the earth and their gradual evolution into higher animals and even human beings.[5] The foundation for this view is the Schellingian one, that nature herself is comparable to a living crea-

5. Johann Wolfgang von Goethe, "Weltseele," in *Werke* (Hamburger Ausgabe), ed. Erich Trunz et al., 14 vols. (Munich: C. H. Beck, 1988), 1:249: "Now all with divine boldness / Strive to advance; / The barren water wants to become verdant, / And every particle of dust lives."

ture. As Goethe put it in fugitive notes of 1805: "She [nature] is subservient to the principle of life, which contains the possibility that the simplest beginnings of phenomena progressively diversify into infinity and in the most variable ways."[6]

In this chapter and in the next two, I will discuss the several theories of life that animated biological discourse during the early Romantic period. I will focus initially on the ideas of Albrecht von Haller, Charles Bonnet, Johann Friedrich Blumenbach, Immanuel Kant, Carl Friedrich Kielmeyer, and Johann Christian Reil. Among them were mechanists, vitalists, and even some who might be described as teleomechanists (though none, save Kant, who fits that category as constructed by Lenoir).[7] They furnished the context, along with Alexander von Humboldt, Goethe, John Brown, and, surprisingly, Erasmus Darwin, out of which Schelling formulated his own theory of the organic. And in epistemological reciprocity, several of these writers—Kielmeyer, Reil, Humboldt, and Goethe—became supporters of Schelling's scientific ideas. Reil, for instance, would incorporate Schelling's conception of an active ego into his own theory of personality development; and Humboldt would endorse the philosopher's notion of an intimate connection between the organic and inorganic forces of nature. Schelling's theories, moreover, complemented and subsequently influenced those of Goethe, and together their biological conceptions formed the embryonic beginnings of ideas that came to maturity during the nineteenth century. I will concentrate more specifically on Goethe's biology in part 3.

Scholars used to maintain that Schelling and Goethe had anticipated the principal evolutionary theories of the nineteenth century, those of Lamarck and Darwin in particular. Darwin himself thought Goethe to have been "an extreme partisan of similar views."[8] More recent historical opinion, however, holds that any attribution of evolutionary ideas to Schelling or Goethe would only betray a lack of understanding of either the theory of Darwin or that of Schelling and Goethe—probably both.[9] I will argue in

6. Goethe, *Physikalische Vorträge Schematisiert* (Schematized lectures on physics, 1805–6), in *Sämtliche Werke,* 6.2:835.
7. See below for a discussion of Timothy Lenoir's ingeniously misleading thesis concerning the concept of teleomechanism.
8. In the third edition (1861) of the *Origin of Species,* Darwin added a historical sketch, in which he listed among the advocates of species descent: Lamarck, Erasmus Darwin, Geoffroy Saint-Hilaire, and Goethe. See Charles Darwin, *On the Origin of Species: A Variorum Text,* ed. Morse Peckham (Philadelphia: University of Pennsylvania Press, 1959), p. 61. Ernst Haeckel had persuaded Darwin that Goethe had advanced transmutational ideas.
9. Lenoir sounds the modern note when he observes: "In truth the works of Goethe and Darwin

chapters 8 and 11 that, on the contrary, the older views were not far wrong: that both Schelling and Goethe were biological evolutionists and that their conceptions straightened the path for German zoologists to advance more quickly and easily to Lamarckian and Darwinian theories than could their counterparts in England and France. In other words, I wish to provide historical substance and bio-Romantic appreciation for the homunculus's erotic couplings with the sea.

Embryology and Theories of Descent in the Seventeenth and Eighteenth Centuries

From the last part of the seventeenth century through the eighteenth, two very different theories of embryological development—each born out of distinct methodological and philosophical considerations—competed with one another for acceptance in the broader community of zoologists and physiologists.[10] One theory came to light in the work of the Dutch insectologist Jan Swammerdam (1637–1680). Swammerdam proposed, in his *Historia insectorum generalis* (1669), that among insects the semen of the female already contained, "in ideas and types according to a rational similitude," a preexisting adult form. The semen of the male acted, he believed, only as a stimulus to the realization of the adult type already encapsulated in the egg. Swammerdam rapidly generalized this theory, tincturing it with a Calvinistic stain: "The entire human race," he concluded, "already existed in the loins of our first parents, Adam and Eve, and for this reason, all of humankind has been damned by their sin."[11] An English reviewer of Swammerdam's volume described the Dutchman's theory as one in which embryological change consisted only "in a gradual and natural Evolution

---

present us with two radically different conceptions of biological science. . . ." See Timothy Lenoir, "The Eternal Laws of Form: Morphotypes and the Conditions of Existence in Goethe's Biological Thought," in *Goethe and the Sciences: A Reappraisal*, ed. Frederick Amrine, Francis Zucker, and Harvey Wheeler (Dordrecht: Reidel, 1987), pp. 17–28; quotation from p. 17.

10. The discussion in this section is based on chapter 2 of my *Meaning of Evolution: The Morphological Construction and Ideological Reconstruction of Darwin's Theory* (Chicago: University of Chicago Press, 1992).

11. Jan Swammerdam, *Historia insectorum generalis*, trans. H. Henninius (Holland: Luchtmans, 1685), p. 45. This Latin translation appeared five years after Swammerdam's death, and it expands the original Dutch passages concerning the theological implications of his theory of generation. This expansion may well have been authorized. It does conform to the remarks in the Dutch that "original sin is supported by this principle." See Jan Swammerdam, *Historia Insectorum Generalis, Ofte Algemeene Verhandeling Van De Bloedeloose Dierkens* (Utrecht: Van Drevnen, 1669), p. 52. (I am grateful to Peter McLaughlin for pointing out the differences in the editions of Swammerdam's work.) The Latin version had a much larger circulation.

and Growth of the parts."[12] The term "evolution" thus became attached to a theory of embryological preformation and thereafter itself would carry certain theological implications. Over the next century and a half, however, the word "evolution" would gradually move from this embryological usage finally to refer to species transformation. The conceptual linkage between embryological development and species development was laid down in the late eighteenth century, and along its tracks the term itself would steadily glide.[13]

The rival conception to embryological evolution was much older. It originated in Aristotle's biological theories and was confirmed in the mid-seventeenth century by William Harvey (1578–1657). Harvey distinguished two modes of gestation: one, *per metamorphosin*, in which all organs became immediately transformed—when, for instance, the caterpillar transmuted into a butterfly; and the other, *per epigenesin*, in which the embryo began as a formless mass and then became gradually more articulate as parts slowly started to take on a definite structure.[14]

During the next century, naturalists weighed the merits of the two theories, evolution and epigenesis, against careful observation and theoretical considerations. Even within the work of one thinker, the scales might dip one way then another—at least this is the story of the great Swiss anatomist Albrecht von Haller (1708–1777). Initially Haller endorsed preformationism, according to which "all the viscera, muscles, and remaining solid parts have already existed in the first beginnings of the invisible human embryo, and that they have at length successively become apparent in those places where they have been slowly dilated by an influxing humor and have become a visible mass."[15] Two versions of the theory of evolution existed, as Haller observed: ovism, espoused by Swammerdam, according to which "some sort of germ or perfect human machine exists in the egg"; and spermism, advanced by his own teacher Hermann Boerhaave (1668–1738), which taught "that man preexists in the little worm and that . . . the fabric

12. Anon., "Review of *Historia insectorum generalis, ofte algemeene verhandeling van de bloedeloose dierkens,*" *Philosophical Transactions of the Royal Society* 5 (1670): 2078. This is a review of the Dutch edition (1669) of the book. Note: the pagination in this volume of the *Philosophical Transactions* must have been set by a printer's devil. There are, for instance, two sequences of pages, each containing a page numbered 2078. The quotation comes on the first of these.

13. I have traced the history of the gradual change in meaning of the term "evolution"—as it moved from describing a theory of embryological preformation to one describing a theory of species descent—in my *Meaning of Evolution*.

14. William Harvey, *Exercitationes de generatione animalium* (London: DuGaidianis, 1651), p. 121.

15. Albrecht von Haller, in Hermann Boerhaave, *Praelectiones academicae*, notes by Albertus Haller, 6 vols. in 3 (Göttingen: Vandenhoeck, 1744), 5.2:489.

of the whole body has been delineated in the earliest embryonic stage and that it is expanded by heat and reabsorbed humor."[16] When Haller offered these descriptions, he had temporarily lost faith in the evolution hypothesis. His extended observational study of fetal development and his pondering of such phenomena as limb regeneration convinced him, for a while, that epigenesis seemed the more likely process of embryogenesis. That theory, though, required the postulation of some formative force that might guide the gradual development of organs. The need for a better angel to watch over embryological formation, however, tempted Haller's north German theology. The requirement was also out of harmony with a more sober Newtonian reluctance to truck with metaphysical entities. In the mid-1750s, Haller began a series of careful examinations of fertilized chicken eggs and the development of their contained embryos.[17] He thought he observed the gradual unfolding of translucent, incipient parts out of what must have been invisible but essentially structured antecedents.[18] He did not, to be sure, think the parental seed to be a miniature adult that would simply balloon out. Rather, the seed and then the fertilized embryo had pre-existing *nascent* parts. These embryonic elements would, during gestation, gradually alter their topology, change shape, solidify, and slowly become identifiable organs. The process of embryological development thus could be understood as a mechanical unfolding and articulating of parts—an evolution that required no mysterious forces to produce out of formless matter a little man.[19]

Haller's final adoption of a refined evolutionary theory received support from a fellow Swiss, Charles Bonnet (1720–1793). Bonnet, in contrast to his friend, set preformationism in a larger theoretical context. He explicitly drew out the implications of the theory and advanced the doctrine of *emboîtement*, or encapsulation. According to this auxiliary conception, God had created a multitude of germs, each encapsulating an embryonic organism, which in turn carried yet smaller organisms within its own germs,

16. Ibid., 5.2:490.

17. Albrecht von Haller's studies were published in his *Sur la formation du coeur dans le poulet; sur l'oeil; sur la structure du jaune &c*, 2 vols. (Lausanne: Bousquet, 1758).

18. Ibid., 1:186.

19. Shirley Roe discusses Haller's wavering positions in her *Matter, Life, and Generation: 18th-Century Embryology and the Haller-Wolff Debate* (Cambridge: Cambridge University Press, 1981). James Larson provides an extended context for understanding Haller's views and those of opponents in his *Interpreting Nature: The Science of Living Form from Linnaeus to Kant* (Baltimore: Johns Hopkins University Press, 1994), pp. 132–69. Jörg Jantzen furnishes a comprehensive consideration of reproductive theories, as well as other physiological theories, that formed the background for Schelling's thought. His monograph *Physiologische Theorien* is included in *Friedrich Wilhelm Joseph Schelling, Ergänzungsband zu Werke Band 5 bis 9: Wissenschaftshistorischer Bericht zu Schellings naturphilosophischen Schriften, 1797–1800* (Stuttgart: Frommann-Holzboog, 1994), pp. 375–668.

Figure 5.2 Albrecht von Haller (1708–1777), physician, anatomist, botanist, etc., of Göttingen, at about age forty-eight. "The supreme rule for him was to be instructed by the voice of nature, and he submitted to this mistress the learned power of his skill; he was a noble friend of truth as well as a friend in a struggle with a foe, and he was a most severe and forthright critic of his own error." (Courtesy Wellcome Institute Library.)

down through ever-smaller encased individuals—whole populations with-in infinitesimal seeds, enough to reach the Second Coming. From these original germs spilled forth lineages of plants and animals, producing, as Bonnet put it, a "natural evolution of organized beings" [*d'Evolution naturelle des Êtres Organisés*].[20] In the course of ages, he suggested, universal catastrophes had swept the earth clean of living creatures, but not their germs, which flowered anew and repopulated the world. Bonnet assumed—since fossils seemed to suggest this—that after each catastrophe, more perfect species had come forth from the kernels of the old and that there had been "a continued progress of all species, more or less slowly, toward a higher perfection."[21] Such an argument presaged the transformation of ideas about embryological evolution into those of species evolution. His theories, of course, did not constitute a naturalistic approach to species change: the environment played no causal role in producing transmutation, and natural forces were not invoked. Yet even Thomas Henry Huxley perceived in Bonnet's ideas "no small resemblance to what is understood by 'evolution' at the present day."[22]

A year after Haller published his study of fetal development in the chicken, he received a dissertation from a young German doctor by the name of Caspar Friedrich Wolff (1734–1794). In *Theoria generationis* (1759), Wolff defended epigenetic theory against Haller's "mechanistic medicine," which explained "the body's vital functions from the shape and composition of its parts."[23] Wolff himself also studied the developing chick embryo, especially its vascular system. He carefully observed the emerging structures of the heart out of fluid antecedents. To explain causally the formation of the various articulations out of homogeneous material, Wolff unhesitatingly postulated "a principle of generation, or essential force [*vis essentialis*], by whose agency all things are effected."[24] A few years later, in his *Theorie von der Generation* (1764), he elaborated on the reasons, aside from lack of visual evidence of preformation, that urged him to adopt epigenesis. He indicated that all analogy was against the idea of generations of homunculi nesting within one another, since "one finds nothing in nature that would be similar to an evolution."[25] Further, the mechanical theory of Haller and Bonnet made all apparent embryological development a prefab-

20. Charles Bonnet, *La Palingénésie philosophique, ou Idées sur l'état passé et sur l'état futur des êtres vivans*, 2 vols. (Geneva: Philibert et Chiroi, 1769), 1:250.

21. Charles Bonnet, *Considerations sur les corps organisés*, 2 vols. (Amsterdam: Marc-Michel Rey, 1762), 2:204.

22. Thomas Henry Huxley, "Evolution," *Encyclopaedia Britannica*, 9th ed. (1878), 8:745.

23. Caspar Friedrich Wolff, *Theoria generationis* (Halle: Hendelianis, 1759), p. 124.

24. Ibid., p. 106.

25. Caspar Friedrich Wolff, *Theorie von der Generation* (Berlin: Birnstiel, 1764), p. 40.

ricated miracle, which depended on a theological foundation. This presumption simply ran counter to the "concept we have of . . . a living nature that undergoes countless changes through its own power."[26]

Haller and Bonnet were the most important representatives of evolution theory during the last half of the eighteenth century. Despite our own estimate that the theory seems rather implausible, its Newtonian abstemiousness—not postulating unnecessary forces—did recommend it to some very shrewd scientists. Even Georges Cuvier (1769–1832), perhaps the most renowned zoologist of the first half of the nineteenth century, held on to the theory, since it obviated the need for German plastic principles.[27] The opposition to evolution, though, grew during the last half of the eighteenth century. John Needham (1713–1781) in England offered microscopical observations of the spontaneous generation of infusoria. This supposed abrupt transition from the inorganic to the organic yielded compelling evidence against the views of Haller and Bonnet. By the end of the century, theoretical weight had moved against evolution. Perhaps the most powerful oppositional force appeared in the form of a small book authored by the young Göttingen physician and physiologist Johann Friedrich Blumenbach (1752–1840). The book bore the title *Über den Bildungstrieb und das Zeugungsgeschäfte* (On the formative force and the operations of reproduction, 1781). This brief treatise would ignite a sequence of small explosions concerning the nature of the force called "life"; the heat of the discussions warmed Schelling for his own effort to reorient the question.

Blumenbach's Theory of the *Bildungstrieb*

Blumenbach studied medicine at the universities of Jena and Göttingen.[28] At the latter, he came to know both Johann David Michaelis, professor of theology and father of Caroline—the woman who became the erotic and

26. Ibid., p. 73.

27. See Georges Cuvier, *Histoire des progrès des sciences naturelles depuis 1789 jusqu'a ce jour*, 4 vols. (Paris: Baudouin Frères, 1829), 1:240–41.

28. For information on Blumenbach's life, I have relied on the memoir by his friend K. F. H. Marx, *Zum Andenken an Johann Friedrich Blumenbach*, 1840. This has been translated, along with several other of Blumenbach's works in *The Anthropological Treatises of Johann Friedrich Blumenbach*, trans. Thomas Bendysche (London: Longman, Green, Longman, Roberts & Green, 1865), pp. 3–45. Timothy Lenoir has discussed in some detail the development of Blumenbach's ideas in several works: "The Göttingen School and the Development of Transcendental Naturphilosophie in the Romantic Era," *Studies in History of Biology* 5 (1981): 111–205; "Kant, Blumenbach, and Vital Materialism in German Biology," *Isis* 71 (1980): 77–108; and *The Strategy of Life: Teleology and Mechanics in Nineteenth-Century German Biology* (Chicago: University of Chicago Press, 1989). I have found Nicholas Jardine's discussion of Blumenbach quite helpful. See his *Scenes of Inquiry* (Oxford: Clarendon Press, 1991), pp. 25–28. I will dissent from Lenoir's interpretation of Blumenbach's major theory. My objections extend to many of the other thinkers that Lenoir discusses: Kielmeyer,

Figure 5.3
Johann Friedrich
Blumenbach
(1752–1840), professor of
medicine at Göttingen.
Mezzotint by F. E. Haid.
(Courtesy Wellcome
Institute Library.)

fluctuating polar force of the Romantic circle—and Christian Gottlob Heyne, the well-known classicist who taught the Schlegel brothers. Heyne hired the student Blumenbach to put order into his newly acquired collection in natural history. It was this experience, as well as his acquaintance with the original owner of the collection, the retired professor Christian Wilhelm Büttner (1716–1801),[29] that led Blumenbach to write his dissertation on an anthropological topic, the races of mankind. Physical anthropology remained a preoccupation throughout his career; and his dissertation, *De generis humani varietate nativa* (On the natural varieties of mankind, 1775), went through three well-spaced editions. Immediately after he received his degree, the young physician attained the status of *Privatdozent* at Göttingen; and within three years he advanced to ordinary professor of medicine.

Über den Bildungstrieb

Blumenbach's *Über den Bildungstrieb* revised and expanded the considerations about generation introduced in his doctoral dissertation and in his highly influential *Handbuch der Naturgeschichte* (Handbook of natural his-

---

Reil, Meckel, Tiedemann, von Baer, et cetera. My own view is closer, on its negative side, to the critique of Lenoir by Kenneth Caneva, in his "Teleology with Regrets," *Annals of Science* 47 (1990): 291–300. My specific objections will be dribbled out below.

29. Büttner was the first person to lecture on natural historical subjects in a German university. See Heinrich Düntzer, "Vorrede," to his edition of *Zur deutschen Literatur und Geschichte: Ungedruckte Briefe aus Knebels Nachlass*, 2 vols. (Nürnberg: Bauer und Raspe, 1858), 1:xv.

tory, 1779–80; and eleven later editions). Initially he had been an advocate
of Haller's evolution theory, which he casually endorsed in the dissertation.[30] In the *Handbuch*, though, he began shifting toward a more neutral
position. He merely described Haller and Bonnet's thesis and the rival epigenetic conception; only in passing did he suggest that evidence indicated
the evolution hypothesis to be more probable.[31] He granted, though, that
the evidence was equivocal. So, for example, in reference to Haller's notion
that essential elements of the embryo were already ensconced in the
mother's egg, Blumenbach observed that "the role of the male semen in the
formation of the embryo is probably greater than is usually assumed."[32] He
then cited the evidence of hybrids, inherited defects, and other phenomena
as pointing to this greater role. By the time he composed *Über den Bildungstrieb*, however, his views had completely changed. In his prefatory remarks to the treatise, he admitted his earlier mistaken endorsement of
evolution, made while still green. He now sought to shrive himself of that
youthful error with the counterproposal of epigenesis.[33]

The occasion for his change of mind was, he said, a certain chance experience that occurred a few years before the publication of his little book,
while he was vacationing in the country. In his leisure, he amused himself
by observing a green, many-armed polyp in a millpond. He then thought to
conduct the classic study of cutting away sections of the hydra's body and
observing the regeneration of parts, which he did over a period of a few
days. He pondered these observations and analogous ones—such as regeneration of flesh after a wound—and was led to conclude that

> there exists in all living creatures, from men to maggots and from cedar
> trees to mold, a particular inborn, lifelong active drive [*Trieb*]. This drive
> initially bestows on creatures their form, then preserves it, and, if they be-
> come injured, where possible restores their form. This is a drive (or ten-
> dency or effort, however you wish to call it) that is completely different

30. Johann Friedrich Blumenbach, *De generis humani varietate nativa* (1775), in *Anthropological Treatises*, pp. 69–70.

31. Johann Friedrich Blumenbach, *Handbuch der Naturgeschichte*, 2 vols. (Göttingen: Johann Christian Dieterich, 1779–80), 1:18.

32. Ibid., 1:20.

33. Johann Friedrich Blumenbach, *Über den Bildungstrieb und das Zeugungsgeschäfte*, 1st ed. (Göttingen: Johann Christian Dieterich, 1781), p. 5. The core of this book was published in subsequent numbers (1780 and 1781) of the journal edited by Georg Christoph Lichtenberg and Georg Forster (Caroline Böhmer's "friend" at Mainz). See Johann Friedrich Blumenbach, "Über den Bildung-strieb (Nisus formativus) und seinen Einfluss auf die Generation und Reproduktion," *Göttingisches Magazin der Wissenschaften und Litteratur* 1, no. 5 (1780): 247–66; and "Über eine ungemein ein-fache Fortpflanzungsart," in *Göttingisches Magazin der Wissenschaften und Litteratur* 2, no. 1 (1781): 80–89. These parts were brought together with added material in his book.

from the common features of the body generally; it is also completely different from the other special forces [*Kräften*] of organized bodies in particular. It shows itself to be one of the first causes of all generation, nutrition, and reproduction. In order to avoid all misunderstanding and to distinguish it from all the other natural powers, I give it the name of *Bildungstrieb* (*nisus formativus*).[34]

Blumenbach insisted that the *Bildungstrieb* not be confused with forces defined by other authors—for instance, Needham's *vis plastica* (an empty word, indicating an occult quality, Blumenbach claimed)[35] or Wolff's *vis essentialis*. The *Bildungstrieb*, according to Blumenbach, was responsible for reproduction, nourishment, and restoration of parts. In these various instances, the force differently expressed itself according to the circumstances in which it operated. These three activities, then, were "merely modifications of one and the same force."[36]

Blumenbach piled up his evidence for the existence of a *Bildungstrieb* from instances analogous to that of polyp regeneration, for example: the restoration of bodily form after an injury; the production of so-called sleep apples in the wild rose from the actions of the gall wasp; the gradual formation of the embryo of larger creatures; the unformed condition of aborted fetuses; and the reproduction through budding in translucent green water moss.[37] (Blumenbach had already mentioned some of this evidence in the first edition of the *Handbuch*, when he had shifted into a neutral position on the question of generation.)

Aside from these positive observations supporting epigenesis and the operations of a formative drive, Blumenbach also marshaled the negative instances, the cases that Haller's evolution theory could not readily handle. The disconfirming evidence, he thought, was rife. For example, freshly fertilized chicken eggs initially showed no trace of blood vessels or blood (vivid traits that ought to be visible from the first, even in a tightly folded miniature chicken); animals of the same species would often produce spermatic organisms of different forms (thus unlikely that the little worms would hold the same kind of miniatures within); hybrids of different varieties or species could be formed (surely impossible if offspring were already preformed in one of the parents); young boys in the Near East often no

34. Blumenbach, *Über den Bildungstrieb* (1st ed.). pp. 12–13.

35. Ironically, Blumenbach would in later editions of *Bildungstrieb* refer to the principle of the *Bildungstrieb* precisely as a *qualitas occulta*, though in the positive sense (pace Leibniz) that might be associated with Newton's principle of gravity. See the text below for examples of this reevaluation of the category of occult quality.

36. Ibid., p. 19.

37. Ibid., pp. 11, 23–24, 40–41, 44, 50.

longer had to be circumcised, since the practice had produced an acquired characteristic (with no likely way for each of the cascading homunculi to be shorn of its infinitesimally small sheath); and most amazingly, a chimera could be produced when half a brown polyp was joined to a half of a green polyp (which seemed unlikely on the basis of any mechanical explanation the evolutionists could imagine).[38]

Though Blumenbach's discussion, with its medley of examples, did have a lethal effect on assumptions of evolution, it still left unclear the status of the principal causal agent that drove the opposing epigenetic process. As Pierre Flourens (1794–1867) put it in his eulogy for Blumenbach, the postulation of the Bildungstrieb, unlike the rival view, did not create any problems, yet it did not remove any either.[39] What, more exactly, was the Bildungstrieb?

In the first edition (1781) of Über den Bildungstrieb, Blumenbach considered the drive an independent vital agency. It caused the formation of the embryo out of homogeneous seminal material and continued to operate in maintaining the vitality of the organism and in repairing its injuries. In this respect the Bildungstrieb differed considerably from Wolff's vis essentialis, which did not of itself produce organic form. Wolff's essential force drove the developmental process of organisms, but the direction and form of that development resulted from local circumstances—that is, the already established parts and external impingements.[40] Blumenbach's Bildungstrieb furnished the architectonic articulations of living matter: it directed the formation of anatomical structures and the operations of physiological processes of the organism so that various parts would come into existence and function interactively to achieve the ends of the species. Kant would have rejected any such force pretending to be constitutive of nature, since a force of this kind would have to operate according to an intellectual plan or an intention, which he believed could only be found in a rational mind but not in a rational, mechanical nature. For Kant, as I will discuss in a moment, the Bildungstrieb could only be a regulative concept, one that helped the naturalist seek out the mechanistic causes really at work. But for Blumenbach, the Bildungstrieb endowed the homogeneous, formless mixture of male and female semen with its most essential character—form, organiza-

---

38. Ibid., pp. 28–29, 33, 69. Blumenbach learned of the supposed effects of circumcision from his friend the theologian Michaelis (p. 69).

39. Marie-Jean-Pierre Flourens, Éloge historique de Jean-Frédéric Blumenbach (1847), in Anthropological Treatises, p. 54.

40. Peter McLaughlin works out these differences between Wolff's vis essentialis and Blumenbach's Bildungstrieb in his carefully detailed article "Blumenbach und der Bildungstrieb," Medizin historisches Journal 17 (1982): 357–72.

tion—and set the various parts so articulated into mutually harmonious operation. This was a teleological cause fully resident in nature.[41] And depending upon the matter on which it operated, the *Bildungstrieb* would produce more or less regular effects, the properties of which might be formulated into laws governing all organisms. So, for instance, Blumenbach asserted as a general proposition that the younger the creature, the "more rapid the growth and the more quickly would form move toward perfection."[42]

In later studies Blumenbach expanded the use of the *Bildungstrieb* to explain other phenomena, most notably the formation of new varieties and subspecies. In the first edition (1775) of *De generis humani*, he had suggested that the varieties of animals and human beings arose from the effects of climate and nutrition on a given stock. He presumed, for example, that cold temperatures effected the smaller stature of Greenland foxes as compared with those animals inhabiting more temperate zones.[43] But in the second edition (1781) of *De generis humani*, he introduced the additional factor of the *Bildungstrieb* to explain the degeneration of an original type into the varieties found populating the world.[44] The concept of degeneration itself and that of its ultimate causes in climate, nutrition, and hybridization were hardly of Blumenbach's own devising. Buffon had earlier argued that degeneration from originally created types, via the aforementioned agencies, had produced the varieties (sometimes called "species") with which we were familiar.[45] The concept, as Buffon formulated it and Blumenbach employed it, did not imply any deterioration, only an alteration from an original condition. What Blumenbach added to this theory was the proposal that extrinsic agencies worked on the *Bildungstrieb* to deflect this "formative force [*nisum formativum*] markedly from its usual path, which deflection is

41. François Duchesneau provides a good account of the functionally integrative character of Blumenbach's *Bildungstrieb*, though he, like Lenoir, suggests (albeit rather vaguely) that the force could have been only a heuristic concept for Blumenbach. The sole evidence for this interpretation seems to be that Kant liked the concept and, well, what else could it be but "subjective." See François Duchesneau, "Vitalism in Late Eighteenth-Century Physiology: The Cases of Barthez, Blumenbach and John Hunter," in *William Hunter and the 18th-Century Medical World*, ed. W. F. Bynum and Roy Porter (Cambridge: Cambridge University Press, 1985), pp. 259–95; especially p. 278.

42. Blumenbach, *Über den Bildungstrieb* (1st ed.), p. 43.

43. Blumenbach, *De generis humani varietate nativa* (1775), in *Anthropological Treatises*, p. 104.

44. Blumenbach, *De generis humani varietate nativa*, 2nd ed. (Göttingen: Vandenhoek et Ruprecht, 1781), pp. 1–2.

45. Georges Louis Leclerc, comte de Buffon, "De la Dégénération des animaux" (1766), *Histoire naturelle*, in *Oeuvres complètes de Buffon*, ed. Pierre Flourens, 12 vols. (Paris: Garnier, 1853–55), pp. 110–44. In *De generis humani*, Blumenbach has many citations to Buffon's *Histoire naturelle*.

the generous origin of degeneration and the mother, properly speaking, of varieties."[46]

Biological Revolution: Blumenbach and Herder

As a complement to his ideas about the production of species varieties, Blumenbach also devised a theory of biological revolution in which the *Bildungstrieb* played the central creative role—it would be the agency for the production of *new* species. In his *Beyträge zur Naturgeschichte* (Contributions to natural history, 1790), he contended that fossils indicated a pre-Adamite creation that was subsequently destroyed in a general, neptunic catastrophe. After a while, according to this supposition, the Creator repopulated the earth, but did so by employing "the same natural powers [*Naturkräfte*] to effect the production of a new organic creation that had filled the same purpose in the preworld [*Vorwelt*]."[47] Thus the *Bildungstrieb* operated to generate—presumably out of the inorganic—a new living world, which would display not only some creatures similar to those of the old creation but also vast kingdoms of entirely new species. The main difference between the operations of the *Bildungstrieb* in the preworld and in the current world was "only that the *Bildungstrieb* had to be applied to a greatly modified matter—after such a total revolution—and through the production of new species had to take a direction differing more or less from the old."[48]

Even in the present dispensation, Blumenbach argued, evidence indicated that some new creations had arisen—for instance, certain kinds of worms in the flesh of domestic pigs, though not in their wild ancestors. He confessed ignorance about the exact process by which such creatures originated but generally attributed the cause "to the great changeability in nature," which itself was but a feature of the activities of the *Bildungstrieb*. Ultimately, though, "this great changeability itself [had to be the result of] the most beneficent and wise direction of the Creator."[49] Such mutability—that is, the fluctuating fortunes of the *Bildungstrieb* under differing conditions—also produced degeneration of existing species, supplying new varieties and subspecies to an ever-changing world.

Though Blumenbach left no overt indications of sources for his theory of

---

46. Blumenbach, *De generis humani varietate nativa*, 3rd ed. (Göttingen: Vandenhoek et Ruprecht, 1795), p. 88.

47. Johann Friedrich Blumenbach, *Beyträge zur Naturgeschichte* (Göttingen: Johann Christian Dieterich, 1790), part 1, pp. 24–25.

48. Ibid., p. 25.

49. Ibid., pp. 31–32.

biological revolution, his ideas harmonized with those authored by Bonnet
(see above) and especially with those of his contemporary Johann Gottfried Herder (1744–1803). In his *Ideen zur Philosophie der Geschichte der
Menschheit* (Ideas for a philosophy of the history of humanity, 1784–91),
Herder constructed a naturalized version of Genesis. He maintained that
the earth and planets had developed out of a nebular chaos, but in obedience to universal laws. He supposed that from a volcanic maelstrom eventually emerged a habitable environment with great varieties of plants and
animals. The first ages proved harsh, winnowing out the most suitable creatures for subsequent times: "Those innumerable volcanoes on the surface of
our earth that once spewed flames no longer do so; the oceans no longer
seethe with the vitriol and other materials that once covered our land. Mil-
lions of creatures have passed away that had to die; those that could survive,
remained and have perdured for thousands of years in great harmony with
one another."[50] Throughout the plant and animal kingdoms, vital powers,
"the fingers of God," operated ontogenetically to form creatures—a *Bildung*
rather than an evolution of preformed parts. And these powers, according
to Herder, drove species to ever more complex development during an early
period of the formation of life on the planet. "Could we but penetrate to
those first periods of creation," he reflected, "we would see how one kingdom of nature was built upon another; what a progression of advancing
forces would be displayed in every development!" These "organic forces,"
he declared, "bud forth in great creations and strive toward new formations." Precisely how nature has achieved these transitions is kept veiled:
"She has laid before us only transitions from the lower realms and only advancing forms in the higher realms; the thousand invisible ways in which
this transition occurs she has reserved for herself; and so no human eye can
penetrate the realm of the unborn, the great ὕλα [matter] or Hades." At
some point, however, "the door of creation was shut." Thereafter, vital
powers continued to produce improvements, but only within established
limits of fixed animal and plant species. The whole development of the
world, in Herder's religious cosmology, arched with deliberate intent toward the crowning achievement, human nature and the perfection of humanity: "The purpose of our present existence," he proclaimed, "is directed
to the formation of humanity [*Bildung der Humanität*], and all the lower necessities of the earth only serve and lead to this end."[51]

Herder wove his history of the earth from threads of sound science, rea-

50. Johann Gottfried Herder, *Ideen zur Philosophie der Geschichte der Menschenheit*, in *Johann
Gottfried Herder Werke*, ed. Martin Bollacher, 10 vols. (Frankfurt a. M.: Deutscher Klassiker Verlag,
1985–), 6:637 [III.15:ii].
    51. Ibid., 6:176 [I.5:iii], 6:178 [I.5:iii], 6:179 [I.5:iii], 6:176 [I.5:iii], 6:187 [I.5:v].

Figure 5.4
Johann Gottfried Herder
(1744–1803). Portrait
(1795) by F. A. Tischbein.
(Courtesy Archiv für
Kunst und Geschichte.)

sonable speculation, Spinozistically tinged theology (in which free human behavior is also regarded as determined by natural law), and a great many colorful strands of poetic musing.[52] Kant, in his extensive review of the first two parts of Herder's *Ideen*, complained about the fanciful and indefinite character of the fabrication.[53] Herder's conception of vital powers investing nature seemed to the philosopher mostly blooms from "the fruitful field of creative imagination," which was nurtured by "a rather dogmatic metaphysics."[54] These speculations, though, were hardly innocent confections. In Kant's estimation, they suggested "ideas so monstrous that reason shudders before them." And with a quaking irritation, he made these ideas explicit: "either one species [*Gattung*] would have arisen out of another and all out of one single original species or perhaps out of a single, productive mother-womb [i.e., the earth]."[55] More generally, these ideas had danger-

52. See Hugh Nisbet's brief but illuminating discussion in his "Historisierung: Naturgeschichte und Humangeschichte bei Goethe, Herder und Kant," in *Goethe und die Verzeitlichung der Natur*, ed. Peter Matussek (Munich: C. H. Beck, 1998), pp. 15–43.

53. Kant reviewed parts 1 and 2 of Herder's *Ideen* (about half the book) in several numbers of Schütz's *Allgemeine Literatur-Zeitung* in 1785–86. See *Rezension zu Johann Gottfried Herder: Ideen zur Philosophie der Geschichte der Menschheit*, in *Immanuel Kant Werke*, ed. Wilhelm Weischedel, 6 vols. (Wiesbaden: Insel Verlag, 1957), 6:781–806 (A 17–22, 309–10, 153–56).

54. Ibid., 6:792 (A 22).

55. Ibid.

ous implications for two fundamental Kantian convictions: the mechanistically necessary structure of scientific law, which would have been compromised by Herder's vitalism; and human freedom, which certainly could not emerge from the interactions of material nature.[56] Kant admitted that it might be unjust to attribute these ideas to Herder, though he did not doubt that they seemed to drift along with his former student's considerations.[57] Kant's estimate of the danger of such monstrous ideas, however, softened during the next five years, when he spied these specters hovering over the more rigidly scientific analyses of Blumenbach. For Kant, the threat of the ideas seemed lessened, since he had found in his *Kritik der Urteilskraft* a way to tame them: he discovered he could render them merely heuristic principles instead of foundational laws. Schelling and Goethe, showing a greater boldness, would adopt these general notions, and with them infuse nature with the necessary energies to produce new organisms and transmute older ones.

Refinements of the Concept of *Bildungstrieb*

Blumenbach's initial theory of the *Bildungstrieb* certainly derived no leads from Herder, rather the reverse: Herder developed his own notion of *Bildung* and vital force in light of Blumenbach's work.[58] But reciprocally, Blu-

56. Beiser argues that the intended outcome of Kant's analysis in the *Critique of Judgment* was precisely this removal of a threat to human rationality and freedom. See Frederick Beiser, *The Fate of Reason: German Philosophy from Kant to Fichte* (Cambridge: Harvard University Press, 1987), pp. 156–57.

57. Herder, Kant's one-time student, bridled at the review he had received from the master. He wrote his friend and Kant's nemesis Johann Georg Hamann to express his deep irritation: "I have heard from several distant quarters that the review has had little success; rather, it has been read with amazement that Herr Kant would mention a shudder of reason. His final preceptorial lectures to me are simply inappropriate: I am 40 years old and no longer sit on his metaphysical school bench." See Herder to Hamann (14 February 1785), in *Johann Gottfried Herder, Briefe*, ed. Wilhelm Dobbek and G. Arnold, 10 vols. (Weimar: Böhlaus, 1977–96), 5:106. The initial vehemence of Kant's reaction to Herder's *Ideen* may have had a decidedly personal cause. Herder's wife Karoline mentions in her biography of her husband that he was told by his publisher that Kant believed his former pupil had, by his criticisms, prejudiced the reception of the first *Critique*. Herder replied: "It's never come into my head to form a cabal against anyone, least of all Kant. Well, his *Critique* is unpalatable to me and contrary to my way of conceiving things; but I have never written or done anything against it. You can assure Kant of that." See Maria Caroline von Herder, *Erinnerungen aus dem Leben Joh. Gottfrieds von Herder*, 2 vols. (Tübingen: J. G. Cotta'schen Buchhandlung, 1820), 2:221. Beiser may have been the first to notice this as a potential cause of Kant's reaction. See Beiser, *The Fate of Reason*, p. 149.

58. Herder frequently cited the second edition of Blumenbach's *De generis humani varietate nativa*, in which the naturalist had introduced his new concept of *Bildungstrieb*. See, for example, Herder, *Ideen zur Philosophie der Geschichte der Menschheit*, in *Herder Werke*, 6:119, 120, 128 [I.4:i], 249 [II.6.vii], etc. Herder's own analysis of *Bildung* followed closely Blumenbach's discussion of preformation versus epigenesis. See ibid., p. 172 [I.5:ii]: "One speaks improperly if one talks

menbach undoubtedly found some inspiration in Herder for applying the concept of *Bildungstrieb* to this new area of inquiry, the history of the earth. In addition to these new applications, Blumenbach continued gradually to alter and refine the core of the concept. These refinements were first introduced into his *De Nisu formativo et generationis negotio* (On the formative drive and the operation of generation, 1787) and in the third edition (1788) of the *Handbuch der Naturgeschichte*. He consolidated these changes a year later in the second edition of *Über den Bildungstrieb*, where they were perspicuously revealed. In that edition he added this paragraph to the general definition of the *Bildungstrieb*:

> I hope it will be superfluous to remind most readers that the word *Bildungstrieb*, like the words attraction, gravity, etc. should serve, no more and no less, to signify a power whose constant effect is recognized from experience and whose cause, like the causes of the aforementioned and the commonly recognized natural powers, is for us a *qualitas occulta*. What Ovid said pertains to all of these forces—*causa latet, vis est notissima* [the cause is hidden, the force is well recognized]. The service rendered by a study of these forces is only that one can more carefully determine their effects and bring those effects into general laws.[59]

Blumenbach secured this Newtonian reconfiguration of his force with a careful footnote to Newton's *Optics* and by emphasizing, in the last pages of the booklet, that one could generalize the various effects of the *Bildungstrieb* into several laws—for instance, that the strength of the *Bildungstrieb* was inversely related to the age of the organism; that it operated more strongly on the young of mammals than on the young of ovipara; that it operated with variable rapidity and strength on different organs of the same creature, and so forth.

It is crucial to note that in this Newtonian rendering of the *Bildungstrieb*, Blumenbach had not suggested that the term referred to nothing, rather that it stood for a force, specified by its effects, but whose cause could not be known directly. As he indicated in the second edition of the *Handbuch*, where he introduced the Newtonian comparison: "It is a proper force [*eigenthümliche Kraft*], whose undeniable existence and extensive effects are ap-

---

about a seed [*Keim*] that only unfolds, or of an epigenesis according to which limbs form by an external power. *Bildung* (genesis) is an effect of inner forces that form the mass prepared by nature and in which it will be manifest."

59. Blumenbach, *Über den Bildungstrieb*, 2nd ed. (Göttingen: Johann Christian Dieterich, 1789), pp. 25–26. See also Blumenbach, *De Nisu formativo et generationis negotio* (Göttingen: Johann Christian Dieterich, 1787), p. x; and *Handbuch der Naturgeschichte*, 2nd ed. (Göttingen: Johann Christian Dieterich, 1788), p. 14.

parent throughout the whole of nature and revealed by experience."[60] In this respect, he thought his use of the term paralleled the way Newton used "attraction." His footnote quotation from Clarke's Latin version of Newton's *Optics* made the point. Newton wrote, "I thus use this term attraction so that it be understood generally to signify any power by which bodies mutually tended toward one another, no matter what cause might be attributed to this power."[61] In comparable fashion, Blumenbach construed the *Bildungstrieb* as a force, deriving from an unknown cause, that could only be characterized by its conspicuous effects. The paradigm employed, then, was a causal chain of this sort: cause (unknown) —produces→ force (the *Bildungstrieb*) —produces→ perceptible effects (e.g., epigenesis). The *Bildungstrieb* thus became a secondary cause yielding immediate effects and itself was the effect of some hidden, primary cause.

Status of the Concept

This cautious Newtonian rendering of the concept of *Bildungstrieb* reduced, though only a little, its metaphysical valence. But did it turn the *Bildungstrieb* into what Lenoir has called a teleomechanistic principle? That is, was the *Bildungstrieb*, as Blumenbach employed it, a principle that Kant could *justly* have adopted, a principle employed to represent a mechanical cause *as if* it were teleological? There is no doubt that Kant appropriated the concept from Blumenbach (see below) and made it a regulative principle, while denying it constitutive status. But did Blumenbach himself hold the principle to be merely regulative? Lenoir argues that this was precisely the case, that Blumenbach and Kant supported "the same program," the program of "teleomechanism."[62] In an otherwise illuminating reconstruction, Lenoir concludes that with the principle of the *Bildungstrieb*,

> Blumenbach adopted what is best characterized as an emergent vitalism: that is to say, the vital force was not to be conceived as separate from mat-

---

60. Blumenbach, *Handbuch der Naturgeschichte* (2nd ed.), p. 14.

61. Blumenbach, *Über den Bildungstrieb* (2nd ed.), pp. 25–26. Newton was often more circumspect about forces than the above quotation suggests. In "Definition VII" of the *Principia*, he declared that he did not wish to consider forces physically, but only mathematically. See Isaac Newton, *Mathematical Principles of Natural Philosophy*, trans. Andrew Motte (1729), ed. Florian Cajori, 2 vols. (Berkeley: University of California Press, 1962), 1:4–6. In the *Optics*, Newton gave way to freer speculation about forces. In any case, it is clear that Blumenbach understood forces to be real phenomena, even if occult. He may have been brought to the Newtonian comparison by an obscure doctoral dissertation (*De respiratione*, by Michaelis Birkholz) that likened his *Bildungstrieb* to Newton's *principium trahens*. Blumenbach cited the dissertation in *Über den Bildungstrieb*, 3rd ed. (Göttingen: Johann Christian Dieterich, 1791), p. 37.

62. Lenoir, *The Strategy of Life*, pp. 23–24.

ter, but matter was not the source of its existence; rather it was the *organization* of matter in certain ways that gave rise to the *Bildungstrieb*. Organization was taken here as the primary given: the presence of organization could not be further explained in terms of unorganized parts.[63]

I will discuss Blumenbach's relationship to Kant in a moment, but I believe one can see immediately why Lenoir's interpretation of the *Bildungstrieb emerging out of organization* is implausible. Blumenbach, it must be remembered, originally developed the concept of the *Bildungstrieb* as an inherent causal principle to explain the possibility of epigenesis, that is, the gradual development of fetal organization out of an unorganized, homogeneous substrate. In the first edition of his booklet, he presented the *Bildungstrieb* straightforwardly as a real cause, a force that produced the phenomenon: "This drive," he said, "initially bestows on creatures their form." Moreover, this drive "shows itself to be one of the first causes of all generation, nutrition, and reproduction."[64] The *Bildungstrieb*, therefore, could not be an *effect* of organization, a property emerging out of organization; it was initially postulated as a *cause* to explain organization. This fundamental employment of the concept is further driven home by Blumenbach's extending the application of the *Bildungstrieb* in order to demonstrate the generation of forms where no like forms could have previously existed—that is, the formation of new species during the pre-Adamite and post-Adamite biological revolutions; and the strange case of gall wasps producing "sleep apples" on rose bushes, which he regarded as a kind of spontaneous generation. In the second edition of his *Bildungstrieb*, Blumenbach did not alter his conception of this fundamental causal relationship, namely, of the formative power causally producing organization. In the second edition, he merely suggested that we could grasp this causal force only through its effects. Even after he had read Kant's endorsement of the concept of the *Bildungstrieb* (see Kant's letter, below), Blumenbach maintained the same causal structure in his account of the *Bildungstrieb*. Thus from the last edition of the *Handbuch* (1830):

> When the ripe, but as yet unformed [*ungeformte*], but organizable [*organisirbare*] seminal matter reaches its time and enters into the required conditions in the place of its determination, then it becomes initially re-

---

63. Lenoir, "The Göttingen School and the Development of Transcendental Naturphilosophie in the Romantic Era," p. 155. See also his *Strategy of Life*, p. 21: "For Blumenbach the *Bildungstrieb* did not exist apart from its material constituents, but it could not be explained in terms of those elements. It was an emergent property having a completely different character from its constituents."

64. Blumenbach, *Über den Bildungstrieb* (1st ed.) pp. 12–13.

ceptive of that same and now teleological [*zweckmäsig*] operative life-force, namely, the *Bildungstrieb* (*nisus formativus*). . . . This power is able to form the variously organizable seminal matter in comparably many, telically modifiable ways into determinate forms.[65]

The *Bildungstrieb* was thus not a Kantian "as if" cause but a real teleological cause (i.e., one acting for ends), which, albeit, was known only through the ends it achieved. The actions of that cause, like causes in physics, could be expressed in general laws, which Blumenbach carefully formulated in the manner of Newton—something Kant would methodologically prohibit in the case of biological principles. And behind this anonymous force, Blumenbach clearly spied the Creator unabashedly pulling the strings, a perception no scientific theory in the Kantian mold would legitimate. Kant did, to be sure, adopt aspects of Blumenbach's conception; but he turned them to his own purposes, which must now be considered.

Kant's Theory of Biological Explanation

The impact of Kant's *Kritik der Urteilskraft* on the disciplines of biology has, I believe, been radically misunderstood by many contemporary historians. It is frequently thought that Kant provided a conceptual framework in terms of which biological science could be conducted.[66] This, I think, is a fundamental misinterpretation of Kant's relationship to the work of biologists during the Romantic period. Those biologists who found something congenial in Kant's third *Critique* either misunderstood his project (Blumenbach and Goethe) or reconstructed certain ideas to have very different consequences from those Kant originally intended (Kielmeyer and Schelling). There were some, of course, who simply and explicitly rejected Kant's analysis of teleology (Reil). These latter two groups seemed to have understood more clearly than the first that the *Kritik der Urteilskraft* delivered up a profound indictment of any biological discipline attempting to become a science.

The third *Critique*'s charge against efforts to make biology into an authentic science (defined as having necessary laws comparable to Newtonian mechanics) can be simply stated, even if its expression lay entangled in the thicket of Kantian distinctions. Kant maintained that in comprehending the organization and operations of living creatures, an investigator had

---

65. Blumenbach, *Handbuch der Naturgeschichte*, 12th ed. (Göttingen: Dieterich'schen Buchhandlung, 1830), p. 15.

66. In addition to differences with Lenoir on Kant's role, I also take exception to the interpretations of James Larson, in his *Interpreting Nature*, pp. 170–82, and Jörg Jantzen, in his *Physiologische Theorien*, pp. 658–64. Both of these historians generally follow Lenoir.

Figure 5.5

Immanuel Kant
(1724–1804). Portrait
(1791) by Gottlieb
Doebler. (Courtesy
Archiv für Kunst und
Geschichte.)

to assume a teleological causality, for no application of merely mechanistic laws could ultimately make biological processes intellectually tractable.[67] One had, from this perspective, to regard the organizational features of animate nature as the result of a kind of causality in which the idea or plan of the whole produced the specific formal relationships of the parts to one another. Yet such assumption of final causes could only be heuristic, an "as if" causality—since no telic causes, which ultimately presumed intentionality, could be validly understood as producing natural phenomena. Natural phenomena, according to Kant, could only be scientifically—that is, properly—explained by appeal to mechanistic laws. Such laws would specify the constituent parts of some entity as the adequate causes of the arrangement of the whole—that was the very meaning, for Kant, of mechanistic cause. He did not assert that natural phenomena were possible only as the result of mechanistic causes—that would presume knowledge of the supersensible world. Indeed, it might be that "in the unknown, inner ground of nature itself, the physical mechanical connections and the purposive connections can be found together in the same things united under one principle."[68] Kant rather claimed that, given the character of our human understanding, we "*ought* always reflectively consider all events and

67. See chapter 3 for a discussion of Kant's theories of teleological and aesthetic judgment.
68. Kant, *Kritik der Urteilskraft*, in *Werke*, 5:502 (A 312, B 316).

forms in nature [e.g., biological occurrences] according to the principle of the mere mechanism of nature, and we *ought* to employ this principle, as far as we can, in research, since without it as a basis for our investigations, we can in no way have proper natural knowledge [*Naturerkenntnis*]."[69] But he yet believed there would be organic phenomena (especially reproduction) that could only be understood *by us* teleologically. We would thus have the methodological imperative to reduce organic phenomena to mechanistic laws as far as we were able, but we would not be ultimately successful. We would have to have recourse to teleological assumptions. Kant thus maintained that biology could not really be a science, but at best only a loose system of uncertain empirical regularities, not a *Naturwissenschaft*, but a *Naturlehre*.[70] Most biologists of the period, by contrast, thought their disciplines could be developed into sciences and could, in that respect, come to stand as certainly on that pinnacle of human accomplishment as Newton's physics. They believed, in part due to Schelling, that teleological processes could be found governing natural phenomena and that valid laws could be formulated to capture such relationships. It is, however, quite understandable how some of them might have been misled into thinking that Kant's system was congenial to biology, especially if they received blandishments from the great philosopher himself, as Blumenbach did.

In August 1790, just after the publication of his *Kritik der Urteilskraft*, Kant wrote an admiring letter to Blumenbach to acknowledge an intellectual debt. The letter read in part:

> I wish to extend my thanks for sending me last year your excellent work on the formative force [*Bildungstrieb*]. I have learned a great deal from your writings. Indeed, in your new work, you unite two principles—the physical-mechanistic and the sheerly teleological mode of explanation of organized nature. These are modes which one would not have thought capable of being united. In this you have quite closely approached the idea with which I have been chiefly occupied—but an idea that required such confirmation [as you provide] through facts.[71]

---

69. Ibid., 5:501 (A 311, B 315). I have added the emphasis in the quotation. For a general discussion of Kant's respective conceptions of mechanism and teleology, see Jardine, *Scenes of Inquiry*, pp. 28–33.

70. Kant, *Kritik der Urteilskraft*, in *Werke*, 5:510 (A 325–29, B 329–33). That Kant excluded biology from the realm of real science (*Wissenschaft*) is, I think, indisputable. But for those who might wish further discussion of this conclusion, see the extensive account of Kant's position offered by Reinhard Löw in his *Philosophie des Lebendigen: Der Begriff des Organischen bei Kant, sein Grund und seine Aktualität* (Frankfurt a. M.: Suhrkamp Verlag, 1980), pp. 130–31.

71. Kant to Blumenbach (5 August 1790), in *Immanuel Kant, Briefwechsel*, ed. Otto Schöndörffer (Hamburg: Felix Meiner Verlag, 1972), p. 466.

Though this letter suggests otherwise, Kant did not really concede that Blumenbach had pulled off the unification of the principles of mechanism and teleology; the *Kritik der Urteilskraft* rather argues that those principles cannot be reconciled, at least by us.

Kant mentioned in his letter that he was having his bookseller send along a copy of the *Kritik der Urteilskraft* so that Blumenbach might see the use to which the concept of the *Bildungstrieb* had been put. In the *Kritik*, Kant introduced the notion of the *Bildungstrieb* at the beginning of a long appendix discussing the "methodology of teleological judgment." As with Blumenbach himself, Kant urged the idea both as a solution to the problem of the origin of organic form and as a way of comprehending how organisms achieved species-specific goals—both perennial concerns of the philosophers of nature.

In his consideration of these latter topics, Kant broached two interrelated conceptions that would come to dominate theories of life through the next century. The first was that of the *archetype*; the second, more dubious, idea was that of a gradual biological development, that is, an evolution (in our sense) of animal forms out of the inorganic and their continued transformation into the multitude of species. In his discussion, Kant admitted that animal species, despite their variety, seemed to display common patterns or archetypes (*Urbilde*). The vertebrates, for instance, all displayed common parts—articulated backbone, ribs, skull, limbs, and so on. On this basis, we might suspect that mechanical transformations of an archetypical pattern could indeed have produced the various species:

> Many animal species resemble one another according to a certain common scheme, which scheme seems to lie at the foundation not only of the structure of their bones but also of the ordering of their other parts, so that the proliferation of species might arise according to a simple outline: the shortening of one part or the lengthening of another, the development of one part or the atrophy of another. This possibility produces a faint ray of hope that something might be done with the principle of mechanism, without which no natural science can generally be constituted. This analogy of forms—insofar as they seem to have been produced, despite their differences, according to common archetypes— strengthens the suspicion of a real relationship of these forms by reason of their birth from a common, aboriginal mother [*Urmutter*].[72]

The archaeologist of nature, according to Kant, might make such an assumption and, on the evidence of fossils, even propose that out of a state of

---

72. Kant, *Kritik der Urteilskraft*, in *Werke*, 5:538 (A 363–64, B 368–69).

chaos, "the maternal womb of the earth (like a large animal) might have given birth initially to creatures of a less purposeful form and these to others whose forms became better adapted to their place of origin and their relationships to each other."[73] In outline this theory of development that Kant mooted conformed to suggestions made both by Herder, in his *Ideen zur Philosophie der Geschichte der Menschheit*, and by Blumenbach, in his *Beyträge zur Naturgeschichte*.[74] But in the *Critique*, Kant moderated what had been his initial reaction to proposals like those of Herder, which he had previously considered as "monstrous ideas." Now he had no conceptual objection to what became a "daring adventure of reason." If one actually dared such a theory, one would, nonetheless, have to refrain from supposing a sheer transition from the inorganic to the organic. One would, instead, have to "attribute to this common mother an organization that purposefully produced these creatures, otherwise one could not at all conceive of the possibility of the purposeful form [*Zweckform*] that exists in the production of the animal and plant kingdoms."[75] Purposeful organization, in Kant's judgment, could only be understood by us as the crafted product of an intentional being—that is, one that acted consciously for ends—ultimately an "*intellectus archetypus*," as he called it.[76] For from mechanism alone, one could not understand the possibility of purpose.[77] This was why he found Blumenbach's principle of the *Bildungstrieb* so attractive—because he interpreted the biologist to be saying that ultimately only organized matter could causally produce organized matter.[78] But in the end, Kant rejected the theory of species transition, even under the aegis of the *Bildungstrieb*, since he did not believe any empirical evidence supported it.

For Kant, the *Bildungstrieb* united, as he mentioned in his letter to Blu-

---

73. Ibid., 5:539 (A 365, B 369–70).
74. See my discussion above.
75. Ibid., 5:539 (A 365, B 370).
76. Ibid., 5:526 (A 346–47, B 350–51).
77. See chapter 2 for an account of Kant's conceptions of mechanism and purpose, and their opposition.
78. Two years before the publication of the *Kritik der Urteilskraft*, Kant published an essay in *Der Teutsche Merkur* (January 1788, pp. 36–52, and February 1788, pp. 107–36) that made this same point. In that essay he declared, citing Blumenbach: "For my part I derive all organization from organic beings (through reproduction) and later forms (of these kinds of natural individuals) according to the laws of general development of such forms from aboriginal dispositions [*Anlagen*]—of the sort that are often found in the transplanting of vegetation. These dispositions often characterize the original stem of these organisms. How the stem itself arose is a question beyond the limits of physics possible for human beings, within whose ambit I must remain." Kant, "Über den Gebrauch teleologischer Prinzipien in der Philosophie," in *Werke*, 5:164. Kant credited Bonnet with the idea that organisms might harbor aboriginal dispositions that would manifest themselves in a changed environment (ibid., note).

menbach, mechanistic considerations with teleological. The biologist himself had likened the *Bildungstrieb* to Newton's gravitational force and discriminated distinct natural laws specified by the force. Because the principle had that Newtonian ring, it sounded mechanistic, and thus could play
a role in scientific judgment about organisms. Yet it also implied an ultimate causality having intellectual features. As Kant put it in the third *Critique*:

> In all physical explanations of this sort [i.e., epigenesis], Blumenbach be-
> gins with organized matter. He rightly declares as unreasonable any pro-
> posals that raw matter has originally formed itself according to mech-
> anistic laws, that out of the nature of the lifeless, life has sprung, and that
> matter could have produced in itself a form of self-preserving purposive-
> ness. Under this principle of an original organization (a principle we can-
> not further explore), he provides an undeterminable but unmistakable
> role for natural mechanism [*Naturmechanism*]. To this ability of matter in
> an organized body (which ability he distinguishes from the commonly
> present, merely mechanistic power of formation), he gave the name *Bil-
> dungstrieb* (and this latter guides and directs the mechanistic power of for-
> mation).[79]

For Kant, the postulation of the *Bildungstrieb* was supported by our experience of the epigenetic properties of organisms, while making "the smallest
possible expenditure of the supernatural" in explanations of phenomena.[80]
Yet the concept of the *Bildungstrieb* did spend the currency of the supernatural (i.e., the nonmechanical) in explanation and thus could not properly
serve as a foundational—that is, *constitutive*—principle of any purported
science of biology. At best the *Bildungstrieb* could only be suggestive and
function as a *regulative* aid for the examination of mechanistic laws involved in the formation and operation of organisms. Kant had met, in the
work of his former student Herder, the "monstrous idea" of vital, organic
development. By the time he confronted the same idea in the scientifically
astute Blumenbach, he thought he had a means to tame it, and so returned
to Blumenbach the now-domesticated idea, apparently declawed of the
threats to science and human freedom that it had initially bared.

Blumenbach was obviously flattered by Kant's endorsement of a principle with which he was so solidly identified. After 1790 he usually added to
his description of the *Bildungstrieb* a parenthesis, stemming directly from
Kant's letter, that indicated this force "united the mechanistic with the pur-

---

79. Kant, *Kritik der Urteilskraft*, in *Werke*, 5:545 (A 374, B 378–79).
80. Ibid., 5:545 (A 373, B 378).

posively modifiable."[81] But aside from this grateful bow to Kant, was the principle as Blumenbach formulated and used it a sign that he agreed with or adopted the Kantian program, which Lenoir calls teleomechanism?

I have already indicated some of the reasons why it would be a mistake to interpret Blumenbach's principle in the fashion of Lenoir and other like-minded historians. Let me now try to indicate that deep divide across which Kant and Blumenbach made overtures to one another but which they never successfully bridged.[82] First, Kant interpreted the theory of the *Bildungstrieb* as implying that we could only understand a particular zoological organization by assuming it had come from matter that already had organization. In drawing this implication, he meant one of two things: he either meant that, for instance, the genital fluid of an animal pair already had organization, which then could produce a further-developed organization— that is, the fetus; or he meant something more general, namely, that only a being with organization (the mother) could produce another being with organization (the child). But neither of these interpretations can really be squared with Blumenbach's general theory. In distinction to Kant, Blumenbach wanted to explain the origin of organization in the first place. If Blumenbach were merely contending that biological organization came from biological organization, he would not have needed to postulate the *Bildungstrieb*. One could hardly claim any originality in asserting *omne vivum a vivo*. Moreover, Blumenbach certainly denied that the genital fluid had any initial organization, though it was organizable—hence the first interpretation of Kant's conclusion is precluded. And so is the second: Blumenbach proposed that a pre-Adamite creation had entirely disappeared—hence there was no residual organized matter—but that it was then replaced by a new creation produced by the *Bildungstrieb,* which operated in materially different circumstances. In both of these examples, new vital forms arose in matter initially bereft entirely of biological organization.

The principal objection, however, to amalgamation of the Kantian and Blumenbachean research projects concerns their respective understanding of the science of life. For Kant, an organism was one in which "every [part] is reciprocally an end and a means."[83] One could not therefore explain why a particular part existed in an organism, except that it was understood as ei-

---

81. See, for example, Blumenbach, *Handbuch der Naturgeschichte* (12th ed.), p. 17.

82. Phillip Sloan has meticulously shown the ways in which, on the related issue of the nature of species, Kant and Blumenbach creatively misunderstood one another. See his "Buffon, German Biology, and the Historical Interpretation of Biological Species," *British Journal for the History of Science* 12 (1979): 109–53.

83. Kant, *Kritik der Urteilskraft*, in *Werke*, 5:488 (A 292, B 295–96).

ther a goal of certain physiological processes or as a means to achieve some other process or structure. In short, one had to conceive of an organism as realizing a *Bauplan*, a network of purposeful design. Yet, Kant insisted, no such concept of a purposeful design could play any *constitutive* role in scientific explanation. If we considered, for example, the structure of a bird's anatomy, we would find its hollow bones, the angle of connection of its wings, the structure of its tail feathers, and so on all directed to the purpose of flight. Without the concept of flight as the end or purpose, we could not understand the necessary unity of configuration found in the bird's anatomy. Kant maintained, in other words, that

> nature, regarded as mere mechanism, could have formed in a thousand different ways without coming to that unity [of organization] according to such a principle [of purpose]; thus one cannot hope to have any foundation purely a priori for such unity, except that we look beyond the concept of nature, not within it.[84]

But in scientific explanation, we had to stay within the bounds of nature, not vault beyond it into the transcendent realm. In Kant's view, only mechanistic principles or laws involving mechanistic causes could really serve to explain natural phenomena, organic or otherwise. Principles that jumped the world to come, leaping over the limits of mechanism, simply landed beyond the range of sober science.

In Kant's scheme such principles as the *Bildungstrieb* could only play a heuristic or regulative role in a discipline; such principles, to put it in the Kantian jargon, resulted from *reflective* as opposed to *determinative* judgment. As he argued in the first introduction to the *Kritik der Urteilskraft:* "A concept [of purposiveness], though it certainly does not objectively determine the synthetic unity [of experience of objects] the way a category does, it nonetheless provides a subjective consideration that serves as a guide [*Leitfaden*] for research into nature."[85] This meant that the biologist could treat organisms *as if* they were teleologically regulated, *as if* the idea of the whole, its design, operated to organize the parts, to cause them to develop toward certain ends. This would be an aid for the discovery of those mechanistic laws that could actually be employed to explain certain operations of creatures. For example, to presume as a regulative idea that the vertebrate eye has the purpose of providing accurate information about the environment and that it does so for the welfare of the entire organism—this allows one to explore how that end is accomplished. The physiologist might then discover

84. Ibid., 5:470 (A 265, B 269).
85. Kant, "Erste Fassung der Einleitung in die Kritik der Urteilskraft," in *Werke*, 5:181n.

that an image on the retina serves that ultimate goal. Further, the researcher might then explore just how an image gets cast onto the retina. Examination might show, then, that the various translucent media of the eye have performed that function because of their refractive properties. The researcher might then apply Snell's law of refraction, a quantitative, mechanistic law, to understand how light rays are bent by the cornea, the aqueous humor, the lens, and the vitreous humor to form an image on the retina. Snell's law then allows the physiologist properly to explain, *given the arrangement or organization* of various refracting media, the mechanisms by which an image is produced on the retina. No mechanistic law, however, can strictly explain—according to the Kantian view—why or how the various media of the eye are so organized. As Kant succinctly phrased it: There could be no Newton of the grass blade.[86] On the other hand, nonmechanistic principles likewise could not *properly* be used to explain biological organization; for in such efforts, "reason is betrayed into poetic swooning."[87]

As part of his explanatory methodology, Blumenbach, of course, made no such distinction between a regulative, reflective principle and a constitutive, determinate principle. He blissfully used the *Bildungstrieb* as part of a constitutively causal account of organization. After 1790 he continued to employ the *Bildungstrieb* in the formation of general laws, comparable to the way Newton used the concept of gravity. He thus conceived of this teleological principle as quite analogous to a mechanistic principle in its explanatory function, something simply unacceptable to Kant.[88] Kant employed Blumenbach's rather loose theory for his own ends. Schelling also scrutinized Blumenbach's principle and, under the guidance of Kielmeyer and Reil, would turn it into something that might even be called a teleomechanistic principle of explanation. I will turn to Schelling's contribution in a moment. We next need to consider those other two late-eighteenth-century theorists whose work closely paralleled Blumenbach's but rejected Kant's arguments about the status of organic principles—the biologists Carl Friedrich Kielmeyer (1765–1844) and Johann Christian Reil (1759–1813).

---

86. Ibid., 5:516 (A 334, B 338).

87. Ibid., 5:529 (A 351, B 355).

88. Larson says (*Interpreting Nature*, p. 178) that "only after Blumenbach had introduced the principle of an original organization was he in a position to 'prove' the theory of epigenesis. His principle was a determinate concept, but its application remained reflective." But there is no evidence that Blumenbach made any such distinction between a determinate and reflective use of a concept, implicitly or explicitly. Moreover, it is unclear what it would mean, even in Kant's terms, to say that the principle was a "determinate" one, but its application was "reflective." A determinate concept would play a role in the constitution of nature—it would cognize particulars as being under a universal notion. A reflective concept is one that arises from examination of particulars; it is not antecedent to such examination.

Chapter 6

KIELMEYER AND THE ORGANIC POWERS OF NATURE

The goal [of the Idealists] is to save the moral freedom of humanity by representing the objective as a product of mind.

—Carl Friedrich Kielmeyer to Georges Cuvier

Carl Friedrich Kielmeyer remains for us an enigmatic figure. Being, perhaps, excessively cautious—serving at the astringent grace of Duke Karl Eugen—he published only a very few articles during his lifetime. His then-contemporary fame rested chiefly on one of those articles and on the direct impact he had on his students and associates. Among the latter he could count Karl Schelling, Friedrich's younger brother, and Georges Cuvier, often thought the greatest zoologist and anatomist of the nineteenth century. Cuvier wrote in 1793 that Kielmeyer was a friend whom "I will always regard as my master, and will admire his genius as much as I love his character."[1] Kielmeyer's larger fame fed on his published lecture "Ueber die Verhältnisse der organischen Kräfte" (On the relationships of the organic forces, 1793), which Friedrich Schelling reckoned "a talk that future ages, without doubt, will regard as an epoch of the new natural history."[2] Alexander von Humboldt concurred; he judged that the essay vaulted Kielmeyer to the rank of "the first physiologist of Ger-

---

1. Georges Cuvier to Friedrich Kielmeyer (25 October 1793), quoted by Kielmeyer's student Georg Jaeger, in his "Ehrengedächtniss der Königl. Würtembergischen Staatsraths von Kielmeyer," *Novorum Actorum Academiae Caesareae Leopoldino-Carolinae Naturae Curiosorum (Verhandlungen der Kaiserlichen Leopoldinisch-Carolinischen Akademie der Naturforscher)* 13, 2nd part (1845): XL. Jaeger discusses Kielmeyer's various ideas based on published works and on manuscripts he received from the widow.

2. Friedrich Schelling, *Von der Weltseele, eine Hypothese der höheren Physik zur Erklärung des allgemeinen Organismus,* in *Schellings Werke,* ed. Manfred Schröter, 3rd ed., 12 vols. (Munich: C. H. Beck, 1927–59), I:633 [II:565]. The volume and page reference in brackets are to *Sämtliche Werke,* ed. K. F. A. Schelling, 14 vols. (Stuttgart: Cotta'scher Verlag, 1857). This latter is the standard though outmoded edition.

many."[3] And Goethe sought him out personally to confer about matters broached in the piece.[4]

Schelling's organic conception of nature, which he detailed in his *Weltseele*, owed much to this paragon of the natural historian, as I will describe in chapter 8. For his part, Kielmeyer, perhaps not fully cognizant of Schelling's debt, thought that the philosopher had advanced an understanding of nature in full conformity to his own, even if they had reached their common view by different means. He expressed this judgment in a note to one of Schelling's critics:

> What my countryman and friend Schelling has up to now produced from the depths of the human mind [*Geistes*] concerning external nature could and should agree with what I have perceived in external nature—the world of appearance; however, since no written communication occurred [between us], this agreement arose quite independently for each.[5]

Kielmeyer's defense of Schelling must, however, be put in the context of his general reaction to idealistic philosophy, a matter I will take up below.

Kielmeyer, born in 1765, entered Duke Karl Eugen's Karlsschule, outside of Stuttgart, in 1774. The Karlsschule, a military academy that trained administrators for civil service in Württemberg, cared for all the necessities of its students, except their need for intellectual and social freedom. Friedrich Schiller, a fellow student at the institute, bridled under the rigidity of the discipline and finally went over the wall. Cuvier, who entered the academy two years after Kielmeyer, seems, however, to have adjusted to the restrictions of the life, perhaps guided by his older friend and mentor. Kielmeyer himself stayed to deliver his famous lecture on the occasion of Karl Eugen's birthday in 1793. He always felt a deep gratitude for his training in the Karlsschule, a training that encompassed languages (ancient and modern), geography, history, philosophy, mathematics, physics, chemistry, and natural history. After receiving his medical degree in 1786, he sought permis-

---

3. Alexander von Humboldt made this declaration in the dedication of his *Beobachtungen aus der Zoologie und vergleichenden Anatomie: Gesammelt auf einer Reise nach den Tropen-Ländern des neuen Kontinents* (Tübingen: F. G. Cotta, 1806).

4. At the beginning of September 1797, on the occasion of his third trip to Switzerland, Goethe stopped at Tübingen and met with Kielmeyer. He previously had asked for some of Kielmeyer's lectures on physiology to be transcribed, and he certainly was familiar with his published essay. See Goethe's record of the meeting in *Goethes Tagebücher*, in *Goethes Werke* (Weimar Ausgabe), 133 vols. (Weimar: Hermann Böhlau, 1888), 2.3:129–30. See also Heinrich Balass's discussion, "Kielmeyer als Biologe," *Sudhoffs Archiv für Geschichte der Medizin* 23 (1930): 271–72.

5. This manuscript is quoted by Jaeger in his "Ehrengedächtniss," p. LXXXI.

sion from the duke to travel to Göttingen to study with Blumenbach and Georg Christoph Lichtenberg, which he did to lasting effect. Dutifully, he returned to the Karlsschule and was appointed in 1790 teacher of zoology and curator of the natural history collection, and in 1792 ordinary professor of chemistry. When the Karlsschule collapsed in financial ruin after the profligate Karl Eugen died, Kielmeyer's growing fame secured for him a call to Tübingen as professor of chemistry in 1796. At Tübingen he extended his teaching domain to materia medica and botany as well. In 1816 he returned to Stuttgart, where he was appointed director for science and art at the state library; his tenure ended after he suffered a series of strokes, which led to his death in 1844.[6] Though Kielmeyer's fame dimmed over the next century, his lecture of 1793 retained enough of a spark to allow a historiographical rekindling in our own time.[7]

6. The details of Kielmeyer's life are provided in ibid., and in his own autobiographical sketch done in 1801–2. The latter is found, along with the bulk of Kielmeyer's unpublished work, in Carl Friedrich von Kielmeyer, *Gesammelte Schriften*, ed. F.-H. Holler (Berlin: W. Keiper, 1938), pp. 7–12.

7. The first ignition occurred with the republication of Kielmeyer's lecture of 1793, accompanied by two hagiographic articles. See Kielmeyer, "Ueber die Verhältnisse der organischen Kräfte unter einander in der Reihe der verschiedenen Organisationen, die Gesetze und Folgen dieser Verhältnisse," *Sudhoffs Archiv für Geschichte der Medizin* 23 (1930): 247–67. The essay is buttressed by Felix Buttersack's "Karl Friedrich Kielmeyer, ein vergessene Genie," ibid., pp. 236–46, and Heinrich Balass's "Kielmeyer als Biologe," pp. 268–88. Dorothea Kuhn has provided an important analysis of Kielmeyer's position on materialism and organicism in her "Uhrwerk oder Organismus, Karl Friedrich Kielmeyers System der organischen Kräfte," *Nova Acta Leopoldina (Abhandlungen der Deutschen Akademie der Naturforscher Leopoldina)* n.s. 36, no. 198 (1970): 157–67. Just prior to the Second World War, a collection of Kielmeyer's unpublished work appeared in a small volume: *Gesammelte Schriften*, mentioned in the previous note. William Coleman, in an illuminating essay, discusses Kielmeyer's caution in accepting recapitulationism. Coleman too easily presumes, I believe, that Kielmeyer simply adopted the Kantian perspective in opposition to that of the *Naturphilosophen*. See William Coleman, "Limits of the Recapitulation Theory: Carl Friedrich Kielmeyer's Critique of the Presumed Parallelism of Earth History, Ontogeny, and the Present Order of Organisms," *Isis* 64 (1973): 341–50. Lenoir elaborates on Kielmeyer's developmental proposals in comprehensive fashion. See Timothy Lenoir, *Strategy of Life: Teleology and Mechanics in Nineteenth-Century German Biology*, 2nd ed. (Chicago: University of Chicago Press, 1989), pp. 37–53. I will, though, continue to dissent from his effort to amalgamate the work of such zoologists as Kielmeyer to Kant's system. Jantzen provides a brief discussion of Kielmeyer's paper. See Jörg Jantzen, *Physiologische Theorien*, in *Friedrich Wilhelm Joseph Schelling, Ergänzungsband zu Werke Band 5 bis 9: Wissenschaftshistorischer Bericht zu Schellings naturphilosophischen Schriften, 1797–1800* (Stuttgart: Frommann-Holzboog, 1994), pp. 511–15. More recently, Kielmeyer's Rede has been published in facsimile, along with a splendid introduction; see Kielmeyer, *Ueber die Verhältnisse der organischen Kräfte*, with an introduction by Kai Torsten Kanz (Marburg an der Lahn: Basilisken-Presse, 1993). Kanz has edited a collection of articles on Kielmeyer: Kai Torsten Kanz, ed., *Philosophie des Organischen in der Goethezeit: Studien zu Werk und Wirkung des Naturforschers Carl Friedrich Kielmeyer (1765–1844)* (Stuttgart: Franz Steiner Verlag,1994). Finally, see Kanz's comprehensive bibliography of works by and about Kielmeyer: Kai Torsten Kanz, *Kielmeyer-Bibliographie* (Stuttgart: Verlag für Geschichte der Naturwissenschaften und der Technik, 1991).

Lecture on Organic Forces

Kielmeyer's lecture on the "Organic Forces" requires careful analysis to recover those features that often attracted his contemporaries and that have just as often eluded ours. In his essay Kielmeyer produced an argument of subtle intent, which piqued Goethe's interest and about which he wanted to talk when they met in Tübingen. Kielmeyer attempted to demonstrate from inductive evidence that teleological laws operated in nature in the same manner as mechanistic laws and that such telic laws explained the balance of faculties throughout the animal kingdom. Additionally, these laws gave an account not only of the development of the embryo, but also the development of species out of one another and, ultimately, out of the inorganic. Under the aegis of the forces described by such laws, nature had, he thought, constituted herself a teleologically balanced and organically connected *cosmos*.

At the beginning of his lecture, Kielmeyer introduced a definition of life that certainly appears to have been gotten from Kant, whose works he knew quite well.[8] An organism, he proposed, was a system in which the organs were "so united that each becomes mutually cause and effect of the other"; any alterations, then, would require a readjustment of the entire system. This causal configuration, he suggested, also characterized the organization of members of the same species (e.g., children depending on adults and vice versa) and the relationships of members of different species to one another, so that "all other species are bound together in a system of effects to produce the great machine of the organic world."[9]

Kielmeyer then formulated an initial and quite subtle argument, which is easy enough to misinterpret. He thought his auditors had to keep in mind, in light of the apparent organic interactions ramifying through the world, this small consideration:

Let us grant that nature had no intention in establishing this artful juxtapositioning of appearances in time, that effects and their consequences were to form no goals that she had wished to achieve; let us grant, it were an empty dream [*Träumerei*] for us to wish to detect some higher goal yet unbeknownst to us; nonetheless, we still must confess that the chain of effects and causes in most cases seems like a chain of means and ends to us and that we would find it advantageous for our reason to assume such a chain; and so we will at least be in a position finally to confirm that na-

8. Jaeger, in his "Ehrengedächtniss," p. LXXX, indicates that Kielmeyer's library held a large number of Kant's writings.
9. Kielmeyer, "Verhältnisse der organischen Kräfte," p. 249.

ture in these instances, no less than in the case of the heavens, is able to convince us of the truth of those observations with which I began.[10]

Kielmeyer's remarks were grammatically cast as counterfactual subjunctives; they granted that nature might not have any intrinsic purposes and that the search for a higher goal might appear to be an empty fantasy. His concessions seem to have had rather direct reference to an argument of Kant. In the third *Critique*, Kant had maintained that any inductions focused on supposed real organic processes in nature (as opposed to mechanical interactions) would lead only to poetic fantasizing (*dichterisch zu schwärmen*), as we conjured up teleological principles supposedly governing such processes.[11] But Kielmeyer's point was that such inductions nonetheless convinced us—and properly so—that nature was teleologically (and thus intentionally) structured; we would be convinced in this case no less than in the case of inductions of mechanistic laws governing astronomical phenomena. Kant, of course, would not grant parity to physical laws and so-called organic laws. The remainder of Kielmeyer's lecture detailed those organic laws so induced—that is, laws of organic adaptation derived via means-ends balancing. Moreover, those laws, he proposed, had a quantitative character—exactly the kind of property Kant believed required for natural scientific laws, but also the kind of property that he maintained could not be found systematically characterizing organic relationships.[12]

10. Ibid., pp. 249–50. Kielmeyer's original manuscript, which served for the oral presentation, delivers these passages with a different rhetorical spin. The implications of the manuscript initially seem more favorable to a Kantian interpretation, but I think it yet admits of the same construal I give above for the published version. In the manuscript, nature herself is personified as addressing the audience: "And if now some of you, whose courage and boldness I must myself at least admire, should say directly to my face that I expressed no intention [*Absicht*] at the time in all this artful interweaving of appearances, that actions and consequences were not purposes [*Zwecke*] which I had wished to achieve—I want to say it myself, I had no intention—yet it must nonetheless be acknowledged that the chain of actions and effects still appears (I want here to flatter your vanity and compare me to you) as if it were a chain of means and intentions of the sort you might form, and that your reason will find it more tractable, when you wish to comprehend these things easier and better, to assume such a chain. Nature could not long continue thus to speak, though she had not yet touched on the more important thing, I mean the wonder of this ability to so represent the situation; and if we ignore the small bits of boasting that are noticeable here and there, we will at least of necessity admit that she has been able to convince us of the truth of those things I spoke of at the beginning of this talk." See Kielmeyer, *Gesammelte Schriften*, pp. 66–67. Lenoir and Coleman assume that these remarks support Kant's conception of teleology as merely regulative. I think they do not appreciate the playful ambiguity of Kielmeyer's remarks—academic lectures need not be lethally flatfooted. I will discuss Kielmeyer's stated objections to Kant at the end of this chapter.

11. Kant, *Kritik der Urteilskraft*, in *Immanuel Kant Werke*, ed. Wilhelm Weischedel, 6 vols. (Wiesbaden: Insel Verlag, 1957), p. 529 (A 351, B 355).

12. In the introduction to the *Metaphysische Anfangsgründe der Naturwissenschaft* (*Werke*, 5:14 [a viii–ix]), Kant declared: "In each particular natural discipline, one meets only so much real science [*eigentliche Wissenschaft*] therein as there is mathematics to be met." He explicitly excluded in

Reil and Schelling, as will be discussed, were in perfect accord with Kielmeyer's challenge to this Kantian orthodoxy.

Kielmeyer distinguished five different organic forces: (1) sensibility, or the ability of the nerves to retain representations; (2) irritability, or the ability of muscles and other organs to respond to stimulation through contraction; (3) reproductive force (*Reproductionskraft*), or the ability of organization to restore injured parts of a creature or to produce a new individual of like kind; (4) secretive force, or the ability to deliver different juices to the right places; and (5) propulsive force, or the ability to move fluids through vessels, especially in plants. Kielmeyer preferred not to treat the last two forces in his lecture, but formulated several laws involving the first three. So, for example, he proposed as a general law governing sensibility that "the number of sensory modalities decreases in the series of organization as the acuity and refinement of the remaining modalities in the limited sphere increases"—thus, worms lack eyes and ears but have an exquisite sense of touch.[13] Some of the laws that Kielmeyer formulated involved the interactions of more than one force. Thus concerning irritability: "The irritability of an organism increases—with respect to the durability of its expression—as the rapidity and frequency or variability of this expression and the variability of sensation decreases"—thus, a turtle with head and heart excised would still slowly move its limbs, unlike the warm-blooded animals.[14] The reproductive force exhibited several laws of still greater complication. For instance, "the more the reproductive force expresses itself in one determinate place by the number of new individuals, the less is the size of the bodies of the new individuals, the simpler the form of the body, the shorter the time in formation in the body of the parent, and the less extended their life-span—or at least some of these latter quantities are diminished"—thus, offspring of worms are more numerous, simpler, and of shorter life than offspring of mammals.[15] As suggested by the last quoted phrase, Kielmeyer recognized that these organic inductions had exceptions and thus held only approximately. This would certainly characterize most biological laws. But this feature did not move him to think biology less of an authentic science than, say, physics. Rather, "the simplicity of laws governing such diverse phenomena," he thought, had to impress one. And indeed,

---

that introduction both psychology (*Seelenlehre*) and chemistry from the realm of authentic science. Kielmeyer had read Kant's book (Jaeger, "Ehrengedächtniss," p. LXXX); but Kielmeyer was a chemist, who, it may be presumed, had thought otherwise about the *wissenschaftlich* character of his discipline.

13. Kielmeyer, "Verhältnisse der organischen Kräfte," p. 254.

14. Ibid., p. 256.

15. Ibid., p. 259.

they all could be summarized in one law: "The more one of these forces, from one side, increases, the more would those on the other side be reduced."[16]

At the end of his lecture, Kielmeyer drew some far-reaching conclusions from his analysis of the shifting equilibria expressed throughout the living world. These conclusions had to do with the origin of species. As part of the analysis of balance, he observed that the embryos of animals higher in the series of organized beings seemed to pass through stages similar to the stages characterizing the organization of extant animals lower in the series.[17] Kielmeyer may not have been precisely the first to enunciate this principle of recapitulation,[18] but he indeed seems to have been the first to draw from it extraordinary implications for species evolution, implications that would shape evolutionary thought through the end of the next century. In the following passage, he sketches out an ancestor, the archetype, as it were, of the theory we recognize as Darwin and Haeckel's conception of evolutionary recapitulationism. Kielmeyer, though, placed emphasis on the recapitulation of forces in embryogenesis and species genesis and only suggested the recapitulation of morphological types, which later formulations of the principle made central. He expressed the principle in his lecture thusly:

> Since the distribution of forces in the series of organizations [i.e., of organized beings] follows the same order as the distribution of forces in the different developmental stages of a given individual, so it can be concluded that the force through which the production in this latter case occurs—namely, the reproductive force—agrees in its laws with that force through which the series of different organizations of the earth were called into existence. And since the lowest classes, in which the individuals are so numerous, are productive of the most numerous species, we are readily permitted to assume that the force through which the series of species are produced, in respect of its nature and laws, is probably one with the force through which the different developmental stages are ef-

---

16. Ibid., p. 261.

17. Kielmeyer reiterated his conviction about recapitulation to an acquaintance, Karl Joseph Windischmann, a physician and physiologist. The letter (25 November 1804) is published under the title "Ideen einer Entwicklungsgeschichte der Erde und ihrer Organisation, Schreiben an Windischmann, 1804," in Kielmeyer, *Gesammelte Schriften*, pp. 203–10. See Coleman's discussion in "The Limits of the Recapitulation Theory."

18. John Hunter, in 1782, was likely the first to suggest in any public way that the embryos of higher animals went through the morphological forms of lower. See my *Meaning of Evolution: The Morphological Construction and Ideological Reconstruction of Darwin's Theory* (Chicago: University of Chicago Press, 1992), p. 18.

fected. Indeed, if this were the place to follow this idea out, it could be shown that one is guided through carefully investigated analogies to assume that the kind of material cause that explains the phenomena of development can also be represented as effective in the first production of organization on our earth.[19]

At one level of analysis, Kielmeyer's postulation of a reproductive force (*Reproductionskraft*) to explain phenomena of regeneration and embryogenesis echoes his teacher's similar postulation of the *Bildungstrieb*. And insofar as Blumenbach wished to formulate authentic laws of biological processes, their endeavors were virtually identical. In the case of Blumenbach, however, one hardly needs a powerful spyglass to observe the Creator standing just behind the force being wielded. With Kielmeyer, however, no trace of the deity is visible. His proposed laws of organic processes were completely naturalistic, something Schelling found quite congenial to his own *Naturphilosophie*.

One striking difference between Blumenbach's and Kielmeyer's understanding of the relationship between embryogenesis and species genesis, despite considerable similarities, was the latter's advancement of the theory of recapitulation. From the observation of embryological recapitulation of the morphological forms of lower species, Kielmeyer concluded that the same force was at work in reproducing the forms at both the individual and species level, and that very likely this same force operated to bring organisms originally out of inorganic matter. Kielmeyer obviously built upon notions of the historical development of species inaugurated by Herder and Blumenbach. He added to their ideas the important conception of recapitulation—that is, the rehearsal by the fetus of the same morphological stages as exemplified in the systematic gradations of extant organisms. He also suggested, though more cryptically and vaguely, that the fetus passed through the same stages as the species historically traversed in its evolution out of lower species. From our perspective, this begins to look much more like a modern theory of species evolution, even if the contours yet lack

19. Kielmeyer, "Verhältnisse der organischen Kräfte," p. 262. Kanz maintains that Kielmeyer's principle of recapitulation only regarded a similarity in physiological states of embryos of more developed creatures and those of the various classes of lower animals found on the earth. He thinks Kielmeyer's principle is thus different from the classical formulation of recapitulation, which regarded *morphological* stages of organisms. Kielmeyer's principle, as here expressed, is ambiguous, but certainly the comparison he draws had to be based on morphology, since the physiological stages—that is, the organic forces to which Kielmeyer referred—can only be expressed in morphology. See Kanz, "Einleitung," to the facsimile reprint, pp. 63–67.

sharp definition. In succeeding years, Kielmeyer would refine his ideas about species origin and development, ideas that he would communicate widely to his students, friends, and associates.[20]

Theory of Species Origin and Transformation

Kielmeyer only cryptically suggested an evolutionary theory in his lecture on organic forces; but some years later, in 1804, he gave a rather extended account of his ideas in a letter to a friend, the Bavarian court physician Karl Joseph Windischmann (1775–1839).[21] Kielmeyer had developed his initial conceptions in advance of Lamarck and Erasmus Darwin, but by the time he wrote Windischmann, he had become familiar at least with Lamarck's theory. In his letter he particularly concentrated on the relationship between the history of the earth—its geological formations and the forces operating on it—and the history of organic forms that populated the planet.

Kielmeyer told Windischmann that he had become convinced of the intimate relationship between these two histories because he considered the force that produced the series of organic bodies to be the same as that which produced the stages in the development of the individual. He explained to his friend that he found the singular force to be quite strongly analogous to magnetism, which was intimately involved in the development of the inorganic world. He thus set up a threefold parallel between the developmen-

---

20. One of Kielmeyer's students, Johann Tritschler, advanced the same theory as his teacher but employed the embryological term "evolution" to characterize also species development. Here we see that gradual transmutation of ideas in which, in this case, a term used to describe one process was shifted to describe a similar process in a different domain. Tritschler wrote: "The series of animals is continuous and directly ascending, with each member distinguished by its grade of evolution and heterogeneity of its organs" [*Seriem animalium esse continuam, recta ascendentem quarum membra modo different gradu evolutionis et heterogeneitatis organorum*]. This line from Tritschler is furnished by Buttersack in "Karl Friedrich Kielmeyer, ein vergessenes Genie," p. 237. I have traced the historical transition of the term and concept of evolution from its first biological use in the description of embryological preformationism to its use in describing species descent in my *Meaning of Evolution*.

21. See Kielmeyer to Karl Joseph Windischmann (25 November 1804), in Kielmeyer, *Gesammelte Schriften*, pp. 203–10. Coleman discusses this letter in his "Limits of the Recapitulation Theory." In his analysis, Coleman stresses what he takes to be Kielmeyer's reservations about a threefold recapitulational relation of geological formations, embryological forms, and species forms. Coleman also attributes to Kielmeyer a commitment to Kantian epistemological restrictions. Coleman, I believe, wanted to save Kielmeyer from the historical clutches of Schelling, and so argued as if Kielmeyer were virtually rejecting the recapitulational theory he in fact advanced. It was a theory, of course, that Kant would not have endorsed. Moreover, Kielmeyer had some deep objections to Kant's basic epistemology; and though he had difficulties as well with the idealists, he certainly inclined more to them than to the Kantians. I discuss Kielmeyer's critique of Kant and the idealists below.

tal history of the earth, the series of organic forms (*Organisationenreihe*), and the organic stages in individual gestation. Just as Charles Darwin and Ernst Haeckel later in the century, Kielmeyer became convinced that these parallel series would reflect one another. Observationally, one could pick out similar aspects of extant species gradations and ontogenic patterns of development, but, as he admitted, the similarities would be considerably less than exact. And their relationship with the history of the earth would be even more problematic, at least as he conceived that history. For by the earth's history, he seems to have meant the geological formations rather than, or at least as much as, the fossil remains.[22] Kielmeyer's conviction of the threefold parallel relied as much on his causal analysis of the identical force behind their respective developmental histories as upon immediate observational evidence. For with changes in the earth, as he explained to Windischmann, changes in the power and direction of its magnetic force ought to be produced and with a consequent effect on the production of organic bodies, "the children of the earth." As he proposed to his friend, we finally come to the conclusion that "the development of our earth and that of the series of organic bodies are precisely connected with one another and that, therefore, their histories must be joined."[23]

Though Kielmeyer remained convinced of the three-part relationship, he doubted that one could empirically recapture the precise parallels.[24] Several obstacles intervened. First, the progressive gradation of extant organic forms had gaps (*Intervalle*), thus making correlation with the geological record difficult. While these gaps—such as those existing even within the forms of one plant (for instance, between leaf and stem)—constituted perfectly normal phenomena in patterns of development, they did not yield to lawlike interpretation. As much as Kielmeyer himself had tried to formulate such laws, he confessed to his friend that he remained in the dark about the explanatory principles at play in gradations of organic forms as well as about those principles governing the earth's magnetism. Second, he suspected that organizational development differed at different times in the earth's history. At one time, one species might have arisen out of another— like the butterfly out of the caterpillar—with the parent species having gone extinct. In this instance, "they would have initially been develop-

---

22. Later in the century, Ernst Haeckel would make central to his argument for evolution the threefold parallelism holding among the series of fossil remains, the stages of embryological development, and the history of species. See Ernst Haeckel, "Ueber die Entwickelungstheorie Darwin's," *Amtlicher Bericht über die acht und dreissigste Versammlung Deutscher Naturforscher und Ärzte in Stettin* (Stettin: Hessenland's Buchdruckerei, 1864), pp. 17–30.

23. Kielmeyer, *Gesammelte Schriften*, p. 206.

24. Ibid., pp. 207–10.

mental stages [of the same creature] and later would have become permanent species." But at another time in the earth's history, the original species might have perdured to the present as "primal children of the birthing earth."[25] Finally, "as Schelling believes," certain current species might have been the result of retrogression from a more advanced form. Or there was even the possibility that a foreign, unearthly type had arrived from a different planet! For all of these reasons, while he stood convinced of developmental progression of species in parallel with the earth's history and that of the individual, he concluded that one would have to give up any thought of precisely articulating the laws governing these correlations.[26]

Kielmeyer's evolutionary views bear direct relationship with those of Schelling. His employment of polar forces, and especially magnetism as a centrally operating principle that might produce organisms out of the bowels of the earth—all of this bore striking resemblance to the more metaphysically grounded ideas of the philosopher—as I will discuss more thoroughly below. And, indeed, Kielmeyer expressed common cause with his fellow landsman in the letter I quoted at the beginning of this chapter. Perhaps because of similarities in their positions and because of Schelling's admiring remarks about Kielmeyer, Cuvier put the question of transcendental philosophy to his one-time colleague.

Critique of Kant and the Idealists

Cuvier wrote Kielmeyer seeking his opinion about the new philosophy sweeping Germany. The inquiry had some bite, since Cuvier believed his friend had really taken up the cause of Schelling's biology.[27] Kielmeyer more than obliged with an extensive essay in the form of a letter, which critically analyzed both Kant's philosophy and the newer idealism. Kielmeyer distinguished an older idealism and a newer idealism. The older sought no explanation of our representations of an apparent world, rather indulged only in skeptical paradox. The newer idealism really did attempt to explain not only how the formal aspects of experience (for instance, space, time, and causality) flowed from the ego but how the material content of the world, the manifold of sensation, also spilled from this primal

25. Ibid., p. 209.

26. Ibid., p. 210.

27. In Cuvier's last lecture before his death, given in May 1832, he observed that "Schelling was the one who first began to develop that idea [of variety in unity]. . . . Thereafter Kielmeyer developed it much more completely and thus he is the true founder of nature philosophy [la philosophie de la nature]." This passage and Cuvier's judgment are discussed by Thomas Bach. See his "Kielmeyer als 'Vater der Naturphilosophie'? Anmerkungen zu seiner Rezeption im deutschen Idealismus," in Kanz, Philosophie des Organischen in der Goethezeit, pp. 232–51; quotation from pp. 237–38n.

source. He recognized two generic kinds of the newer idealism: one that supposed the not-I arose as an impulse (*Anstoss*) of our consciousness; the other that supposed the opposition of an I and not-I stemmed from some latent, more fundamental condition, or simply left the opposition as an unexplained primitive fact (*Urfaktum*).[28] This division of idealisms apparently referred to the positions, respectively, of Fichte and of Schelling. None of these positions, Kielmeyer thought, adequately accounted for the projection of an objective nature. Mostly they only assumed what they ought to have explained.[29] Yet Kielmeyer did not wish to deny the worthiness of the attempt "to bring unity to all human knowledge" and "to provide a complete genealogy of our knowledge and an illustration of the descent of nature [*eine Stammtafel der Natur*]." Moreover, the idealists had a high purpose— "to save the moral freedom of humanity by representing the objective as a product of mind."[30] In summing up his view of the new philosophy, Kielmeyer ticked off to Cuvier some four other of its advantages. First, it emphasized that much of our presumed objective knowledge of nature derived from the subject. Second, it provided a dynamic interpretation of matter as arising out of polar forces, an interpretation that simply put mechanism and atomism "in the shadows." Third, it did not really depend on abstruse metaphysics, rather on the shrewdness of the idealists themselves, to exhibit the dualism of forces that gave rise to organic bodies. Kielmeyer thought he had discovered the explanatory power of polar forces quite independently of the idealists, but concurred in their view that such organic forces were likely transformations of magnetism and electricity. Finally, the new philosophy "accustoms one more than before to the idea that nature at large should be regarded as an organism, living in all its actions, and that particular organized bodies ought to be thought of as individualizing representatives of the larger nature—an idea that is already to be found in the old opposition between the microcosm and macrocosm, between organism and universe."[31] In this last, Kielmeyer, of course, focused on the central concept of Schelling's identity philosophy.

In contrast to the idealists, Kant's position seemed initially to have the advantage, in that he relegated the content of experience, the particulars of the manifold of sensation, to an independent objective sphere, the thing-in-itself. Yet on closer inspection, in Kielmeyer's estimation, the critical philosophy had the sort of weaknesses that made the idealist response inevitable: If mind were active in respect to a passive, unreachable thing-

28. Kielmeyer, *Gesammelte Schriften*, pp. 237–38.
29. Ibid., pp. 238–39.
30. Ibid., pp. 239–40, 240.
31. Ibid., pp. 249–51.

in-itself, then perceptions must originate effectively in the subjective sphere—which was only really a short step away from full-blown idealism; if matter were only the synthesis of repulsive and attractive forces, then differences in matter could only be quantitative—which meant that matter also ought to be derivable a priori, in the way all quantitative relations were in the Kantian system; and if a limiting manifold were a requirement of knowledge, then it would be a subjective condition of the possibility of knowing—it would not differ from temporal, spatial, and causal conditions in epistemological status. (These objections closely paralleled those of Schelling.) Kielmeyer's most serious objection to Kantianism, however, was its inability really to make a separation of the objective from the subjective. After all, the subjective sphere of the categories could only be attained through experience, which meant that it too was only appearance, not the mental apparatus in itself. Thus "the subjective remains always and eternally unknown in itself, like some x, just as the object in the Kantian system remains unknown in itself."[32] Ultimately, then, to distinguish the object from the subject in Kant's philosophy would be an impossible task.[33]

In the early nineteenth century, Kielmeyer's reputation as an exact and creative scientist had few rivals. He developed a conception of organic powers, much in the tradition of his own teacher Blumenbach, that closely allied his ideas with those of the *Naturphilosophen*, especially Schelling. Even his epistemological and metaphysical notions approached those of Schelling and were certainly in deep sympathy with the latter's aims. Kielmeyer himself did not hide his affinity for the philosopher, which undoubtedly strengthened the general impact of Schelling's ideas on other of his contemporaries. From our historical distance, however, that impact has become only faintly discernible, so faint as to be denied by most historians who have recently considered a possible relationship. There are two related historiographic impulses that have produced these nervous denials. First is the general motive of virtually every conscientious historian: the desire to render the subject under the most just and appreciative light. But, second, historians tend to turn up the wattage of that light, obliterating unhappy shadows, by connecting it to the best source of contemporary power, modern science. Certainly someone like Kielmeyer, who appears to have been one of those harbingers of our enlightenment, must likewise have shared our contemporary disdain for the kind of Romantic metaphysics Schelling infused into biology. Yet the denial of the relationship between Schelling and Kielmeyer suggests a lack of negative capability, an unwillingness to sus-

32. Ibid., p. 246.
33. Ibid., pp. 241–49.

pend disbelief. When, on the other hand, such capacity is exercised by the historian, much can be revealed—certainly the insight, learning, passion, and sheer genius that electrified virtually all of Schelling's ideas, which would have made them quite attractive to the likes of Kielmeyer. Yet if the connection between these thinkers seems too tenuous to strengthen the case for the considerable influence of Romantic philosophy on nineteenth-century biology, then one needs consider the unmistakable relationship between Schelling and another of his illustrious contemporaries, the physician Johann Christian Reil. Reil initially conceived of living organisms in a way completely antithetic to that of Schelling, but he came dramatically around to embrace what he before had disdained. Some have attributed this dizzying turn to war; I think it rather due to romance.

Chapter 7

JOHANN CHRISTIAN REIL'S ROMANTIC
THEORIES OF LIFE AND MIND,
OR RHAPSODIES ON A CAT-PIANO

> Over all of this, a sublime group of speculative *Naturphilosophen*
> soars like an eagle. They assimilate their earthly booty into the
> purest ether and return it again as beautiful poetry.
>
> —Johann Christian Reil, *Rhapsodieen*

Johann Christian Reil (1759–1813), one of the most famous medical theorists and physicians of his time, followed an intellectual path paralleling
that of Kielmeyer (though he strode a social path of considerably greater elevation). He initially formulated a conception of living forces that made
them logically indistinct from standard forces of physics and chemistry—
and thus adopted a position antithetic to Kant's assumption that organic
phenomena had to be, as a regulative measure, conceived as teleological.
The weight he gave to materialism began to shift, however, after the turn of
the century, as he perceived the intellectual advantages offered by the new
Romantic thought that had recently taken wing in Jena and Berlin. Two
kinds of experience helped sharpen his perception in this respect. The first
sort of experience arose from distinct areas of his medical practice, psychiatry and gynecology. Reil's study of the insane opened him to a more decidedly organic understanding. In the mad he found revealed, by its absence,
the kind of subtle psychological organization that passed unobserved in
normal life. The other medical area that altered his perspective was not the
pathological but the magically and creatively healthy—the production of
new life in human pregnancy. In both of these medical domains, Reil perceived Schelling's ideas to fit neatly with his own observations. Indeed, he
shifted his epistemological orientation and gave the new categories of transcendental philosophy a specifically physiological construction. The second major influence on his thought was cultural. The Romantic movement
had spread to Halle, where Reil held a longtime professorship, and it enveloped him in its characteristic poetry, music, and philosophy. In that atmosphere, the literary impulse that must have been natively his grew to

powerful expression in various regions of his intellectual life. His metaphorical turn of mind swung sweetly to those Romantic strains, casting its similes now into poetry—surreptitiously published—and now into highly original and inventive medical tracts—of the sort that would recommend for therapy the wonderful instrument of the cat-piano (*Katzenclavier*). And in this new cultural dispensation, he became intimate friends with recently arrived colleagues at Halle, who came trailing clouds of Romantic sentiment—especially Henrik Steffens, Schelling's disciple, and Friedrich Schleiermacher, the divine whose religious thought fed on the poetic transcendentalism cultivated in Jena. Under the inspiration received from this creative matrix of professional and cultural forces, Reil stepped into the new century in the increasingly more intimate embrace of a Schellingian *Naturphilosophie*.

Early Training and Practice

Anna Reil (née Jansen-Streng), wife of the Lutheran pastor Johann Julius Reil, gave birth to her first child, Johann Christian, on 20 February 1759 in the little village of Rhaude (Kreis Leer) in the far northwestern part of the German lands.[1] In 1770 the family moved to the costal town of Norden, where Reil matriculated at the Ulrichsgymnasium. His valedictory address to the school revealed the direction of both his intellectual and imaginative aptitudes—he poetized in long strings of alexandrines, singing the praise of medicine and cleverly, if exhaustively, taking leave of his king, town council, mayor, rector, fellow students, sisters, parents, and his land.[2] In fall of 1779, this gifted, though stubborn and anti-authoritarian, Frisian began his university studies at Göttingen, but quickly became dissatisfied and left for Halle the next year. At Halle he came under the influence of the

1. I have found the following works helpful in reconstructing Reil's life and thought: Henrik Steffens, *Johann Christian Reil: Eine Denkschrift* (Halle: Curtschen Buchhandlung, 1815); Max Neuburger, *Johann Christian Reil: Gedenkrede* (Stuttgart: Verlag von Ferdinand Enke, 1913); Günter Engelberg, *Aus dem Leben des Dr. J. C. Reil: 19 Beilagen des General-Anzeiger Westrhauderfehn* (Westrhauderfehn, 22 November 1958–28 March 1959); and Hans-Heinz Eulner, "Johann Christian Reil: Leben und Werk," *Nova Acta Leopoldina* (*Abhandlungen der Deutschen Akademie der Naturforscher Leopoldina*), n.s. 20 (1960): 7–50. The most comprehensive and provocative work on Reil is Reinhard Mocek's *Johann Christian Reil (1759–1813)* (Frankfurt a. M.: Peter Lang, 1995). Mocek orients his study around the question of when Reil became a follower of Schelling—a question to which I will attend below. Reil's position within the larger context of German medicine is lucidly recounted in two other studies: Cheryce Kramer, "The Psychic Constitution: Psychiatry in Early 19th Century Germany" (master's thesis, University of Cambridge, 1991); and Thomas Broman, *The Transformation of German Academic Medicine, 1750–1820* (Cambridge: Cambridge University Press, 1996), especially pp. 86–88, 120–22, 182–85.

2. This address is transcribed in Engelberg, *Aus dem Leben des Dr. J. C. Reil*, pp. 17–20.

great anatomist Philipp Friedrich Meckel (1755–1803), who poignantly proclaimed his enthusiasm for anatomy by requesting that upon death his body be preserved for scientific delectation. Johann Friedrich Goldhagen (1742–1788), in charge of the natural history cabinet of the university, became a special friend, introducing his student to the mysteries of Free-masonry and sponsoring him as a lodge brother just after Reil had received his medical degree in 1782.

Later in that same year, Reil traveled to Berlin for further medical study. He carried with him a letter of introduction written by Goldhagen to a for-mer student of his, Marcus Herz. During the young physician's brief time in the city (1782–83), he became a regular guest at the Herz salon and there frequented the lecture sessions in physiology, medicine, and philosophy or-chestrated by his host. Herz had also been a student of Kant and had be-come an enthusiast for the new critical philosophy under the guidance of the master himself.[3] The *Kritik der reinen Vernunft* (1781) had appeared just prior to Reil's arrival, and so he and other visitors to the salon were treated to seminars on the ideas buried under the very solid covers of the book. On the periphery of Herz's demonstrations and lectures stood his beautiful, young wife, Henriette, who was but eighteen years old at the time of Reil's visits. She had only begun cultivating the artists, writers, philosophers, and statesmen who would find her charms more compelling than those of her husband.[4]

After a brief stint as a practicing physician in East Friesland, Reil re-ceived a call back to Halle in 1787, arranged by his former teacher Gold-hagen, who also obliged by dying shortly after Reil's arrival. With the new vacancy, Reil swiftly ascended from extraordinarius professor to ordinarius and then to director of the clinical institute there (founded by Goldhagen) in the space of a few months. In 1789 he was named chief physician of the city and had many of its leading citizens as his patients, as well as those, like Goethe, who would travel some distance to seek his help.[5]

Reil's practical experience, his freethinking ways, his deep knowledge of science and philosophy, as well as his difficult and contrarian personality—to which his colleague Henrik Steffens would lovingly attest—all of this

---

3. Marcus Herz had been the respondent for Kant's inaugural dissertation at Königsberg—*De mundi sensibilis atque intelligibilis forma et principiis* (Concerning the forms and principles of the sen-sible and intelligible world, 1770)—which established preliminary critical principles that would be further developed ten years later in the *Kritik der reinen Vernunft* (1781). Herz's own *Betrachtun-gen aus der spekulativen Weltweisheit* (Considerations drawn from speculative worldly wisdom, 1771) closely mirrored his teacher's dissertation.

4. See chapter 2 for a depiction of the Herz salon.

5. Steffens, *Johann Christian Reil*, pp. 14–15; Engelberg, *Aus dem Leben des Dr. J. C. Reil*, pp. 27–38.

gave him a taste for revolution, at least in medicine.[6] In 1795 he founded the journal *Archiv für die Physiologie*, which became the herald of the movement he led, namely, the effort to make medicine a thoroughgoing *Wissenschaft*.[7] He intended physiology to serve as the scientific foundation for medicine. But that discipline itself first had to be reformed, since, as Reil believed, physiology had made less progress than the other sciences. He diagnosed the problem, in the spirit of Herz and that greater teacher Kant, as that of not having properly observed the boundaries of human knowledge. Inattention to those limitations allowed, he insisted, the obfuscating ideas of metaphysics to be smuggled into the realm of the physical sciences.[8] The conception that had caused most of the problems in physiology was that of a *Lebenskraft*, of a life force. He offered his monograph "Von der Lebenskraft," which formed the lead article of the new journal, as a specific for that malady.

Lebenskraft

In his tract Reil accepted the epistemology of Kant's first *Critique*, though rejected the regulative biology of the third. In the spirit of the former, he agreed that we had access only to phenomena, which in the external realm had to be scientifically understood in causal and spatial terms. Accordingly, he thought Blumenbach's concept of the *Bildungstrieb* conveyed misleading implications, since *Trieb* (drive) suggested that some feeling or mental representation functioned in the interstices of purely material operations. Physiological processes, he maintained, were determined by "blind necessity," not by any ghostly intentions.[9] Reil did not, however, reject a mental

---

6. Steffens, *Johann Christian Reil*, pp. 13–14.

7. In the preface to the first number of his new journal, which took the form of an open letter to professors Friedrich Albrecht Carl Gren (1760–1798) and Ludwig Heinrich Jakob (1759–1827), his chemistry and philosophy colleagues, respectively, Reil made explicit his intention to bring medicine into the "form of a science [*Wissenschaft*]." He would further this aim by publishing in his journal articles on physiology that adopted the basic principles he laid down in his lead essay "Von der Lebenskraft." See Johann Christian Reil, "An die Professoren Herrn Gren und Herrn Jakob in Halle," *Archiv für die Physiologie* 1 (1795): 3. For a discussion of Reil's efforts to make medicine *wissenschaftlich*, see the instructive article by Thomas Broman, "University Reform in Medical Thought at the End of the Eighteenth Century," *Osiris*, 2nd ser. 5 (1989): 35–53.

8. Even before the publication of Kant's first *Critique*, Marcus Herz was proclaiming the doctrines of his master. In his *Betrachtungen aus der spekulativen Weltweisheit* (Königsberg: Hohann Jakob Kanter, 1771), Herz laid it down as a rule that "the fundamental principles of our sensible knowledge must never overstep their limits and become mixed with those that our pure rational knowledge has prescribed. . . . It is indeed rather clear that those things that can in no way become known intuitively are unthinkable and therefore impossible" (pp. 105–6).

9. Reil, "Von der Lebenskraft," *Archiv für die Physiologie* 1, no. 1 (1795): 66–67. (The title page of the first volume carries the date 1796.)

Figure 7.1
Johann Christian Reil
(1759–1813), who
established German
psychiatry from a
Romantic perspective.
Lithograph by L. Noel.
(Courtesy Wellcome
Institute Library.).

principle tout court—since he himself, at this early stage of his reflections, embraced a kind of Kantian dualism. The understanding of cognitive processes, therefore, might require principles other than those of physics and chemistry, but the construal of basic physiological activity did not. Reil insisted that he would seek

> the foundation of all phenomena of animal bodies, which are not representations or not connected with representations [i.e., mental operations] as cause or effect, in animal matter, in the original differences of their basic material and in the composition [*Mischung*] and form [*Form*] of that material.[10]

Sensibility, digestion, growth, reproduction—all the physiological operations of the body had thus to be understood purely as functions of the chemical composition and structural form of the material elements making up the animal organism. These basic material elements could also be found in nonliving bodies, though in combinations differing from those found in living ones.

10. Ibid., p. 11.

According to Reil, the conventional name for the property of matter that allows it to be perceived by human senses is "force" [*Kraft*].[11] In itself, though, "force is a subjective concept, the form according to which we think the connection between cause and effect."[12] Talk about life forces, then, ultimately referred only to the necessary causal interactions of the material elements of an organic body and their outcomes. "If it were possible for us to think clearly of each body as it is—simultaneously of the nature of its constituent elements and their connection, of their composition and form—then we would not find," Reil contended, "the concept of force necessary, a concept that produces so many erroneous conclusions."[13]

Reil's notion of what might constitute *Lebenskraft* differed considerably, for example, from Alexander von Humboldt's early conception of *Lebenskraft* and Blumenbach's comparable theory of the *Bildungstrieb*. Humboldt had supposed that a vital force operated in living bodies to suspend the usual affinities of chemical elements in animal matter and to prevent those elements from freely uniting as they would in inanimate compositions.[14] Blumenbach, who had a comprehensive theory of vital force, held the *Bildungstrieb* to be an intermediate cause determined by some hidden, more ultimate cause (God); consequently it acted as a secondary cause to produce more immediately the phenomena of life—generation, nutrition, and repair.[15] Reil, by contrast, thought of the life force as a subjective concept; its objective referent, he argued, was merely the causal relationships between the chemical elements that produced the phenomena of life. He con-

---

11. Ibid., p. 19.

12. Ibid., p. 46. For a discussion of Reil's concern with the subjective side of medical knowledge, see Broman's brief but clear discussion in his *Transformation of German Academic Medicine*, pp. 86–88.

13. Reil, "Von der Lebenskraft," p. 46.

14. Humboldt hardly had an extensive theory of vital force, though he did have some presentiments. In the appended *Aphorismi* to his *Florae Fribergensis*, he remarked: "We call those bodies animated and organic that, though they tend constantly to change into new forms, are contained by some internal force, so that they do not relinquish that form originally introduced. . . . That internal force [*vim internam*] which dissolves the bonds of chemical affinity and prevents the elements of bodies from freely uniting, we call vital." See Alexander von Humboldt, *Florae Fribergensis specimen, plantas cryptogamicas praesertim subterraneas exhibens* (Berolini: H. A. Rottman, 1793), pp. 133–35 (nos. 1 and 2). Humboldt dressed up this idea in a metaphorical story, "Die Lebenskraft oder der rhodische Genius, ein Erzählung," that Schiller published in his journal *Die Horen*, part 5 (1795): 90–96. Humboldt reprinted his fable in the second and third editions of his *Ansichten der Natur*. See Alexander von Humboldt, *Ansichten der Natur*, 3rd ed. (Stuttgart: Cotta'schen Buchhandlung, 1849), pp. 315–24. Humboldt, responding to Reil's criticisms, later reformulated his conception of vital force in decidedly more neutral and ambiguous terms. For further discussion, see the appendix to part 2.

15. See Johann Friedrich Blumenbach, *Über den Bildungstrieb und das Zeugungsgeschäfte* (Göttingen: Johann Christian Dieterich, 1781).

cluded, therefore, that life forces pretending to be something other than causal functions of matter should not be invoked in a truly scientific physiology. The so-called life force was only another physical force of nature.[16]

In his monograph Reil advanced a quite materialistic and mechanistic conception of organic processes. The fundamental operations of assimilation, growth, and reproduction occurred by way of external matter (in the form of food), first being electively attracted to the right parts of an animal body and then being chemically altered through a kind of "crystallization of animal matter."[17] In the case of reproduction, Reil did not wish to pronounce definitively on the debates between evolutionists (i.e., preformationists) and epigeneticists.[18] "How does the seed [Keim] arise? How it is formed? Whether it contains the entire organic individual in miniature or only a part of it, and which part? This we do not know."[19] The kind of analysis he undertook, however, required him to reject a thoroughgoing epigenesis, for that would suggest the operation of an extrinsic force, a *Lebenskraft*, that brought form out of strictly homogeneous matter. But whether the paternal germ only stimulated the maternal seed or actually contributed a part to it, the formation of the fetus after conception had to be analogous, Reil thought, to crystallization.[20]

Reil's metaphor of crystallization spread through all parts of his treatment of biological phenomena. Since crystallization was a clear example of inanimate assimilation and growth, it served well his fundamental assumption that organic life derived from basic chemical and physical forces. This virtue provided the metaphor with an intellectual momentum that would propel it through the next century, allowing such biologists as Theodor Schwann (1810–1882) and Ernst Haeckel (1834–1919) to hitch to it their own theories of organic formation.

In Reil's case, however, some conceptually uneven patches jarred the otherwise smooth carriage ride to the concluding legislative summary of his monograph. First, he felt compelled to assume that, in addition to the grosser, palpable matter of living bodies, there had to be a finer, less visible matter that provided a necessary, interactive ingredient in the mixture of life. After all, from a crude perspective, there seemed little physical or chemical difference between a living body and one that had just died. And

16. Theodore Ziolkowski—who provides a sharply analytical, if brief, account of Reil's work in psychiatry—believes the "Lebenskraft" article to be an example of Romantic *Naturphilosophie*. I think he was greatly misled by the title of the piece. See Ziolkowski, *German Romanticism and Its Institutions* (Princeton: Princeton University Press, 1990), p. 182.

17. Reil, "Von der Lebenskraft," p. 67.

18. See the discussion of "evolution" and epigenesis in chapter 5.

19. Ibid., p. 79.

20. Ibid., p. 80.

certain other considerations—for example, the rapidity of changes in animate processes—suggested to him the presence of this finer, more labile matter. He conjectured that this special element had to be added to the composition of bodies in order to "perfect" the more obvious material constituents of an organism.[21] The fine matter might be caloric fluid, electricity, oxygen, or something else. We simply did not have sufficient experience, he believed, to know exactly the nature of this ingredient.[22] He did resolutely believe, though, that life arose as a force out of a determinate range of chemical interactions of certain gross and fine matter, of the sort that could also be found, though in different mixtures, in nonliving objects.[23]

Though animal and plant life were completely material phenomena, Reil did not think matter could compose itself spontaneously into living organisms. He would not, for instance, have attempted the fanciful experiment of Goethe's Wagner. The elements of organic bodies, while they could be found in dead nature, required a catalyst to organize them into a living being—and the catalyst could only be life itself. Some seed or stock, therefore, was needed to allow the process of food assimilation to occur and thus to crystallize living out of nonliving matter. In short, only living creatures could produce flesh and bone out of inorganic elements.[24] But this last requirement did not lead Reil to adopt anything like Kant's solution to the problem of the organic. Indeed, he explicitly rejected Kant's analysis in the third *Critique*.

Kant also maintained that we had to conceive of life as coming only from previous life, not because we thought some spark was needed, but because life itself seemed the only source for an organization that elevated a creature to a higher level of being than that of a mere chemical-mechanical contrivance. He believed we properly conceived an organism as a "natural end" [*Naturzweck*], that is, as an individual in which each part was related to all the other parts reciprocally as means and ends, and in which the whole determined the parts and the parts the whole.[25] After citing Kant's analysis of our necessary understanding of the organic, Reil dismissed it.[26] "Indeed," he contended, "each part forms itself and maintains itself through its own energy; its connection with the other parts is only the external determination whereby its force can be effective."[27] Thus no power or idea of the

21. Ibid., pp. 30–31.
22. Ibid., p. 31.
23. Ibid., p. 40.
24. Ibid., pp. 25–26.
25. Immanuel Kant, *Kritik der Urteilskraft*, in *Immanuel Kant Werke*, ed. Wilhelm Weischedel, 6 vols. (Wiesbaden: Insel-Verlag, 1957), 5:483–88 (A 285–91, B 289–95).
26. Reil, "Von der Lebenskraft," pp. 54–57.
27. Ibid., p. 55.

whole was required for the organization of the parts. As a good physician, Reil knew that many important organs of the body could fail—for instance, particular muscle groups, this or that sense organ, the higher faculties of the brain—without endangering the maintenance of the whole. He sagely observed that "we often find in an animal one part that is principally good or bad, in opposition to the character of the other parts. With scholars, not infrequently, we find all the organs ill or failing, except the brain."[28] The individual parts of the body, he maintained, existed fairly independently of one another. In a metaphor that would resonate with many biologists throughout the nineteenth century, Reil likened the parts of an organic body to a republic:

> The animal body is like a large republic, which consists of many parts. These parts, of course, stand in a determinate relationship with one another and they contribute to the maintenance of the whole. But each part operates through its own force and possesses its own perfections, deficiencies, and failures independently of the other branches of the body.[29]

The force that each organ exhibited came not from other organs or the whole, but from the form and composition of the matter of that particular part.[30] The phenomena of life, according to this analysis, stemmed from the constituent, citizenlike parts in their external, causal interactions with one another. Reil simply denied the teleological relationship that Kant prescribed for organisms.[31] If Prussian imperialism could not prevail on the battlefield, it should not reign in the laboratory.[32]

Through the 1790s Reil remained convinced that the organization of the animal body, in sickness and in health, would arise from the bottom up, that is, would emerge from the interaction of lower forces. He pressed this conviction through vigorous and often scathing reviews of books that pos-

28. Ibid., pp. 105–6.
29. Ibid., p. 105.
30. Ibid.
31. Timothy Lenoir regards Reil, like Blumenbach and Kielmeyer, as an advocate of a teleomechanist view derived from Kant. Lenoir, *The Strategy of Life: Teleology and Mechanics in Nineteenth-Century German Biology* (Chicago: University of Chicago Press, 1989), pp. 35–37. If Reil can be so easily assimilated to a position he expressly denied, then nothing is less fixed than the past.
32. The potency of this republican metaphor—exercised just at the time that the French had humiliated the Prussian king and forced him into neutrality (1795)—could not fail to excite a liberal readership sympathetic to the Revolution. The theory of the cell state and its rather different political associations in the latter part of the nineteenth century, especially in the science of Rudolf Virchow and Ernst Haeckel, has been explored by Paul Weindling. See his "Theories of the Cell State in Imperial Germany," in *Biology, Medicine and Society, 1840–1940*, ed. Charles Webster (Cambridge: Cambridge University Press, 1981), pp. 99–156.

tulated a force of life as a cause independent of the forces of physics and chemistry.[33] Reil's basic assumption had extreme inertial power, continuing on as a fundamental principle for Germany's leading medical authority at midcentury, Rudolf Virchow (1821–1902). Reil himself, though, did not remain fixed to his original assumption. By the turn of the century, his experience and considerations ripened him for a keener appreciation of a top-down view, a view taken from cultivated Romantic heights. He exhibited this new perspective in his writings on mental illness.

Studies of Mental Illness

The first of Reil's systematic considerations of various forms of psychological disturbance came in his book *Fieberhafte Nervenkrankheiten* (Feverish nervous illness, 1802), volume 4 of his *Ueber die Erkenntniss und Cur der Fieber* (On the understanding and cure of fever, 1799–1815).[34] The dedicatory page of this fourth volume must have arrested the eye of his readers then as it does today: "To the Supreme Consul of the French Republic, Buonaparte [*sic*], a knowledgeable friend of the sciences." (One is tempted to believe that Reil crafted his dedication ironically, namely, to the healer of a feverish and mentally disturbed nation.)[35]

Reil treated mental illness as part of his five-volume work ostensibly because derangement often accompanied fevers; but his own experience with friends who suffered various disturbances, which he mentioned among his examples, must also have stimulated his interest in these phenomena.[36] He augmented his experience in this area by considerable reading in the pertinent literature, including Pinel's just published *Traité sur l'aliénation mentale*. Kant's first *Critique* also played a significantly informing role.

---

33. For example, Reil in quick succession reviewed Johann David Brandis's *Versuch über die Lebenskraft* (Hannover: Hahn'schen Buchhandlung, 1795) and Christoph Wilhelm Hufeland's *Ideen über Pathogenie und Einfluss der Lebenskraft auf Entstehung und Form der Krankheiten* (Jena: Akademische Buchhandlung, 1795). The review of Brandis was mildly approving of the particular features of the book, though skeptical of its appeal to a *Lebenskraft*; in the case of Hufeland, however, he dished sarcasm and ridicule into the lap of his friend, certainly confirming Steffens's estimate of his difficult character. See Reil, "Rezensionen," *Archiv für die Physiologie* 1 (1796): 178–92; 2 (1797): 149–52.

34. Reil, *Ueber die Erkenntniss und Cur der Fieber*, 5 vols. (Halle: Curtschen Buchhandlung, 1799–1815). Volume 4, *Fieberhafte Nervenkrankheiten*, was published in 1802.

35. Kramer surmises this as well. See her "Psychic Constitution," p. 51.

36. Reil had attended his friend and mentor Goldhagen during his final illness, which involved hallucinations and other mental aberrations. Reil left a history of this illness in an essay composed in 1788, his "Krankheitsgeschichte der Oberbergraths J. F. G. Goldhagen," reproduced in Reil, *Kleine Schriften wissenschaftlichen und gemeinnützigen Inhalts* (Halle: Curtschen Buchhandlung, 1817), pp. 3–29.

Reil thought of mental illness as a disruption of the normal functioning of the powers of the soul, which he glossed explicitly in Kantian terms.[37] The very basic powers were the typical Kantian ones of consciousness, understanding, reason, imagination, and sensibility. In his application, though, Reil distinguished three main areas of representational understanding, malfunctions of which might produce illness: representations of "common sense" [*Gemeingefühl*], representations of sensibility, and representations of imagination.[38] Common sense, in Reil's account, consisted in a perception of the well-being of the different parts of the body,[39] and its abnormal activity could result in hypochondria, melancholy, and vertigo, as well as in nonspecific dreaminess. Problems with sensible representations might produce fantasies and hallucinations.[40] And, finally, the classic cases of insanity would arise from an energetic imagination, in which the sufferer could not distinguish manufactured images from reality. This last condition produced those archetypal examples in which "the sick person plays a king, a general, or soldier."[41] Reil thought that Saint Theresa and Emanuel Swedenborg undoubtedly suffered from this kind of derangement.[42]

In his analysis Reil supposed the powers of the soul expressed the more fundamental forces arising out of the composition and form (his basic biochemical categories) of the brain and nerves. "The powers of the soul," he argued, "stand in exact relationship to the structure of the nervous system, which is extended throughout the whole organism and has in each part a particular function."[43] To the soul itself he accorded only a kind of Kantian phenomenal existence, since what it really might be and what connections it might actually have to both the supersensible and corporeal worlds remained "totally unknown to us."[44] The entire direction of Reil's analysis implied that the powers called psychic could ultimately be reduced to forces of the nervous system. In this respect his treatise on mental illness conformed closely to the principles laid down in his monograph "Von der Lebenskraft."

During the year of the publication of his volume on psychiatric fevers, a new world dawned for Reil. Early in 1803 he saw through the press a

37. In *Fieberhafte Nervenkrankheiten*, 4:253, Reil explicitly refers to Kant's *Kritik der reinen Vernunft* as the source for his analysis of the powers of the soul.

38. Ibid., 4:261. Reil had written an extensive essay on the "common sense" earlier in 1794; his "Ueber das Gemeingefühl" was reprinted in his *Kleine Schriften*, pp. 34–112.

39. Reil, *Ueber die Erkenntniss und Cur der Fieber*, 4:261–62.

40. Ibid., 4:265–73, 274–78.

41. Ibid., 4:279.

42. Ibid., 4:288.

43. Ibid., 4:259.

44. Ibid., 4:253.

most unusual work, also on the subject of mental illness, his *Rhapsodieen über die Anwendung der psychischen Curmethode auf Geisteszerrüttungen* (Rhapsodies on the application of psychiatric methods of cure to the mentally disturbed).[45] This book differed fundamentally, however, from his just-completed volume 4 in his series on fevers.

Reil's *Rhapsodieen* became perhaps the most influential work in shaping of German psychiatry before Freud. The conditions of its origin were complex, involving medical practice, philosophical reorientation, and the new cultural environment in which Reil found himself. As before, Reil's experience in treating many in his regular practice who suffered from various mental derangements, including some of his friends (such as Goldhagen and Marcus Herz), obviously played a role.[46] He also wished to contribute, as his preface made clear, to the new movement that sought more humane and rational treatment for the insane, a movement in Halle led by his friend the preacher Heinrich Wagnitz, to whom his book was dedicated. Unlike Reil's previous work, the model of mind that he developed in the *Rhapsodieen* went considerably beyond Kantian boundaries. Quite clearly he had been reading Schelling, whose Romantic idealism, I believe, fundamentally reoriented Reil's understanding of the root causes of mental illness. In the light of this new philosophical conception, Reil came to regard insanity as stemming from the fragmentation of the self, from an incomplete or misformed personality, and from the inability of the self to construct a coherent world of the non-ego—all of which resulted from the malfunctioning of self-consciousness, that fundamentally creative activity of mind postulated by the Romantic philosophers.

Though various particular events could precipitate a breakdown in the self, the undeniable progressive advantages of civilization, Reil thought, in-

45. Reil, *Rhapsodieen über die Anwendung der psychischen Curmethode auf Geisteszerrüttungen* (Halle: Curtschen Buchhandlung, 1803). The translation of *"psychischen"* by "psychiatric" is not inappropriate. In 1808 Reil invented the name *"Psychiaterie"* to designate the new discipline he described in his book. See Reil, "Ueber den Begriff der Medicin und ihre Verzweigungen, besonders in Beziehung auf die Berichtigung der Topik der Psychiaterie," *Beyträge zur Beförderung einer Kurmethode auf Psychischem Wege* 1 (1808): 161–279. See also the brief discussion of Reil's priority in naming, in "Historical Notes: The Word Psychiatry," *American Journal of Psychiatry* 107 (1950–51): 868–69.

46. It seems a historiographic commonplace that Reil had actually little experience with mentally ill patients. But this assumption cannot stand against patent evidence. In his *Rhapsodieen*, Reil specifically recounted the case of his beloved teacher Goldhagen, who, at the end of his life, would wander through his own house to find the sick person, only finally to realize he was searching for himself (ibid., p. 70). He also described the problems of Marcus Herz, who suffered from delusions about his location during his illness (ibid., p. 86). In the *Rhapsodieen*, Reil referred to other patients who had comparable ailments. His good friend Henrik Steffens, who would have known, pointedly declared that Reil "was a very successful psychiatric [*psychischer*] physician, and many who suffered from obvious madness were restored by him." See Steffens, *Johann Christian Reil*, p. 48.

evitably had their dark side. The pressures of advancing culture could also fracture the integrity of the self. In the *Rhapsodieen*, he sharply diagnosed civilization and its discontents:

> Nature has endowed us with so many divine impulses toward lofty and noble deeds, the drive for fame, for one's own perfection, the power of self-determination and perseverance, and the passions, which through their storms guard against the deadly desire for sleep. Yet nature, through these very same inclinations, has also planted in us as many seeds for madness [*Narrheit*]. By equally measured steps, as we advance on the path of our sensible and intellectual culture, we fall back ever nearer to the madhouse.[47]

Thus the same powers of nature that lead us to construct a shining city on a hill can also shatter the self that mirrors the world, so that the city might be duplicated in a frenzy of distorted images. That, at least, is the metaphor with which Reil dramatically opened his volume:

> It is a remarkable experience to step from the whirl of a large city into its madhouse. One finds here repeated still the same scenes, though as in a vaudeville performance; yet, in this fool's system, there exists a kind of easy genius in the whole. The madhouse has its usurpers, tyrants, slaves, criminals, and defenseless martyrs, fools who laugh without cause, and fools who torture themselves without cause. Pride of ancestry, egoism, vanity, greed, and all the other idols of human weakness guide the rudder in this maelstrom, just as in the ocean of the large world.[48]

Aside from the contemporary ring of these portrayals of the troubled soul in its doubles, these passages indicate two significant respects in which the *Rhapsodieen* differed from Reil's previous work, as well as from that of his predecessors. The first difference is in expression. The literary style of the book made it glitter as a quintessential exemplar of the burgeoning Romantic movement. Indeed, its metaphors unleashed a menagerie of wild and wonderful images, and its examples and proposals exuded the macabre and the ironic. As a consequence, the book stood at some distance from the usual run of scientific and medical treatises of the time. But there is an even

---

47. Reil, *Rhapsodieen*, p. 12. Reil had been convinced for some time of the Rousseauean thesis that civilization and its pressures contributed to mental illness. He worked out this thesis in a series of articles published in *Wöchentliche Hallische Anzeigen* in 1788 and 1789. See Michael Hagner's discussion in *Homo cerebralis: Der Wandel vom Seelenorgan zum Gehirn* (Berlin: Berlin Verlag, 1997), p. 160. See also Ziolkowski's lucid description of Reil's account of madness in *German Romanticism and Its Institutions*, p. 184.

48. Reil, *Rhapsodieen*, pp. 7–8.

deeper, if more elusive respect in which the book remained apart from other treatments of madness: Reil suggested that the world of the mad had its own logic, one that duplicated the logic of our own world, albeit with some crucially different premises. The enlightened mind stood close to the brink of a dark abyss.

In the *Rhapsodieen*, Reil again proposed a medical and quasi-physiological interpretation of mind, identifying mental powers quite closely with underlying forces of the brain and nervous system. In "Von der Lebenskraft," Reil had argued that, like a republic, the forces that held an organism together arose from its constituent citizen-elements. Now, however, he represented the nervous system as imperial, a kind of Napoléon that instituted a dynamic ordering for the various lower centers of force. But this Napoléon came up from the people, as it were; that is, the forces expressed at lower levels of interaction gave rise to higher syntheses, to transmutations of organization and powers. This "dynamische Evolution," to use Schelling's terms for the process, would lead ultimately to a controlling force of nervous integration.[49] The nerve complex, Reil urged, was "the knot that ties organization together so as to lift it above lifeless nature as natural purpose [*Natur-Zweck*]."[50] "The brain," in his altered metaphor, "may be conceived as a synthetic product of art, composed of many sounding bodies that stand in a purposeful relationship (i.e., in rapport) with one another."[51] Any change in the brain's components from external sources would then change the orchestration of the whole.

> The ordering of these relations of the parts of the soul's organ is grounded in a determined distribution of forces in the brain and the whole nervous system. If this relationship is disturbed, then arise dissociations, volatile character, abnormal ideas and associations, fixed trains of ideas, and corresponding drives and actions. The faculties of the soul can no longer express the freedom of the will. This is the way the brain of a mad person is produced.[52]

---

49. In *Von der Weltseele*, which Reil read (see below), Schelling argued that the same principle lay at the root of nonliving and living nature, and that principle itself was organic. In his *Erster Entwurf*, Schelling described the further syntheses of the fundamental organic principle, as it passed through higher stages of the inanimate and animate, as a process of "dynamische Evolution." See Friedrich Schelling, *Von der Weltseele, eine Hypothese der höheren Physik zur Erklärung des allgemeinen Organismus* (1797) and *Erster Entwurf eines Systems der Naturphilosophie* (1799), in *Schellings Werke*, ed. Manfred Schröter, 3rd ed., 12 vols. (Munich: C. H. Beck, 1927–59), 1:418, 2:19. For a further discussion of Schelling's concept of dynamic evolution, see chapter 8.

50. Reil, *Rhapsodieen*, p. 112.

51. Ibid., p. 46.

52. Ibid.

Quite obviously Reil now conceived of the nervous system as an integrating force designed to achieve a "natural purpose," precisely the conception of organic activity rejected in his earlier "Von der Lebenskraft."

Like Freud after him, Reil would have preferred an array of medical specifics to reorder defective nervous centers more directly. Little, though, was available. Opium might immediately quiet someone in a frenzy, but the outcome would only be a calmer madman.[53] The *Rhapsodieen* instead proposed an indirect method of cure, which would leap past the obstacles to direct intervention. Reil believed psychological means could be effectively employed to alter deficient ideas and abnormal emotional states, at least of the curably insane.[54] If psychological manipulations were successful, then the underlying nervous connections would be properly readjusted and the rational operations of mentality restored.[55]

It would be a mistake to think of Reil as introducing, via the mind, an indirect means of altering the pathological brain. In his construction, brain and mind became inextricably joined. Indeed, while not worrying about theoretical problems of the mind-body relationship, he treated them as virtually identical, as if mind were completely instantiated in the nervous system.[56] Hence, an altered mind was an altered brain. This practical identity justified, to some extent anyway, his use of a mixed mental and physical vocabulary—though even today it is hardly the style of medical writers scrupulously to avoid verbal and conceptual promiscuity when discussing the operations of the brain or mind. The close identification of brain and mind would have been more offensive had Reil retained Kantian ties. But his mental model, launched just the year before in his volume on fevers, slipped its Kantian moorings and sailed away on currents flowing directly from idealist philosophy.

In the *Rhapsodieen*, Reil distinguished three chief forces of the soul, whose disruption could produce pathology. These were self-consciousness (*Selbstbewusstsein*), prudential awareness (*Besonnenheit*), and attention (*Aufmerksamkeit*). Though he had mentioned the latter two powers in his earlier work on mental illness, he devoted most of his effort in the *Rhapsodieen* to the analysis of a force now considered the most crucial for understanding

---

53. Ibid.

54. Reil knew that some people were irretrievably mad and could only be humanely cared for, not cured. See ibid., pp. 20–21.

55. Ibid., p. 150.

56. Reil periodically asserted in the *Rhapsodieen* that the relationship between mind and body simply remained unclear (e.g., ibid., p. 111). Hagner has also interpreted Reil's theory of mind-brain as asserting a practical identity between the two. See Hagner, *Homo cerebralis*, pp. 157–70.

pathologies, that of self-consciousness and its attendant powers of temporal and spatial perception.

Reil conceived of self-consciousness as a distinctively active force, much in the manner of Schelling, who, unlike Kant, made self-consciousness do real work in the construction of the self and its world. "The essence of self-consciousness," Reil held, "seems chiefly to consist in joining the manifold into unity and assimilating the representations as one's own."[57] Self-consciousness wove together disparate representations (*Vorstellungen*) into a coherent whole and constituted them as *ours*. Without this force, each of us would become "an empty likeness in the mirror of a sea that simply reflects floating objects but cannot hold fast to the received images, cannot make them one's property." Moreover, without self-consciousness, we would lack personality; our history would remain disconnected, scattered shards without solidifying temporal relationships. Self-consciousness "synthesizes the mental man, with his different qualities, into the unity of a person."[58] The child, by contrast, having an incompletely developed power of self-consciousness, would be like the madman whose faculty had deteriorated. The child thus displayed through its deficiencies the vital ties between self-consciousness and personhood:

> The child also observes; it observes itself and the world, but without connection. Its ideas flow disconnectedly, as the images in a brook. It plays with its own limbs, as with trinkets. It feels something, namely, itself, and does so with pleasure or pain, which stimulates it to laugh or cry. But it does not recognize that it is the person who represents the world and through its own self is affected either pleasurably or painfully.[59]

As the individual emerges from childhood, the self-conscious activities that constitute the riper intuitions of time and space create a personality existing within history's horizons. Those intuitions endow us with "a consciousness of the past and connect everything that has seeped into our awareness from the world—all the catastrophes of our bodily and psychic existence, which are reproduced in our memories and fantasies, are joined to the governing ego, to which also our present condition is linked."[60] And the world, too, comes into view from the vantage of our personality. Reil captures this transcendental posture with another of his vivid images:

---

57. Reil, *Rhapsodieen*, p. 54.
58. Ibid., pp. 53–54, 55.
59. Ibid., p. 57.
60. Ibid., p. 58.

The mind, in self-consciousness, rolls up the immeasurable thread of time into a ball, reproduces the long dead centuries, and gathers into the miniature of one representation mountain ranges, rivers, woods, and the stars strewn through the firmament, all stretching into endless space. The mind senses itself as if it were in each representation, it relates what is represented to itself, as the creator of the same, and maintains thereby a special rule over the world outside of itself, insofar as it is representable.[61]

When self-conscious action falters, when pathology of the ego strikes, then personality fragments and the world becomes incoherent. Some people will not be able to distinguish real objects from phantoms of their imaginations. Or they might react in horror to the commonplace, and with indifference to the horrible, as when a patient can endure a needle stuck through an arm without seeming to register pain. The fractured personality might set up a double, in which one part speaks to the other as to a separate individual. At times the deranged person will be convinced that he or she is someone else. A woman of Reil's acquaintance so identified with the refugees from the French Revolution, that she thought herself one of them.[62] When the faculty of prudential awareness (*Besonnenheit*), which keeps mental focus fixed on an object or project, becomes weakened, then attention shifts with the wind, and patients live in another world. But what could be done for such patients? Psychological methods, Reil believed, could be employed.

Rudimentary theories of psychological methods of cure had already begun to spread in Europe and the United States during the Romantic period.[63] Reil made himself quite familiar with the pertinent literature, which drifts through his footnotes. The classic work of the genre, Philippe Pinel's *Traité médico-philosophique sur l'aliénation mentale, ou la Manie*, left, however, a larger wake than the rest. Reil had cited Pinel previously in his book on fevers, but the *Rhapsodieen* made more generous use of the tract, perhaps because a German translation had recently come into his hands.[64] Reil thought there were enough madmen in France after the Revolution to sup-

61. Ibid., p. 55.
62. Ibid., pp. 75–78.
63. See Martin Schrenk's *Über den Umgang mit Geisteskranken* (Berlin: Springer, 1973). Jan Goldstein offers the most detailed analysis of the methods and philosophical framework behind Philippe Pinel's development of the "moral cure" of the insane. See her *Console and Classify: The French Psychiatric Profession in the Nineteenth Century* (Cambridge: Cambridge University Press, 1987), especially pp. 64–119.
64. Philippe Pinel, *Philosophisch-Medicinische Abhandlung über Geistesverirrungen oder Manie*, trans. Michael Wagner (Vienna: Carl Schaumburg und Compagnie, 1801). Though Reil had Pinel's volume at hand when he wrote his work on fevers, he seems not to have had the German translation, which is extensively cited in the *Rhapsodieen*.

ply Pinel with sufficient cases for an extensive study; but he judged the Frenchman's work, while "excellent in particular parts, ill in systematic treatment, and without principles or originality."[65] Reil, one of the most famously accomplished men of his profession, may have been negatively reacting to what Goldstein calls the "populist" character of Pinel's treatment regimens—that is, his borrowing of methods from nonprofessional empirics and quacks, who nonetheless had some success in treating the insane.[66] One might suspect, though, that these more superficial differences between Pinel and Reil disguised a deeper unity of conception, since, after all, they both advocated a kind of psychological therapy. But as one descends to more profound levels, the rifts appear even larger. Pinel had cast his notions about therapy into a Condillacian sensationalism, which supposed that ideas streamed rather directly from the empirical world into the mind; madness, according to this interpretation, resulted from the improper arrangement and ordering of those ideas. Reil, by contrast, assumed in his Schellingian way that consciousness created both the self and the world whence ideas supposedly came. Madness stemmed not from a lack of logic in the ordering of ideas; rather, it flowed in broken rivulets through the cracks of a fragmented personality. Pinel's diagnostic categories—melancholia, mania without delirium, mania with delirium, dementia, and idiotism— were quite general and depended on the kinds of symptoms displayed. Reil's categories were more refined and based on the faculties of mind assumed to be affected.[67] Pinel's recommendations about the sort of moral treatment to be administered lacked specificity; they had no ground, as his German translator judged, "in any established and determinate principle."[68] Pinel, in justifying his admittedly vague suggestions, argued that it was unwise to make general prescriptions for all classes of people—the Jamaican slave surely had to be treated differently than the "well-bred Frenchman."[69] Reil's therapeutic suggestions drew inspiration from Pinel, but also from other authors—such as the Englishmen Francis Willis (1718–1807) and Alexander Crichton (1763–1856), and the Germans Johann Langermann (1768–1832) and Johann Benjamin Erhard (1766–1827). But as compared to the prescriptions of his predecessors, Reil's were considerably more numerous and

65. Reil, *Rhapsodieen*, p. 31.

66. See Goldstein, *Console and Classify*, pp. 72–89.

67. Ziolkowski suggests some of the similarities between Reil and Pinel, but he stops there. He holds that Reil had not made one step beyond Pinel in his analyses—a position, of course, from which I completely dissent. See Ziolkowski, *German Romanticism and Its Institutions*, pp. 185–87.

68. Wagner, translator's introduction, in Pinel, *Philosophisch-Medicinische Abhandlung*, p. ii.

69. Philippe Pinel, *Traité médico-philosophique sur l'aliénation mentale, ou la Manie* (Paris: Richard, Caille et Ravier, 1801), pp. 64–65.

more specific to the disease symptoms and the faculties that produced them. Finally, at the most comprehensive level of intellectual constitution—in the orbits of imaginative construction—Reil and Pinel lived worlds apart. Against the empiricistic and prosaic formulations of Pinel's work stands the rhapsodic and poetic character of Reil's book. The psychiatric tradition that Reil spawned could well have grown to produce a Freud, the French tradition of Pinel, only a Charcot.[70]

The psychological methods that Reil prescribed ranged from the commonsensical to, from our perspective, the bizarre.[71] For example, one might bring a patient to a sense of well-being by exposing him or her to quite normal surroundings and a good diet (even spiking the wine with a bit of opium to produce a warm, contented glow). Plenty of sunshine could also yield positive results. Gymnastic exercises and dancing might harmonize the mind just as they brought the body into balance. Sexual intercourse (Beischlaf), perhaps with a prostitute, could well reduce accumulated lascivious energy that might contribute to mental disturbance.[72] The mad should not be denied reading, learning poetry by heart, and practicing sums. Any well-built asylum, Reil proposed, would include a theater, in which patients would have "their imaginations strongly excited in a purposeful way; their prudential awareness awakened; contrary passions elicited; fear, terror, amazement, anxiety, and tranquility excited; and the fixed ideas of the mad confronted."[73] Reil strongly recommended the use of music, to speak directly to the heart; for "music quiets the storm of the soul, chases away the cloud of gloom, and for a while dampens the uncontrolled tumult of frenzy."[74] These humanizing measures, of course, stood in stark contrast to the wretched conditions endured by the institutionalized insane in much of

70. Schrenk indicates the impact of Pinel and Reil on subsequent generations of psychiatric physicians. See his Über den Umgang mit Geisteskranken, part 3. Doris Kaufmann shows how Reil's considerations reverberated through Prussian social and institutional policy concerning the insane. See especially her Aufklärung, bürgerliche Selbsterfahrung und die 'Erfindung' Psychiatrie in Deutschland, 1770–1850 (Göttingen: Vandenhoeck and Ruprecht, 1995), especially pp. 283–85.

71. Reil, Rhapsodieen, pp. 182–217.

72. Ibid., pp. 185–86.

73. Ibid., p. 210. Kramer discusses in some interesting detail Reil's employment of the theater as a therapeutic tool. See her "Psychic Constitution," pp. 34–36.

74. Reil, Rhapsodieen, p. 207. In the last part of his book, Reil discussed the administrative and architectural requirements for an ideal asylum. Among the features he thought necessary for the accomplishment of cure was an auditorium for plays and musical performances (p. 462). When Christian Roller (1802–1878) set the plans for the first psychiatric institution in Germany built for specific therapeutic purposes—Illenau in Baden—he followed Reil's recommendation and included a musical theater. Cheryce Kramer discuss this and other features of the construction of Illenau in her "Illenau, Château de Plaisir," a paper delivered at the Fishbein Workshop in the History of Human Sciences, 1994, University of Chicago.

the world. They also provided the kind of stimuli that a Romantic personality would regard as deeply restorative.

Reil also recommended what we would call aversive conditioning and even primitive shock therapy. His descriptive accounts of these latter, which build toward a dissonant crescendo, do abruptly remind one of the work's eighteenth-century character. For example, withdrawing food or applying hot wax to the body would restore control to the hitherto uncontrollable, without real damage being inflicted. Hysterical mutes, he assured his readers, had been brought to speak by the application of a strong irritant to the soles of their feet. Cold baths seemed therapeutic for the willfully convulsive. To place an unsuspecting madman in a tub of live eels must, Reil thought, rather strongly "work on his emotions through the torturous play of imagination."[75] For those whose attention could not be easily tamed, Reil recommended the amazing device of a *Katzenclavier*—indeed, a piano made from cats. One would first voice the instrument with suitable animals, which would, then,

> be arranged in a row with their tails stretched behind them. And a keyboard outfitted with sharpened nails would be set over them. The struck cats would provide the sound. A fugue played on this instrument—particularly when the ill person is so placed that he cannot miss the expressions on their faces and the play of these animals—must bring Lot's wife herself from her fixed state into prudential awareness.[76]

But, with his own imagination stimulated, Reil went even further. He observed that "the voice of the jackass is even more heartbreaking." Too bad, he reflected, that despite its small talent, the animal nonetheless suffered from "artistic caprice."[77]

Reil's *Rhapsodieen* marked a new phase in his own intellectual development. The book not so much abandoned the reductive metaphysics of "Von der Lebenskraft" and the Kantian analysis of consciousness of the *Fieber-*

75. Reil, *Rhapsodieen*, p. 190.

76. Ibid., p. 205. The story of the *Katzenclavier* is just the sort of tale Reil might have heard from the musicians at Giebichenstein (see the text below for a description of this salon). Thomas Hankins has briefly traced the history of this amazing instrument, which he supposes was actually constructed—a most improbable event, I believe. Caspar Schott in his *Magia universalis naturae et artis* (1657–59) attributed it to his teacher Athanasius Kircher's *Musurgia universalis* (1650), though Hankins was not able to find it there. Hankins discovered two other descriptions of the device: one in 1725, by Louis-Bertrand Castel, who understood it as a kind of joke; and the other in an anonymous article in a French journal (1883). See Thomas Hankins, "The Ocular Harpsichord of Louis-Bertrand Castel; or, The Instrument that Wasn't," *Osiris* 9 (1994): 141–43. Reil provides one more sighting of this wonderful instrument.

77. Reil, *Rhapsodieen*, p. 205.

Figure 7.2 An early emblem of a *Katzenclavier* (with a dog or two added), from Johann Theodor de Bry, *Emblemata Saecularia Mira et Lucunda* (1596). The verse at the top of the illustration reads: "Come all, gather 'round and see who the most beautiful might be, / But count me in as well, for I am the bass of the choral company." The inscription at the bottom reads: "There is no music sweeter to Midas's ears."

hafte Nervenkrankheiten, as it raised their perspectives to a new and more finely integrated level, to a heightened plane on which Romantic conceptions began to flower. The active and constructive role of self-consciousness became the central principle of the new position. In his previous book on mental illness, Reil considered the insane to have made errors in their understanding of the world. In the *Rhapsodieen*, he depicted the deranged as having succumbed to more than logical mistakes: their metaphysics had gone bad. Usually because of some psychological trauma, the mad faltered in the dynamic construction and continuous reconstruction of their very selves, with the consequence that their personality crumbled and external nature, as it emerged from their egos, withered into phantasms.

Reil's new philosophical perspective on the insane occurred in an atmosphere much altered from that which enveloped his previous studies. The

very tone of the *Rhapsodieen*, its metaphors, examples, and delicious iro-
nies, suffused the work's philosophical and medical considerations with a
deeper appreciation of the poetry of life—its beauty and its tragic features,
to which Reil's description of the delusional last days of his teacher Goldhagen gave poignant illustration. But it is his generous exercise of Romantic irony that has made the work a puzzle to modern readers (and to some of
his contemporaries). Typically, prescriptions such as the cat-piano have led
these critics simply to dismiss Reil as someone who might himself have ben-
efited from therapy. But this attitude misses the strategic uses to which Romantic thinkers put irony. Friedrich Schlegel conceived irony as a galvanic
jolt to heart and brain; it poetically expressed the absolutely required in the
face of the relatively attainable. "Irony springs," he proclaimed, "out of the
union of an artistic sense of life [*Lebenskunstsinn*] and the scientific spirit
[*wissenschaftlicher Geist*], out of a collision of the most complete nature philosophy and the most complete aesthetic philosophy." Schlegel thought it
"a very good sign when harmonious dullards simply do not know how to
take these constant self-parodies; in a vertiginous whirl, they are by turns
credulous and incredulous; they take the joke as serious and the serious as a
joke."[78] In comparable confusion, Reil's recent critics simply have not noticed the new and marvelous territories into which his thought moved by
the end of 1802. In a rhapsodic moment in his book, where he praised efforts at humane treatment of the insane, he more pointedly indicated that
larger environment in which his hopes breathed life for a better dispensation for the mentally ill:

> A bold race dares a tremendous idea, which only excites to vertigo men
> of more common stripe: to eradicate from this globe one of the most di-
> sastrous of plagues. And we seem to be entering the vicinity of the harbor.
> Over all of this, a sublime group of speculative *Naturphilosophen* soars like
> an eagle. They assimilate their earthly booty into the purest ether and re-
> turn it again as beautiful poetry.[79]

The Romantic Movement in Halle

The beautiful poetry to which Reil alluded, he likely heard orally descanted
in the gardens of Johann Friedrich Reichardt (1752–1814), Kapellmeister
to Friedrich the Great and Friedrich Wilhelm II, director and writer of for-
gettable operas for the Berlin stage, but the composer who set to evocative

78. Friedrich Schlegel, *Lyceum Fragmente*, in *Kritische Friedrich-Schlegel-Ausgabe*, ed. Ernst Behler et al., 35 vols. to date (Paderborn: Verlag Ferdinand Schöningh, 1958–), 2:160 (fragment
108).

79. Reil, *Rhapsodieen*, pp. 52–53.

music the lyrics of Schiller, Goethe, and many of the Romantic poets, the editor of the republican journal *Deutschland*, and the author of five beautiful and talented daughters. Reichardt's estate Giebichenstein was located just outside of Halle, and from 1794 to its destruction by Napoléon's troops in 1806, it was the gathering place for some of the most talented poets, philosophers, and musicians of the period. Schiller, the Schlegels, Novalis, Tieck, Jean Paul Richter, Joseph von Eichendorff, and Clemens Brentano all spent a few days to a couple of weeks as guests at the estate. Fichte and Schelling also visited, and during several seasons Goethe heard his own poetry put to music there.[80] And from the university itself, select professors made the guest list. In its wild and wonderful gardens—which a friend of Reichardt's described as "the most beautiful composition of his life and spirit"[81]—the Kapellmeister would arrange ravishing musical performances. On a balmy summer evening, Reichardt might take up his violin or sit down at the piano, and then, accompanied by other visiting musicians and singers, play a new composition for one of Novalis's lyrics. And in these performances, he was always joined by his spirited and highly gifted daughters. Frequently, though, the musicians would sequester themselves in some hidden location, so that their songs would seem to waft from the very trees around the garden. The poet Joseph von Eichendorff (1788–1857) recalled his own evenings at Giebichenstein:

> How completely mysterious appeared much of the garden that Reichardt had laid out at Giebichenstein, as did his striking and beautiful daughters, one of whom composed Goethean songs and another who was Steffens's fiancée. Often on a mild summer evening, out of the mysteriously obscuring boskets, as from a distant magical island, songs and sounds of strings would drift over. How many a young poet there would look in vain through the gates or sit on the garden wall among the budding branches, dreaming up wonderfully crafted stories half the night.[82]

Exactly when Reil became a member of this Romantic community is difficult to determine; but his scientific reputation, his social standing, his ap-

80. The atmosphere at Giebichenstein is artfully described by Walter Salmen in his *Johann Friedrich Reichardt: Komponist, Schriftsteller, Kapellmeister und Verwaltungsbeamter der Goethezeit* (Freiburg i. Br.: Atlantis Verlag, 1963), pp. 75–147. Goethe recorded his visits to Giebichenstein in his diary (22–24 May, 17–19 July 1802, 6–8 May 1803, 3 April 1805). See *Goethes Tagebücher*, in *Goethes Werke* (Weimar Ausgabe), 133 vols. (Weimar: Hermann Böhlau, 1888), 3.3:57, 60, 72–73, 111.

81. Wilhelm Dorow, Reichardt's nephew, as quoted by Salmen in *Johann Friedrich Reichardt*, p. 78.

82. Joseph von Eichendorff, *Erlebtes*, in *Joseph von Eichendorff Werke*, ed. Hartwig Schultz, 6 vols. (Frankfurt a. M.: Deutscher Klassiker Verlag, 1993), 5:427–28.

preciation of the female form, and his own poetizing—anonymously published—made him exactly the sort of cultivated intellectual that Reichardt would have eagerly sought out.[83] Reil must have been an intimate of Giebichenstein's society by the time he wrote the *Rhapsodieen*, for shortly thereafter he was arranging for Reichardt's son-in-law Steffens to return from Denmark to become a member of the faculty at Halle. Reil seems to have been enjoined by Reichardt to retrieve his daughter, Johanna, from her exile in Copenhagen, where her husband took her immediately after their marriage in September 1803.[84] At the same time that Reil wrote Steffens, a known disciple of Schelling, with the invitation to return to Germany, the university also moved to get another individual deeply engaged in the Romantic movement, Friedrich Schleiermacher, who had simultaneously received a call from Würzburg. Since the theologian's sister was a close friend of Johanna Steffens, and since Henriette Herz, who was warmly attached to Reil (as his own wife nastily observed)[85] and even more heatedly attached to Schleiermacher—well, it seems obvious that social lubrication made it possible to slide Schleiermacher's intellectual qualities more easily before the authorities at the university. Schleiermacher's abilities, though, hardly needed much of a boost; his sublime intellect, as lofted in his *Über die Religion* and *Monologen*, already floated high above the German lands for all to see. Moreover, the competing call he received from Würzburg would have made him even more attractive (according to the calculus still operative at universities). Shortly after his arrival at Halle, Schleiermacher formed a close intellectual alliance with Steffens, Friedrich August Wolf (1759–1824)—the great philologist who had taught Schleiermacher when he was a student at

---

83. Reil wrote a screed against corsets ("Ueber den nachteiligen Einfluss der Schürbrüste auf Schönheit und Gesundheit"), since they artificially and unhealthily distorted the most beautiful form in the animal world. The article is republished in Reil, *Kleine Schriften*. His poetry appeared anonymously in the journal *Mannigfaltigkeiten* between 1790 and 1800. Several of these verses are reprinted in Engelberg, *Aus dem Leben der Dr. J. C. Reil.*

84. In his letter to Steffens making the offer, Reil asked him not to mention the negotiations to his father-in-law until the position had been fully secured. Steffens also asked Reil to keep it quiet until it had been approved in Berlin, lest his father-in-law intercede on his behalf. The father's desire for his daughter's return must, however, have been obvious to Reil, otherwise these solicitations of secrecy would not have occurred. See Henrik Steffens, *Was Ich Erlebte*, 10 vols. (Breslau: Josef Max, 1840–44), 5:102–4.

85. Ludwig Börne, a student living with the Reils, quoted for the delectation of Henriette Herz a remark by Reil's wife: "Mrs. Reil related that her husband had stayed for a long time in your house, when he followed his curriculum in Berlin. Among other things she said 'Yes, I am certain that were I to die today, my husband would marry the Herz woman tomorrow; he can tolerate her very well.'" Ludwig Börne to Henriette Herz (8 October 1803), in *Briefwechsel des jungen Börne und Henriette Herz*, ed. Ludwig Geiger (Leipzig: Schulzesche Hof-Buchhandlung, 1905), p. 93.

Halle[86]—and Reil. Around them gathered an outer circle of students who would take their classes and join with them against the more conventional members of the faculty.[87] Under their inspiration, the university became a cauldron of youthful enthusiasm for the new philosophy and science. Indeed, the pot often bubbled over. According to one who was a student at the time, two groups of Schelling devotees, while agreeing on matters of polarity, yet clashed over the concept of the "indifference point," and as a result blood flowed in the city streets.[88] Metaphysics proved a dangerous game. In this exhilarating intellectual atmosphere, Reil could not but be borne aloft with the soaring eagles of the Romantic movement.

Reil may well have been fledged in Romanticism along with an eaglet of the party, Löb Baruch (1786–1837), a Jewish medical student who would later gain fame, under the name Ludwig Börne, as a writer and journalist of precious though biting wit. The sixteen-year-old Börne (as I will refer to him) had originally been sent to Marcus Herz for tutoring in medicine, and in fall of 1802 he came to reside in the Herz household. But he had other things on his mind than medicine. He spent most of his time mooning over Henriette Herz, his senior by twenty-two years. After Marcus died in January 1803, Henriette decided she had gently to dispose of the lovesick boy, of whom she herself was becoming inordinately fond.[89] She asked her friend

86. Wolf, who originally studied with Heyne at Göttingen, advanced at Halle his program of *Altertumswissenschaft*. He argued that the texts of the ancients could not really be understood until the historian had mastered the details of everyday life and thought in the ancient world. The historian had to think like an ancient to give the most perspicuous rendering of any text at issue. In 1795 he published the fruits of his program in a book that would achieve considerable fame and notoriety, his *Prolegomena ad Homerum*, which maintained that the Homeric epics were not of unified and single authorship, but the work both of troubadours, who improvised on poems that perhaps came from an ancient bard, and of later editors, who shaped the works into unified wholes. For an appreciation of the extent of Wolf's originality, see the lucid essay by Anthony Grafton, "Prolegomena to Friedrich August Wolf," in his *Defenders of the Text: The Traditions of Scholarship in an Age of Science, 1450–1800* (Cambridge: Harvard University Press, 1991), pp. 214–46.

87. See Steffens, *Was Ich Erlebte*, 5:158. Steffens and Reil became fast friends. When Reil moved to Berlin, he promised to help Steffens gain a call to the new university, though his own untimely death prevented this. Steffens's first encounter with Reil, during an excursion as a student in 1799, undoubtedly was framed by Schelling's critical analysis of Reil's materialism. Steffens wrote Schelling of his impressions of the various professors he met: "Shall I describe to you the lamentable, famous philosophers with whom I have spoken on my trip? . . . Reil looks even as insignificant as his philosophy and, if it were possible, speaks more stupidly than he writes." See Henrik Steffens to Friedrich Schelling (26 July 1799), in F. W. J. Schelling, *Briefe und Dokumente*, ed. Horst Fuhrmans, 3 vols. to date (Bonn: Bouvier Verlag, 1962–), 2:176.

88. The student was Ludwig Börne, who related the story in his "Die Apostaten des Wissens und die Neophyten des Glaubens" (1828), in *Sämtliche Schriften*, ed. Inge and Peter Rippmann, 5 vols. (Düsseldorf: Joseph Melzer Verlag, 1964), 1:600–1.

89. The sixteen-year-old knew he had to leave Henriette, but his farewell plunged him into an emotional maelstrom: "I am a man.—You have rendered my sentence: I cannot stay. You pour oil

Reil to take him in. Börne eventually sidled away from medicine, but during his four years in Halle, he lent an adolescent élan to the Reil household: he kept Reil apprised of his beloved "dear mother" (as Börne took to calling Henriette); he carried himself, in relation to her, as a young Werther; he enthused over Schleiermacher, who conducted his courses "like Socrates";[90] he became smitten with Reil's older, charmingly blond daughter; and though he had an initially rocky relationship with the doctor, he came to admire this medical researcher who would "begin and intermix his lectures on therapy and diseases of the eye with poetry from Schiller and Goethe, so that the delicious fruits of his research were hidden among flowers."[91]

The spring flowers of Romanticism, whose seeds had blown over from Jena and Berlin, had taken deep root in Halle. Music, poetry, and literature flourished. On this fecund ground, the university plowed under much traditional academic conservatism and stood ready to cultivate fresh ideas, especially in the disciplines of theology, philosophy, philology, the natural sciences, and medicine. The famous phrenologist Franz Joseph Gall (1758–1828), for instance, came to lecture in 1805, an event that attracted even Goethe from Weimar.[92] On that memorable July occasion, with an auditorium filled to overflowing, Gall, surrounded by the skulls of animals and men, gestured to Goethe—or rather to his head—as he illustrated the evenly developed contours of universal genius. The genius, though, must have harbored an enlarged bump for irritation, since he seems not to have been very pleased to become a specimen in Gall's lecture—at least such was the judgment of Steffens, who was also in the audience. Gall next turned to Reichardt, seated at Goethe's left, as he indicated that sublime musical talent would produce bulges in the temporal regions of the skull, which the Kapellmeister's perfectly bald and powdered pate exemplified. Gall then prepared to discuss the organ of language. But before he could even smile toward Wolf, on Goethe's right, the great philologist took off his glasses and swiveled his head in all directions. "For the moment, he became transmuted into the skull bones turned by the hands of a demonstrator," or so

---

on the flame, and it consumes my heart. I will die if I can no longer remain in your presence." Ludwig Börne to Henriette Herz (March 1803), in *Briefwechsel des jungen Börne*, p. 57. During his stay with Reil, Börne and Henriette exchanged letters frequently, sometimes almost daily, with both parties complaining when the other was tardy with a missive. Quite clearly Henriette felt deep affection for the boy.

90. Börne, "Die Apostaten des Wissens und die Neophyten des Glaubens," in *Sämtliche Schriften*, 1:598.

91. Ibid., 1:598–99.

92. This episode is related by Steffens in *Was Ich Erlebte*, 6:48–52.

imagined Steffens.[93] The behavior broke up the skeptical philosopher, who even in recollection could not suppress his sense of the ridiculous. At the time, though, Steffens felt compelled to counter the nonsense with a set of his own lectures. The next day, in the same room, he began; and Eichendorff, then a student at Halle, thought he completely won over his auditors with the "animated and fiery force of his enthusiasm."[94] Steffens did allow, however, some good effect had come of Gall's lecture—it made Reil determined to begin what eventually became a magnificent series of studies of the brain and nervous system.[95] These studies would occupy him during a dark time, when the university was shuttered by invading French forces.

In fall of 1806, Napoléon's armies met the Prussian troops just outside Halle. As the citizens gazed out from the city walls onto the Saale valley, they saw the pomp and glory of the German forces wither before the onslaught of the seasoned French companies, whose uniforms were still bloodstained from their recent victory at Jena. During the subsequent occupation, Steffens and his wife had to move in with Schleiermacher, but all had to make room for quartered French officers. Before the battle, Reil was forced to share the great house he had built near Reichardt's at Giebichenstein with the German duke Karl Eugen von Württemberg; and then after the rout, Marshal Jean-Baptiste Bernadotte moved in. Reichardt fled the city, since he was implicated in the publication of a book that defamed Napoléon. His magnificent house was razed. Napoléon entered Halle and, because of the threat he perceived in the congregation of young men, closed the university and scattered its students, who had to return to their homes.[96]

The Romantic Naturphilosoph

During the two years the university remained closed (1806–8), Reil continued to do research, care for patients, and edit his journal, though now with the help of a colleague. In 1807 he published in the Archiv a hundred-page monograph that quite unequivocally revealed the transformed state of his thinking.[97] The article described the results of his investigations of the

93. Ibid., 6:52.

94. Eichendorff, Tagebücher, in Werke, 5:124.

95. Steffens, Was Ich Erlebte, 6:62.

96. The battle at Halle and the occupation are described by Steffens in ibid., 5:183–227.

97. Reil, "Ueber das polarische Auseinanderweichen der ursprünglichen Naturkräfte in der Gebärmutter zur Zeit der Schwangerschaft, und deren Umtauschung zur Zeit der Geburt, als Beytrag zur Physiologie der Schwangerschaft und Geburt," Archiv für die Physiologie 7 (1807): 402–501. Mocek regards this as the first work of Reil really in the "romantic-naturphilosophisch vein." See his Johann Christian Reil, p. 163. Below I will discuss further and dissent strongly from Mocek's analysis.

physiological changes undergone by the human uterus during pregnancy and birth, and the relationship of these changes to the rest of female anatomy and physiology. His research included a considerable amount of first-hand observation of pregnant women and experiments conducted on gestating animals. His general conclusions concerning the nature of organic processes transcended those of his "Lebenskraft" monograph. The dynamic metamorphosis of the uterus after conception and its altered relationship to the body of the mother could not, he now asserted, be understood as a merely passive and mechanical process by which the organ simply became stretched out—as if it were a piece of gold hammered thin. The whole structure and its relationships during pregnancy underwent a profound alteration. In particular, the uterus itself became miraculously "transubstantiated": "the hard, white, and structureless substance transmuted into a soft, reddish, vascularly rich, and fibrous one." These large structural changes proclaimed a singular purpose, quite obviously the production, retention, and nurturing of the fetus. Moreover, at birth, unlike a chick that mechanically breaks its shell, the human fetus is transported by the uterus to the outside in a purposeful fashion. All of this, Reil now concluded, "says to us loudly that there is something more fundamental here than dead mechanism."[98]

Mechanism and chemistry alone now seemed to Reil insufficient to explain the purposeful stages that the pregnant uterus passed through. He was not, however, circling back to Kant. In an extensive footnote, he made clear that the purposive activity of the process hardly became clarified by simply uttering the Kantian slogan that "in organism everything is mutually related as means and ends."[99] For, after all, the uterus has purposive relationships extending beyond the mother of which it is a part—its end is the species itself. Moreover, we see during the process of pregnancy and birth a transformation from one organic individual during gestation (the mother and fetus whose interactions seem as one) to the separate existence of two individuals after the birth of the child.[100] This is but an example, Reil urged, of a larger principle that governs "all organisms on the entire earth, the balance of kinds, of sexes, of origins and extinctions, of epidemics and contagious illnesses, and of hunger and war—a connection that lies over all reality in an unconditioned and all-powerful spiritual region."[101]

In his fine-grained and extensive analyses of the pregnant uterus, Reil

98. Reil, "Gebärmutter," pp. 437, 409.
99. Ibid., p. 411n.
100. Ibid., p. 464.
101. Ibid., p. 412n. Reil here echoes Kielmeyer's principle of balance in the organic world. See chapter 6.

made considerable use of Schellingian conceptions. For instance, the fundamental phenomenon he wished to investigate—namely, the reversal of polar forces governing the dynamics of the uterus—he conceived in terms drawn directly from *Naturphilosophie*. After conception, the "force of expansion increases in relation to that of contraction." The two forces come to govern opposite poles of the uterus—the body of the uterus now expands (to accommodate the fetus), while the neck of the uterus contracts (to retain the fetus), so that "the axis of the uterus is like a magnetic line with different poles." As the time of birth draws near, these polar forces begin to reverse, until during birth the contractive power resides in the body of the uterus (to expel the child) and the expansive force opens the passageway for the child to slip out.[102] Were the forces postulated mechanical or vital? Reil argued that during gestation, mechanical and vital forces worked in such harmony that one would have to conclude they expressed a more fundamental, underlying power.

Reil employed his Schellingian conceptions to do real work, not to serve as mere decoration. They became woven through the descriptive analyses of the intricate changes in the uterus and mutually related changes in other parts of the female anatomy (such as the swelling and engorgement of the breasts). These studies were of striking detail and precision, and included some exacting experiments.[103] For instance, strong opinion (Albrecht von Haller's among others') held that birth contractions were due to the muscles of the abdomen and diaphragm. Reil demonstrated that the contractions, on the contrary, were chiefly due to new musculature developed in the uterus itself, which worked in coordination with the fibers of the stomach and diaphragm. He showed this by sacrificing a good many pregnant, near-term rabbits and, after dissection, attaching a galvanic apparatus to their uterus, which would contract violently when a current was passed through it, expelling the babies.[104]

In the concluding two sections of his monograph, Reil sought to expand more generally the framework that allowed him to articulate his study of pregnancy. In "Von der Lebenskraft," he demurred from the concept of a

---

102. Ibid., p. 416.

103. Reil's observational and experimental acuteness displayed in his monograph elicited from a hard-nosed professor of medicine in our century—a Dr. Dr. Joachim-Hermann Scharf, who had little patience for Reil's Romantic *Naturphilosophie*—the admission that "a modern anatomist and endocrinologist must recognize that this study grounded the entire clinical-gynecological anatomy of the second half of the nineteenth century." See Joachim-Hermann Scharf, "Johann Christian Reil als Anatom," *Nova Acta Leopoldina* (*Abhandlungen der Deutschen Akademie der Naturforscher Leopoldina*), n.s. 20 (1960): 51–97; quotation from p. 53.

104. Reil, "Gebärmutter," p. 434.

Bildungstrieb, since it suggested something intelligent or willful about bio-logical processes. Now he fully embraced the concept, emphasizing just those features he had dismissed before. "The whole sensible world," he pro-claimed, "is its work [that of the Bildungstrieb], from the array of stars that stretch from one pole to another through the immeasurable spaces of the universe to the crystal that imbibes water." The Bildungstrieb forms every in-dividual as "not only an inclusive totality for itself, but also a member of the universal organism of the world structure, in which the worm is as necessary as Orion." "One does not know," he confessed, "whether one should won-der more at the beauty or the purposiveness of their formation." Both beauty and purpose, he thought, melted together in the artful constructions of nature. The principal difference between the human artist and nature, he urged, was that the human artist took his model from nature and im-pressed form on recalcitrant material. But "nature, like a Proteus, brings everything from herself; she herself is the material, tool, craftsman, and ar-chetype. Then she breathes spirit into her forms, since they are both one. Pygmalion's beautiful Elise, however, remains without feeling, a mute piece of marble."[105]

In Reil's monograph, one moves from a surface layer of analysis, in which the phenomena of pregnancy and birth are given an exacting empirical ac-count, to a deeper layer, wherein the phenomena receive a scientific expla-nation in terms of the Bildungstrieb, to the metaphysical foundation, in terms of which the Bildungstrieb attains its real significance. At the deepest level, Reil invoked Schelling's Spinozistic absolutism. Every form, he main-tained, lay sequestered in absolute substance. This substance was, thus, "the mother of all finite things and insofar as the sensible world as the work of ideas develops out of her, she must carry under cover the idea of the sensible universe, like the seed carries the future plant." The Bildungstrieb, then, as expressed in the "metamorphosis of material is nothing other than a striv-ing of the ideal, which expresses itself in the forms it creates, to become ob-jective and for the objective really to represent what it ideally is."[106] Right through to his last works, Reil articulated his science explicitly in these Schellingian terms.[107]

---

105. Ibid., pp. 477–79.
106. Ibid., pp. 479, 491.
107. For example, in Reil's posthumously published study of general pathology, he explicitly laid out "the system of absolute identity, which has been chiefly most excellently articulated by Schelling," as the foundation for understanding nature and her aberrations in sickness. See Reil, Entwurf einer allgemeinen Pathologie, 3 vols. (Halle: Curtschen Buchhandlung, 1815–16), 1:11–95; citation from 1:13–14.

Final Years: War and Romance

In 1808 the university at Halle reopened, but it was like an aged parent who had suffered the ravages of prolonged illness: the body ached with decrepitude and the mental faculties had stiffened up. Steffens, upon his return to Halle, thought the whole experience like coming back after a great fire had reduced all of one's possessions to ashes.[108] Little remained among the ruins—social life had disappeared; few students had regathered (about one-quarter of the original number); the university building (which was rented from the city) reeked of desolation; the once-Prussian Halle had been incorporated into the new kingdom of Westphalia, ruled by a puppet king, Napoléon's brother Jérôme Bonaparte; and Steffens's family and close friends had scattered. His father-in-law, Reichardt, stayed out of reach of the Corsican; Wolf and Schleiermacher remained in Berlin in anticipation of the opening of a new university. Reil, however, returned, though not for long.

After Napoléon's armies departed the city of Berlin in November 1808, planning began in earnest for a new university, one that would realize the muted desire of the dispossessed Friedrich Wilhelm III to have the state "replace through intellectual force what it has lost in physical force."[109] The new university would reflect the philosophical dispositions of Kant, Fichte, Schelling, and most especially Wilhelm von Humboldt, its chief intellectual architect. Humboldt wished to elevate philosophy and the *Geisteswissenschaften* to a more prominent position in relation to the professional faculties of law, medicine, and theology. He quickly assembled an extraordinary faculty: Fichte and later Hegel in philosophy, Wolf and Franz Bopp in philology, Schleiermacher in theology, and in medicine Christoph Wilhelm Hufeland and Reil.[110]

Reil received his call in 1810 to become dean of the medical faculty; and given the respective prospects of Halle and Berlin, he quickly accepted. In his departing speech from Halle, he recalled the great changes in science that had occurred just during the time of his own professional life. The direction of those changes, at least in his estimate, was clear:

---

108. Steffens, *Was Ich Erlebte*, 6:1–2.

109. The quotation is taken (with some changes) from Elinor Shaffer's insightful article "Romantic Philosophy and the Organization of the Disciplines: The Founding of the Humboldt University of Berlin," in *Romanticism and the Sciences*, ed. Andrew Cunningham and Nicholas Jardine (Cambridge: Cambridge University Press, 1990), pp. 38–54; quotation on p. 38.

110. For a discussion of the contrasting ideals of medical education that Hufeland and Reil brought with them to Berlin, see Broman's *Transformation of German Academic Medicine*, pp. 182–85.

Figure 7.3

Christoph Wilhelm Hufeland (1762–1836), professor of medicine at Jena and later founder with Johann Christian Reil of the medical faculty at the new University of Berlin. Lithograph after F. A. Tischbein. (Courtesy Wellcome Institute Library.)

The period of my present teaching position coincided with the most noteworthy time in which the study of medicine, as well as that of the entire natural sciences, underwent an almost complete revolution. It is unbelievable how far more real is the present instruction than that which I enjoyed. The effort at explanation has made place for living intuition [*lebendige Anschauung*]; the idea has entered the arena of the mechanical principle; and observation has achieved a standpoint from which to view things in their natural relations. Indeed, the dead have been resurrected to life; the machine of the heavenly bodies has been animated [*vergeistet*]; science has penetrated into the depths of the earth; and natural behaviors are being reduced to laws that are one with the laws of thinking mind [*denkende Geister*]. This revolution has already delivered a great boon, which will grow with the times. This will occur when first the storm of the initial agitation has passed and minds divided are subsequently reunited, so that they will have achieved mutual understanding. Only the German scholars have given birth to this renaissance of science; and without vanity, I might boast that I might count myself among those who, by reason of the various kinds of ideas which they were the first to circulate, have helped to prepare for this upheaval.[111]

111. Reil, "Abschiedsrede," in *Kleine Schriften*, pp. 318–19.

The revolution in thought that Reil described was, of course, that wrought by the Romantics, especially Schelling. With Schelling, "the idea entered the arena of the mechanical principle," and it was he who reduced the laws of natural activity to those of "thinking mind." Reil's claim that he participated in this Romantic revolution obviously did not prove a hindrance to his appointment in Berlin. Likely it comported well with the spirit that Humboldt wished to create at his new university.

Reil did not long enjoy his prestigious position in Berlin. After Napoléon's catastrophic retreat from Moscow during the fall and early winter of 1812, Prussia broke from its French masters and allied itself with Austria, Russia, and Britain. Friedrich Wilhelm III called upon his people to form voluntary guards. Reil agreed to establish and manage field hospitals, which quickly came into great demand. Through spring and summer of 1813, battles raged throughout the German lands, with the French forces being worn down even in victory. In the fall, the allies cut off the French at Leipzig. The battle raged for four days (16–19 October 1813), and the slaughter was fierce. In Thomas Hardy's lines:

> Five hundred guns began the affray
> On next day morn at nine;
> Such mad and mangling cannon-play
> Had never torn human line.[112]

Napoléon's forces were routed and effectively cast out of the Germanies. Some thirty thousand French corpses and only a slightly smaller number of allied bodies littered the plains around the city. The wounded and sick, the latter suffering mostly from dysentery and typhus, numbered also some thirty thousand. The casualties were transported to field hospitals, which Reil himself directly managed. He worked unceasingly in those hospitals around Leipzig, but finally succumbed to the kind of threat that made the medical profession in those times a heroic undertaking. He contracted typhus. Reil believed he had actually been infected just prior to tending the sick and wounded around Leipzig. Before he left Berlin, he visited an old physician friend, Karl Grappengiesser (1773–1813), who himself had fallen to the disease. When Reil walked through the door, Grappengiesser, in his mad delirium, leapt out of bed and flung his arms around his friend. Reil, taken aback, simply uttered: "I have been infected."[113] Though believing the embrace fatal, he nonetheless rushed to his duty at Leipzig.

---

112. Thomas Hardy, "Leipzig," in *The Essential Hardy*, ed. Joseph Brodsky (Hopewell, N.J.: Ecco Press, 1995), p. 71.

113. Steffens recounted this story from witnesses in his *Johann Christian Reil*, pp. 62–63.

When he started feeling ill and thought himself infectious, he made his way to his sister's house in Halle, and there spent his last days. Knowing full well the prognosis, he quickly wrote his family in Berlin. His sister reported to his friend Steffens that only one fear darkly intruded on his otherwise calm state: he believed he might lose his mind as had his friend. But it did not happen. On his last day, extremely weak, he lay on the sofa and, with some lingering pleasure, smoked a pipe. He bade his sister to sit by his side. And then, in what became his last request, he asked if she would "`place on the table some beautiful flowers, so that my eye may be pleased, and next to my seat place a goldfinch who can sing well, so that I might hear something pleasant.'"[114] Reil died on 22 November 1813.

In trying to understand Reil's intellectual trajectory, a significant problem remains, one that has embarrassed older scholars but supplied a puzzle for more recent ones.[115] It is this: What explains Reil's transition from a thoroughgoing physiological materialist—whose anti-vitalistic program outlined in "Von der Lebenskraft" became marshal music for many biologists through the nineteenth century—to a Romantic *Naturphilosoph*, who reconceived his empirical work in terms drawn from Schelling? Neither group of scholars thinks the transition internally generated—certainly it could not be the product of Reil's deeper understanding of nature and biological processes.

Among recent scholars, Reinhardt Mocek has provided his explanation of the transition in two stages. He first maintains that Reil did not adopt any notions of *Naturphilosophie* until the "Gebärmutter" monograph in 1807, during the time when the university was closed by Napoléon.[116] He then argues that the instantaneous adoption of Schelling's language occurred *because* of Napoléon's invasion of Prussia. He thinks Reil decorated the "Gebärmutter" monograph and consequent studies with the trappings of *Naturphilosophie* as an attack on the French, pitting Romanticism as the intellectual representation of the German nation against the arrayed armies of the Republic. Reil recognized, according to Mocek, "the necessity to mobilize every spiritual means against the Napoleonic domination of Prussia—even that of science."[117] Mocek's neo-Marxist hypothesis suffers from several difficulties, not the least of which is that the "Gebärmutter" treatise whispers not a hint of political sentiment. Moreover, even a cursory examination of Reil's *Rhapsodieen* reveals that his theory of consciousness, as

---

114. Ibid., p. 65.
115. For example, Hans-Heinz Eulner ("Johann Christian Reil," p. 29) remarks that one often has to translate Reil's *naturphilosophische* nomenclature into more common terms.
116. Mocek, *Johann Christian Reil*, p. 163.
117. Ibid., p. 168.

well as the tone of the whole study, took wing with the eagles of *Naturphilosophie* at least by late 1802, shortly after the time he had dedicated his *Fieberhafte Nervenkrankheiten* to Napoléon and more than three years before the outbreak of war. Moreover, by 1803 Reil had begun incorporating Schelling's *Naturphilosophie* into his medical lectures. As one of his students remarked: "Reil is, in much of his thought, himself a disciple of Schelling, and constructs everything on a foundation of *Naturphilosophie*. . . . He begins his colloquium thus: 'Medical treatment is the application of *Naturphilosophie* to the cure of human illness.'"[118] Further, in 1804, after the publication of Reil's book on the training of physicians (his *Pepinieren zum Unterricht Artzlicher Routiniers*), Andreas Röschlaub (1768–1835), the well-known doctor in Bamberg who had become a follower of Schelling, wrote his mentor that the famous Reil had become one of them: "Reil has shown in his most recent writing (his *Pepinieren*) that he would happily take over your ideas and wishes to apply them to medicine."[119] Thus, the evidence that Reil had become well disposed to Schelling's thought prior to the outbreak of hostilities with the French is quite patent. And even if we prescind from the question of timing, for Mocek's hypothesis to work, Reil had to have adopted Romantic *Naturphilosophie* only for extrinsic reasons. But the "Gebärmutter" work exudes from every pore the fundamental principles of *Naturphilosophie*—Reil certainly did not simply adopt such principles to serve as spiritual armor for the moment, to be removed after the battle. Finally, why would Reil, or anyone, suppose Romantic *Naturphilosophie* to be the authentic representative of the German spirit? Rudolph Haym, in his comprehensive survey *Die romantische Schule*, did argue that the early Romantic movement represented "the spiritual heritage" of Germany.[120] But even in 1870, when he published the book, that was an unconventional notion. It certainly was not a common view in 1807.[121] The

118. The student was Adolph Müller, who attended Reil's lectures. The cited passage is from a letter of 14 January 1804, quoted in Werner Gerabek, *Friedrich Wilhelm Joseph Schelling und die Medizin der Romantik* (Frankfurt a. M.: Peter Lang, 1995), p. 89.

119. Andreas Röschlaub to Friedrich Schelling (30 October 1804), in Schelling, *Briefe und Dokumente*, 3:132.

120. See Rudolf Haym, *Die romantische Schule: Ein Beitrag zur Geschichte des deutschen Geistes* (Berlin: Rudolf Gaertner, 1870), pp. 4–5.

121. Mocek (*Johann Christian Reil*, pp. 170–71) portrays Steffens as one who claimed that Schelling's thought represented the true German spirit. What Steffens actually said in the article that Mocek cites is: "No writer in Germany has had so decisive an influence on the scientific mind of his fatherland as Schelling. There is hardly a now-living natural researcher (except for the oldest)—including his decided opponents—whose thought has not oscillated, to a greater or lesser degree, to Schelling." See Henrik Steffens, "Schellingsche Naturphilosophie" (1805), in his *Schriften, Alt und Neu*, 2 vols. (Breslau: Josef Max, 1821), 1:85–86. It is quite one thing to say that a figure has had a considerable influence; quite another to contend on that basis that he represents

Romantics commanded no spiritual divisions with which to combat Napoléon; they had rather to marshal all of their intellectual forces against their own countrymen.

There is a simpler explanation. Reil became introduced to Schelling's philosophy undoubtedly by 1797 or 1798, when he would have read Schelling's *Weltseele*, which contained a sustained critique of the materialism of his "Von der Lebenskraft."[122] Schelling, like Reil, disputed the need to postulate a separate vital principle in addition to the mechanical and chemical processes governing life. Thus, like Reil, he regarded physical and biological processes as of a piece, a position they both adopted in opposition to Kant. Schelling, however, simply turned the relationship around: he argued that organic principles grounded the physical, that the laws of chemistry derived from higher organic laws. This kind of monism would, upon reflection, have appealed to Reil. After all, he initially rejected Kant's analysis of biological organisms because of the artificial way in which teleological considerations were secreted into physiological research. Yet certainly as a medical investigator, he met time and again—especially in his studies of pregnancy—the overwhelming and real telic processes that operated therein. The various ways in which a woman's organs teleologically functioned in relationship to each other and to the fetus were amply manifested in the processes of life. Reil displayed no inclination to drag the Creator in to explain such goal-oriented organization, as Blumenbach and Kant surreptitiously did, and Darwin's theory was not yet available to suggest any alternative approach to understanding that teleology. Schelling, however, offered a powerful way of understanding it. Many other medical and biological researchers found his approach congenial—as did Goethe. Moreover, initially stuck as he was in a kind of Kantian dualism when considering the nature of consciousness and its aberrations, Reil continued to look for ways of conceiving the brain and mind as one. Schelling again offered that means. In the *System des transscendentalen Idealismus*, Schelling demonstrated how the laws of mind became isomorphic with the organic realm, an idea that Reil himself mentioned in his farewell address at Halle.[123] And, of course, Schelling's arguments for idealism were extremely powerful—and, even from our perspective, irrefutable. (Rather, what has happened in our

---

the "spirit" of a nation. In any case, Steffens's sentiments would have hardly been widely shared, except by those who believed that the philosopher had created many negative thought oscillations.

122. Schelling, *Weltseele*, in *Schellings Werke*, 1:564–75.

123. See my quotation from his farewell address above. Reil, in his *Entwurf einer allgemeinen Pathologie* (1:36), reiterated the identity of life and matter, though with "pure mass as such only the bearer (the framework) for the higher stages of life."

time is not that we have demonstrated the falsity of these arguments, merely that we have judged them best forgotten.)

The quite substantial intellectual advantages to Reil's research program that derived from Romantic *Naturphilosophie* would have been augmented and endorsed in the warm and encouraging environment created during the Romantic florescence at Halle, especially in the company he kept at Giebichenstein. Steffens judged that Reil lacked the sort of unbridled imagination that could give wing to his own poetry.[124] And in that art he undoubtedly lacked the creative sensitivity of a Novalis or an Eichendorff. Yet his imagination had the power to pull his empirically grounded reason away from the sliced-up brains and uteruses that lay in profusion on his dissecting table, so as to raise his science to new heights and there to observe connections hitherto unknown. Romantic *Naturphilosophie* held powerful appeal for those whose imaginations exalted in the fresh, bright air in which the poets dwelt. Reil breathed in some of that atmosphere, which animated his empirical work sufficiently to inspire a century of medical and biological researchers, even those who might not be disposed to utilize in their practice Reil's wonderful, ironic device, the cat-piano.

124. Steffens, *Johann Christian Reil*, p. 60.

Chapter 8

SCHELLING'S DYNAMIC
EVOLUTIONISM

The alteration to which the organic as well as the inorganic nature
was subjected . . . occurred over a much longer time than our small
periods could provide measure.

—Schelling, *Weltseele*

Schelling's conception of the organic, which lay at the heart of his idealism,
undoubtedly served as the magnet that initially attracted such researchers
as Reil, Kielmeyer, and Goethe to labor over his quite abstruse philosophy.
That Schelling's own ideas about the organic had been forged in a dialectical consideration of the conceptions of the aforementioned (along with
those especially of Kant, Humboldt, John Brown, and Erasmus Darwin) undoubtedly added to their initial piquancy.

In summer of 1797, in the third installment of his long review of philosophical literature done for Niethammer and Fichte's *Philosophisches Journal*
(see chapter 3), Schelling briefly introduced his conception of the organic.
He maintained that the human mind produced out of itself a chain of representations, such that the very activity of producing the series constituted
the self and brought conscious mind into existence. The human mind thus
reproduced itself. For Schelling, self-reproduction instantiated the basic
meaning of the organic. The human mind was, therefore, organic in its fundamental operations. He further argued that, as a consequence of its organicity, the mind established (and in codependency) was established by an
objective world whose very structure also displayed this property:

> Since our mind [*Geist*] strives endlessly to organize itself, the external
> world must reveal a universal tendency toward organization. And that is
> the case. The world system is a kind of organization that has structured it-
> self out of a common center. The forces of chemical matter are already
> beyond the border of the merely mechanical. Even raw matter, cut from a
> common medium, includes regular figures. The universal *Bildungstrieb* of

nature loses itself ultimately in an infinity, which even the aided eye cannot measure. The continuous and steady process of nature toward organization betrays clearly enough an active drive that, struggling with raw matter, at times conquers, at times is suppressed, now breaking through into more open, now into more limited forms. It is the universal mind [Geist] of nature that gradually structures raw matter. From bits of moss, in which hardly any trace of organization is visible, to the most noble form, which seems to have broken the chains of matter, one and the same drive governs. This drive operates according to one and the same ideal of purposiveness and presses forward into infinity to express one and the same archetype [Urbild], namely, the pure form of our mind [Geist].[1]

Kant discovered within the human mind mechanistic categories that provided the necessary framework for the phenomenal world of nature. For him, organicity served only as a regulating conception, one that might help guide a researcher to discover mechanistic laws actually governing the natural world. The principle of the organic itself, because it harbored notions of intentionality, he deemed unfit to be a constitutive principle of nature. While Kant's world of nature clanked on in mechanical fashion, yielding to a jejune understanding, it yet obscured a mysterious, dark Urwelt, about which only rumors could be floated. Schelling, by contrast, excavated the mind more deeply and discovered at its foundation the principle of organism. This principle operated not simply to bind together with categorical relations a Potemkin world, but functioned to create a real world—at least as real as the human mind itself. And this world lived; even the most seemingly inert matter stirred with life. "To philosophize about nature," Schelling contended, "means to lift her out of dead mechanism, wherein she seems trapped, and to animate her with freedom and reposition her in a free development."[2] For Kant, the world of natural necessity and the world of human freedom existed disjointly and with apparent incompatibility. One would have to examine each of these worlds separately, with the eye for the other closed. Schelling strove to unite them stereoscopically. In this effort, he produced a theory of evolution that, while differing from Lamarck's and Darwin's, yet made the way easy for the conceptions of these biologists to advance on German soil. Schelling's developmental ideal arose not simply from metaphysical speculation; he worked it out within the complex of the

1. Friedrich Schelling, Abhandlungen zur Erläuterung des Idealismus der Wissenschaftslehre, in Schellings Werke, ed. Manfred Schröter, 3rd ed., 12 vols. (Munich: C. H. Beck, 1927–59), 1:310–11 [I:386–87]. The Abhandlungen ordinally appeared in installments in the Philosophisches Journal (1797–98), under the title "Allgemeine Uebersicht der neuesten philosophischen Literatur."

2. Schelling, Erster Entwurf eines Systems der Naturphilosophie, in Schellings Werke, 2:13 [III:13].

empirical biology of his day and cast it against the evolutionary ideas of Charles Darwin's grandfather Erasmus Darwin.

Biological Treatises

In quick succession, Schelling composed three biological treatises in which emerged, in ever-greater detail, his notions about the organic: *Von der Weltseele, eine Hypothese der höheren Physik zur Erklärung des allgemeinen Organismus* (On the world soul, a hypothesis of the higher physics for the clarification of universal organicity, 1798); *Erster Entwurf eines Systems der Naturphilosophie* (First sketch of a system of nature philosophy, 1799), and the introduction to the latter, *Einleitung zu dem Entwurf* (1799).[3] Schelling wrote these lengthy and often arcane tracts against the various problematics established by Kant, Reil, and Blumenbach on the negative side, and Kielmeyer and Goethe on the positive. The views of Alexander von Humboldt, Erasmus Darwin, and John Brown both negatively and positively charged the ideas of the philosopher with a kind of underground current, enough to be felt through the nineteenth century.

As explained in chapter 3, Schelling's *Weltseele* represented a wideranging inductive examination of the latest empirically based theories regarding organic as well as (seemingly) inorganic nature. These inductions, he believed, supported the quite general principle that external phenomena derived from polar forces in nature: magnetism and electricity, for instance, exemplified such positive and negative forces; and light resulted from a kind of combustion in the atmosphere, in which positive and negative forces were torn asunder. Ultimately these opposing forces had to be understood as expressions of the two fundamental activities of self-consciousness, for which his *System des transscendentalen Idealismus* provided the deductive justification. In this respect, the *Weltseele* did not represent in Schelling's mind an independent, self-contained analysis of natural phenomena; rather, it required a transcendental idealism for its completion. In the *Erster Entwurf*, however, he began to formulate a deductive system of nature that he believed required no further justification. Principles induced from the latest experimental and theoretical work could, he argued, be repositioned within a framework that was both deductive and yet reliant only on natural philosophical concepts. The *Einleitung* made this presumption quite explicit and brought Schelling to the brink of his "identity philosophy." All three of the tracts, written at great speed (the *Erster Entwurf* really consists of a series of lectures delivered at Jena), catch

---

3. I have discussed the basic structure of these works in chapter 3. Here I will deal more substantively with the details of their content.

Schelling's thought on the wing—in these tracts he struggled to make his ideas clear to himself, as well as to his readers. They thus have an insistent but inchoate quality: just as by running up and down the bank of a swiftly flowing stream, one might detect well enough some of the creatures moving along the bottom; but then something will jump up in the distance, leaving one with only a fleeting impression. So with Schelling's elusive thought.

Critical Analysis of the Biological Theories of Contemporaries

Schelling worked out the details of his organic theory against the conceptions of Kant, Reil, Blumenbach, and Kielmeyer in particular. I have already suggested the nature of his advancement on Kant. Reil and Blumenbach posed for him antithetical positions on the nature of life. In his early tract "Von der Lebenskraft," Reil maintained that the life force emerged, just as all natural forces, out of chemical elements combined in certain ways. He needed no additional or special *Lebenskraft* to explain organic phenomena. Schelling had several different arguments to refute Reil's contention, but they all came down to a fundamental objection: that animal matter—that is, the particular combination of chemical elements that productively caused the activities of life—already assumed some prior cause of the combination and organization of elements and thus could not explain such organization. "Life," Schelling urged, "is not a property or product of animal matter, rather the reverse; animal matter is a product of life."[4] The processes of life required some parts to react upon themselves— thus the heart must furnish blood to the parts, including itself, so as to be able to pump blood; others parts need to perform homeostatic functions— such as producing fever during illness and thereafter returning temperature to normal; and still others must construct the same forms for one part (in the growth of tissues, for instance) and different forms for others (bone produced instead of flesh).[5] By the lights of his age, Schelling was quite correct—simple mechanism could not explain organicity. For in any natural, mechanical relationship,

> we recognize (insofar as it is not regarded as a whole that returns upon itself) a mere series of causes and effects, none of which in itself can be a principle of establishment, endurance, or governance—nothing, in short, that forms its own world. Such elements appear as mere phenomena that arise out of determined law and disappear again according to another law.[6]

4. Schelling, *Von der Weltseele, eine Hypothese der Höheren Physik zur Erklärung des allgemeinen Organismus*, in *Schellings Werke*, 1:568 [II:500].

5. Ibid., 1:568–69 [II:500–1].

6. Ibid., 1:583 [II:515].

Reil's mechanistic theory failed to account for the phenomena of life. But Schelling did not repair directly to Blumenbach's or Humboldt's vitalism. He understood them both to have introduced into the life sciences something *ab extra*, something foreign to natural processes.

Blumenbach added the *Bildungstrieb* to the natural forces resident in physical and chemical processes to yield biological phenomena. Alexander von Humboldt did the same, though his principle of *Lebenskraft* had the negative function of preventing naturally attractive elements in the animal body from uniting and thereby returning living flesh to dead corruption.[7] Schelling thought the unwarranted addition of such forces prevented a general systematization of the science. When naturalists, he observed, "assume a particular vital force, which as a magical force cancels all effects of natural laws in living beings, they likewise eliminate the possibility of explaining a priori the physical organization of those bodies."[8] The concept of *Lebenskraft* actually obscured biological processes; it became a barrier "for progressive reason and serves only as a chair overstuffed with dark qualities designed to allow reason to rest."[9] Even Kant, Schelling maintained, permitted his reason to recline in the third *Critique* on that comfortable idea of *Lebenskraft*.[10] The idea suggested, at least to Schelling, that nature could act freely, without constraint of natural law. Her behavior would thus be rendered, as Kant admitted, scientifically intractable. By contrast, Schelling sought to construct an autonomous and foundationally complete science. He was "thoroughly convinced that it was possible to explain organizing natural processes from natural principles."[11]

Schelling's own position would come as a kind of synthesis of materialism and vitalism. He would even appropriate the term *Bildungstrieb* to ex-

7. I have briefly sketched Humboldt's early views in chapter 7. When Humboldt reprinted his allegory about vital forces in his *Ansichten der Natur*, a story that originally appeared in Schiller's *Horen* (1795), he added a set of clarifying notes. He confessed in his reprint that he had become quite dubious about so-called vital forces. He adopted instead a position that came very close to that of Schelling. He indicated that we should call that entity alive whose parts were held in dynamic equilibrium, in which each part was reciprocally means and ends: "The equilibrium of elements maintains itself in the living matter insofar as the elements are parts of a whole. One organ determines another, one gives to the other a corresponding temperature, a tone, in which these and no other affinities operate. So in an organism, everything is mutually means and ends." See "Die Lebenskraft oder der rhodische Genius, eine Erzählung," in Alexander von Humboldt, *Ansichten der Natur*, 3rd ed. (Stuttgart: Cotta'schen Buchhandlung, 1849), pp. 315–24. The quoted note comes from p. 323. I will return to Humboldt's shifting conception about life in the appendix to part 2.

8. Schelling, *Weltseele*, in *Schellings Werke*, 1:594 [II:526]. This objection was pointed directly at Humboldt and Blumenbach.

9. Ibid., 1:595 [II:527].

10. Ibid., 1:597 [II:529].

11. Ibid.

press this new synthesis (though allowing thereby some confusion of his position with that of Blumenbach). What his construction sustained, he thought, were the legitimate though seemingly antithetical convictions that nature had to be understood as lawful and as free. "Nature," he urged, "must neither act simply without law (as the defenders of the *Lebenskraft*, if they are consistent, must maintain), nor act simply lawfully (as the chemical physiologists hold); rather, she must be lawless in her lawfulness and lawful in her lawlessness."[12] Nature quite generally could have these properties, Schelling intimated, only under two conditions. She could be lawfully free only if *organism* became the fundamental concept not only in biology, but in chemistry and physics—thus reversing Reil's kind of reductionism. The second condition flowed from the first, namely, that organism could be the basic constructive concept of nature only if nature herself ultimately issued from organic self-consciousness. This latter condition—of consciousness giving law to itself in the production of a world—established freedom at the core of nature's activity. Nature would, therefore, be rendered lawful in her activities; but because such law issued from her own deepest core, it was law freely imposed upon herself. Thus only Schelling, not any of Kant's other supposed disciples, constructed an authentic teleomechanistic understanding of nature.

Nature as a Dynamically Shifting Balance of Forces

Schelling's organic conception of nature owed much, of course, to Kant's general analysis of teleological judgment, which, after all, seemed to be a judgmental function similar to that grounding the categories of cause, substance, and the rest. More generally, his view expressed the Romantic conviction that all of nature lived, yet not merely lived, but breathed as one with man. He sought, with Wordsworth, "To look on nature, not as in the hour / Of thoughtless youth, but hearing oftentimes / The still, sad music of humanity."[13] The narrower gauge of Schelling's theory of the organic, however, achieved stability through the ideas of Kielmeyer, whose own inductions discovered a balance of power pervading all of living nature.

In the *Weltseele*, Schelling followed the lead of Kielmeyer and derived from the most recent experimental work in physiology—especially the galvanic studies of Alexander von Humboldt in his *Versuche über die gereizte*

12. Ibid., 1:595 [II:527].

13. From William Wordsworth's "Lines Composed a Few Miles above Tintern Abbey," in *The Essential Wordsworth*, ed. Seamus Heaney (Hopewell, N.J.: Echo Press, 1988), p. 46.

Muskel- und Nervenfaser (Experiments on stimulated muscle and nerve fibers, 1797)[14]—evidence of balancing forces throughout the plant and animal worlds. In Schelling's conception, however, these were dynamic forces: an initial disturbance would cause a realignment and recalibration of forces to restore balance—thus, the more irritable an organism (i.e., the more its muscular movement), the more it required for nutrition, and consequently the more oxygen uptake it needed for assimilation of elements, which finally would reestablish balance.[15] This dynamic balancing of forces occurred, as Schelling comprehended it, throughout the organic world. So, for example, plants take in water, but through combustion break it down into its components, exhaling oxygen. Animals, in turn, breathe in oxygen and through productive synthesis release water into the atmosphere.[16]

As discussed in chapter 3, Schelling perceived even nonliving matter's origin in a balance of forces held in dynamic tension.[17] In chemical reactions—particularly combustion of various types—the balance would be upset, but quickly a new equilibrium would be attained, resulting in a new kind of matter. The qualities of matter, detailed in the *Erster Entwurf*, thus displayed themselves as expressions of variously combined oppositional forces.[18] In this way, organicity—the dynamic rebalancing of forces—constituted the fundamental property of all natural bodies. Reil had reduced living phenomena to the actions of dead matter. Schelling elevated dead matter to the actions of lethargic life. For him, matter was but life reduced to a minimum.

According to Schelling, life, as conventionally manifested, displayed a defining set of properties that fell under the general rubric of excitability (*Erregbarkeit*). Excitability encompassed the organic powers of sensibility, irritability, and reproduction. Schelling had both physiological and metaphysical reasons for holding excitability to be the fundamental property of life. On the physiological side, his considerations followed a long trail of previous argument concerning irritability, sensibility, reproduction, and their relationship (see the appendix to part 2 for a short history of ideas about excitability and sensibility). Schelling drew the vital waters for his

14. Alexander von Humboldt, *Versuche über die gereizte Muskel- und Nervenfaser, nebst Vermuthungen über den chemischen Process des Lebens in der Thier- und Pflanzenwelt*, 2 vols. (Berlin: Heinrich August Rottmann, 1797–99).

15. Schelling, *Weltseele*, in *Schellings Werke*, 1:609 [II:541].

16. Ibid., 1:581 [II:513].

17. See also ibid., 1:566 [II:498].

18. The qualities of matter, in Schelling's scheme, expressed different combinations of elementary forces, but these forces themselves, glimpsed from behind the transcendental veil, were modes of the acts of self-consciousness. See Schelling, *Erster Entwurf*, in *Schellings Werke*, 2:23, 34 [III:23, 34].

own theory of physiological processes from a pool of ideas fed by strong currents from the various works of Kielmeyer, Alexander von Humboldt, and the Scots physician John Brown. Schelling, it will be recalled, held Kielmeyer's essay on the organic powers in particularly high esteem. And in his *Weltseele* and *Erster Entwurf*, he had frequent recourse to Humboldt's *Versuche*, as well as to Brown's *Elements of Medicine* (German translation, 1795).[19] In the light of these works, he maintained that sensibility was the fundamental power through which we discriminated external stimuli. The surface reaction to felt stimuli would be irritability—the expansion and contraction of muscles and organs. This reactivity would give sensation the quality of "coming from without," thus virtually identifying it with a supposed external stimulus. Finally, reproduction, in Schelling's scheme, included the activities of growth and maintenance, as well as instincts (*Kunsttriebe*) and generation of other individuals of like kind. Reproduction, he thought, naturally flowed from both the organism's sensibility and its intrinsic drive to advance beyond its present state.

Schelling's larger metaphysical theory neatly accommodated this basic physiological scheme. Excitability, from this philosophical perspective, was not a simple product of nerves or muscles. Indeed, according to Schelling, sensibility did not fundamentally depend on the brain and nerves; rather, brain and nerves depended upon organic feeling for their initial formation in the organism.[20] At the deep, metaphysical level, excitability represented a kind of continuous self-reproduction, which, at the phenomenal level, appeared to be the organism's response to the incessant stimuli from the external world. Each stimulation constrained an organism's naturally positive expansion, producing distinct stages of growth and development. For the inorganic, continuous stimulation would eventually only destroy the object; but for organisms, continuous stimulation produced advancement.[21]

In animal organisms, according to Schelling, physiological processes occurred through continuous disturbance and then complementary reestablishment of balance between positive and negative forces. Oxygen served as a principal negative force in irritable reaction. Within the animal body, oxygen united with other chemical elements to produce just the composition that constituted a specific kind of organic matter. The negative force of oxygen was balanced by a positive force that operated through the nerves. The positive force, which Schelling referred to as the *Bildungstrieb*, functioned to modulate irritable reactions in organisms. The balance of such

19. These works are discussed in the appendix to part 2.
20. Schelling, *Erster Entwurf*, in *Schellings Werke*, 2:155 [III:155].
21. Ibid., 2:23, 34 [III:23, 34].

forces allowed Schelling to formulate lawful relationships much in the manner of Kielmeyer. For example:

The positive principle operates by means of the nerves on the irritable organ. The fewer the nerves going to an organ, the less its capacity for oxygen, and the less its capacity for oxygen, the more compelling (i.e., less under the direction of the will) the deoxidation process in it, and the more reactive its irritability. . . . The more and larger the nerves to an organ, the greater its capacity for oxygen, and the greater its capacity for oxygen, the less compelling and arbitrary its irritable expression.[22]

The positive principle, to which Schelling rather mysteriously referred, functioned in ways comparable to Blumenbach's *Bildungstrieb*; and the philosopher appropriated that very term to characterize the force of life which he perceived streaking through all organic phenomena. He differed from his predecessor, however, in the extent to which he embedded his empirical theories in a comprehensive metaphysical and epistemological system.

The fundamental idea of Schelling's *Naturphilosophie* was simply that nature strove to achieve the absolute. At the transcendental level, this nisus reflected the constant effort of self-consciousness to achieve perfect expression. By the time he wrote the *Entwurf*, however, Schelling was attempting to understand such basic natural striving independently of its transcendental complement. In natural-philosophical terms, nature's striving amounted to this: "She seeks the most general (or perfect) proportion in which all actions of individuality can, without injury, be united." This effort could never be realized within the temporal, empirical dimension, so natural striving could be "nothing other than a formation [*Bildung*] into infinity." In this developmental process of nature, certain stages, or apparently fixed points, might be reached—the formation of a particular species, for instance—but the process would be that "of continual evolution, constant transformation, which would only seem played out at a particular stage."[23] Nature, in short, had to be conceived as a progressive evolution, achieving ever-new productive moments, never at rest, but striving toward perfection.[24]

---

22. Schelling, *Weltseele*, in *Schellings Werke*, 1:618–19 [II:550–51]. Such formulations as these Schelling likely derived from Humboldt, who in his *Florae Fribergensis* cites the Swiss electrophysiologist Christoph Girtanner (1760–1800) on the relationship between oxygen in a plant and its irritability: "The irritability of an organic body is always in direct proportion to the quantity of oxygen that it contains" (ibid., 2:155n).

23. Schelling, *Erster Entwurf*, in *Schellings Werke*, 2:43 [III:43], 2:42 [III:42], 2:19 [III:19].

24. Ibid., 2:42 [III:42]: "The development [*Entwicklung*] of an absolute product, in which the natural activity would be exhausted, is nothing other than a formation [*Bildung*] into infinity."

The very terms Schelling employed to name this process of development—"dynamic evolution" [*dynamische Evolution*]—suggested the entire theory. The term "evolution" had originally referred to the embryological doctrine of preformation, according to which the embryo unfolded (evolved) out of an inchoately formed adult. In Schelling's view, the species was preformed in the archetypal ideal, but yet was dynamically realized in time through gradual transition in form. Thus there would be real historical metamorphosis, a temporal development and alteration. Was Schelling, then, an evolutionist in our sense of the term?

Theory of Dynamic Evolution

Kuno Fischer, professor and dean of the arts faculty at Jena in the 1860s and early 1870s, admonished his friend Ernst Haeckel for contending that no one before Lamarck and Darwin had considered the origin of species. He pointed to a passage in the preface to the *Weltseele* where "Schelling was the first to enunciate with complete clarity and from a philosophical standpoint the principle of organic development that is fundamental to the Darwinism of today."[25] In the passage to which Fischer referred, Schelling took direct aim at Kant. He asserted that it was a "vintage delusion" to hold that "organization and life cannot be explained from natural principles." He contended that only courage was lacking in the effort to do so.

> One would at least take one step toward explanation if one could show that the stages of all organic beings have been formed through a gradual development of one and the same organization.—That our experience has not taught us of any formation of nature, has not shown us any transition from one form or kind into another (although the metamorphosis of many insects . . . could be introduced as an analogous phenomenon) —this is no demonstration against that possibility. For a defender of the idea of development could answer that the alteration to which the organic as well as the inorganic nature was subjected . . . occurred over a much longer time than our small periods could provide measure.

"One and the same principle," Schelling argued in the main part of the *Weltseele*, "unites inorganic and organic nature."[26] That principle, or at least one version of it, was his dynamic evolutionism. The principle proclaimed: "Every product that seems now fixed in nature exists only for a mo-

25. Kuno Fischer, *Friedrich Wilhelm Joseph Schelling*, 2 vols.; vol. 6 of *Geschichte der neuern Philosophie* (Heidelberg: Carl Winter's Universitätsbuchhandlung, 1872), 2:448.
26. Schelling, *Weltseele*, in *Schellings Werke*, 1:416–17 [II:348–49], 1:418 [II:350].

ment, and is in the process of continual evolution [*Evolution*], a constant transformation, which would only seem played out at a particular stage."[27]

Schelling's specific remarks in the *Weltseele*, his opposition to Kant, his great admiration for Kielmeyer—who had an explicit evolutionary theory—and his general dynamics of nature, all suggest that Fischer's historical judgment was sound enough: Schelling does appear to be among the very first to advance an evolutionary conception of nature, even down to using the cognate term that would come to name Darwin's theory. Yet it was not quite a Darwinian-like evolutionism that he advocated. Only about a year after he made those suggestive proclamations in the *Weltseele*, he offered a demurring clarification. In the *Erster Entwurf*, he allowed that

> several naturalists seem to have harbored the hope of being able to represent the source of all organization as a successive and gradual development of one and the same original organization. This hope, in our view, has vanished. . . . The belief that the different organizations are really formed through a gradual development out of one another is a misunderstanding of an idea that really lies in reason.[28]

Some contemporary historians have latched on to these last remarks to urge that "Schelling is no forerunner of Darwin." Dietrich von Engelhardt, for instance, has maintained that Schelling proposed no actual transition from one fixed organic stage to another—"no real descent"—rather only "a metaphysical ordering of plants and animals."[29] If von Engelhardt were entirely correct that Schelling only postulated an ideal development of nature and not a real, historical development in time, then one would have the puzzle of a grand cleavage in Schelling's philosophy, a split that occurred in approximately a year's time: from the *Weltseele*, where he quite unmistakably referred to a temporal transformation of species, to the *Erster Entwurf*, where, under this interpretation, he denied a real evolution. Precisely how are we to understand this apparent opposition, and what theory did this restless thinker finally settle upon?

I believe Schelling did shift his thought on the question of organic transformation, but the change amounted, I think, to a further specification and not a rejection of the idea that in nature morphological transitions have occurred, that over time new species have appeared on the earth in dependent

27. Schelling, *Erster Entwurf*, in *Schellings Werke*, 2:19 [III:19].

28. Ibid., 2:62–63 [III:62–63].

29. Dietrich von Engelhardt, "Schellings philosophische Grundlegung der Medizin," in *Natur und geschichtlicher Prozess: Studien zur Naturphilosophie F.W.J. Schellings*, ed. Hans Jörg Sandkühler (Frankfurt a. M.: Suhrkamp Verlag, 1984), pp. 305–25; quotations from pp. 312–13.

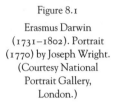

Figure 8.1

Erasmus Darwin
(1731–1802). Portrait
(1770) by Joseph Wright.
(Courtesy National
Portrait Gallery,
London.)

relation to those that preceded them. The principal event that sparked the
further refinement of Schelling's evolutionary dynamism was, I believe, the
appearance of Darwin's theory of evolution—Erasmus Darwin's theory,
that is.

The two volumes of Erasmus Darwin's *Zoonomia, or the Laws of Organic
Life*, published in England in 1794–96, were quickly translated into German, with the first volume appearing in two parts during 1795.[30] Schelling
approvingly referred to the first part of this German edition in the *Weltseele*,
where he deferred to Darwin's view that every bud of a plant constituted a
new individual.[31] Darwin's own evolutionary theory occurs, however, only
in part 2 of volume 1, which Schelling seems not to have read before he finished the *Weltseele*. In part 2, Darwin described the evolutionary process as
one in which, at the beginning, God endowed a simple organic filament
with life and certain powers—"irritations, sensations, volitions, and associations." These powers enabled this simple life "to improve by its own inherent activity, and . . . [to deliver] down those improvements by generation to

30. Erasmus Darwin, *Zoonomia, or the Laws of Organic Life*, 2 vols. (London: Johnson, 1794–
96); and *Zoonomie, oder Gesetze des organischen Lebens*, trans. J. D. Brandis, 3 vols. in 5 (Hannover:
Gebrüder Hahn, 1795–99).

31. Darwin's discussion occurs in *Zoonomie*, 1.1:185; Schelling mentions this in *Weltseele*, in
Schellings Werke, 1:603 [II:535].

its posterity, world without end!"[32] Thus, after a divine shove, organic nature might, under its own power and through genealogical transmission, produce ever-improved species into eternity.

Darwin's evolutionary theory—or, at least, theories like it—suffered in Schelling's estimation from a principal defect. Such theories attempted to explain the unity of organic structure genealogically, that is, through a physical transmission of developing form. It was this conception to which Schelling strongly objected. According to the genealogical theory, all species gradually emerged from a single aboriginal form until the earth became populated with the myriad of species familiar to us now. Schelling raised three crucial difficulties with this conception of how species appeared on earth. These difficulties concerned: (1) the nature of species and their boundaries; (2) the function of sex in fixing species at a particular morphological stage; and (3) the role of the archetype in both the epistemology of understanding dynamic evolution and the metaphysics of its production. Schelling incorporated these three concerns in an objection, the expression of which is not a little enigmatic and requires some explication. About genealogical theories of evolution, he complained:

> The belief that the different organizations are really formed through a gradual development out of one another is a misunderstanding of an idea, which really lies in reason. Namely: all particular organizations together should constitute only one product. This would be thinkable only if nature had before her eyes for every organism one and the same archetype [*Urbild*]. This archetype would be the absolute, the sexless condition that is neither the individual nor the species, but both together, in which the individual and the species are conjoined. This absolute organization cannot be represented through a particular product, but only through an infinity of particular products, which particulars deviate from the ideal in infinite ways, but in the aggregate are congruent with the ideal.[33]

First, let us consider the species theory lurking in the background of this passage. Schelling used the term "species"—"*Gattung*" in the German—in a very broad and rather vague way, so that a hyena, dog, and wolf would be regarded as varieties of the same species. He construed the logic relating characters of the individual to those of the class as inclusive, or as parts to the whole that comprised them, a logic rather different from the reductive approach that Linnaeus took. But this difference needs a few words of clarification.

32. Darwin, *Zoonomia*, 1:505 (xxxix, 4); *Zoonomie*, 1.2:458 (xxxix, 4). I have discussed Erasmus Darwin's theory in *Darwin and the Emergence of Evolutionary Theory of Mind and Behavior* (Chicago: University of Chicago Press, 1987), pp. 31–34.

33. Schelling, *Erster Entwurf*, in *Schellings Werke*, 2:63–64 [III:63–64].

In the last part of the eighteenth century, three notions of species could be found in pertinent literature: a reductive concept, a perfective concept, and an inclusive concept. One might, according to the reductive concept, consider every group of individuals, even of rather different varieties, as members of the same species if each exhibited a basic or reductive set of species characteristics—each individual would be regarded fully and completely as a member of the species. So, for example, the hyena, dog, and wolf might be regarded as one species if they shared a common set of traits—a particular shape of skull, tail of a certain character, clawed paws, and so on. Or one might think, in light of the perfective concept, that only certain varieties, the most perfect, captured the full meaning of the species (or of a higher taxonomic group)—then only the most developed individuals would be taken as representatives of the species—thus, the human being as the crown of the vertebrates. Or, finally, like Schelling, one might think inclusively of individuals of several varieties as members of the same species if each realized a different aspect of the ideal for that species—no one individual would fully express the species; only the aggregate of individuals would embody the ideal.[34] An evolutionary theory like Darwin's and his grandson's would attempt to explain unity of organization reductively. And, indeed, the theory of the archetype as it developed in Britain (especially by Richard Owen and taken over by Darwin) construed, for example, the vertebrate archetype in reductive fashion, that is, as merely a string of vertebrae. Schelling, by contrast, found no individual, certainly not a genealogically primitive individual, embodying all the possibilities of which the species ideal was capable. The reductive concept of species seemed to suggest, at least to Schelling, that a fixed and determinate structure already existed in the Ur-progenitor, a structure that would abolish the freedom of expression intrinsic even to nature. Though he allowed that the *Anlage*, or hereditary deposit, would fix the generic boundaries of a large species group, he insisted that environmental contingencies would foster individuals of infinite variety, both in their external morphology and internal character —so that only in infinite time and in full freedom would the possibilities of

34. Ibid., 2:54 [III:54]: "The community that no particular individual expresses, but all individuals together do, is called the species [*Gattung*]." This inclusive concept of species trails through Herder's *Ideen zur Philosophie der Geschichte der Menschheit*, which may have been the inspiration for Schelling's theory. According to this inclusive notion, individuals would be related to their species as parts to a whole—a conception of species given prominent play in the contemporary literature by David Hull and Michael Ghiselin, both of whom would shudder at the association with Schelling or Herder. Goethe's notion of the archetype is somewhere between an inclusive and perfective conception (see chapter 11). The perfective and inclusive conceptions differed quite markedly from the reductive theory of the archetype advanced by Richard Owen in England.

the species be realized.[35] What must be stressed in this conception, however, is that the notion of species (*Gattung*) encompassed a great variety of what we, in the present day, might call separate species. When Charles Darwin began his own work on species change, he originally proposed that out of the Ur-slime some four or so archetypal groups arose (e.g., vertebrata, articulata, mollusca, and radiata); and he supposed that morphological change occurred only within those groups.[36] And this seems very close to what Schelling had in mind: namely, independent species groups arising out of the earth—through a dynamic evolution—and then great morphological change within the groups, with one variety arising out of the other, so that all the varieties might realize the potential of the species.

Schelling argued that all developmental "operations of nature in the organic world are a continuous individualization [*Individualisiren*] of matter."[37] The apex of individualization, according to Schelling, "is reached in the formation of the reproductive power of the opposite sexes."[38] Once sexual dimorphism has been achieved, then a species would simply reproduce its general form (yet with great varietal change) ad infinitum. In the dynamics of evolutionary development in nature, certain fixed points seem to have been reached—new species groups appeared—and sex has frozen these places in the flow of natural becoming.[39] The internal opposition generated by sex, accordingly, determined the boundaries of species, which could not be transgressed.[40] But for Schelling, the fixation of species limits certainly did not imply a cessation in nature's continued active and free construction of varietal forms over time. Later Goethe would show that the giant Megatherium, whose fossil remains had been discovered in the late

35. Schelling, *Erster Entwurf*, in *Schellings Werke*, 2:56 [III:56].

36. For a discussion of Darwin's conception, see my *Meaning of Evolution: The Morphological Construction and Ideological Reconstruction of Darwin's Theory* (Chicago: University of Chicago Press, 1992), pp. 125–26.

37. Schelling, *Weltseele*, in *Schellings Werke*, 1:600 [II:532].

38. Ibid., 1:602 [II:534]; and Schelling, *Erster Entwurf*, in *Schellings Werke*, 2:49n [III:49n].

39. Schelling *Erster Entwurf*, in *Schellings Werke*, 2:53 [III:53]: "The difference of sexes, therefore, we maintain is the real and only reason why [organic] natural products seem generally to be fixed. (Though they are not really fixed. The individual passes away, only the species remains; and nature, for that reason, never ceases to be active.)"

40. Why the eruption of sex in natural evolution? Schelling had no theoretical explanation, something he mentioned in a footnote: "That such a bifurcation at each developmental stage is necessary and that the production is to be constrained—this we have clearly shown. But we have not explained the bifurcation itself. It is thus a necessary assumption for us; it is, in connection with our present investigation necessary, although we cannot clarify it. This clarification must necessarily be given later, when our science should be complete" (ibid., 2:52 [III:52]). In chapter 3 I suggested a historical explanation for Schelling's concern with sex and its "retarding" functions. His preoccupation arose after his introduction to the group for whom sex, in the abstract and in the act, became of paramount—one is tempted to say of paramour—importance.

eighteenth century, spawned the lowly sloth that roamed our world—the same *Gattung* had large possibilities of varietal metamorphosis. The idea of evolution, then, for Schelling, included great morphological change within a species form, as well as a sequence of species forms that progressively realized in time the more general archetype of organicity.

Schelling supposed that the *Bildungstrieb*, that is, the nisus of creative force, would determine new seeds or *Anlagen*, whence sprang particular species groups. Every new organic product, he maintained, "that seems fixed to us, must begin from the beginning, that is, with a wholly new *Anlage*."[41] The *Anlage* would take form as the result of a new level of oppositional forces appearing in the dialectic of nature. In this respect, "all products lie hidden in the original productivity of nature; and as soon as a determinate constraining point in nature has been reached, they arise out of identity."[42] Our detection of the appearance of new species varieties might be hampered, as Schelling suggested in the *Weltseele*, by the vast periods of time over which this might occur.[43] The researcher thus had to be careful that he not mistake a newly appearing variety of an established species for a completely novel one.[44] The appearance of new species—that is, those establishing new taxonomic groups—would occur, he believed, in a regular sequence, such that higher, more developed forms would appear in the wake of less-developed forms.[45] The force that produced this epigenetical production of species varieties would be identical, he claimed in concert with Kielmeyer, to the force operative in the dynamic *evolution* of the fetus.[46] He thus concluded with Kielmeyer that the more developed forms would go through ontogenetic stages in which they recapitulated the lower stages of species transition.[47] Hence, new, more fully evolved species would

41. Ibid., 2:63 [III:63].

42. Ibid., 2:63n [III:63n]. The notion that new species would suddenly appear in the world was not, of course, original with Schelling. Blumenbach, as we have seen above, advanced a similar view. The idea also found its way to Charles Lyell, who made the same supposition. See Charles Lyell, *Principles of Geology*, 3 vols. (London: Murray, 1830–33), 2:124. The difference between Schelling, on the one hand, and Blumenbach and Lyell, on the other, is that the philosopher conceived of the introduction of new species not as an instance of divine foresight (as Lyell did; see ibid., 2:42), but as a natural and necessary consequence of the forces of nature.

43. Schelling, *Weltseele*, in *Schellings Werke*, 1:416–17 [II:348–49]; see this passage quoted above.

44. Schelling, *Erster Entwurf*, in *Schellings Werke*, 2:63 [III:63]: "It thus remains a task for the natural researcher precisely to locate the original *Anlagen*, so that he might not mistake something that is merely a subspecies of a given *Anlage* for a different species."

45. Ibid., 2:63n [III:63n].

46. Ibid., 2:61 [III:61].

47. Ibid., 2:53n [III:53n]: "Nature is only one activity—thus her product is only one. She seeks through individual products to represent the one, the absolute product. She can differentiate her products only through the differentiation of stages. But several products are already constrained at

emerge over time; and in the aggregate, they would progressively realize the general ideal of the organic, just as each individual variety of a species would come to realize more fully the potential of the species.

With these considerations, we are now in a position to appreciate Schelling's principal objection to a genealogical theory of evolution. A theory like Erasmus Darwin's could not explain, to Schelling's mind, unity of organization, either at the level of species or more generally, at the level of organization per se. In this respect, he agreed with Kant that the possibility of organism in general and of individual organisms in particular could not be understood without assuming an archetype as a creative force; he differed from Kant by maintaining that such an assumption could be scientifically and philosophically justified. When Schelling complained that genealogical evolutionists had misunderstood "an idea, which really lies in reason," he meant they assumed that an actual primitive organism—Darwin's living filament—could fully embody the archetype of organism itself (a "mistake" that Charles Darwin would explicitly make, thus turning the idealist archetype into a biologically real entity).[48] Rather, the only way in which archetypes could be realized in nature was through dynamic evolution—the temporally gradual appearance of new species forms and the free development of those forms within definite boundaries. Through infinite time, species in the aggregate would thus evolve toward the full realization of the absolute archetype of organism per se. But if one simply examined actual species forms unreflectively, without attending to the archetypal ideal in whose light they had to stand, then "that experience itself would be as if the forms were originally only different developments out of one and the same organized being"[49]—which was Darwin's mistake. From the standpoint of empirical observation, there would be little difference between a Darwinian theory of evolution and a Schellingian theory. At their causal roots, however, they would differ considerably. Darwin presumed that organic unity could be explained by an efficient *vis a tergo*. Schelling rather conceived it as arising teleologically, as a *vis a fronte*, as it were. Species were not shoved from behind toward an already determined future; rather they rationally and thus freely moved toward the realization of the ideal of ab-

the lowest stages. Those at the higher stages must necessarily pass through the lower in order to achieve the higher." Though this last sentence seems to imply recapitulation in Kielmeyer's sense, it does remain ambiguous enough (as much in Schelling does) to sustain an alternative interpretation. He might only have meant that in the progressive display of species over time, more advanced forms would appear only after more primitive ones had been realized.

48. See the epilogue for a discussion of Darwin's theory of the archetype.

49. Schelling, *Erster Entwurf*, in *Schellings Werke*, 2:64 [III:64].

solute organism. Schelling's was a theory of evolution, *dynamic evolution*. It was just not the particular theory of Darwin, grandfather or grandson.

Schelling did not develop, to any great extent, his evolutionary ideas beyond those found in the *Ernster Entwurf*. But he seems never to have abandoned them, either. In an appreciative letter to Goethe, Schelling indicated how his mentor's morphological doctrine became the scheme for his own evolutionary theory:

> The metamorphosis of plants, according to your theory, has proved indispensable to me as the fundamental scheme for the origin of all organic beings. By your work, I have been brought very near to the inner identity of all organized beings among themselves and with the earth, which is their common source [*gemeinschaftlicher Stamm*]. That earth can become plants and animals was indeed already in it through the establishment of the dynamic basic organization, and so the organic never indeed arises, since it was already there. In the future we will be able to show the first origin of the more highly organized plants and animals out of the mere dynamically organized earth, just as you were able to show how the more highly organized blooms and sexual parts of plants could come from the initially more lowly organized seed leaves through transformation.[50]

Schelling's gratitude may have too easily persuaded him of the role of Goethe's morphology in his own ideas, though the fit of his theory with the poet's scheme is comfortable. As I will try to show in chapter 11, Goethe also felt a debt to Schelling for his own conception of species transformation, which only differed from his young friend's by reason of the latter's deep metaphysical grounding.

---

50. Schelling to Goethe (26 January 1801), in F. W. J. Schelling, *Briefe und Dokumente*, ed. Horst Fuhrmans, 3 vols. to date (Bonn: Bouvier Verlag, 1962–), 1:243.

Chapter 9

CONCLUSION:
MECHANISM, TELEOLOGY,
AND EVOLUTION

From antiquity to the present, scientific thinkers have postulated special forces to explain the distinctive properties and activities of the living and nonliving worlds. Aristotle and his disciples, for instance, thought the fundamental properties of bodies, whether animate or inanimate, sprang from an activating principle (substantial form) investing a potential principle (matter). Particular kinds of substantial form and their respective matter, as co-entative principles, constituted different substances, from the water of the seas and the clay of the earth to plants, animals, and human beings. According to this view, a special sort of substantial form, called *psyche*, or soul, breathed life into matter. Aristotle's living world displayed an ordered hierarchy of animating principles. The most basic kind of soul, the vegetative soul, characterized plant life, with specifically different souls giving rise to the various species of plants. Animals were distinguished from plants by their ability to sense, and each grade of animal—from sessile clams to the highly intelligent elephant—had their particular traits realized through the entelechy of soul activating the potentialities of matter. Human beings possessed the highest kind of mundane soul, which made matter able to think. Aristotle's world thus displayed a progressive ontological series of inanimate and animate beings.

Aristotle's set of principles, like his universe, was complex and adjusted to the variegated experiences of the research naturalist. Particular plant and animal souls, as well as the bodies they animated, fell into the general explanatory categories of formal and material causes. These served as intrinsic principles by which to understand the natures of individuals. Aristotle conceived efficient cause as usually extrinsic, something that moved or altered a body from without (for instance, a spear piercing a chest). Efficient cause, in later science, became the foundation of mechanistic explanation and has survived in present-day scientific discourse. The most important cause for Aristotle, the one that bore much of the burden of explanation, at least in the biological sciences, was the final cause, "that for the sake of which." For example, we understand, according to Aristotle, the existence of lungs in higher vertebrates in terms of their telic properties—thus, the lungs exist *in order to* cool the heat of the heart.

With the advent of the Enlightenment, Aristotle's conception of the world began to appear by turns both too circumscribed and too complex. It was too circumscribed for the likes of Copernicus, Kepler, Galileo, and Newton, under whose mathematical operations the spatially and temporally constrained universe of the ancients and medievals began to expand into infinity. At the same time, Aristotle jammed his limited world with too many grades of being and too many principles; it was an overripe garden that needed pruning. Cartesian mechanics reduced this luxuriant growth to two kinds of being—*res cogitans* and *res extensa.* Everything that was not mind could be understood simply in terms applicable to matter in motion. The nonmental sphere ran like a machine, and only efficient causes were necessary to explain its behavior. Descartes thus transformed biology into a branch of physics. The real power of this reductive strategy, though, only became manifest as a consequence of the work of Newton. In his wake, even human mind became an engine for producing ideas, or at least this was the conceit of Julien Offray de La Mettrie's *L'Homme-machine* (1747), a work, albeit, reflecting perhaps as much the ironic temper of the Enlightenment as a serious scientific proposal. In a more sober effort, David Hume exploited the mechanical model of mind, arguing that the manufacture of ideas was subject to the same causally efficient laws believed to characterize the external world. The Enlightenment thus enshrined mechanism as the model to explain all phenomena of matter, life, and mind. Teleological causes had been unveiled, in Bacon's terms, as "barren virgins." Any work they supposedly accomplished could be turned over to efficient, mechanical causes.

Curiously, though, Hume's analysis allowed the machinery of the world to slip gears; the causality according to which it operated lost necessary power. Cause became a relationship of mere subjective association: one billiard ball caused another to move across the table—at least this had been the common experience of those who played the game. Yet, according to Hume, analysis of the linkage between cause and effect revealed no necessary bond. Only fluctuating experience joined them, and future experience might undo them. At the next moment, colliding billiard balls might produce a bouquet of flowers. Reason could not gainsay this possibility. No rational principle, from the Humean perspective, could guarantee the universal applicability of previously established causal relationships. Mechanism was supreme, but the engine clanked and groaned, threatening to break down at any moment.

Kant, an enthusiast for the Newtonian design of the world, reinstalled the machinery of that world within the subject, not as a contingent conse- quence of the subject's past experience, but as an integral part of rational mind itself. Indeed, the Newtonian framework virtually came to define hu- man rationality. Under this reconstruction, however, the world of experi- ence—the phenomenal world—lacked depth. The Newtonian machinery, to use a crude but justifiable metaphor, projected an image onto the surface of external reality, providing a stable picture of nature amenable to scien- tific investigation. Beneath the phenomenal surface, though, lay an active but unknowable substructure—the noumenon or thing-in-itself—that fur- nished, or so Kant implied, the sensual data that mind illuminated accord- ing to the projected categorical principles of unity, substance, causality, and the rest. Kant's model of the interaction of mind and nature perhaps pro- vided—if one did not inquire too deeply—a sufficient protection against the psychological subjectivism of Hume, at least in the area of the physical sciences. But even Kant found mechanical principles insufficient to explain the activities of living matter. The Königsberg sage, not a great naturalist but a great reader of natural history, concluded that, in a sense, Aristotle had been right. The natural researcher had to employ teleological consider- ations, since organisms appeared to behave as if their structures and actions had been intelligently planned. Human mind could thus confront nature's products only if they appeared to result from an *intellectus archetypus*. Al- though mindful of the Enlightenment's rejection of the barren virgins of teleology, Kant relegated final causes to the category of regulatory ideas. They might modestly serve as heuristic principles, necessary handmaidens to true science, which ought to employ only mechanistic causes in explana- tion. For Kant, this meant that biology, which seemed to depend radically on teleological principles to organize its inquiry, could never really be a science.

Kant's extremely powerful analysis of the foundations of scientific knowledge became the temple in which Fichte and Schelling originally worshiped, only to find that the beauty of its promise began to crumble when any pressure was applied to its buttress, universal mechanism. In or- der to save that promise, they had to destroy the temple.

Kant's mechanistic analysis of nature required a given: the noumenal realm had to provide the contingent factors of sensible qualities, which then would be organized through the activities of reason. Fichte and Schelling rejected the postulation of a noumenal sphere; no evidence or ar- gument could sustain it. Either such postulation was a necessary require- ment of understanding—but then no different than the categories—or it made its presence felt through an assumed causal interaction—but then that would entail an illegitimate use of the category of causality, which, by Kant's own strictures, was applicable only to the phenomenal realm. Moreover, as Schelling urged, how could something independent of mind—a kind of raw givenness with its own necessities—anticipate the operations of reason? How could the two be made to mesh without a supernatural harmonizer?

Aside from considerations drawn from Fichte, Schelling's understanding of nature had other sources that directed him against the Newtonian conception of a mechanical universe. Blumenbach, one of the great biological authorities of the late eighteenth century, had recognized the need to postulate a nonmechanistic force to explain the telically structured actions of organisms. And Kielmeyer, greatly admired within a smaller circle of cognoscenti, had discovered biological relationships, governed by principles of balance, that defied mechanistic reconstruction. In light of the work of these naturalists—and mindful of Kant's persuasive argument that biological explanation could not dispense with teleological notions— Schelling advanced the proposal that organism, not mechanism, must serve as the grounding concept to explain the activities of nature, both in the animate and inanimate spheres. Indeed, the deep harmony of law and content, form and matter, convinced him that not only the categorical structures but also the contingent contents of nature must flow from a common source. This source, he thought, could best be comprehended as something like an artistic intelligence. Only creative mind could account for the harmonious patterns actually found in nature. Blind mechanism certainly could not. If the categorical structures that Kant postulated, as well as the material content of those structures, flowed from mind, then nature and mind could be conceived as merely two aspects of one absolute reality. Here falls Newtonian mechanism and the mind-body dualism that covertly sustained it. Here arises a new primacy for biology as the foundational science.

The trail of reasons that led Schelling to restructure biological science was smoothed by his deeply aesthetic response to nature, nicely captured in these lines from his long poem *Epikurisch Glaubensbekenntniss Heinz Widerporstens* (1799):

Ascending up to thought's youthful force,
Whereby nature continually renews its course,
There is one power, one pulse, one life,
A continuous exchange of resistance and strife.[1]

1. Friedrich Schelling, *Epikurisch Glaubensbekenntniss Heinz Widerporstens*, in F. W. J. Schelling, *Briefe und Dokumente*, ed. Horst Fuhrmans, 3 vols. to date (Bonn: Bouvier Verlag, 1962–),

From our perspective, peering back over the two centuries since Schelling wrote these lines, we might be as ready to ascribe the youthful élan he felt more to the revolutionary times—to the Romantic fever of the period and to the particular circle in which he moved—than to nature. Yet at other times and other places, say, in the jungles of South America (with Humboldt and Darwin) or the Mediterranean coasts of Sicily (where Haeckel sketched, poetized, and biologized)—or even on the island of Lesbos (where Aristotle worked some two thousand years before)—scientists have directly experienced a pulsating, active, and value-charged nature. Schelling's vitalization of nature was hardly unique, and it was based, despite the fancies of critics who hold the contrary, on immediate experience, the anchor of good scientific theory.

There is, though, a residual difference separating Schelling's naturalism from that of those who would follow in his wake. Later naturalists of a Romantic turn tended to be evolutionists. Neither Schelling, Goethe, nor Humboldt, however, was an evolutionist, at least not in the genealogical mode of Charles Darwin. I have argued in the preceding chapter that Schelling was indeed an evolutionist—in his terms, a dynamic evolutionist. After all, his metaphysical idealism had to be realized in time, which meant that the archetypal relationships discernable as logically uniting classes of organisms (the vertebrates, for example) had to unfold gradually over the course of ages. Schelling appears to have been initially receptive to the idea of descent of creatures out of one another, that is, until he understood the full meaning of the genealogical theory of evolution. That theory had been most completely worked out during the late 1790s by Erasmus Darwin, who generally inclined to empiricism and mechanism. His evolutionary constructions simply seemed too mechanical, too contingent to be trusted. Schelling understood the forces of nature not to impel species from behind, as one billiard ball might bang into another, but magnetically to draw them forward, toward the realization of ever more complete organic forms. He did not, as it were, dispute the empirical facts of evolution—that is, the appearance of more advanced creatures over time. Armed with a capacious notion of species (more like the archetypal categories of, say, quadrupeds or fish), he supposed that the *Anlagen*, or seeds, of such groups would develop sequentially, from those that realized less perfectly the ideal of organism to those that realized it more perfectly. "The jump from polyp to man," he acknowledged, "seems indeed enormous; and the transition from the former to the latter would be unclear if intermediate stages did not

---

2:212: "Hinauf zu des Gedankens Jugendkraft, / Wodurch Natur verjüngt sich wieder schafft, / Ist Eine Kraft, Ein Pulsschlag nur, Ein Leben, / Ein Wechselspiel von Hemmen und von Streben."

intervene between them. The polyp is the simplest animal and likewise the stem from which all other organizations sprout."[2] Within each archetype, however, significant morphological transitions might occur genealogically. According to the theory, which Schelling adapted from Kielmeyer, the same forces at work in individual development operated in species development. For both thinkers, the similarity of forces implied recapitulation: that is, species at "the higher stages must necessarily pass through the lower in order to achieve the higher."[3] The German Romantic tradition had thus begun to excavate a path that would lead to genealogical evolutionary thought in the latter part of the nineteenth century. That path, however, had branching trails that momentarily led in other directions. Karl Ernst von Baer, for instance, adopted Schelling's notion of the archetype and would utilize its resources first to oppose genealogical evolution but then to propose something like Schelling's own dynamic evolutionism. More proximately, the theory of the archetype would become central to Goethe's biological conceptions. The great poet advanced along the path that his protégé had cleared, as we will see in the next two chapters.

2. Schelling, *Erster Entwurf*, in *Schellings Werke*, ed. Manfred Schröter, 3rd ed., 12 vols. (Munich: C. H. Beck, 1927–59), 2:54n [III:54n].
3. Ibid., 2:53n [III:53n].

Appendix

Theories of Irritability, Sensibility, and Vital Force from Haller to Humboldt

Various theories of vital force that exfoliated at the turn of the eighteenth century usually developed out of or by reference to ideas planted earlier in the century, particularly in the work of Albrecht von Haller. Haller proposed, much like Aristotle, that animal life had to be understood in terms of two fundamental powers, irritability of tissues (a property shared with plants) and sensation. Disputes about these traits of animal organisms, culminating in the work of electrophysiologists, form the conceptual background for the history related in part 2. It seems advisable, then, to sketch these antecedent ideas in order to achieve greater definition for the history of Romantic conceptions of life.[1]

Albrecht von Haller

Albrecht von Haller (1708–1777) had distinguished sensibility and irritability as the fundamental forces in organisms. "Sensation," according to Haller, "arises when an impression of some sensible object, which has impinged on a nerve of the human body, arrives in the brain, through that nerve's connection to the brain, so that it is thus represented to the soul."[2] Only nerves, he held, were sensitive; consequently, parts without nervous involvement, like tendons, should feel no pain when pricked. Hearts and other muscles reacted to nerve stimulation by contracting. But, in Haller's view, they could also directly respond to other irritants: for instance, a freshly excised heart (thus with nervous connections severed) might contract when touched with a probe or jolted with electricity.[3] Irritability, thus, was not essentially a nervous phenomenon, but a property of the muscle fiber itself. Haller proposed that the irritable reaction ensued from a *vis insita*, or resident force.[4] Even when nerve impingement caused a muscle to contract, the shortening of fibers resulted, he thought, from nervous energy activating the *vis insita*. In the case of non-nervous stimulation, contraction would be caused by some object (e.g., a metal probe) or substance (e.g., blood in the chambers of the heart) evoking the *vis insita* directly.[5] During the next half century, considerable controversy arose over Haller's distinction between sensibility and irritability, and over the physiology behind the distinction.

---

1. Joan Steigerwald has discussed the experimental foundation of ideas about *Lebenskraft*, especially that stemming from Albrecht von Haller and the electrophysiologists. I have been oriented to this literature through the lucid treatment in chapter 2 of her forthcoming *Lebenskraft in Reflection: German Perspectives of the Late Eighteenth and Early Nineteenth Centuries*.

2. Albrecht von Haller, *Primae Lineae Physiologiae in usum Praelectionum Academicarum*, 4th ed. (Lausannae: Grasset et Socios, 1771), p. 206 [CCCLXVI].

3. Ibid., p. 50 [CI].

4. Ibid., p. 225 [CCCC].

5. Ibid.

Robert Whytt

Robert Whytt (1714–1766), the Scotsman who first demonstrated the reflex character of the spinal cord, took decided issue with Haller. Haller had performed vivisectional experiments to show that internal organs stimulated with a probe gave no indication of communicating pain, though they might retain an irritable response. Whytt thought, however, that Haller's experiments artificially masked the sensitivity of organs. A dog's misery in having its thorax cut and separated would surely overwhelm any feeling of pain from a prick to its heart or liver.[6] Whytt's own experiments convinced him that all the body's organs were sensible. Further, "if sensibility be a sure mark of the existence of nerves in any part of the body, then there is none without them, altho' anatomists will never be able to demonstrate them in every part." Concerning irritability, Whytt held that "it always infers some degree of sensibility, yet sensibility does not infer irritability, unless the part be, by its structure, fitted for motion, i.e., in other words, unless it be what we call muscular."[7] Accordingly, sensibility was the fundamental force and irritability the derived one. Whytt was convinced that the soul, either in perception or in muscular contraction, had to act through the "nervous power" that surged through nerves to all organs of the body.

Erasmus Darwin

Just prior to the publication of Schelling's *Weltseele*, another author contributed to the debate on animal forces, and his steely ideas made several surprisingly deep dents in the philosopher's theories. This was the English physician Erasmus Darwin (1731–1802). His *Zoonomia* appeared in 1794 and was quickly published in German translation the next year.[8] Darwin held that the "spirit of animation" ("*Geist der Belebung*" in the German) consisted in a tenuous matter that had the "power" [*Kraft*] of causing fibrous contractions in the muscles or sense organs. "Irritations" [*Reitzungen*] engaging the power might be excited by external objects or by feelings of pleasure and pain, volitions, and habits of association.[9] Darwin held, along with Whytt and in opposition to Haller, that

6. Robert Whytt, *Observations on the Sensibility and Irritability of the Parts of Men and other Animals Occasioned by the Celebrated M. De Haller's late Treatise on those Subjects*, in *The Works of Robert Whytt, M.D. Published by his Son* (Edinburgh: Becket and DeHondt, 1768), pp. 257–306; reference to pp. 260–61. In response, Haller objected that he never denied the heart, liver, spleen, and kidneys to have nerve connections, and thus the capability of pain; he simply maintained that they had few nerves, and thus could feel little pain. Haller replied to his critics in a lecture (*De partibus corporis humani sensibilibus et irritabilibus*) delivered to the Royal Society of Science at Göttingen on 22 April 1752. It was translated under Haller's supervision some twenty years later. I have used a reprint edition: Albrecht von Haller, *Von den empfindlichen und reizbaren Teilen des menschlichen Körpers* (Klassiker der Medizin), ed. Karl Sudhoff (Leipzig: Verlag von Johann Ambrosius Barth, 1922). The reference is to pp. 29–30. Ultimately, both Whytt and Haller believed sensibility required nerves; the question became what organs had proper nerve connections. Whytt also maintained, against Haller, that irritability was a function of sensation, and thus also required nerve connections. Haller held to the idea that muscle fibers themselves could react with irritable contraction in the absence of nerves.

7. Whytt, *Observations*, pp. 268, 280.

8. Erasmus Darwin, *Zoonomia, or the Laws of Organic Life*, 2 vols. (London: Johnson, 1794–96); and *Zoonomie, oder Gesetze des organischen Lebens*, trans. J. D. Brandis, 3 vols. in 5 (Hannover: Gebrüder Hahn, 1795–99). In *Von der Weltseele*, Schelling refers to the first volume, parts 1 and 2 (1795) of the German edition.

9. Darwin, *Zoonomia* (English), 1:32; *Zoonomie* (German), 1.1:51.

nerves were necessary for any fibrous motion in the animal body, since the spirit of animation only operated through the nerves that invested the muscles and sense organs.
For Darwin, then, contractility of muscle or sense fibers, in virtue of the actions of the
spirit of animation, constituted the foundation for life's processes. In a general way, this
was the physiological position that Schelling elaborated. But his understanding of the
fundamental physiological processes more likely derived from the ideas of the Scots
physician John Brown (1735?—1788), whose medical theories had established a strong
front in the Baden region of Germany,[10] and from Alexander von Humboldt, whom he
knew and greatly respected.

John Brown

In his *Elementa medicinae* (1780)—published in German translation in 1795—Brown
maintained, comparable to Darwin, that "excitability" constituted the basic property of
living bodies.[11] Excitability encompassed reactions of "sense, motion, mental action,
and the passions."[12] The force or power of excitability, according to Brown, resided in
the nervous system and in the muscle fibers, insofar as he regarded the latter as exten-
sions of nerve fibers.[13] Plants, animals, and human beings came into the world with a
certain quantity of excitability, which differed from individual to individual, from time
to time, as well as from organ to organ. For Brown, this meant that the well-being of the
creature could be assessed in terms of the proportion existing between the exciting stimuli an individual faced and the amount of excitability resident in the body. Good health
would be maintained when the level of excitement remained within certain boundaries.
Were a stimulus either too strong or too weak in relation to the degree of excitability,
disease would result. The physician, then, had the task of either lowering or increasing
the power of the stimulus in order to achieve the just balance and consequent good
health.

Brown said he discovered these causes of disease when treating his own painful gout.
He discovered that strong drink and opium, two potent stimulants (as he thought), restored the health of his big toe, a feat of cure he proudly demonstrated on one occasion
to his surprised dinner companions.[14] Brown generalized his results to other diseases. In
the *Elements of Medicine*, he spent a fair number of pages analyzing various maladies into
either diseases of excess: "sthenic diseases," in which the stimulus surpassed the quantity
of excitability in a person, or diseases of deficiency—"asthenic diseases"—in which the
stimulus could not release sufficient excitability. Typhus, for example, he determined
to be an asthenic disease, resulting from "impure air"; balance could only be restored

10. The pervasive influence of Brown on German medicine is discussed by Markwart Michler
in *Medizin zwischen Aufklärung und Romantik: Melchior Adam Weikard (1742–1803) und sein Weg in
den Brownianismus*, in *Acta historica Leopoldina*, no. 24 (Halle: Deutsche Akademie der Naturforscher Leopoldina, 1995). See also the lucid discussion of Thomas Broman in his *Transformation
of German Academic Medicine, 1750–1820* (Cambridge: Cambridge University Press, 1996),
pp. 143–58.

11. John Brown, *The Elements of Medicine; or, a Translation of the Elementa Medicinae Brunonis
by the author of the Original Work*, 2 vols. (London: Johnson, 1788), p. 14 [xxiii]. Though Schelling
and the rest of the German medical community would have had Brown's original Latin treatise
available, the popularity of his work demanded a German translation: John Brown, *Grundsatze der
Arzeneylehre*, trans. M. A. Weikard, 2 vols. (Frankfurt a. M.: Andreaischen Buchhandlung, 1795).

12. Ibid., p. 5 [xv].

13. Brown, *Elements of Medicine*, p. 38 [xlviii].

14. Ibid., pp. 7 [xviii], XI.

through stimulants, such as opium (something he also found rather good for his gout).[15] The opium treatment for typhus was, it will be recalled, the specific prescribed for Auguste Böhmer by the Brunonian physician in Bamberg, a treatment that Schelling moderated and for which he received the obloquy of his enemies.

Schelling accepted Brown's general account of health and disease (not without some refining criticisms) but recognized that the Brunonian conception of excitability yet remained awfully vague. He also objected that excitability and exciting stimuli could only constitute a negative, restraining condition of life; Brown lacked a positive principle that could account for the progressive, developmental character of vital processes.[16] Schelling sought enlightenment and confirmation in this regard from Alexander von Humboldt (1769–1859), who had conducted a series of experiments that gave considerably greater physical content to the notion of excitability. Humboldt also had some explicit proposals about the positive principle governing living organisms. The positive principle achieved prominence in Humboldt's initial studies of plants.

Alexander von Humboldt

Humboldt had composed his *Florae Fribergensis* while serving as inspector in the Prussian Department of Mines, a position he obtained just after completing his studies at the Bergakadamie in Freiberg (1792). The work sketched out a conception of the natural history of plant distribution, a notion that he would more fully develop as a result of his five-year voyage to the New World, which he undertook in 1799. To this composition he appended a series of *Aphorismi*, or short position statements (replete with citations to the contemporary literature) on the physiology of plant and animal life. He introduced his more detailed chemical analyses in these aphorisms with a general consideration of the distinction between life and lifeless matter.

> We call that matter inactive, brute, or inanimate if its elements are combined according to the laws of chemical affinity. We call those bodies animated and organic that, though they tend constantly to change into new forms, are constrained by some internal force, so that they do not relinquish that form originally introduced.... That internal force [*vim internam*] which dissolves the bonds of chemical affinity and prevents the elements of bodies from freely uniting, we call vital.[17]

The phenomenon of life, in both plants and animals, that gave fundamental expression to the internal vital force was excitability. Some parts of plants and animals simply obeyed laws of chemical affinity and thus were, strictly speaking, inanimate (for instance, bones in animals or epidermis in plants); but others, the vital organs, responded with irritable contraction to stimulation. Such excitability clearly distinguished vital organs from those nonvital parts; the elements of the former were held in a dynamic tension that opposed the chemical affinities that threatened to dissolve their animate form. On the question of the relationship between irritation and sensation, Humboldt sided with Haller: irritative parts did not necessarily also enjoy sensation.[18] But to further secure the conception of irritability as the fundamental response of life, Humboldt under-

15. Ibid., p. 129 [cxlvi].

16. Schelling, *Weltseele*, in *Schellings Werke*, ed. Manfred Schröter, 3rd ed., 12 vols. (Munich: C. H. Beck, 1927–59), 1:573–74 [II:505–6].

17. Alexander von Humboldt, *Florae Fribergensis specimen, plantas cryptogamicas praesertim subterraneas exhibens* (Berolini: H. A. Rottman, 1793), pp. 133–35 (nos. 1, and 2).

18. Ibid., pp. 137–38 (nos. 3 and 4).

took a series of exacting experiments in electrophysiology, which ultimately led him to reconsider the nature of vital force.

While visiting Vienna in October 1792—as part of a tour undertaken for the Prussian Department of Mines—Humboldt witnessed an exciting spectacle, which had been repeated in learned gatherings all over Europe, namely, the galvanic revivification of frogs. Throughout Italy, France, and Germany, ponds had been seined to supply the myriad of creatures required to re-create the entertaining experiments performed by the Bolognese physician Luigi Galvani (1737–1798).[19] Galvani discovered that he could make a decapitated frog contract its leg muscles by exposing its crural nerve in the vicinity of a sparking electrostatic machine. Even more startling and theoretically imposing were the experiments he performed to establish that animals carried an endogenous electricity that operated through the nerves. In these experiments, he prepared his frogs by dissecting out as a unit the spinal cord, crural nerve, and posterior limbs. He found that if he placed a metal hook in the spinal cord and touched one end of a metal arc to the hook and the other end to the leg muscle, the muscle would vigorously contract, as if the frog were perfectly alive. This and other well-executed experiments convinced him that animal organisms contained an innate electrical fluid, held in the muscle fibers as in so many miniature Leyden jars: positive electricity was stored in the core of the fibers and negative electricity was retained on their surface.[20] When a metal arc touched the nerve and the outer surface of the muscle, accumulated electricity in the core of the muscle would, he conceived, discharge back through the nerve and metal bridge to the negative surface, producing a contraction. He suggested that due to an act of will, electrical current flowed from the brain through the conducting nerves to muscles; equilibrium would be maintained as long as no excessive amount of positive electricity flowed into the muscle condenser. However, when a large quantity of positive electricity did flow into the core of the muscle, say, through a jolt of voluntary action, the muscle would discharge its stored electricity to the negative surface, causing a contraction.[21]

Galvani's theory of animal electricity evoked a strong challenge from the physicist Alessandro Volta (1745–1827). Volta was quite impressed with Galvani's experiments and initially accepted the idea of animal electricity, though with some reservations. Upon further consideration and experiment, however, he quickly came to believe that the electrical current causing muscle contraction in prepared frogs flowed not from en-

19. Galvani published his discoveries in *De viribus electricitatis in motu musculari commentarius* (1791). This treatise is conveniently available in facsimile, along with an English translation, in Luigi Galvani, *Commentary on the Effects of Electricity on Muscular Motion*, trans. Margaret Foley, ed. I. Bernard Cohen (Norwalk, Conn.: Burndy Library, 1953).

20. Ibid., pp. 38–39 (facsimile).

21. Ibid., p. 47 (facsimile): "Concerning how voluntary motion occurs: it is likely that the soul, by its remarkable power, which resides either within the brain or outside it, affects, as it were, the nervous-electrical fluid, which flows strongly from the reacting muscle to that part of the nerve to which it was recalled by the impulse. When it will have arrived, with the increased forces having overcome the non-conducting part of the nerve substance, it is then drawn away from that place, leaving either by the extrinsic moisture of the nerve, or by the membrane, or by other contiguous conducting parts, and through these it is restored to the muscle from which it departed as through an arc. I believe the restoration occurs in order that, according to the law of equilibrium, the electrical part might copiously flow to the negative fiber of the muscles, just as it had previously effluxed from the positive electrical part of the muscle by reason of the impulse in the nerve." This passage constitutes only one sentence in the Latin. Typically it gives only a rather vague impression of the physiological operations and theory Galvani supposed in order to explain his experiments, which themselves are clearly enough described.

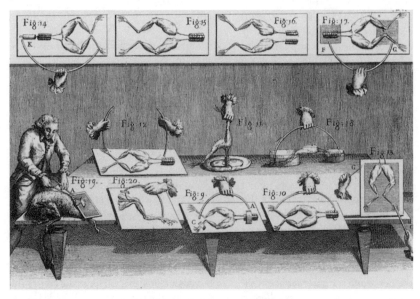

Figure A.1   Plate from the second edition (1792) of Luigi Galvani's *De viribus electricitatis*, showing the endogenous electrical production of muscle movements in excised frog legs, a chicken leg, and the leg of a living sheep.

dogenous sources but from the bimetallic connections (i.e., from the hook and the metal arc) or from impure metallic arcs (thus essentially heterogeneous metals) that Galvani had used in his experiments. The dissimilar metals, not the internal constitution of the frog, produced the electricity. Volta ingeniously demonstrated this likelihood by constructing what became known as a voltaic pile, a column of some dozen or so alternating metal disks (e.g., silver and zinc) separated by a disk of moist pasteboard. This apparatus (a modern battery) would supply a continuous number of shocks, very like a constantly recharging Leyden jar. Thus, all evidence for an internal animal force evaporated with the artifacts of Galvani's experiments.[22] Or so Volta contended. Humboldt, however, thought Galvani had been essentially correct and that his theory provided an account of the nature of irritable contraction of muscles, and, ultimately, a general solution to the problem of vital forces.

After he had witnessed the galvanic phenomena in Vienna, Humboldt began preparing his own frogs and conducting his own experiments during the several years thereafter, often carting his instruments and materials along with him on his many travels for the Prussian Department of Mines. On one extensive trip—actually a vacation— he visited Volta at his villa on Lake Como.[23] From 1794 through 1797, he conducted a variety of experiments, variations on those executed by Galvani and Volta. In many of these, he would construct voltaic piles but would use animal muscle interspersed with plates of metal to test the existence of currents and the force with which they might

22. The most extensive treatment of the controversies between Galvani and Volta is provided by Marcello Pera in his *The Ambiguous Frog: The Galvani-Volta Controversy on Animal Electricity*, trans. Jonathan Mandelbaum (Princeton: Princeton University Press, 1992).

23. Hanno Beck, *Alexander von Humboldt*, 2 vols. (Wiesbaden: Steiner Verlag, 1959–61), 1:74.

cause contraction in frog preparations. In March 1797 Humboldt returned to Jena, where his brother Wilhelm and his family were living. He wrote his friend Carl Freiesleben, suggesting a calmer engagement with study there than would be possible for most other mortals: "I have been living since March 1 in Jena wholly occupied with my book, chemical experiments, and anatomy. I have really retreated into the life of a student, since my sphere is narrow and limited totally to myself. . . . Since I am now preparing very earnestly for a West Indian trip and intend there chiefly to concern myself with organic forces, anatomy is now my principal study."[24] In fact, Humboldt met often with Goethe and Schiller, conferred with August Carl Batsch about invertebrates, took a seminar on anatomy with Justus Christian Loder (often accompanied by Goethe), discussed galvanism with Johann Wilhelm Ritter, and did not neglect his brother's family.[25] All the while, he was testing instruments (barometers and theodolites) in the local countryside in preparation for his intended trip. Yet during this time in Jena, Humboldt somehow managed to complete and see to press the first volume of his book *Versuche über die gereizte Muskel- und Nervenfaser* (Experiments on stimulated muscle and nerve fibers, 1797), which detailed his galvanic studies; and he continued frantically to work on the second volume, which expanded his consideration of the chemical and anatomical conditions of animals and plants that regulated the galvanic response.[26]

In his experiments, Humboldt constructed various kinds of voltaic piles, usually using muscle tissue interspersed between metallic elements. He demonstrated that the strength of muscle contraction of a prepared frog depended not only on what metals were used in the generating pile, the way they were stacked with the tissues, and the general moisture of the elements of the pile, but also on the gender, age, health, and a myriad of environmental conditions that affected the excitability of muscle fibers in the once-living frog.[27] His principal discovery was that a pile that consisted only of nerve and muscle, without any interspersed metals, could generate an electrical current, causing contraction and thus that the metals thought necessary by Volta played only a secondary role.[28] This meant that electrical currents might have an internal source, likely in the brain and nerves, as Galvani maintained. Electrical fluid would flow easily through organic matter, with greater resistance through metal, and with greatest difficulty through piles of heterogeneous metals. Greater resistance could be overcome only by a larger current, which thus explained the vigorous contractions Volta got with his bimetallic piles.[29] Humboldt theorized, on the basis of his experiments, that muscle fibers had (as Haller supposed) an innate power of contraction, which an alteration of the composition of the fibers could influence. Many exogenous stimuli—such as a metal

24. Alexander von Humboldt to Carl Freiesleben (18 April 1797), in *Die Jugendbriefe Alexander von Humboldts, 1787–1799*, ed. Ilse Jahn and Fritz Lange (Berlin: Akademie-Verlag, 1973), p. 574.

25. Goethe frequently joined Humboldt in conducting physiological experiments. See chapter 11.

26. Alexander von Humboldt, *Versuche über die gereizte Muskel- und Nervenfaser nebst Vermuthungen über den chemischen Process des Lebens in der Thier- und Pflanzenwelt*, 2 vols. (Berlin: Heinrich August Rottmann, 1797–99).

27. Humboldt seems to have followed Volta in observing that the strength of reaction depended on aspects of the physiology of the frogs used.

28. Ibid., 1:379–80.

29. Ibid., 1:418: "The greater the resistance and the later the breakthrough, the more effective must that breakthrough be. The galvanic fluid seems to flow most easily through animal matter, with greater difficulty through metal, and with greatest difficulty through a heterogeneous composition."

Figure A.2

Humboldt's experiments
showing (in figs. 3 and 5)
that excised muscle and
nerve tissue alone,
arranged in a circuit,
could produce a current
and generate muscle
contraction. Plate from
Alexander von
Humboldt, *Versuche über
die gereizte Muskel- und
Nervenfaser* (1797).

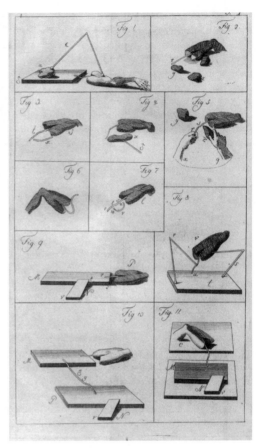

probe or acid—might evoke a contraction in muscles; but within the ordinary animal
economy, endogenous electrical stimulation produced contraction. As he conceived it,
following Galvani, "in the case of voluntary muscle movement, a secretion of [galvanic]
fluid in the brain or nerves occurs simultaneously with the idea of the will; the fluid becomes greater and causes a current or sudden overflow into the muscle," resulting in
contraction.[30]

In the second volume of his work, published in 1799, Humboldt focused on the internal physiology of animal and plant organisms. His experiments led him to rethink the
conception of vital force that he had advanced a few years earlier. Under the barrage of
criticism he had received, especially from Reil and Röschlaub, he felt compelled to reformulate what before he had regarded as mere facts of observation. To be theoretically
safe, he said he would no longer refer to a unique vital force, which might simply be the
interaction of known physical forces. He would put it this way:

> I call that stuff animated whose parts, when arbitrarily separated from the whole,
> change their composition while maintaining their previous external relationships.

30. Ibid., 1:417.

> . . . The balance of elements in living matter maintains itself only so long as every part is part of a whole and exists as such.[31]

The parts of inanimate objects would, by contrast, retain their same composition after separation from their original whole (thus, parts of broken glass are still glass). This definition of life, Humboldt expressly pointed out, depended "directly on the idea of that immortal thinker 'that in an organisms every part is mutually means and end.'"[32]

Humboldt's physiological analysis of the organic and his reconception of the nature of vitality strongly appealed to Schelling. Humboldt had shown that a fundamental force of animal and plant activity—electricity—was not a mystical power distinct from the ordinary forces operating in nature. Yet, obviously this force was intimately related to life, being generated under specifically organic conditions. Life was merely a more conspicuous array of forces—such as electricity and contractility—that maintained a certain balance. Moreover, Humboldt thought the balance was dynamic:

> Animal and plant fiber must, as I believe, be regarded not merely as susceptible to irritation but as continuously irritated. The secretions, whose composition differs in every species, keep them constantly moist. Light, heat, electricity and the other components of the atmosphere, in which all creatures are bathed, operate upon them at every moment. What one calls the natural tone of the fibers—or that condition which responds to the stimulus—never exists in the strict sense. With the vital activity of the organs, with the incessant alteration of their component parts, the idea of rest vanishes. What we take as the absence of stimulation is only a lesser grade of stimulation.[33]

Schelling's thesis that life was a dynamic balance of forces, forces that could be found rife within nature—this thesis was thus shared by a number of physiologists and experimentalists at the end of the eighteenth century. His conception was neither unique nor the spillage of a metaphysically overheated brain.

31. Ibid., 2:433–34.
32. Ibid., 2:434. The immortal thinker, of course, is Kant.
33. Ibid., 2:59–60.

Part Three

GOETHE, A GENIUS FOR POETRY, MORPHOLOGY, AND WOMEN

•◆•◆•◆•

Figure 10.1 Johann Wolfgang von Goethe (1749–1832). Chalk portrait (1791) by Johann Heinrich Lips. (Courtesy Archiv für Kunst und Geschichte.)

Chapter 10

THE EROTIC AUTHORITY
OF NATURE

The beautiful is a manifestation of secret laws of nature, which
without its appearance would have remained forever hidden.

—Goethe, *Maximen und Reflexionen*

After Schiller's death in 1805 and his own sudden illness that year, Goethe
became increasingly preoccupied with the history of his own mental development. Presentiments of mortality turned him toward the rebirth of himself in memory. He wished to explain to himself and to the world the
formation of an individual who had achieved a remarkable cultural position; and thereby, he undoubtedly hoped as well, to solidify that position.
The tendency to use his various artistic and scientific works as media for
self-reflection had begun quite early, though it sharply accelerated during
these later years. In the principal construction site for memory, his autobiography *Dichtung und Wahrheit* (Poetry and truth, 1811–14, 1833), Goethe
remarked "that I could never overcome this impulse, namely, to transform
whatever delighted or pained me—or whatever I was engaged with—into
an image, into a poem, and by that means to come to terms with myself."[1]
Throughout his longer autobiographical writings, through the essays that
depicted the course of his thought about scientific subjects, into the novels
that portrayed barely transformed events in his life, in the conversations
that conjured up ideas and episodes of long ago, and especially in the poetry that transmuted significant experiences—through all of these vehicles
of self-consideration, he sought to discover that strand of interpretation
holding together a life of great complexity and abrupt change. No history
of an individual, of course, can be tied up in a neat epitome, and certainly
not Goethe's, as the libraries written about him attest. "Our lives," he be-

---

1. Goethe, *Aus meinem Leben: Dichtung und Wahrheit* (pt. 2, bk. 7), in *Sämtliche Werke nach Epochen seines Schaffens* (Münchner Ausgabe), ed. Karl Richter et al., 21 vols. (Munich: Carl Hanser Verlag, 1985–98), 16:306.

lieved, "like the whole of nature in which they are contained, are composed in an incomprehensible manner out of freedom and necessity."[2] The lived reality of his life simply escaped the confines of even his own efforts at self-recognition. Four tightly plaited threads do, however, become quite visible in his autobiographical writings and letters, and they provide strategic guides through the labyrinth of his life. These are the variously hued themes of art and science, of nature and women.

Goethe's many intense relationships with women—what Friedrich Schiller referred to as "the *Weiberliebe* that plagues him"[3]—became a passional means by which he explored the aesthetic and the universal in nature, yielding with each encounter, as well, a different modality in his art. In his poetry, he expressed the joys and sorrows of a youthful embrace of nature, whether in the figure of a golden grove or a golden girl. And in the autobiography, he poetically represented his approach to both nature and women in similar fashion: at first from a distance, with fertile imagination, then more immediately and intimately, and finally in the fabrications of memory. His many relationships with women—beginning with Gretchen, the first, stinging love of the fifteen-year-old boy and the melancholic re-creation of the sixty-year-old poet, and extending through the numerous real and imaginatively elaborated affairs thereafter—these became for him emblematic of the hopes and disappointments, the joys and sorrows of his equally long scientific affair with nature. The young Gretchen broke his heart, and the aging memorialist recalled it poignantly and ruefully, rapping the truth it contained within the poetic creations of reminiscence. With each subsequent relationship, his swelling desires penetrated to an ever-deeper investigation of nature both poetically and scientifically. In Italy he finally achieved physical satisfaction in love and not accidentally, I believe, in morphology—and in the poetry coming from that period, he wove them sensually together. The female became for him the symbol of creativity—of the generative power of the artist and of nature—and in her embrace Goethe's poetry was joined to his science. This man, who confessed his life to have been led in closer accord to the heart than the brain, had even the rhythms of his science orchestrated by the pulse of that vital organ.[4]

2. Ibid. (pt. 3, bk. 11), 16:511.

3. Friedrich Schiller to Christian Gottfried Körner (1 November 1790), in *Schillers Werke* (Nationalausgabe), ed. Julius Petersen et al., 55 vols. (Weimar: Böhlaus Nachfolger, 1943–), 26:55.

4. In his autobiography, just after recalling the joy, humiliation, and disappointment of his first love, he admitted that, compared to more intellectual matters, "the affairs of the heart always seem to me the most important. I never tired of reflecting on the fleeting moments of affections, the changeability of human nature, moral sensitivity, the sublime and the mundane—all of those elements joined in our nature that can be considered the riddle of human life." See Goethe, *Dichtung und Wahrheit* (pt. 2, bk. 7), in *Sämtliche Werke*, 16:310–11.

The comparison between Goethe's passional life and his scientific life might appear too tenuous, too circumstantial to bear much interpretive burden, even if he himself encouraged the comparison. He attested that "affairs of the heart" had to "animate a person's life" before that person could affect the larger circle of peers; and such experience, he further observed, led to greater communication with nature, at least in his case.[5] In nature he found the form and seduction of the female, and in women the very nature that he sought. Faust in desperation uttered Goethe's incessant question: "Where can I grasp you, infinite nature? / You breasts, you sources of all life, where?"[6] Despite, however, Goethe's own depiction, only a more detailed exploration can possibly justify the use of love and poetry as threads by which to follow the course of his scientific relations with nature and to demonstrate the erotic authority that nature exercised over him. Let me begin, preliminarily, with an assessment of three chords of the four-part theme, Goethe's science and his poetry, and both as ways of comprehending nature. I am aided in this by essays from the great physicist-physiologist, Hermann von Helmholtz (1821–1894).

In 1853, two decades after Goethe's death, Helmholtz, who had just become professor of anatomy at Königsberg, delivered an evaluation of the poet's contributions to science.[7] The young Helmholtz observed that Goethe had, in biological science, advanced two singular and "uncommonly fruitful" ideas.[8] Goethe recognized, first, that the anatomical structures of various kinds of animals revealed a unity of pattern or type underlying the superficial differences arising from variability of food, habit, and locality. His second lasting achievement, in Helmholtz's estimation, was the related theory of the metamorphosis of organisms: the theory that the various articulations within an organism developed out of a more basic kind of structure—that, for instance, the different parts of plants were metamorphosed leaves or that the various bones of the animal skull were but transformed vertebrae. These two general morphological conceptions,

5. Ibid. (pt. 3, bk. 12), 16:554–55: "How shall young individuals find the highest interest, how shall they arouse the interests of their peers, if love does not animate them, and if affairs of the heart, of whatever kind, are not alive within them? I had a lost love to mourn silently. . . . My spirit could only take rest out under an open sky, in the valleys and mountains, in the fields and forests. . . ."

6. Goethe, *Urfaust* (ll. 102–3), in *Sämtliche Werke*, 1.2:136. At the time he was working on this first draft of *Faust*, Goethe composed a poem that again employed the metaphor of sucking at nature's breasts—"Auf dem See" (in ibid., 3.2:21).

7. Hermann von Helmholtz, "Ueber Goethes wissenschaftliche Arbeiten. Ein vortrag, gehalten in der deutschen Gesellschaft in Königsberg, 1853," *Allgemeine Monatsschrift für Wissenschaft und Literatur* (1853), pp. 383–98; republished in *Populäre wissenschaftliche Vorträge*, vol. 1 (Braunschweig: Friedrich Vieweg und Sohn, 1865).

8. Ibid., pp. 34–38; quotation on p. 34.

according to Helmholtz, grounded the biology of his own time. Goethe came to these ideas, Helmholtz quite shrewdly maintained, as the result of a poetically intuitive conception (*anschauliche Begriffe*).[9] He described, for instance, Goethe's immediate recognition, while playfully tossing around a sheep's skull on the Lido in Venice, that the fused bones of the battered cranium consisted of transmuted vertebrae. This experience resulted in his vertebral theory of the skull, which became a standard conception in later morphology. Though I will question the reliability of the story of Goethe's supposed discovery of the vertebral skull, Helmholtz's depiction remains illuminating: for Goethe, poetic intuition liberated ideas initially embedded in matter and made them available to the analytic understanding of the scientist.

While Goethe's poetic approach to nature yielded fruit in biology, it had, in Helmholtz's estimation, less happy consequence in physical science, especially in his theory of light and color. Goethe stubbornly rejected the results of Newton's prism experiments, which demonstrated that white light was composed of several homogeneously colored rays. Goethe, in the view of Helmholtz, could not get behind the phenomena of nature to detect the wires, pulleys, and gears that made her operate. The young researcher suggested that Goethe's poetic inspirations bore fruit in science only, as it were, by chance—while walking on the Lido, fortune smiled, but in conducting experiments with prisms, luck turned away.

Forty years later, in 1892, at the meeting of the Goethe Society in Weimar, Helmholtz returned to reexamine the poet's scientific accomplishments, and, it would seem, implicitly his own; for by the end of his career, Helmholtz himself had achieved a position in German culture only a few steps below that of Goethe. He reiterated his earlier judgment about the significance of Goethe's morphological ideas and now indicated how those ideas led to an accelerated acceptance of Darwin's theory of evolution, particularly in Germany. His evaluation of Goethe's achievements in physical science, however, was now more complex than his earlier assessment had been. While allowing that Goethe too rapidly dismissed Newton's analyses, Helmholtz admitted the considerable difficulty in experimentally finding one's way to an adequate theory of light and color. And remarkably, in this second essay, he conceded that Goethe was intuitively right to have rejected Newton's particulate theory of light.[10] Had Goethe but known of Christiaan Huygens's (1629–1695) wave theory, Helmholtz suggested, he

9. Ibid., p. 43.
10. Hermann von Helmholtz, *Goethes Vorahnungen kommender naturwissenschaftlicher Ideen. Rede, gehalten in der Generalversammlung der Goethe-Gesellschaft zu Weimar den 11 Juni 1892* (Berlin: Verlag von Gebrüder Paetel, 1892), pp. 30–33.

might well have moved toward a more satisfactory conception. Moreover, seventeenth- and eighteenth-century physicists had needlessly confused their science with unanchored, abstract notions of, for instance, a bare matter having in itself no properties or forces. Goethe, he thought, had been essentially correct in the effort to ground his considerations concretely and to reduce multiple phenomena to an Ur-phenomenon.[11]

After fifty years of a career that ranged from physics to physiology, from optics to theories of artistic representation, Helmholtz was able to assess more sensitively Goethe's aesthetic approach to nature. In this second essay, he emphasized a principle operative in Goethe's work that I have, in earlier chapters, represented as a fundamental organizing principle in the philosophy of the Romantics, particularly that of Schelling. This was the aesthetic-epistemic tenet of the complementarity of the poetic and scientific conceptions of nature. Helmholtz held that "artistic representation" provided another way into the complexities of nature.[12] Both aesthetic intuition and scientific comprehension drove down to the type, to the underlying force that gave form to the surface of things. Exercising aesthetic intuition within the realm of science, therefore, would not introduce anything foreign, but only aid the scientist in comprehending the fundamental structures and forces of nature.

In attributing to his predecessor the principle of complementarity, Helmholtz implicitly joined Goethe with the Romantics. He made the linkage even stronger by his recognition that under Schiller's tutelage, Goethe had come to appreciate the Kantian conceptions of art and nature. Kant's third *Critique*, which animated many of Goethe's discussions with Schiller, helped him to understand better both the constructed aspects of nature and the structural isomorphism between aesthetic and scientific comprehension. Prior to his association with Schiller, Goethe had regarded the Kantian epistemology as too subjective, quite insensitive to the "rights of nature."[13] It was Schiller, however, who initiated that course of considerations that gradually moved Goethe toward a critical idealism and led him later to remark that his friend had convinced him that he himself was

11. Ibid., p. 33: "In this respect, however, contemporary physics has already taken the course along which Goethe wished to lead it."

12. Ibid., p. 7.

13. Goethe described his initial encounter with Schiller as one in which "he preached the gospel of freedom, while I wanted to make sure the rights of nature were not neglected." His frequent discussions with Schiller thereafter "gradually accustomed me to a language that had been completely foreign, but one which I found more congenial as I considered the higher representation of art and science, which such language well served." See Goethe, "Einwirkung der neueren Philosophie," in *Sämtliche Werke*, 12:97.

indeed a Romantic.[14] While reluctantly admitting the justice of the Kantian epistemology, Goethe nonetheless lamented the loss of what seemed an immediate congress with nature. It would take a principal figure of the Romantic circle, Friedrich Schelling, to lead him back into a justified relationship with his erstwhile companion.

Goethe did not come to his ideas in biology, aesthetics, and epistemology in a flash. He matured into these ideas through the formative impact of multifarious experience. I would like to show in this chapter and the next how Goethe came to these positions and, as suggested by my opening paragraphs, how his various intimate experiences with women fostered and became emblematic of those achieved insights—how, in short, Goethe became a Romantic biologist.

Growing Up in Frankfurt

"I came into the world in Frankfurt am Main on 28 August 1749, as the clock struck twelve noon," so begins Goethe's *Dichtung und Wahrheit*.[15] He feigned that the moon, exerting astrological power, resisted his birth. Otherwise, his entrance into the world was propitious. Certainly good fortune assisted his arrival into a free city and a prosperous house. His grandfather, Friedrich Georg Goethe (1657–1730), a well-to-do tailor, had established the family fortune through a second marriage to a rich Frankfurt widow, Cornelia Schellhorn (1668–1754), who owned a successful inn, Zum Weidenhof. Their only surviving son and Goethe's father, Johann Caspar Goethe (1710–1782), inherited the business and the small family fortune. Goethe's father became a lawyer and adviser to the Frankfurt city council; but, with modest political prospects, he mostly indulged his own educa-

14. See my introductory chapter for a brief discussion of this admission and the next chapter for an elaboration.

15. Goethe, *Dichtung und Wahrheit* (pt. 1, bk. 1), in *Sämtliche Werke*, 1:13. In reconstructing Goethe's life, I have used his autobiography, though with caution. His letters (in various collections) have provided a tempering influence on his recollections, though they themselves have distinctively artful elements. I have also found generally helpful: Baker Fairley, *A Study of Goethe* (Oxford: Oxford University Press, 1947); and Karl Otto Conrady, *Goethe: Leben und Werk*, 2 vols. (Frankfurt a. M.: Fischer Verlag, 1988). Nicholas Boyle sets the standard by which all biographies of Goethe will be measured with his *Goethe: The Poet and the Age*, vol. 1: *The Poetry of Desire, 1749–1790*; vol. 2: *Revolution and Renunciation, 1790–1806* (Oxford: Oxford University Press, 1991–2000). Indispensable for following Goethe's activities in detail is Robert Steiger, *Goethes Leben von Tag zu Tag*, 8 vols. to date (Zurich: Artemis Verlag, 1982–). The range of Goethe's activities has received comprehensive and authoritative treatment in the *Goethe Handbuch*, ed. Bernd Witte et al., 4 vols. (Stuttgart: J. B. Metzler, 1996–98). Also of extreme value are the various commentaries in the *Werke* (Hamburger Ausgabe), ed. Erich Trunz et al., 14 vols. (Munich: C. H. Beck, 1988), and in the *Sämtliche Werke nach Epochen seines Schaffens* (Münchner Ausgabe). Other secondary literature on particular subjects will be listed in the footnotes below.

tional and leisurely pursuits. Intending, perhaps, to boost his political fortunes as much as to satisfy his amorous needs, he married, at age thirty-nine, the seventeen-year-old Katharina Elisabeth Textor (1731–1808), daughter of Johann Wolfgang Textor (1693–1771), who was secretary (*Schultheiss*) of the city council and the emperor's representative in the city during the period 1747 to 1771. They named their first child after the maternal grandfather, and the father's political connections seemed thereby to have modestly improved.

Goethe's large childhood home—in which he occupied a well-lighted room on the third floor—sheltered not only him, his parents, and a small staff of servants, but his paternal grandmother and his slightly younger sister Cornelia (1750–1777), with whom he formed a very strong attachment.[16] In his autobiography, Goethe fondly recalled childhood play and tutorial lessons with his sister. She was not particularly good-looking, and rather sensitive about that.[17] Later she would enjoy a marriage, short-lived though it was, with a friend of her brother, Johann Georg Schlosser (1739–1799). She died at age twenty-seven, just after the birth of her second child. None of Elisabeth Goethe's other five children survived beyond age six. The mother herself, though, lived through Goethe's later years, dying in 1808 at age seventy-seven. The son did not visit his mother during the last decade of her life nor did he attend her funeral. Goethe's astrological fable about his birth may thus conceal some deeper truth, or perhaps a guilty conscience, or merely the fact that his mother was in labor for three days prior to his birth.[18]

---

16. In his autobiography, Goethe intimated that his relation with his sister verged on the erotic. They shared, he said, youth's "amazement at the awakening of sensuous drives"; and he suggested that their desires were forestalled by "the holy dread of the close relationship." His sister offered him consolation when his first serious love affair went sour, and they "regarded themselves utterly unhappy, the more so since in this particular case the confidants could not transform themselves into lovers." See Goethe, *Dichtung und Wahrheit* (pt. 2, bk. 6), in *Sämtliche Werke*, 16:250, 254. The strongly tinctured character of this relationship has given the psychoanalyst Kurt Eissler much to speculate about. One can derive insight from Eissler's considerations without adopting his conclusions. Particularly about books like Eissler's, Pliny hit the mark: "Nullus est liber tam malus, ut non aliqua parte prosit." See Kurt R. Eissler, *Goethe: A Psychoanalytic Study, 1775–1786*, 2 vols. (Detroit: Wayne State University Press, 1963).

17. Goethe, *Dichtung und Wahrheit* (pt. 2, bk. 6), in *Sämtliche Werke*, 16:253.

18. The writer Bettina Brentano (1785–1859) became quite friendly with Goethe's mother, and at the mother's death conveyed to him the tale of his birth, just as he was writing his autobiography: "She was at the time 18 years old and one year married. Three days—think about it—before you saw the light of the world, you caused your mother some difficult hours. Out of anger you were driven from your native dwelling place, and through the mishandling of the nurse you came completely black and without signs of life." See Bettina Brentano to Goethe (4 November 1810), in *Briefe an Goethe* (Hamburger Ausgabe), ed. Karl Robert Mandelkow, 3rd ed., 2 vols. (Munich: C. H. Beck, 1988), 2:60.

Despite these astrological and psychological signs, Goethe's youth was routine for someone of his upper bourgeois class. He contracted the usual childhood diseases, including a mild case of smallpox (without lasting effect). His education began at home, where he received instruction at the hands of his father and some eight tutors over the years, with special focus on languages—Latin, Greek, Italian (his father's delight), French, English, and even Yiddish and some Hebrew. Johann Caspar Goethe's splendid library and fine collection of paintings by local artists provided opportunity for the son to cultivate eclectic tastes. As Goethe grew into early adolescence, he engaged in those activities meant to refine manners and habits appropriate to his station: he studied some music, practiced fencing and riding, and began dancing lessons. He would take up these pursuits again at university, but then with the seriousness of a young man wishing to make a noticeable entrance into the larger social world.

In his autobiographical recollections of childhood, Goethe centered on two kinds of events that would echo through other of his works: his experience of the theater and his delight in emotionally driven poetry. One Christmas his grandmother arranged for a puppet show for the children, "which made a very strong impression on the boy, reverberating with long-lasting effect."[19] The children got to keep the puppets and they later staged many plays for which Goethe wrote the scripts. In his great novel *Wilhelm Meisters Lehrjahre* (1795–96), a similar childhood experience with puppets would propel the hero into the life of the theater. For the real boy, interest in the theater intensified when, during the Seven Years' War, a French general was billeted in the Goethe house. The father had favored the Prussians and so resented the intruder. The boy, by contrast, became quite friendly with the visitor, improved his French, and began attending the French theater. For this latter pursuit, his father rebuked him, much as Wilhelm Meister's father would, declaring that "the theater is of no utter use and can lead to absolutely nothing."[20]

Johann Caspar Goethe, a cultured though conservative man, also disdained the new poetry of Friedrich Klopstock (1724–1803), whose verses exuded powerful emotions through an unconventional meter and free form. The son, however, felt a deep affinity for this new kind of poetry. In a memorable scene from Goethe's *Leiden des jungen Werthers* (Sorrows of young Werther, 1774), Werther and Lotte, on their initial meeting, stare out a window after a thunderstorm: "She looked up at the heavens and then

19. Goethe, *Dichtung und Wahrheit* (pt. 1, bk. 1), in *Sämtliche Werke*, 16:18.
20. Ibid. (pt. 1, bk. 3), 16:116. In *Wilhelm Meister*, the father remonstrates about the theater: "Of what use is it! How one can so waste his time." See *Wilhelm Meisters Lehrjahre* (bk. 1, chap. 2), in *Sämtliche Werke*, 5:11.

at me, I saw that her eyes were full of tears, she laid her hand on mine and said: 'Klopstock!'"[21] Shortly after he finished writing the novel, the young author began a correspondence with Klopstock and soon met him; the friendship, however, quickly dissolved over political differences. Goethe yet gently re-created in his autobiography a charming scene of his child-hood that involved smuggling Klopstock's poetry into his father's house. While his father was being shaved by the visiting barber, Goethe and his lit-tle sister had been warming themselves behind a stove and reading aloud Klopstock's *Messias*. His sister declaimed in a loud voice, "Before I could hate you, with terrible, hot hatred! But now I can't anymore! This is a wretched mess!" And then she fairly shouted: "Oh, how I am crushed!" This caused the barber to upset the shaving bowl on their father's chest. The older memorialist concluded this poetic tale with: "Thus children and ordinary people make a game of greatness, of the sublime. They turn it into a farce; for otherwise how should they be able to stand it, to bear it!"[22]

Through Goethe's autobiography, the descriptions that sound the most lyrical moments of memory, that achieve the most delicate emotional pitch, playing softly the chords of desire, regret, and longing for youth, are those dealing with his various love affairs. His first infatuation came as he turned fifteen. The girl, whom he called Gretchen, was about seventeen and a waitress at an inn Goethe and his friends frequented. She seems to have been the kind of girl who today would be wearing a tank top, sporting a light tan, and displaying a small tattoo on her shoulder. Throughout their relationship, she stayed his amorous advances but behaved so sweetly and solicitously that his passions were continually stoked. The affair, if it can be called that, ended sadly. Because of some minor trouble with the authori-ties, Gretchen had to return to her village, leaving the boy in a state of deep despair. His desperate love, though, turned to humiliation and remorse when he learned she thought him a mere child. Despite his deep hurt, he could not keep the image of her from running through his mind—"her form, her manner, her behavior"—and he spent many a night "given over to his pain."[23] As he perceived the event almost half a century later, he de-tected in it a lever that would allow him to pry up social surfaces to study the psychological realities crawling beneath.[24] Undoubtedly his teenage experience gave some imaginative substance, a decade later, to the young

---

21. Goethe, *Die Leiden des jungen Werthers* (1774), in *Sämtliche Werke*, 2.2:369. (One can per-haps imagine a corresponding modern scene that would evoke from two contemporary lovers the exclamation "Larkin!")
22. Goethe, *Dichtung und Wahrheit* (pt. 1, bk. 2), in *Sämtliche Werke*, 16:87–88.
23. Ibid. (pt. 2, bk. 6), 16:241–42.
24. Ibid. (pt. 2, bk. 7), 16:308.

Gretchen whom Faust loved and then seduced, and whom the poet condemned and then saved.[25]

University Education

Goethe had wanted to attend the university at Göttingen in order to study with Christian Gottlob Heyne and Johann David Michaelis—figures, along with Johann Friedrich Blumenbach, who formed the university's reputation during the last part of the century. Goethe's father, however, was of a more practical mind. He insisted his son attend his alma mater in Leipzig and there take up law, a profession that provided social respectability, political advancement, and money.

In October 1765, at age sixteen, Goethe matriculated at Leipzig, and desultorily pursued a path toward a law degree. He heard lectures on law, legal history, and legal institutions—and they bored him. He tried philosophy, but the scholasticized versions of Leibniz and Wolff "simply refused to become clear to me."[26] He had high hopes for lectures on literary subjects, and to pursue those interests he attended the courses of Johann Christoph Gottsched (1700–1766) and Christian Fürchtegott Gellert (1715–1769). Gottsched followed the already antiquated tradition of Wolff, and his students thought him hopelessly old-fashioned. Gellert, himself a novelist and an imposing intellect, took as his duty the moral welfare of his students. Goethe observed, however, that the professor appeared to believe his aristocratic students more in want of solicitous advice than his middle-class students.[27] When Goethe offered Gellert his fledgling poetry for judgment, his teacher recommended he avoid the genre altogether.[28] Goethe leavened his disappointment and loneliness with sourdough, writing his friend Johann Riese (1746–1827) a poetic account of his experience with the great men of Leipzig:

> As I saw the fame of great men,
> And first perceived what it means
> To earn a reputation,
> I finally understood that my own sublime flight,

---

25. In *Faust, Part One*, Faust falls hard for Gretchen and gets her with child. In her delirium, she drowns her baby and is to be executed. In an early draft (about 1775), the play ends with Mephistopheles saying, "She is condemned." The published version (1806), however, follows this with "a voice from above: 'She is saved.'" See *Urfaust* and *Faust: Erster Teil*, in *Sämtliche Werke*, 1.2:188, 6.1:673.

26. Goethe, *Dichtung und Wahrheit* (pt. 2, bk. 6), in *Sämtliche Werke*, 16:271.

27. Ibid. (pt. 2, bk. 7), 16:318–19.

28. Ibid. (pt. 2, bk. 6), 16:279.

Or so it seemed to me, was only the effort
Of a dusty worm that sees an eagle
Soaring toward the sun,
And desires to rise as it does.
The worm stretches up, wriggles,
And anxiously strains all its nerves,
But yet remains in the dust.
Suddenly a wind blows up,
That lifts the dust in a whirl,
And the worm along with it.
The little creatures believes itself
To be mighty, like the eagle,
And rejoices in its ecstasy.
But suddenly, the wind becomes placid,
The dust falls back, and so the worm,
Now creeps along as before.[29]

The professors at Leipzig smothered Goethe's aspirations under encrusted attitudes and imperious indifference. He consequently sought consolation in the traditional refuge of university students—comrades, self-education, and love. Among his new friends was Johann Schlosser, whom he had known slightly in Frankfurt as a practicing lawyer and who was now visiting in Leipzig. Schlosser's own companions provided Goethe his first coterie of acquaintances in the city. Among these was Ernst Wolfgang Behrisch (1738–1809), a tutor in Leipzig and, as Goethe remembered, "a most amazingly queer fellow." Behrisch read and spoke French quite well, constantly wore gray, and had a talent for "wasting time."[30] But unlike Gellert, he encouraged the young poet. He copied out a selection from Goethe's lyrics in a fine, practiced hand and had it bound in a beautiful volume, which the poet treasured. His solicitude and positive criticisms endeared him to Goethe's heart, and the affection spanned the years.

Goethe's self-education took several forms. One kind consisted in the

29. Goethe to Johann Riese (28 April 1776), in *Goethes Briefe* (Hamburger Ausgabe), ed. Karl Robert Mandelkow, 4th ed., 4 vols. (Munich: C. H. Beck, 1988), 1:26: ". . . als ich den Ruhm, / Der Grossen Männer sah, und erst vernahm, / Wie viel dazu gehörte; Ruhm verdienen. / Da sah ich erst, dass mein erhabner Flug, / Wie er mir schien, nichts war als das Bemühn, / Des Wurms im Staube, der den Adler sieht, / Zur Sonn' sich schwingen, und wie der hinauf / Sich sehnt. Er sträubt empor, und windet sich, / Und ängstlich spannt er alle Nerven an, / Und bleibt am Staub. Doch schnell entsteht ein Wind, / Der hebt den Staub in Wirbeln auf, den Wurm / Erhebt er in den Wirbeln auch. Der glaubt / Sich gross, dem Adler gleich, und jauchzet schon / Im Taumel. Doch auf enimal zieht der Wind / Den Odem ein. Es sinckt der Staub hinab, / Mit ihm, der Wurm. Jetzt kriecht er wie zuvor."

30. Goethe, *Dichtung und Wahrheit* (pt. 2, bk. 7), in *Sämtliche Werke*, 16:320.

philosophical-religious discussions carried on with friends. As a result, the constricted religious beliefs of his youth began to loosen under the skeptical and irreverent gibes of companions; he finally broke from formal church ties during his Leipzig years. Goethe remained religious, though in a more generous sense. In the older Greek writers—Homer, Aeschylus, Sophocles —he found a poetically realized religion that cultivated the kind of feeling for the divinity of nature that he began to recognize as his own. Another sort of education—that of the senses—came from his lessons in drawing and painting. His study of art while at Leipzig helped develop, he thought, his talent for being able to perceive nature through the eyes of this or that artist, that is, to come to feel and live through different aesthetic perceptions.[31] Likely it was the just-published *Laokoon* (1766) of Gotthold Ephraim Lessing (1729–1781) that encouraged him to regard the ineffable ideas of the artist as a particular kind of knowledge of the natural world. Lessing, by the elder Goethe's estimation, led him and his contemporaries "from the region of pedestrian views into the open field of thought."[32] Lessing furnished those of Goethe's generation with a revolutionary perspective, teaching them the distinction between painting, an expression of visual beauty, and poetry, a stimulus to the imagination. Lessing's conceptions differed from those of the standard authorities, and so they became even more attractive; but Goethe himself would quickly abandon these notions. He would come to maintain that all art—and science—required the creative play of the imagination. It was, rather, Lessing the dramatist that left the deepest impact on the young writer. As an ironic homage, Goethe has his hero Werther reading Lessing's drama *Emilia Galotti* just before he shoots himself.

Goethe had arrived in Leipzig in October 1765. By the following spring, he had become fully inducted into Schlosser's community of friends. The group regularly took their afternoon meals at an inn owned and run by Christian Gottlob Schönkopf, himself an émigré from Frankfurt. In a repetition of the immediate past, the young poet lost his heart to the innkeeper's daughter. Anna Katharina Schönkopf, known to Goethe and his friends as Käthchen (and to the poet's alter ego as Ännchen and Annette), was an attractive, lively girl of nineteen, again slightly older than the lad of seventeen. But Goethe, by this time, had a reputation that enhanced his charms, giving them a patina of maturity; and she, as the elder Goethe remembered, was all the smitten fellow could desire: "She was so young, pretty, cheerful,

---

31. Ibid. (pt. 2, bk. 8), 16:346. In his *Italienische Reise*, he refers to this as an "old gift." See Goethe, *Italienische Reise* (Venice, 8 October 1786), in *Sämtliche Werke*, 15:101.

32. Goethe, *Dichtung und Wahrheit* (pt. 2, bk. 8), in *Sämtliche Werke*, 16:341.

lively, and oh so nice that she probably deserved to be installed during that time in the shrine of my heart as a little saint, in order to devote the kind of worship to her that is often more pleasant to impart than to receive."[33] The romance kept Goethe emotionally exhausted for over two years. It was encumbered by Käthchen's lower social status, his friend Behrisch's traumatic departure, and his own anxiety-driven maneuvers of approach and avoidance. He would pursue her ardently; and when she would respond, he would cool to the point of indifference. She would then seek other company, and he would be driven into wild despair and brash acts. A series of letters to Behrisch, who had departed to Dresden for a new tutorial position in fall 1767, offers some brief scenes of turmoil.[34]

Goethe begins one of the most telling of his letters on Tuesday evening, 10 November:

> (7 o'clock in the evening) Ha, Behrisch this is one of those times. You're away, and this paper is only a cold substitute for the refuge of your arms. O God, God.—I've got to get hold of myself. Behrisch, love be damned. Oh, if you could see me, if you could see how sorrow makes me rage, and I don't know against whom I should rage—Oh, you would be upset. Friend, friend! Why do I have only one? (8 o'clock) My blood runs more calmly, and I'll speak with you more quietly. Will it be rational? God only knows. No, no, not rational. How should a madman speak rationally? That's what I am.[35]

After another deluge of anguished expression, the boy finally tells his friend of the event that unleashed the passion. He had gotten into such a row with Käthchen on the previous Sunday that the next evening he had broken out in fever (which probably was due to an incipient attack of tuberculosis that would soon send him back to Frankfurt for parental care). The next day he asked his beloved to look after him. She replied, in icy tones, that she and her mother would be attending the theater that evening. She left, and his fever spiked. The delirious fool then followed mother and daughter to the theater and spied on them in their box. He saw a one Herr Ryden seated behind Käthchen, whispering in her ear. In a fit, he snatched someone's opera glasses to see better what was going on. He felt relief when she remained unresponsive to Ryden's solicitations. But Goethe left the theater sick and despairing, and came home to write his friend: "I have tried in vain the whole evening to cry, but my teeth are chattering, and when that happens, one

33. Ibid. (pt. 2, bk. 7), 16:307.
34. These letters may be found in *Goethes Briefe*, 1:54–66.
35. Goethe to Ernst Wolfgang Behrisch (10 November 1767), in ibid., 1:57.

cannot cry." He added: "But I love her. I believe I have drunk poison from her hand"[36]—a remark that later may well have been transformed into one of the most trenchant love epigrams in literature: "Bin so in Lieb zu ihr versunken / Als hätt ich von ihrem Blut getrunken."[37] By the next spring, the young lovers had let their relationship cool, though Goethe continued to cultivate in his heart its possibilities and became increasingly more wretched as her wedding to a young lawyer in 1770 came ever closer.

In his autobiography, as an introduction to his affair with Käthchen, Goethe remarked:

> And so began that path, from which I have not been able to stray during my whole life, namely, that inclination to turn whatever delighted, tortured, or otherwise occupied me into an image, a poem, and thereby to come to terms with myself so that I might as much justify my conception of external things as well as calm my inner self about them.[38]

There is little doubt that Goethe performed just this transformation of his experience with Käthchen Schönkopf. During his Leipzig period, he wrote a number of anacreontic poems—in shades of the erotic and the ironic—that traced the changing face of his relationship with the girl. They run from the playfully amatory—"To Sleep" is a light poem that entreats the god to bring the beloved's mother into a deep slumber lest she hear them carrying on in the next room—to the more vividly sensual "Inconstancy":

> I lay upon the bed of a brook, how splendid,
> And there await an arriving swell, arms extended.
> Against me she wantonly presses a beckoning breast,
> Then carries her playfulness down into the stream
> But soon brings the second to brush against my chest,
> And joy of mutual pleasure makes real my dream.
> O youth, be wise, do not cry in vain
> Over the happiest hours of life's sad refrain.
> When a maiden forgets you in her fickleness,
> Go, recall the earlier time of tempting bliss,
> When you kissed the second breast so sweetly,
> As if you forgot the first completely.[39]

---

36. Ibid., 1:60.

37. Goethe, "Bin so in Lieb," in *Sämtliche Werke*, 2.1:81: "I have fallen so in love with her / It's as if I had drunk her blood." This epigram was included in a letter to Charlotte von Stein (23 January 1782), but Goethe called it an "old verse." See ibid., 2.1:592.

38. Goethe, *Dichtung und Wahrheit* (pt. 2, bk. 7), in *Sämtliche Werke*, 16:306.

39. Goethe, "Unbeständigkeit," in *Sämtliche Werke*, 1.1:126–27: "Auf Kieseln im Bache, da lieg ich, wie helle, / Verbreite die arme der kommenden Welle, / Und buhlerisch drückt sie die sehnende Brust. / Dann trägt sie ihr Leichtsinn im Strome darnieder, / Schon naht sich die zweite

In the poetry of this period—as well as in his letters—Goethe vented strongly fluctuating emotions—of the sort that would later receive more extended expression in his novel *Die Leiden des jungen Werthers*. Being self-reflective, often in the extreme, he could not help but be aware that his own interior feelings were coloring the external world of nature—an emotional preparation for accepting the kind of idealism that would come to ground his scientific epistemology.

In July 1768 Goethe began to suffer chest pains, likely the result of tuberculosis. His condition became exacerbated when he fell from a horse, and the resulting violent hemorrhaging brought him close to death. His slow convalescence, attended to by friends, left a lingering melancholy that eventually drove him back, in September, to his family in Frankfurt and the care of his sister. During his recovery, his mother introduced him to Susanna Katharina von Klettenberg (1723–1774), a woman of deep Moravian piety (who inspired a similarly mystical character in his novel *Wilhelm Meister*). Upon the recommendation of his physician, Goethe began reading certain alchemical treatises—including those of Paracelsus, Helmont, and Starkey—as well as more orthodox chemical works, such as those of Boerhaave. The compound of disease, mystical piety, and alchemy had an intoxicating effect on Goethe's soul, leading him further away from the orthodox religious path and easing him toward "neo-Platonic" byways.[40] In spring of 1769, having recovered sufficiently and having been sufficiently irritated by his father, he set off for Strasbourg, where he intended to pursue, in relaxed fashion, the law degree that might secure his professional future.

The Law, Herder, and Lotte

Goethe took a very pragmatic view of his legal education, studying just enough to keep up. With the help of a handbook on law from his father's library and with a little concentration, he did pass his exams; but the law faculty refused to accept his dissertation, in which he defended the right of a sovereign to determine church worship. The dean of the law faculty recommended in place of a dissertation that he propose some theses to defend in a regular academic disputation, which would win him a bachelor's degree. So with some of his friends and drinking mates taking the opposition, Goethe did satisfactorily defend a set of theses, received his licentiate, and

---

und streichelt mich wieder. / Da fühl ich die Freuden der wechselnden Lust. / O Jüngling sei Weise, verwein nicht vergebens / Die fröhlingsten Stunden des traurigen Lebens / Wenn flatterhaft dich ja ein Mädgen vergisst; / Geh ruf' sie zurüche, die vorigen Zeiten! / Es küsst sich so süsse der Busen der Zweiten, / Als kaum sich der Busen der ersten geküsst."

40. Goethe, *Dichtung und Wahrheit* (pt. 2, bk. 8), in *Sämtliche Werke*, 16:376.

called himself, according to the custom of the time, "Doctor Goethe." During his brief stay in Strasbourg (April 1770 to August 1771), Goethe found other occupations more compelling than law.

While at the university, he took classes in chemistry and anatomy, and his appetite for these would continue to grow in the immediately following years. The city offered various social pleasures in which the young law student indulged, from raucous drinking parties to social dances. He took instruction in dance, pursuing not only the latest steps but the dancing instructor's daughters as well. He fell in love with the younger of the two, while the older fell for him. This complicated triangle dissolved into kisses, tears, and curses. Among the very middle class, Goethe's wit and charm smoothed every social situation. In the higher social strata of the city, though, he felt a bit out of place because the French were always correcting his language. A person would be tolerated, he allowed, but "in no way would he be taken up into the bosom of the one linguistically blessed church." What really nettled him and his fellow countrymen, though, were the suggestions by the French that Germans "lacked taste."[41]

At Strasbourg Goethe met the man who would become the greatest intellectual and emotional force of his early life, Johann Gottfried Herder (1744–1803). Herder, former student of Kant and an ordained clergyman, had already achieved some fame for his essays on aesthetics and for championing the idea that out of the genius of the culture, the Germanies might produce a powerful, indigenous literature. He had recently abandoned a tutorial position and had come to Strasbourg in September 1770 for an eye operation—he had a blocked tear duct—which in the end offered hardly any relief. The two men chanced to meet at the entrance of an inn, under the portentous sign "Zum Geist," and within a short time they became friends. Goethe would later bring this austere clergyman to take charge of ecclesiastic administration in Weimar. But during Herder's four months in Strasbourg, they spoke mostly of literature—rather, the older man displayed his immense learning and, with faint regard for the adolescent ego, dispensed dismissive, if instructive, criticism of the young poet's opinions, or so Goethe recalled:

> The older individuals with whom I had associated until now had attempted to instruct me with kind solicitude, perhaps even to have indulgently spoiled me. But from Herder, one could never expect approval, however one presented oneself. Thus on the one side there was my great inclination toward him and my respect, on the other the discomfort that

41. Ibid. (pt. 3, bk. 11), 16:513, 515.

he awoke in me—these were continually at war with each other. So there arose a split in me, the first of its kind that I had experienced in life. Since his conversations were at all times significant, whatever his questions, answers or other communications, he inevitably brought me to new points of view daily, indeed, hourly.[42]

Herder certainly did not serve as a nurturing guide for the younger man. Goethe's early poetry, itself much like a scene of natural beauty, evoked a direct and immediate emotional response. Herder pointedly lectured the poet about the formal mode in which poetry should achieve its effects. The prickly preacher could only stimulate in this negative way. The two, nonetheless, became quite friendly during the Strasbourg period, so that Herder felt easy enough to introduce him to his fiancée, Karoline Flachsland (1750–1809), venturing something of a chance thereby. For Goethe took an immediate liking to this woman, and she to him.[43] Despite his infatuation, he lent his friends money so that they could get married (1773), though Herder's tardiness in repaying the debt stuck as a barbed irritant in the poet's memory. During the years of their association, Herder's temper would often shatter social constraints, and the swelling tide of his irascibility would, frequently enough, push friendship hard against the rocks of their differences.

While in Strasbourg, Herder completed a small monograph on the origin of languages, which won a prize competition in Berlin. He argued that languages originated not from the instructions of God—as was the presumption of conventional theology—but from men imitating natural sounds of particular regions: "the entire, many-voiced, divine nature is the language teacher and muse," he wrote. Different languages would be indicative of different locations and historic traditions. What to our ears may be even more startling, coming as it did from a clergyman, was Herder's insistence that language and human reason developed simultaneously and

---

42. Ibid. (pt. 2, bk. 10), 16:437.

43. Flachsland obviously became quite attracted to her "good Goethe," though she observed the proprieties in writing to her husband: "I love Goethe like my soul. But how can I and should I express it to him? There is nothing in the world I can do for him. Indeed, I don't know what I would not do for him." See Karoline Flachsland to Johann Gottfried Herder (12 December 1772), in *Goethe in vertraulichen Briefen seiner Zeitgenossen*, ed. Wilhelm Bode and Regine Otto, 2nd ed., 3 vols. (Berlin: Aufbau-Verlag, 1982), 1:41. Goethe would remain good friends with Karoline, writing her often. When Goethe left Weimar in 1786 for Italy, she felt his absence keenly: "How lonely we are since his absence! We have shared our whole life only with him. . . . He is one of the few mortals who has learned the wisdom of life and with whom one so happily travels only a little way with him." Karoline Herder to J. G. Müller, in *Goethes Gespräche*, ed. Flodoard von Biedermann, expanded by Wolfgang Herwig, 5 vols. (Munich: Deutscher Taschenbuch Verlag, 1998), 1:396.

naturally, so that "the first, most primitive use of reason could not occur without language."[44] "Man," Herder thus concluded, "is, in his distinctive features, a creation of the group, of society: the development of a language is thus natural, essential, and necessary for him."[45] This conception of a natural evolution of language and reason would later provide a leading motif for the likes of Fichte, Schelling, and Hegel, the architects of German idealism.

Goethe read Herder's treatise, but even his positive evaluations were not gratefully received by his friend.[46] Initially the poet did not detect the dangerous metaphysics that formed the soul of Herder's little book. He rather reacted to the notion of a language having specific roots in the formation and experience of a people. To Herder's confident call for a re-creation of a national literature, especially by recovery of traditional songs, Goethe responded in the way he best could, in poetry. During this period he adapted some folk songs, most notably "Heidenröslein" (Little rose on the heath), which Herder saw to publication, and set to composing other poems in the mode of the traditional song.

While a student in Strasbourg, Goethe became involved with Friederike Brion (1752–1813), the charming and beautiful daughter of a clergyman who had his parish in Sesenheim, a village outside of Strasbourg. He had been introduced to the family by a mutual friend in fall of 1770. In his autobiography, Goethe wrote poignantly and dramatically of his affair with Friederike, but with so artful a touch that the contours of the original relationship can only be vaguely discerned.[47] At least this much is fairly clear: she was alluring (she had other suitors); the young pair experienced fits of love; he left her for murky reasons; and she inspired some of Goethe's most moving poetry. But even in the poet's artful descriptions, a lived reality shimmers through:

> Slim and ethereal, as if she were light as a feather, she entered the room. It seemed as if her neck were too delicate for the large blond braids of her

44. Johann Gottfried Herder, *Abhandlung über den Ursprung der Sprache* (1772), in *Johann Gottfried Herder's Sprachphilosophie, aus dem Gesamtwerk ausgewählt*, ed. Erich Heintel (Hamburg: Felix Meiner, 1975), pp. 32, 28.

45. Ibid., p. 67. Herder's thesis about language creating human mind would later be reinforced by the linguistic studies of August Schleicher, who, in turn, exerted a powerful influence on Ernst Haeckel and Charles Darwin. See my "The Linguistic Creation of Man: Charles Darwin, August Schleicher, Ernst Haeckel, and the Missing Link in 19th-Century Evolutionary Theory," in *Experimenting in Tongues: Studies in Science and Language*, ed. Matthias Doerres (Stanford: Stanford University Press, 2002).

46. Goethe's description of his interactions with Herder is in *Dichtung und Wahrheit* (pt. 2, bk. 10), in *Sämtliche Werke*, 16:434–47.

47. See Conrady, *Goethe: Leben und Werk*, 1:122.

pretty, dainty head. From lively blue eyes, she looked sharply around, and she turned her nose up into the air so freely, as if she had not a care in the world; her straw hat hung from her arm—and thus I had the pleasure to see and recognize at the first glance her full grace and loveliness.[48]

Goethe initially seems to have fallen hard for this eighteen-year-old, and the autobiography paints the course of their two-year romance as idyllic, with a chaste passion fueling the relationship—"as opportunity offered, I did not hesitate to kiss my tender love with warm embrace, and even less did I deny myself the repetition of this joy."[49] The few letters that survive from this period do not, however, suggest exactly the same idyll. In a note to his friend Johann David Salzmann (1722–1812), written from Sesenheim in late spring of 1771, Goethe lamented that his soul turned like a weather cock, now this way, now that. He asked himself: "Wasn't this the fairy garden that you desired? Yes, yes it is! I feel that it is, dear friend, and I feel that one is not one hair happier when he achieves what he wishes."[50] The rockier course of their affair might also be measured by the kind of poetry Goethe produced at this time.

The poetry became increasingly personal. It ranged from the tender and whimsical "Ich komme bald, ihr goldnen Kinder" (I'm soon coming, you golden children), which he wrote for the sisters Brion, to the exuberant and lovely "Maifest" (May celebration), which identifies the force of creative nature with that of creative love. In "Maifest" the two forces of nature and love inspire the poet in his song:

As the lark loves
A joyful song and air,
And morning flowers,
The sent of heaven fair,
So I love you
With blood at the boil,
You who give me youth
And set heart to toil
To compose the new dance
And sing the new song!
So be happy forever,
And love me so long![51]

---

48. Goethe, *Dichtung und Wahrheit* (pt. 2, bk. 10), in *Sämtliche Werke*, 16:466.

49. Ibid. (pt. 3, bk. 11), 16:493.

50. Goethe to Johann David Salzmann (12 and 19 June 1771), in *Goethes Briefe*, 1:122.

51. Goethe, "Maifest," in *Sämtliche Werke*, 1.1:163: "So liebt die Lerche / Gesang und Luft, /

The very poignant "Willkommen und Abschied" (Welcome and farewell) reveals the transformation of his initial feeling of joyful and creative love; it depicts lovers first meeting and then, in ecstasy and pain, parting:

> As the sun arose that day,
> The departure shrank my heart.
> What joy your kisses convey!
> What pain your eyes impart!
> I left and you stayed, cast down,
> With eyes staring through a mist.
> Yet, to be loved, what bliss profound!
> And to love, O ye gods, what bliss![52]

This is the much-reworked 1810 version of the poem. In the 1775 version, perhaps catching more of the original situation, the departure is reversed: the woman walks away from the poet, who has his moist eyes downcast. The 1810 version yet has the aesthetic merit of making love a true perplexity—the male figure abandoning that which brings supreme happiness.

Whoever initiated the parting, Goethe did leave Friederike. He provided no explanation of this decision in the autobiography—perhaps at that far temporal distance the exact reason eluded even him. Nicholas Boyle, in his account of the episode, suggests that Goethe realized that marriage would impede his life as a poet; for Goethe's poetry, according to Boyle, fed on desire, not fulfillment.[53] Goethe, to be sure, used his emotions, often self-consciously, as the staging for poetry that flowed ever more vitally, poignantly, and effectively from his pen—and the principal emotion that drove his verse was that of longing, the staple of much love poetry through the ages. That Goethe hesitated to commit himself to a woman,

---

Und Morgenblumen / Den Himmels Duft, / Wie ich dich liebe / Mit warmen Blut, / Die du mir Jugend / Und Freud und Mut / Zu neuen Liedern, / Und Tänzen gibst! / Sei ewig glücklich / Wie du mich liebst!" David Wellbery shows that this lyric and others, though apparently arising from a moment of spontaneous interaction with nature, fall within an already established tradition of nature poetry, though one that Goethe creatively alters to achieve novel results. See David Wellbery, *The Specular Moment: Goethe's Early Lyric and the Beginnings of Romanticism* (Cambridge: Harvard University Press, 1996), chap. 1.

52. Goethe, "Willkommen und Abschied," in *Sämtliche Werke*, 3.2:16: "Doch ach shon mit der Morgensonne / Verengt der Abschied mir das Herz: / In deinen Küssen, welche Wonne! / In deinem Auge, welche Schmerz! / Ich ging, du standst und sahst zur Erden, / Und sahst mir nach mit nassem Blick: / Und doch, welch Glück geliebt zu werden! / Und lieben, Götter, welch ein Glück!"

53. See Boyle, *Goethe: The Poet and the Age*, vol. 1, especially pp. 107–13. In his interesting, if highly imaginative psychoanalytic study of Goethe, Eissler argues that in these early days Goethe suffered from premature ejaculation, usually brought on by dancing and kissing. When Friederike urged sexual intercourse, he believes, Goethe then bolted. The evidence for this scenario approximates that for the reality of Mephistopheles. But see Eissler, *Goethe: A Psychoanalytic Study*, 2:1064.

seems clear enough—that was a constantly repeated story. Just why he was reluctant, well, Boyle gives a plausible explanation, even if a rather precious one. Yet Goethe's powers as a poet did not noticeably wane when he later achieved sexual fulfillment or had a stable relationship and family; nor did he ever hint that his art would suffer in a regular love relationship. He might indeed have sensed his poetic ability would have declined were he to become domesticated and rooted in family life at age twenty-one; but then, poetry aside, he would have languished in every way. He already had a taste for travel and the possibilities of a larger life, one not confined to the narrower range he found in his father's house or in the staid existence he sampled at Sesenheim. And for an individual of charm and talent, which Goethe certainly had in abundance, the world awaits, and the charming and talented carry it away. Finally, the idealized and poignant affairs Goethe described in his autobiography appear more emotionally tumultuous in his contemporary letters. Thus the question as to why he did not settle with one woman during the first part of his life may be artificially urgent: chance may simply not have provided the kind of satisfying relationship that stills the need to explore the world and experience adventure. In any case, Goethe left Strasbourg two days after receiving his degree on 6 August 1771. He wrote Friederike a farewell letter, and she responded with one that, as Goethe recalled, "tore my heart into pieces."[54]

Goethe returned to Frankfurt, where, to his father's considerable satisfaction, he began the practice of law. More than law, however, he practiced poetry and drama. He began reading Shakespeare, and recognized, even through poor German translations and his own reading of the English, an artist of unparalleled genius. In an essay he wrote at this time, he exclaimed: "After reading the first page, I became his for life, and when I finished the first play, I stood like a man born blind who has his sight restored in a moment by a magic hand."[55] He then himself undertook to write a drama, something in the mode of Shakespeare, but with a distinctively German cast. Through November and December 1771, he composed with fantastic speed a work that would propel him before a wider public: *Götz von Berlichingen mit der eisernen Hand* (Götz von Berlichingen with the iron hand, 1773). He based the play on a history he had read while in Strasbourg, that of a sixteenth-century robber-knight. The sobriquet for the hero comes from having lost a hand in battle, which was replaced with an iron prosthesis. The play sprawls over its canvas—a crude, disunified, and, for audiences used to more constrained conceptions, wildly exciting piece.

54. Goethe, *Dichtung und Wahrheit* (pt. 3, bk. 12), in *Sämtliche Werke*, 16:555.
55. Goethe, "Zum Shakespeares-Tag (1771), in *Sämtliche Werke*, 1.2:411–12.

Into its construction Goethe threw more passion than theatrical form. Herder criticized it as having been influenced too much by Shakespeare, and as merely "thought," as opposed to carefully structured.[56] Its publication in 1773, nonetheless, stirred many of Goethe's contemporaries to ecstasy. Wrote one: "The knight with the iron hand—What a piece! I can hardly restrain my enthusiasm. How shall I repay the author for my delight." The reviewer in Der Teutsche Merkur called it a "most beautifully interesting monster, for which [he] would be happy to exchange a hundred of our tragi-comic theatrical pieces."[57] With this work Goethe helped give birth to the Sturm und Drang movement in literature, the predecessor of the Romantic movement that would come to birth two decades later.

For the practical part of his training as a new lawyer, Goethe's father, in May 1772, sent him to Wetzlar (thirty miles north of Frankfurt) to inscribe his name in the lists of advocates in the Reichskammergericht, the highest court of appeals in the Empire. There the twenty-three-year-old novice quickly joined a company of young lawyers, among whom was the Hanoverian legation secretary Johann Georg Kestner (1743–1800). Goethe also met Kestner's fiancée Charlotte Buff (1753–1828) when they chanced to attend the same spring party. He remembered her in his autobiography as one "who, if she was not created to cause a tide of great passion, yet did excite a general delight."[58] Duty frequently called Kestner away, and Goethe began spending considerable time with Lotte—walking through meadows and gardens, engaging in domestic chores, enjoying the simple pleasures of each other's company. Quickly enough, though, his general delight turned to specific passion, and Lotte became indispensable to his being. In Die Leiden des jungen Werthers, her alter ego, also called Lotte, thoroughly captures the heart of Werther and then drives him to deep despair, since she too has a fiancé.

The novel is epistolary in form, which helps create its considerable psychological intensity. In the first several letters, Werther describes to his friend Wilhelm his experience of the countryside, just beyond the city, where the sublimity of nature overwhelms the artistic self, making that self into the very art created by nature:

> I am so happy, my friend, so completely sunk in the feeling of a calm existence that my art suffers. I now cannot sketch, not a line, and yet I have

56. Herder's letter to Goethe expressing his criticisms has not survived. We know of his reaction because of Goethe's reply. Herder's letter to his fiancée, though, makes clear his admiration for the play. See the selection of responses to Goethe's play in Goethe, Werke, 4:486–87.

57. Ibid., 4:488, 489.

58. Goethe, Dichtung und Wahrheit (pt. 3, bk. 12), in Sämtliche Werke, 16:576.

never been a greater painter than in these moments. When the lovely valley surrounds me in mist, and the sun at noon lingers on the surface of the impenetrable darkness of my woods, and only a few rays steal into the inner sanctum, and I lie by the rushing brook in the high grass, . . . then things grow dim before my eyes, and the surrounding world and the entire heavens come to rest in my soul, like the form of a beloved; then I often well up in desire and think: Oh, could you but repeat this, breathe on to paper that which lives in you so completely, so warmly, that it would become the mirror of your soul, just as your soul is the mirror of the infinite God. My friend—but I am overcome, I lie under the power, the glory of these apparitions.[59]

Werther's soul becomes the canvas on which nature inscribes her art, since "she alone forms the great artist." His soul—now the refuge of a nature that steals into that inner sanctum like a lover in the night—becomes an even more receptive medium when nurtured by love of Lotte: "Never before have I been happier," he writes his friend, "never before has my sensitivity to nature—yea, even to the rocks and grass under my feet—been richer or deeper."[60] The intimate connection between the artist, nature, and the significant woman will be a constant theme of Goethe's literary work and, indeed, the work of his life.

But frustrated love can sully the canvas that is the artist's soul, blotting out as well the sustaining relationship to nature. With frustration, increasingly volatile emotions come to dominate Werther's interior life, and ultimately his exterior life:

When I have sat with her for two or three hours and have devoted myself to her form, to her bearing, to the heavenly way she speaks, then gradually my sensibilities become tense, and my vision becomes darkened, and I can hardly hear anything; and it's as if an assassin were choking me; and then my heart, in wild beating, attempts to get some air into my head, but only increases my confusion. . . .[61]

In final despair over a futile love—and even as a test of that love—Werther commits suicide. The self-murder is preceded by a growing insensitivity to nature, as if when that umbilical to nature is severed, all the blood of life must gush out. The raw effectiveness of the novel, which usually splits stu-

---

59. Goethe, *Die Leiden des jungen Werthers*, in *Sämtliche Werke*, 1.2:199. This passage remains unchanged in the second edition.

60. Ibid., 1.2:205, 228.

61. Ibid., 1.2:242.

dents into the extremely sympathetic and the violently antipathetic, depends on the vivid expression of authentic and powerful feeling.[62]

Werther consists of many elements—composites of characters, emulations of other literature, and imaginative reconfigurations. Yet many features of the novel, especially the transformed sensitivity to nature and the singular expressions of passion, have a firm anchor in the reality of Goethe's relationship with Kestner and his fiancée Lotte Buff. In a diary entry for the end of June 1772, Kestner recorded that he caught Dr. Goethe and Lotte in the garden and that he recognized that Goethe loved his girl. The tension between the three friends grew until Lotte made clear her commitment to her fiancé. Kestner entered a note in his diary for 16 August: "Goethe received a talking to by Lotte. She declared that he could hope for nothing more than friendship. He became pale and crestfallen."[63] Goethe left Wetzlar to return to Frankfurt in mid-September 1772, and his condition grew more erratic as the marriage of Kestner and Lotte approached. Yet, even after the event (11 April 1773), he kept up a correspondence with the newlyweds—still professing his love of Lotte and even detailing for her husband dreams and reveries he had of the young wife. That fall the tolerant Kestner wrote to tell Goethe the story of a mutual acquaintance, Karl Wilhelm Jerusalem (1747–1772), who had been involved with the wife of a friend and who, disappointed in love, committed suicide, using pistols borrowed from Kestner. When Goethe got the detailed description of events, he began formulating the plan for *Werther*, or so he recalled in his autobiography.[64] He actually started writing the book some time later, during February 1774, and finished it in four weeks. Another love gone sweetly sour seems to have served as the proximate cause of his putting pen to paper. He became infatuated with the daughter of his friend Sophie von La Roche (1731–1807), a writer of some modest repute. Her daughter Maximiliane had married, rather unhappily, an older man; and Goethe spent some time in her company just as he began composing *Werther*. The Lotte of the novel reflects back Maximiliane's dark eyes and comely features. But there can be little doubt that Charlotte Buff remained Goethe's inspiration and still fer-

62. In my classroom experience, the reaction to the novel divides the sexes. For the most part, male students empathize completely with the hero, while women usually regard Werther as a jerk and Lotte as bubble-headed flirt. What is clear, though, is that the two-hundred-year-old novel still has enormous force.

63. The passage is from *Kestner's Tagebuch*, selections in Goethe, *Werke*, 6:517–18.

64. Goethe, *Dichtung und Wahrheit* (pt. 3, bk. 13), in *Sämtliche Werke*, 16:618–19. Even prior to learning of Jerusalem's suicide, Goethe sketched out a program for a work that would detail an "apprenticeship of the heart," one that sounds like a rehearsal for the novel. See Gert Mattenklott, "Die Leiden des jungen Werthers," in *Goethe Handbuch*, 3:57.

vent love.[65] During the work on the novel, he confessed that he himself often thought of suicide.[66] In June 1774 he wrote Lotte to tell her of his composition and more. The letter could have been taken from the novel itself:

> Last night I wanted to write you, but it wasn't possible. I paced back and forth in my room, and spoke to the shadow silhouette of you, and even now it is difficult to write.—Am I never again, never again to hold your hand, Lotte? . . . Adieu, dear Lotte, I am sending you shortly a friend who is quite similar to me; I hope he is well received. He is called Werther, and is and was—well, he will explain this himself.[67]

Götz von Berlichingen brought Goethe fame. *Werther* brought him the celebration of genius. The poet Johann Jakob Wilhelm Heinse (1746–1803), in a long journal review, expressed one kind of popular reaction:

> Who has felt and feels what Werther felt—that person's rational thoughts disappear like a fine mist before a blazing sun if he should merely mention the name. The heart of such a person is simply full of that feeling and his head fills with tears. O humanity, what a flood of pain and ecstasy you are capable of. There he lies in the church graveyard, under the two linden trees in the high grass. Deep is his sleep, and his kisses now dust. Oh, when will it be morning in this yard to bid the sleeping: Awake, poor Werther! And even more unhappy Lotte![68]

The novel penetrated deeply into the culture of Europe during the subsequent decades. At the beginning of the nineteenth century, Napoléon himself told Goethe that he read *Werther* several times.[69] Even Frankenstein's monster received an education in the passions by reading the novel.[70] And at the end of the century, Ernst Haeckel would adopt the persona of Werther—when he was sixty-eight years old—and would call his young fe-

---

65. Conrady (*Goethe: Leben und Werk*, 1:178–79) warns against taking *Werther* as a literal description of Goethe's relationship to Lotte Buff—generally a sound kind of advice. But both the evidence and Goethe suggest the close parallel. See, for example, the letter quoted next in the text.

66. Goethe, *Dichtung und Wahrheit* (pt. 3, bk. 13), in *Sämtliche Werke*, 16:618.

67. Goethe to Charlotte Kestner (15 and 16 June 1774), in *Goethe-Briefe*, ed. Philipp Stein, 8 vols. (Berlin Wertbuchhandel, 1924), 1:288–89.

68. Johann Jakob Heinse, "Über *Die Leiden des jungen Werthers*," originally published in *Iris: Vierteljahrsschrift für Frauenzimmer* (December 1774), pp. 78–81. The review is collected in Karl Mandelkow, ed., *Goethe im Urteil seiner Kritiker*, 5 vols. (Munich: C. H. Beck, 1975), 1:23.

69. In early fall of 1808, Napoléon summoned Goethe to Erfurt and spoke to him about the novel, even slightly criticizing what he thought some inauthentic representation in the text. See Goethe's notes on the meeting, "Unterredung mit Napoleon," in *Sämtliche Werke*, 14:578.

70. Mary Shelley, *Frankenstein; or the Modern Prometheus*, in *The Complete Frankenstein*, ed. Leonard Wolf (New York: Penguin, 1993), p. 176. Mary Shelley's parents, William Godwin and Mary Wollstonecraft, read the novel together on the morning before their daughter's birth.

male friend "Lotte," gracing her with all of the passion and poetry that testify to the transforming power of Goethe's creation. Not every sensibility, though, was comparably touched. The physicist and epigrammarian Georg Christoph Lichtenberg (1742–1799) reacted with less tender regard. He wrote a friend thanking him for the gift of the novel: "For the sorrows and joys and madness of the young Werther, I give you much thanks. It is true that a young fellow from Lüttichow shot himself over the book, which makes him a true citizen of Lüttichow. I believe the smell of a pancake is a stronger motivation for remaining on this earth than all the enormously vulgar conclusions of the young Werther for the leaving of it."[71]

Johann Heinrich Merck (1741–1791), poet, literary critic, and friend of Goethe, rendered, I believe, the soundest judgment about that still-affecting portrayal. In a review, in which he considered the novel, its imitators, and parodies, he remarked:

> The inner feeling that all his compositions exhibit, the living presence that accompanies the art of his representation, the felt details that reside in all the parts of the work and their particular selection and ordering— all of this shows a universal and powerful master of his material. How much truth is included in the story of the young Werther and how much he has drawn out of his horn of plenty, we leave to the judgment of present and future justifiers, falsifiers, and bumbling imitators. Anyone who knows what it is to write will easily understand that no event in the world, with all its circumstances, can make a dramatic composition except that the hand of the artist has infused it with another disposition.[72]

Merck's conclusion must, I think, be ours. Goethe's literary compositions are singular works of art that an inventive imagination constructed according to its own principles. Yet they did have instigating circumstances in his life; and in light of evidence and guided by informed judgment, those circumstances might be recovered. To adopt a more abstemious attitude that rejects a biographically motivated interpretive practice—thinking such must wallow in the "backwaters of literary-historical research"—is the critical equivalent of endorsing the virgin birth; it is a modernist theology that must founder on a more severe study of generation.[73]

---

71. Georg Christoph Lichtenberg to Johann Christian Dieterich (1 May 1775), in Mandelkow, *Goethe im Urteil seiner Kritiker*, 1:530.

72. Johann Heinrich Merck, "Sammelrezension über *Die Leiden des jungen Werthers* und Werther-Schriften," originally published in *Allgemeine deutsche Bibliothek* 26 (1775). In Mandelkow, *Goethe im Urteil seiner Kritiker*, 1:59.

73. Wellbery rejects, soundly enough, a particular biographical approach to Goethe's poetry. One benighted author he chastises had interpreted some poems as expressions of Goethe's own immediate, youthful responsiveness to nature since the poems themselves celebrate that kind of im-

Though Goethe completed *Werther* with quick dispatch, he began and then abandoned many other dramatic works and poems during the years immediately surrounding the publication of the novel. He did see to near completion a drama based on the Faust legend, which would become known to the scholarly world as *Urfaust*. Goethe read portions to friends during this period and later, but he brought this drama to full publication, after many additions and revisions, only in the first decade of the nineteenth century. The turmoil of his creative productions matched that of his emotional attachments. Within the first few months of 1775, Goethe had moved to a new love.

Though all of his amorous connections left their mark—both on his psyche and his poetry—Goethe recalled his relationship with the sixteen-year-old Lili Schönemann as the most blissful of his life. She was the daughter of a Frankfurt banker, whose fortunes would come to naught within a decade. Goethe met her at a ball in the family house. She was remembered by a friend, who knew her when she was in her early thirties, as the "ideal of femininity": "tall, with a slim frame, a slight melancholic expression on her somewhat faded but still charming features, and above all, a sublime dignity, of which her whole being gave evidence."[74] Goethe spent the spring of 1775 on the Schönemann country estates in Offenbach am Main (just outside of Frankfurt), whiling away the hours in long walks, heartfelt conversations, evening songs, and poetry reading—an interlude of sweet existence that Goethe extensively described in his autobiography.[75] By Easter they had become engaged. In his later years, he revealed to Eckermann, in some reverie, the measure of his love:

> I see again the charming Lili, in all her liveliness, before me, and it's as if I could feel again the breath of her happy presence. She was, in fact, my

---

mediacy. Wellbery indicts this approach by pointing out that Goethe's early lyric poetry conformed in theme to certain already established literary models. But this is hardly an argument for the bankruptcy of biographical interpretation that Wellbery assumes. It only means the instigating causes for Goethe's poetic expression were, perhaps, particular kinds of reading rather than (or in addition to) particular experiences of nature. Moreover, Goethe's constantly reiterated theme suggests that he, as a matter of fact, held such immediacy to nature as an ideal—a judgment supported by a great deal of other evidence, as will be discussed below. Unless those literary models that Wellbery mentions came to Goethe as divinely inspired, they too were instances of biographically relevant experience; and the poems indicate their impact. See Wellbery, *The Specular Moment*, p. 7.

74. After Goethe discussed this relationship with Johann Peter Eckermann in 1830, he inquired of his friend Henriette, countess of Beaulieu-Marconnay, what she recalled about Lili's later days. See Henriette von Beaulieu-Marconnay to Goethe (3 December 1830), in *Briefe an Goethe*, 2:565.

75. Goethe, *Dichtung und Wahrheit* (pt. 4, bk. 17), in *Sämtliche Werke*, 16:730–49.

first deep and true love. And I can say, she was the last—since all the small attachments, which subsequently followed in my life, were compared with this first, and were in that comparison slight and superficial.[76]

This is the judgment of an elder Goethe reliving his youth. Other later attachments, especially his decade-long devotion to Charlotte von Stein, would seem to have had considerably more emotional and erotic weight, at least at the time. And one must not forget, as Goethe apparently did, Christiane Vulpius, his longtime mistress and later his wife. But the yet-green memory of Lili proved powerful. Unlike Vulpius, Lili had a lively intellect, and unlike von Stein, she reciprocated Goethe's love generously and explicitly. According to a friend, she had been willing to give herself sexually to him. Amazingly, he refused her sacrifice—and thus was able to preserve a sacred memory of honor not despoiled.[77] If his past relationships with women give indication, this refusal of sexual favors may have been less high-minded than he afterward suggested. Goethe, it seems, was too frightened for such direct intimacy. The poetry of this period did transform those libidinous desires into poignant emotional release. In lived experience, though, he simply could not pull the trigger, even when the target obliging moved right in front of him.

During this period of barely contained passion, the celebrated author of *Werther* penned several plays and large quantities of poetry: he finished *Claudine von Villa Bella*, he began *Egmont* (and continued to add and reconfigure it over the next decade), and he saw *Stella* published. His "Lili" poems tremble with anguished desire: "Could I but be filled just once / With you, O eternal love / Oh, this long, penetrating suffering / How it endures on the earth."[78] The final poems, though, change the emphasis to speak virtually of desired anguish: "Pure Lili, you were for so long / All my passion and all my song. / But now, Oh, all my pain / Yet, all my song you do remain."[79]

Within a few weeks of his betrothal to Lili, Goethe, in what might appear an almost pathological reaction, escaped from Frankfurt with a group of friends who were traveling to Switzerland. Therewith he effectively broke off his relationship with the woman he would later describe as his one

---

76. Johann Peter Eckermann, *Gespräche mit Goethe in den letzten Jahren seines Lebens*, 3rd ed. (Berlin: Aufbau-Verlag, 1987), (15 February 1830), p. 621.

77. Beaulieu-Marconnay to Goethe (3 December 1830), in *Briefe an Goethe*, 2:566: "If his great-heartedness had not, with propriety, refused the sacrifice that she wanted to offer, she would not have looked back, robbed of her dignity and social honor, on the past with the happy memories that she now does."

78. Goethe, "Sehnsucht," in *Sämtliche Werke*, 1.1:273.

79. Goethe, "Holde Lili," *Werke*, 1:104.

true love.[80] In the poetry written during this first of his Swiss excursions, the lingering love of Lili ("If I didn't love you Lili / What would my happiness be?") is replaced by a desire for reanimation through nature ("I suck on my umbilical / To draw nourishment from the world, / And all around so wonderful is Nature / Who holds me to her breast.")[81] Goethe's high-spirited traveling companions through Switzerland, the brothers Christian (1748–1821) and Friedrich Leopold (1750–1819) zu Stolberg-Stolberg and Christian August von Haugwitz (1752–1831), were poets and counts on the loose. Goethe had an intimate relationship with these young men, which he characterized in his autobiography as "the sort of love peculiar to the period." Looking back, he regarded the relationship as merely an expression of the exuberance of youth.[82] At the outset of their journey, his friends, after visiting at court in Darmstadt, went bathing naked in a nearby fishpond, causing a "scandal."[83] And frequently during the journey, this affectionate and lively group of friends would not be able to resist, as they passed through the alpine countryside, the sound of fresh, cold, rushing water. Just outside of Basel, some local bluenoses came upon the boys frolicking in the water and pitched stones at them, which caused another "scandal."[84] Goethe remembered these events vividly in his autobiography, and, as he mentioned, he also wove them into a story that he later published in Schiller's journal *Die Horen*.

Goethe probably wrote the story "Briefe aus der Schweiz" sometime after his Italian journey, in the early 1790s. In his tale he employed the conceit that a packet of letters, written by Werther before meeting Lotte, had been found. In these letters to a friend, Werther describes a trip through Switzerland. The journey quickens his enthusiasm for works of art, "if they are true, if they are direct expressions of nature." Up to that point, Werther admitted, he never had the experience of the singular masterpiece of nature, the unadorned human form. In order to gain that model for his imagination, he reports that he caused his friend Ferdinand to bathe naked in a nearby lake: "How wonderful is my young friend built! What a balance of all the parts! What a richness of form, what a radiance of youth, what a prize for my imagination."[85] He then desired to see the female form and in Geneva hired a

80. Eissler thinks that it was Goethe's love for his own sister Cornelia—and her belief that the marriage of her brother and Lili would fail—that led him to break off with the girl whom, at one level, he certainly loved. See Eissler, *Goethe: A Psychoanalytic Study*, 1:128–29.

81. Goethe, "Tagebuch, Schweizerreise 1775," in *Sämtliche Werke*, 1.2:544, 543.

82. Goethe, *Dichtung und Wahrheit* (pt. 4, bk. 18), in *Sämtliche Werke*, 16:763: "This sort of relationship, which indeed looked like a confidence, was taken for love, for true affection."

83. Ibid., 16:766: "To see naked young men in bright sunshine seemed in this region to be quite unusual; in any case, it produced a scandal."

84. Ibid. (pt. 4, bk. 19), 16:796.

85. Goethe, "Briefe aus der Schweiz," in *Sämtliche Werke*, 4.1:638, 640.

prostitute under the pretext that he was a painter seeking a model. The meeting was arranged; and as he waited in a room, a woman, who seemed not to notice him, walked in. She had

> a beautiful, regular form, brown hair with many large locks that rolled down off her shoulders. She began to undress. As one piece of clothing after another fell to the floor and nature was disrobed of foreign drapery, an amazingly strange sensation came over me and, I might say, almost gave me a shudder. Oh, my friend, is it not thus with our opinions, prejudices, institutions, laws, and whims? Are we not frightened if one of these strange, unsuitable, and untrue garments is cast away and some part of our true nature stands naked? We shudder, we are ashamed, though we do not feel the least aversion before any astounding and tasteless mode [of dress] that disfigures us through an external force. . . . What do we see in women? What kind of woman pleases us and how do we confuse all ideas? A small shoe looks good, and we remark: What a beautiful, small foot! A narrow corset has something elegant about it, and we love a pretty waist. . . . She was enchanting when she disrobed, beautiful, wonderfully beautiful, as the last garment fell.[86]

The woman then climbed into bed, and lay there, seemingly asleep. She called out as if to a lover, and then beckoned to Werther. Werther assured his correspondent that nothing untoward happened, but "my imagination was aflame, my blood was boiling. Oh, but could I have stood next to the greatest mass of ice to cool myself down again."[87] In Italy Goethe had ample opportunity to experience the naked form of many artist's models; though he, like Werther, likewise refrained, mostly from fear of the French disease.[88]

In the story, and presumably in the originating adventure with his young friends, Goethe identified nature with its masterpiece, the human form—in its most erotic posture. In the "Briefe," all the elements melt into a coherent, lyrical expression: art, a love abandoned (Lili), a love renewed (the lusty counts), and the naked form of nature and nature's young gods and goddesses. For Goethe, nature evoked and emblematically meant the human, naked form, which inspired poetry, art, and, as I will discuss shortly, science.

86. Ibid., 4.1:643–44. The descriptions of the nude model bear resemblance to the Faustine, about whom the poet sung in his *Römische Elegien*, as well as to Christiane Vulpius, whom he met on his return from Italy in 1788 and who quickly became his mistress.

87. Ibid., 4.1:642.

88. See below for the discussion of Goethe's experiences in Italy.

The Weimar Councillor and the Frustrated Lover

In December 1774 Prince Carl August (1757–1828) arrived in Frankfurt. The next year, on his eighteenth birthday, he would succeed to the duchy of Saxe-Weimar-Eisenach. Goethe was introduced to the prince, and the two spent some time speaking about literature and affairs of the day. Goethe had occasion to meet the young noble again in Karlsruhe (Baden) in May 1775, while on his way to Switzerland. Carl August was then preparing for his marriage in the autumn to Princess Luise of Hesse-Darmstadt (1757–1830). In September, on the way to his wedding, the prince stopped again in Frankfurt, and on seeing Goethe invited him to come to Weimar, the capital of his modest lands.[89] In November Goethe, weary of living under his father's roof, acted upon the invitation; and thus began his long career in the service of the duke, who, boy and man, would remain a close friend and confidant.

Goethe was not only a sporting and carousing companion to the young nobleman—and he was surely that—but he formally entered into governmental service. In June 1776 he became a member of the privy council, with the title of *Geheimer Legationsrat;* and three years later, in August 1779, he was promoted to *Geheimrat.* On the occasion of his thirty-third birthday, in 1782, Goethe, at the duke's request, received from the Holy Roman Emperor Joseph II a certificate of ennoblement, allowing him to add "von" to his name. Herder, who had been brought by Goethe to administer the duchy's ecclesiastical affairs in 1776, animadverted to his friend Johann Georg Hamann about Goethe's power and influence, and his recent ascendance to the presidency of the ducal chamber after the dismissal of the previous holder of that office:

> After this fellow's honorable sacking had been dictated in the council, Goethe was named to the presidency of the chamber, though without being precisely so called, which would be for him too small an appendix to attach to his name. He is really, though, privy councillor, president of the chamber, president of the war council, overseer of buildings and mines,

---

89. Weimar was a town of approximately 6,000 individuals (Frankfurt had at the time about six times that), with around 100,000 in the entire duchy. The inhabitants consisted of court ministers, small tradespeople (apothecaries, saddlers, blacksmiths, bakers, shoemakers, farmers, etc.). In Ilmenau, a town thirty miles southwest of Weimar, the duchy had silver and copper mines, for which Goethe became responsible. The university at Jena, about fifteen miles from the capital, was jointly administered by the Saxon duchies (Saxe-Weimar-Eisenach, Saxe-Gotha-Altenburg, Saxe-Meiningen, Saxe-Coburg), though Weimar had the principal say, guided often by Goethe's shrewd judgment in academic affairs.

Figure 10.2
Carl August
(1757–1828), duke of
Saxe-Weimar-Eisenach.
Portrait by an unknown
artist. (Courtesy
Inspektorhaus,
Botanische Garten, Jena.)

and also directeur des plaisirs, court poet, orchestrator of beautiful festivals, of court operas, of ballets, of masks, of writing, of art, etc., director of the drawing academy, in which during the winter he holds lectures on osteology, and himself, principally, the first actor, dancer and, in short, the factotum of the house of Weimar and, by God, soon the majordomo of all of the Ernestine [Saxon] houses, in which he will receive the prayers of all concerned. He has become a baron. . . .[90]

Goethe's power in the duchy had no equal, and that power extended far beyond the formalities of court, well into the boudoir.

In the first few years at court, Goethe wrote, produced, and acted in several plays, which seem to have had the usual kind of erotic effect. In fall 1776, for instance, he quickly composed *Die Geschwister* (The siblings), in which he acted opposite Amalia Kotzebue, sister of August Kotzebue (1761–1819), later playwright and diplomat. He took to calling her Mariana, the name of a character in the play whom the hero Wilhelm loves and also the name of a woman in *Wilhelm Meisters theatralische Sendung* (Wilhelm Meister's theatrical mission), whom another Wilhelm loves but discovers is the mistress of a local merchant.[91] Goethe also seems to have had

90. Johann Gottfried Herder to Johann Georg Hamann (11 July 1782), in *Johann Gottfried Herder, Briefe,* ed. Wilhelm Dobbek and G. Arnold, 10 vols. (Weimar: Böhlaus, 1977–96), 4:226.
91. Goethe worked on the novel *Wilhelm Meisters theatralische Sendung* during this first Weimar period, but never published it. In the revised version it became *Wilhelm Meisters Lehrjahre,* which appeared in the collection of Goethe's works in 1795–96.

a brief flirtation with Corona Schröter, an actress he helped bring from Leipzig and who, in spring of 1779, played opposite him in the first version of his *Iphigenie auf Tauris*. Carl August, however, had a more commanding interest in her. The duke's usurpation, though, seems to have had little effect on Goethe.[92] Carl August's later evaluation may have been correct, at least about these minor affairs, namely, that Goethe read too much into the women to whom he was attracted. "He loved his own ideas in them," the duke believed, "and never really felt grand passion."[93] This judgment, made just before the poet's death, was, however, no fair measure of the one woman on whom Goethe, over the course of a decade, expended great love in all its denominations, save one. This was Frau Charlotte von Stein (1742–1827).

She was born Charlotte Albertine Ernestine von Schardt and raised by a pious mother to those accomplishments—reading, sums, music, and dance —befitting a daughter of the minor nobility. At age fifteen she became lady-in-waiting to Duchess Anna Amalia, mother of the duke, and married, at twenty-two, Gottlob Ernst von Stein, a fairly wealthy landowner who became master of the stables in the duchy. When the twenty-six-year-old Goethe met her, she was thirty-three, had been married eleven years, and had borne seven children, four girls who died immediately after birth and three boys who survived. Just prior to meeting Goethe, her physician Johann Georg Zimmerman (1728–1795) described this fascinating woman to a friend:

> The lady-in-waiting, mistress of the stables, and Baroness von Stein of Weimar. She has large black eyes of the greatest beauty. Her voice is soft and low. Every man, at first glance, notices in her face earnestness, tenderness, sweetness, sympathy, and deeply rooted, exquisite sensitivity. Her courtly manners, which she has perfected, have been ennobled with a very fine simplicity. She is rather pious, which moves her soul with a quiet enthusiasm. From her easy, Zephyr-like movements and her theatrical accomplishment in artful dancing, you would not suspect—something that is certainly true—that a hushed moonlight and a quiet evening fill her heart with the peace of God. She is thirty-one years old, has quite a few children and weak nerves. Her cheeks are very red, her hair completely black, and her skin, like her eyes, is Italianate. Her figure is slim, and her entire being is elegant in its simplicity.[94]

92. See Boyle, *Goethe: The Poet and the Age*, 1:303.
93. Quoted in ibid., 1:265.
94. Johann Georg Zimmerman to Johann Kaspar Lavater (1774), quoted in Wilhelm Bode, *Charlotte von Stein*, 3rd ed. (Berlin: Mittler und Sohn, 1917), p. 63.

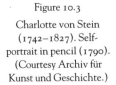

Figure 10.3

Charlotte von Stein
(1742–1827). Self-
portrait in pencil (1790).
(Courtesy Archiv für
Kunst und Geschichte.)

When von Stein had heard that the famous author of *Werther* might be
coming to live at Weimar, she wrote Zimmermann to ask what he knew
about the man. Her friend, who had not yet met Goethe, related what he
had heard from a "woman of the world." "Goethe," he reported, "is the most
handsome, liveliest, most original, fieriest, stormiest, softest, most seductive, and for the heart of a woman, the most dangerous man that she had
ever seen in her life."[95] And for von Stein, he would prove to be all of that.

The relationship between Goethe and von Stein grew to be quite complex during the decade of their closeness (1776–86). She was obviously alluring, coquettish, sensitive, sexually attractive, if not a great beauty, and
had a lively intellect, at least one that might appreciate Spinoza, the decidedly arcane philosopher whom they read together. Goethe swooped into
her dull existence like a frenzied swan from Olympus. He was maddeningly
impetuous and foolish with her, trampling on her sensitivities, and shock-
ing her with his ribaldness and vulgarities. His genius was matched by the
size of his ego. She would frequently remonstrate with him about his behavior, and he would hangdoggedly become apologetic, only shortly to do
something else that upset her. Yet he loved her for her efforts at taming him;
and, in time, she—and the responsibilities of the court—did rein in his ac-

95. Johann Georg Zimmermann to Charlotte von Stein (19 January 1775), as quoted in ibid.,
pp. 69–70.

tions. Goethe's initial hyperbolic behavior serves as an index, partly at least, of his frustrated desire for her. Typical of the early letters is this, from spring 1776:

> Why should I plague you! Lovely creature! Why do I deceive myself and plague you, and so on.—We can be nothing to one another and are too much to one another.—Believe me when I speak as clear as crystal to you; you are so close to me in all things.—But since I see things only as they are, that makes me crazy. Good night angel and good morning. I do not wish to see you again—Except—You know everything—I have my heart—Anything I could say is quite stupid.—I will look at you just as a man watches the stars—think about that.[96]

Despite her occasional lapses into peevishness, she was strongly attracted to him. Together they discussed the gossip of the court; he gave her instruction in English and read his poems and works-in-progress to her. She was his confidante, support, and, for the early years, his passion.

During the decade of their relationship, Goethe wrote her at least once every other day, though they lived but a few minutes apart and he saw her several times a week. In those letters he made love to her—to use an older meaning of that phrase. The love could be expressed without being "distorted by my madness and foolishness," as so often happened when they were together.[97] Those letters describe his projects and the affairs of court, but mostly explore his own reactions to people, especially to her, and to other matters churning deep within his ego. They often have a decidedly artful turn, and one suspects the writer of trying out different modes of expression, rehearsals for his dramas, novels, and poetry. Yet it would be a mistake, I think, to suppose that Goethe simply donned a guise to hone his art, as if the man and the artist were not one. The letters are constantly leavened with lines too insistent to be other than authentic expressions of love, as for instance: "Dearest, I love only you—something I feel when I'm interacting with other women"; "I know that you love me, I sense that, since I love you so much."[98] In June 1780 she sent him a ring, which seemed to seal for him the magic of their relationship.[99] By spring of the next year, that relationship had become ever richer, if not quite as sensational. Their love seems to have worked a profound effect on his entire being: "I cannot say and should not even try to understand the kind of upheaval your love has produced in my inmost self. It is a condition which, as old as I am, I have

96. Goethe to Charlotte von Stein (1 May 1776), in *Goethes Briefe*, 1:213.
97. Goethe to von Stein (6 September 1777), in ibid., 1:235.
98. Goethe to von Stein (6 and 13–16 September 1777), in ibid., 1:235, 237.
99. See Goethe to von Stein (14 June 1780), in ibid., 1:306.

Figure 10.4　Apartments of Charlotte von Stein in Weimar. The family of the chief equerry, Baron Josias von Stein (1735–1793), occupied the upper part of the building; horse stalls were originally on the lower level. (Photograph by the author.)

never known."[100] Goethe's declarations suggest that the sacrament that he so desired to bless their union should anoint a physical bond:

> My soul is grown intimately into yours. I won't waste words, you know that I am inseparable from you and that neither heights nor depths can cut me off from you. I should wish that there were some oath or sacrament that bound me obviously and legally to you—how I would love that.[101]

Charlotte von Stein remained during their long relationship mistress of his soul, though very probably not of his body. The Weimar court, a small, ingrown, gossiping community, knew of Goethe's devotion to her, but seems never to have suspected a physical relationship of the sort their duke enjoyed with a number of women, save his wife. When Schiller visited the court in summer of 1787, a year after Goethe had left for Italy, he heard no hint of scandal. He wrote his friend Körner a note that suggests he perhaps did expect to hear some: "This woman possesses perhaps over a thousand letters from Goethe, and he has written her every week from Italy. They say

100. Goethe to von Stein (23 March 1781), in *Goethe-Briefe*, ed. Philipp Stein, 8 vols. (Berlin: Wertbuchhandel, 1924), 2:231.
101. Goethe to von Stein (12 March 1781), in *Goethes Briefe*, 1:350.

that their relationship is completely pure and blameless."[102] From the ardent outpourings of his missives, Goethe undoubtedly wished the relationship to have become less chaste. Even after he had given up hope, his ghostly desires yet recalled to her that for which he had so desperately longed:

> I hang on to you with every fiber of my being. It's awful how those recollections so often tear me apart. Oh, dear Lotte, you don't know what violence I've done to myself and will do, and the thought that I will not possess you, no matter how I consider it, reconsider it, lay it out—that drives me wild and tears me apart. I would like to give form to my love for you, which I wish, forever and ever—Forgive me for wanting to say again once more what has for so long stuck in my throat and has made me dumb.[103]

During the Weimar years, von Stein stoked Goethe's passions; but restrained by her piety, by her reputation, and by thoughts of her children (hardly, it seems, by thoughts of her husband), she would immediately pull back when he came too close. He deeply valued her not only as a sexual being but even more specially as a friend; and he certainly would not have wished to destroy that friendship by a reckless attempt at seduction. One can believe that his physical desire for her, though fluctuating, never completely died—until, that is, his journey to Italy. In that land, the days of golden sun and nights of sensual delight conspired to make his physical desires overwhelming, and he finally found the satisfaction for which he longed. Von Stein may have learned of this or at least suspected. When he returned to Weimar two years later and quickly brought a young woman into his home, von Stein was devastated; and despite his efforts to retain her friendship, they unalterably broke from one another, with only occasional, very polite meetings thereafter. She returned all of his letters—some eighteen hundred—and demanded hers in return, which she then burned. If the pitch of their relationship, that love in all its complexity, should be measured by the poetry she inspired, then one must judge their love the most significant of his life. Indeed, subsequent generations would look to Goethe's relationship with von Stein as modeling the very trope of unrequited love's ability to inspire great poetry.

During his first decade at Weimar, Goethe wrote a very large number of poems—certainly some of his most beautiful—that were initially penned for her. For instance, on 12 February 1776, as they were becoming seriously

---

102. Friedrich Schiller to Christian Gottfried Körner (12–13 August 1787), in *Schillers Werke*, 24:131.

103. Goethe to von Stein (21 February 1787), in *Goethes Briefe*, 2:50.

involved with one another, he sent her the plangent "Wandrers Nachtlied" (Wanderer's evening song), which he composed along the slopes of the Ettersberg (just northeast of Weimar):

Der du von dem Himmel bist,
Alles Leid und Schmerzen stillest
Den der doppelt elend ist,
Doppelt mit Erquickung füllest,
Ach! Ich bin des Treibens müde
Was soll all der Schmerz und Lust?
Süsser Friede!
Komm ach komm in meine Brust![104]

Such poetry, of course, becomes completely flat in translation, when rhythm, rhyme, and allusion are traduced in another language. Certainly this holds true for perhaps the best known of Goethe's poems, one that also came to be called "Wandrers Nachtlied." It arose out of a journey to the mining region of Ilmenau in September 1780. There Goethe climbed the Hermannstein and found the cave where four years before he had chiseled an "S" in the rock; and now, as he wrote Charlotte, he kissed and kissed that stone till it seemed to breathe a sigh.[105] He then spent the night in a small hunter's cabin by himself and on its wall scratched in pencil a poem whose delicate feeling might even coax love out of a rock.

Über allen Gipfeln
Ist Ruh,
In allen Wipfeln
Spürest du
Kaum einen Hauch;
Die Vögelein schweign im Walde.
Warte nur, balde,
Ruhest du auch.[106]

Aside from compositions that have justly become famous, Goethe penned a myriad of other verses in that occasional, effortless way that made

---

104. Goethe, "Wanders Nachtlied I," in *Sämtliche Werke*, 2.1:13: "You who are from heaven, / You quiet all sorrow and pain / Him who is doubly suffering / You fill with double refreshment / Oh! I am tired of this striving / What does all this pain and joy mean? / Sweet peace! / Come, Oh, come into my breast!"

105. Goethe to von Stein (6 September 1780), in *Goethes Briefe*, 1:314–15.

106. Goethe, "Wandrers Nachtlied II," in *Sämtliche Werke*, 2.1:53: "Over all the peaks / There is quiet, / Through the treetops, / You feel / Hardly a breath; / The small birds remain silent in the woods, / Only wait, soon now / You will rest as well."

Schiller think of him as a sheer phenomenon of nature. So, for instance, in fall of 1776 he dropped this into a letter sent to von Stein from Dornburg, in the Saale valley (just north of Jena):

> I am nowhere safely hidden
> Far along the Saale River here.
> Many cares follow unbidden;
> But my love for you is also near.[107]

One poem in particular clearly exhibits the web of feeling holding close the poet, von Stein, and nature. It appears to have been inserted in a letter to von Stein of November 1777. The letter contains the remark: "That I am always envisioning the phenomena of nature and my love for you in reverie, you will see from the enclosed."[108] In the poem "An den Mond" (To the moon), the poet compares the soft, haunting light of the moon to his lover's watchful eye, which holds his fiery heart like a ghost is held on a riverbank.

> Your gaze softly spreads
> Over my field
> Like the eye of my beloved
> Gently watching over my fate.
> You know it to be so voluble,
> This heart on fire,
> You hold it like a ghost
> Close by the river.[109]

In these remarkable lines, the poet's burning heart and the ghost—perhaps it is a lonely spirit that comes from the earth—melt into one, as does the moon's light and the lover's look. For Goethe, woman and nature flow into a single being. The poem ends:

> Blessed is the person who, without hate,
> Retreats from the world,
> Takes a man to the breast,
> And enjoys that of which
> Humans are unaware

107. Goethe, [To Charlotte von Stein], in ibid., 2.1:28: "Ich bin eben nirgend geborgen / fern an die Saale hier / verfolgen mich manche Sorgen / Und meine Liebe zu dir."

108. Goethe to von Stein (11 August 1777), in *Goethes Briefe an Charlotte von Stein*, ed. Julius Petersen, 2 vols. (Leipzig: Insel Verlag, 1923), 1.1:77.

109. Goethe, "An den Mond," in *Sämtliche Werke*, 2.1:34: "Breitest über mein Gefild / Lindernd deinen Blick / Wie der Liebsten Auge, mild / Über mein Geschick. / Das du so beweglich kennst / Dieses Herz im Brand / Haltet ihr wie ein Gespenst / An den Fluss gebannt." The first "you" is in the familiar form (*du*), the second "you" (*ihr*) is plural, and seems to mean both the moon and the lover.

Figure 10.5 The river Ilm, which runs near Goethe's garden house in Ilm park. (Photograph by the author.)

> Or perhaps despise,
> And which in the night wanders
> Through the labyrinth of the heart.[110]

What the lovers share flows through their hearts like a stream through moon-blessed fields. These last lines suggest a bond of lover and beloved drawn tight beyond friendship.

After Goethe's return from Italy, von Stein redacted this poem, changing and adding some verses, such as:

> Your gaze softly spreads
> Over my field,
> Where my friend no longer
> Gently watches over me.
> Extinguish that image from my heart
> Of my departed friend,
> For whom inexpressible grief
> Cries silent tears.[111]

---

110. Ibid., 2.1:34: "Selig wer sich vor der Welt / Ohne Hass verschliesst / Einen Mann am Busen halt / Und mit dem geniesst, / Was den Menschen unbewusst / Oder wohl veracht / Durch das Labyrinth der Brust / Wandelt in der Nacht."

111. Charlotte von Stein, "An den Mond nach meiner Manier," in ibid., 2.1:561: "Breitest

This was the woman who for a decade held Goethe in thrall, and now, because he could wait no longer, she sadly and bitterly rejected even his friendship.

The Science of Goethe's First Weimar Period

When Goethe came to Weimar, he fell deeply in love with Charlotte von Stein. Simultaneously, his love for nature became transformed. In his early years, he poetically captured nature in vigorous, urgent, and cascading images, in the quicksilver emotions of youth. In his Weimar years, his relationship to nature struck a lyrical cord, with powerful emotions restrained by a more complex expression, as that relationship became entangled with his love of von Stein. He wrote to her on 9 April 1782: "For me you are transubstantiated into all objects. I see everything clearly and yet see you everywhere."[112] He most frequently saw her, paradoxically, when he left Weimar and traveled through the local region, usually to visit the mines in Ilmenau and to scout out other areas for possible mineral deposits, particularly around the Ettersberg. It was on two such occasions that he composed the poems later called "Wandrers Nachtlied I" and "Wandrers Nachtlied II."

Goethe's work as superintendent of mines required him to read in mining literature and to observe quite carefully the geology of the different areas of the duchy. In this work he engaged the help of Johann Carl Wilhelm Voigt (1752–1821), a young student of the great Freiburg geologist Abraham Gottlob Werner (1750–1817) and a researcher who would in time become a leading geologist in the Germanies.[113] Goethe commissioned Voigt to make a geological survey of the territory, and they both would often ride out together for particular investigations. But Goethe was ready, during these years, to turn any journey to the countryside into a geologizing effort. When, for instance, he made an excursion with Carl August to Switzerland (September 1779—January 1880), his letters back to von Stein were filled with reactions to the natural beauty of the land, as well as specific speculations about geological processes.[114] His letter of 3 October 1779 is typical:

über mein Gefild / Lindernd deinen Blick, / Da des Freundes Auge mild / Nie mehr kehrt zurück. / Lösch das Bild aus meinem Herz / Vom geschiednen Freund, / Dem unaugesprochner Schmerz / Stille Träne weint."

112. Goethe to von Stein (9 April 1782), in *Goethes Briefe,* 1:391–92.

113. Voigt was the brother of Goethe's later colleague in the court administration, Christian Gottlob Voigt.

114. With some modifications and embellishments, he turned these letters into a small travel book, initially meant for his friends. Parts were later published in Schiller's *Horen* (1796), and the entire manuscript in Goethe's collected works of 1806. See Goethe, "Briefe aus der Schweiz 1779," in *Sämtliche Werke,* 2.2:595–647.

At the end of a ravine I dismounted and alone ventured back a way. A deep feeling of pleasure grew in me as I looked around. One ruminates darkly about the life and origin of these strange forms. Whenever they arose, these masses took form according to their weight and the similarity of their parts. Whatever revolutions thereafter moved, separated, and split them, they were only particular shudders, and even the thought of so enormous a movement gives the elevated feeling of eternal solidity. Time, bounded by eternal laws, has also affected them, sometimes more, sometimes less.[115]

Many years later Goethe reflected back on how his geologizing activity grew out of "subjective" concerns, as did all of science.[116] In this instance, his aesthetic and utilitarian interests combined to push him deeper into the investigations of geology; but he moved through this realm while trailing out an emotional thread that always led him back to von Stein. He thus passed swiftly from that "consideration and characterization of the human heart, the most recent, most voluble, most changeable, most quaking part of creation to the observation of the oldest, most stable, deepest, most unshakeable son of nature."[117]

In the notes from this period (1784–85), Goethe ventured to formulate more explicitly some of the shaded considerations he had broached in his letter to von Stein. For instance, he tried to work a compromise between the Neptunists, who believed as Werner that different rocks precipitated out of a primitive ocean when the waters receded, and the Vulcanists, like Voigt, who thought that primitive rocks, such as granite and basalt, solidified from lava flows during ancient times. Goethe himself admitted that the process of granite's deposition remained shrouded in mystery and that one could not with confidence say whether this backbone of the earth was formed by water or fire.[118] He did, though, sketch out for himself a basic history of the formation of the earth that, while it included elements from both theoretical sides, inclined more to the Wernerian view.[119] These cosmological speculations undoubtedly gained focus from his discussions with

---

115. Goethe to von Stein (3 October 1779), in *Goethes Briefe*, 1:275–76.

116. Goethe, "[Verhältnis zur Wissenschaft, besonders zur Geologie]," in *Sämtliche Werke*, 13.2:236: "As one moves deeper into a subject, one sees how much the power of the subjective operates, even in the sciences; and one does not advance until one begins to know one's own self and character."

117. Goethe, "[Granit II]," in ibid., 2.2:505.

118. Ibid., 2.2:504.

119. Goethe, "[Epochen der Gesteinsbildung]," in ibid., 2.2:509–11.

Herder, who at this time was working on his own theory of the earth's formation for his *Ideen*.[120]

Goethe supposed that the earth was initially in a fluid condition, with an internal fire keeping minerals in solution. Gradually the core of the earth crystallized; and then granite, precipitating out of a boiling sea, crusted over the core in large and many-layered, blocklike crystals. The cooling seas deposited other rock forms (such as mica) against the granite; and these minerals sometimes ran through the granite, indicating a rapid sequential deposition. After a long period, parts of these original mountains decomposed and the detritus formed secondary mountains (as in the Harz range). This Neptunist position spoke to Goethe of gradually maturing formations, while Vulcanism suggested youthful eruptions of fire. So just as von Stein was teaching him calmness in his life, so did calmness enter his science.[121] After the turn of the century, as Vulcanism (especially as advanced by Cuvier) gained force, Goethe yet remained faithful to his brand of Neptunism, as these lines written at the death of Werner (1817) indicate:

> Hardly had the noble Werner turned around,
> Than they destroyed the realm of Poseidon.
> Though all now before Hephaestos do bow down,
> I am not among those to be relied on.[122]

The Unity of Biological Nature: Goethe's Discovery of the *Zwischenkiefer* in Human Beings

Goethe first met Justus Christian Loder (1753–1832), professor of medicine at Jena, in 1780. He attended several of the physician's anatomical demonstrations during the year of their meeting and the next. They also belonged to the same Masonic lodge, advancing through the various grades

120. Herder thought the earth had gone through many different epochs in which the forces of water, fire, and wind operated to change its surface dramatically. See Herder, *Ideen zur Philosophie der Geschichte der Menschheit* (pt. 1, bk. 1, chap. 3) in *Herder Werke*, 6:29–32 [I.ii:4].

121. Shortly after the turn of the century, when he learned that limestone deposits in France had yielded fossils with various portions of living representatives, Goethe had the inspired notion that fossil strata might be used to date different rock formations. Unfortunately, the mix of fossil strata near Weimar made such a project then impossible. He prophesied, however, that "the time will soon come, when you will not toss fossils randomly together, but will arrange them according to their epoch." The quotation comes from George Wells, "Bergbau in Ilmenau und Anschten über Gebirgsbildung," in *Goethe Handbuch*, 3:658–59.

122. Goethe, "Zahme Zenien," in *Sämtliche Werke*, 13.1:224: "Kaum wendet der edle Werner den Rücken / Zerstört man das Poseidaonische Reich, / Wenn alle sich vor Hephaestos bücken / Ich kann es nicht sogleich." This was probably written at the time of Werner's death in 1817.

Figure 10.6
Justus Christian Loder
(1753–1832), physician
to Duke Carl August and
collaborator with Goethe
on his early
morphological studies.
Engraving by J. G.
Mueller after a portrait
(1801) by F. A. Tischbein.
(Courtesy Wellcome
Institute Library.)

together.[123] In fall of 1781, from late October through early November, Goethe visited Loder almost daily in order to study anatomy with him, since the poet wished to perfect his artistic rendering of the human form. He made use of that knowledge at the Weimar Drawing Academy, where he gave lessons in anatomical drawing during the winter of 1781–82. But his interest in anatomy quickly advanced beyond its utility for artistic expression. He continued to work with Loder, but now in more general areas, particularly in comparative anatomy. Then in winter of 1783–84, his anatomical researches became more focused as a result of his discussions with Herder, who at that time was working on the first part of his *Ideen zur Philosophie der Geschichte der Menschheit*. Their mutual interest centered on a rather obscure anatomical question, namely, whether human beings, like other animals, had an intermaxillary bone—*Zwischenkiefer* in German. In most apes, dogs, rodents, and other vertebrates, this bone runs along the nasal opening of the upper jaw and has the incisor teeth set in it. Contemporary authorities—such as Pieter Camper (1722–1789), Samuel Thomas von Soemmerring (1755–1830), and Johann Friedrich Blumenbach (1752–1840)—had denied the existence of a *Zwischenkiefer* in human beings.

123. Carl August was also a lodge brother. The Masons were suppressed in mid-1780 throughout the Germanies because of alleged republican sympathies.

They offered up this bone as a natural sign of man's radical separation from the animals (just as a half century later Richard Owen would argue that the hippocampus of the human brain provided the same kind of distinction). At the end of March 1784, Goethe resolved the question differently than these experts. He had found the presence of the bone in the human embryo and believed he could trace its faint residue in an adult skull. He quickly wrote von Stein of his wonderful observation, which produced such pleasure as to "move all my innards."[124] He believed that Herder would be particularly delighted, since the discovery supported his friend's belief in the unity of nature. Goethe wrote him on the same day, 27 March 1784:

> According to the instruction of the Evangelist, I must quickly let you know of a happiness that has befallen me. I have found—neither gold nor silver, but something that makes me unspeakably glad—the *os inter-maxillare* in man! While comparing human and animals skulls with Loder, I came on its tracks and saw it there. Now please don't mention any of this, since it must be kept secret. It should also make you right glad as well, since the capstone for man isn't lacking, it's there. Indeed, I have thought about it in relation to your whole [book], how wonderful it will be.[125]

From fall of 1784 through spring of 1785, Goethe did extensive research on the skulls of a variety of animals—elephant, hippopotamus, bear, anteater, horse, rabbit, walrus, armadillo, ape, lion, and, of course, man.[126] During this same period, Herder had been consulting with him—often seeking his consolation and encouragement—over his long contemplated *Ideen*.[127] According to Herder's conception, the earth went through a vast series of developments—a scheme that Kant thought fraught with evolutionary dangers (see chapter 5). That Kant was correct can be inferred from a remark that Charlotte von Stein made to a friend: "Herder's new work," she wrote on 1 May 1784, "makes it probable that we were first plants and animals. What nature will further stamp out of us will remain well un-

---

124. Goethe to Charlotte von Stein (27 March 1784), in *Goethe-Briefe*, 3:17.

125. Goethe to Johann Gottfried Herder (27 March 1784), in *Goethes Briefe*, 1:435–36.

126. His notes on this research are in *Sämtliche Werke*, 2.2:545–62. Some of the material he got from collections at the university in Jena; other skulls he borrowed from acquaintances, such as Soemmerring.

127. Herder wrote his friend Hamann, at the end of his labors on the *Ideen*, that "if my wife, the author of the author of my writings, and Goethe, who happened to see the first book, had not unceasingly encouraged and driven me, everything would have remained in the Hades of the unborn." See Johann Gottfried Herder to Johann Georg Hamann (10 May 1784), in *Herders Briefe*, ed. Wilhelm Dobbek (Weimar: Volksverlag, 1959), p. 231.

Figure 10.7 Remains of the *Anatomieturm* in Jena, where Goethe did research on the *os intermaxillaris* while working with Loder. Originally the tower was part of a thirteenth-century fortress, which was converted into an anatomical theater in 1750; it was destroyed in 1860. (Photograph by the author.)

known. Goethe now messes around [*grübelt*] thoughtfully in these things, and anything that first has passed through his imagination becomes extremely interesting."[128] Herder argued that these transitions in form and the patterns they exhibited revealed an underlying plan, a unity of organization running throughout nature. "Now it is undeniable," he observed, "that given all the differences of living creatures on the earth, generally a certain uniformity of structure and a *principal form* [*Hauptform*] seem more or less to govern, a form that mutates into multiple varieties. The similar bone structure of land animals is obvious: head, rear, hands, and feet are generally the chief parts; indeed, the principal limbs themselves are formed according to one prototype and vary in almost infinite ways."[129] This similarity, he asserted, even extended to the *os intermaxillare*, "the final feature of the human countenance" shared with animals.[130] Herder did not men-

128. Charlotte von Stein to Karl Ludwig Knebel (1 May 1784), in Heinrich Düntzer, ed., *Zur deutschen Literatur und Geschichte: Ungedruckte Briefe aus Knebels Nachlass*, 2 vols. (Nürnberg: Bauer und Naspe, 1858), 1:120. Knebel (1744–1834) was a court administrator and friend also to Goethe.

129. Herder, *Ideen zur Philosophie der Geschichte der Menschheit*, in *Herder Werke*, 6:73 [I.ii:4].

130. Ibid., 6:119 [I.iv:1].

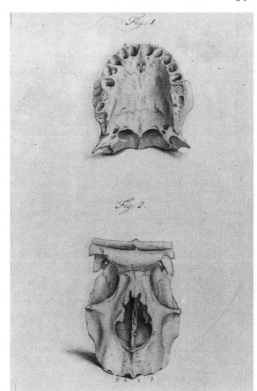

Figure 10.8
Wilhelm Waitz's
illustrations prepared for
Goethe's essay on the
Zwischenkiefer; they show
the faint sutures of the
intermaxillary bone
(which holds the incisors)
in a human skull. From
Goethe, *Die Schriften zur
Naturwissenschaft:
Morphologische Hefte*.

tion that his friend had made this discovery, but the acknowledgment and the implicit endorsement of the finding meant much to Goethe.

During spring of 1784, Goethe, with the help of Loder, composed a small tract that would later be given the title "Dem Menschen wie den Tieren ist ein Zwischenknochen der obern Kinnlade zuzuschreiben" (An intermaxillary bone of the upper jaw is ascribed to man as well as to animals).[131] In this essay he noted that recent authors, such as Camper and Blumenbach, had denied the presence of the intermaxillary in man, but that older writers had recognized it. He indicated that Galen (c. 131–c. 210), for instance, had described the bone and that Andreas Vesalius (c. 1514–1564) had included a woodcut in his *De humani corporis fabrica* (1543) that depicted sutures outlining the bone.[132] He admitted, though, that Vesalius believed the sutures

131. Goethe, "Dem Menschen wie den Tieren ist ein Zwischenknochen der obern Kinnlade zuzuschreiben," in *Sämtliche Werke*, 2.2:530–45.

132. See Galen, *de ossibus*, in *Claudii Galen Opera Omnia*, ed. D. Carolus Kühn, 20 vols. (Leipzig, 1821–33), 2:749. Galen does not name the bone; he merely describes the sutures in the palate and around the nasal area. Likely, as Vesalius maintained, he was using an ape skull as a model for the human. See the next note.

did not penetrate deeply enough to form a separate structure, as is the case in animals.[133] Goethe asserted that the bone could be seen more distinctly in the human embryo, though he certainly believed it was discernable in the adult. In his tract he also traced out its different configurations in a variety of animals but noted that in some the bones of the jaw had so completely fused that the intermaxillary appeared absent. For his essay he commissioned a young Weimar artist, Johann Christian Wilhelm Waitz (1766–1796), to illustrate the bone in several animal skulls, as well as in man; and he asked Loder to translate the essay into Latin. The tract, however, was not published until 1820, when it appeared as part of Goethe's multipart collection *Zur Morphologie* (1817–24).

Goethe sent his essay to his friend Johann Heinrich Merck, who, at Goethe's request, forwarded it to Soemmerring and Camper. Goethe anxiously sought the approval of these experts. He wrote to Soemmerring: "Since my little treatise is not intended for publication and should be regarded merely as a proposal to consider, I would be greatly pleased to learn whatever you wish to impart concerning my project."[134] If the authorities approved, he undoubtedly would have sought to have his treatise published, despite his disclaimer; since, after all, he went to the trouble and expense of hiring an illustrator, of having the treatise translated into Latin, and of having a fair copy made. The authorities at first seemed reluctant to alter their opinion about the bone's absence in human beings; and Soemmerring even tried to talk him out of the thesis, perhaps partly out of his disdain for Loder.[135] Anticipating these responses, Goethe wrote Merck: "I am confident that a learned man of the profession will deny his own five senses. They seldom have to do with the living concept of the thing, but only what is said about it."[136] This attitude came to characterize Goethe's relationship to professional scientists. They often politely deferred to him, without taking him too seriously, which gravely offended him. This sort of

133. For the woodcut, see Andreas Vesalius, *De humani corporis fabrica* (Basel: Ioannus Oporinus, 1543), p. 38 (fig. 2, chap. 9, bk. 1). Vesalius thought Galen based his description on an animal skull, not a human. He yet did admit that sutures outlining the bone in man could be discerned. He wrote: "Lateral to this foramen [nerve opening in the palate] sutures appears, or rather lines, which in the young [*pueris*] are filled by cartilage; they stretch across to the canine teeth. But truly nowhere do they penetrate so that by reason of these sutures the fourth area of the jawbone might be regarded as being divided, which (as I will indicate shortly) occurs in dogs, apes, and pigs" (ibid., p. 41). From this passage it seems that Vesalius recognized that in the young this bone appears more distinctly.

134. Goethe to Samuel Thomas Soemmerring (6 March 1785), in *Goethe und Soemmerring: Briefwechsel 1784–1828*, ed. Manfred Wenzel (Stuttgart: Gustav Fischer Verlag, 1988), p. 51.

135. Goethe to Johann Heinrich Merck (13 February 1785), in *Goethes Briefe*, 1:471: "I have received a gentle letter from Sömmerring. He wants to talk me out of it. Oi!"

136. Goethe to Merck (8 April 1785), in ibid., 1:475.

patronizing dismissal became especially pronounced in relation to his later studies in optics. But in this instance, the reaction was not as negative as Goethe anticipated (and is commonly believed). Camper, for instance, admitted Goethe's discovery of the *Zwischenkiefer* in the walrus, and Soemmerring added a note to the second edition of his *Lehre von den Knochen* (Treatise on the bones, 1800) that seemed to endorse Goethe's theory of the existence of that bone in man.[137] Blumenbach also made some concession. In his *De generis humani varietate nativa* (1775), one of the tracts that originally stimulated Goethe to think about the *Zwischenkiefer*, Blumenbach had simply observed that this bone "man is without, although all the apes and most of the other mammals have it."[138] Goethe communicated his findings to Blumenbach and initially got a rather accommodating response. The letters are lost, but Goethe referred to them in a note to Soemmerring: "Professor Blumenbach's letters were very welcome to me. You can easily believe they have strengthened me in my fixed idea. I am amazed that he let himself be so easily sidetracked from the path that he was once on."[139] Likely as the result of Goethe's essay, Blumenbach revisited the question of the *os intermaxillaris*. In a book published the year after he had read Goethe's essay, his *Geschichte und Beschreibung der Knochen des menschlichen Körpers* (History and description of the bones of the human body, 1786), he now admitted that in the human fetus or the young child "and often in adults . . . a faint track of the *os intermaxillaris*, which is found in other mammals, is indicated." He also noted that he had in his possession a small ape skull that gave hardly any sign of intermaxillary sutures.[140] And in the new

---

137. To an observation in his text that a certain area of the human skull was similar to the *Zwischenkieferbein* in animals, Soemmerring appended a note remarking that while the bone did not exist in a separate state in the adult human, he had noticed it in a three-month embryo. He went on to say: "Goethe's empirically rich investigation, which concludes that the intermaxillary of the upper jaw is common to man and the other animals—an investigation based on a comparative treatment of the bones, written in 1785 and with very precise illustrations—deserves to be published." See Samuel Thomas Soemmerring, *Vom Baue des menschlichen Körpers. Erster Theil. Knochenlehre*, 2nd ed. (Frankfurt a. M.: Varrentrapp und Wenner, 1800), pp. 201–2. Later, in preparing a new edition of this work, Soemmerring struck out the passage about Goethe in his own copy of the book. See *Samuel Thomas Soemmerring Werke*, ed. Jost Benedum and Werner Kümmel, 21 vols. (Stuttgart: Gustav Fischer Verlag, 1997–), 1:474–76.

138. Johann Friedrich Blumenbach, *De generis humani varietate nativa*, in *The Anthropological Treatises of Johann Friedrich Blumenbach*, trans. and ed. Thomas Bendysche (London: Longman, Green, Longman, Roberts & Green, 1865), p. 92.

139. Goethe to Samuel Thomas Soemmerring (6 March 1785), in *Goethe und Soemmerring*, p. 51.

140. Johann Friedrich Blumenbach, *Geschichte und Beschreibung der Knochen des menschlichen Körpers* (Göttingen: Johann Christian Dieterich, 1786), pp. 194–95. Blumenbach also, in a note, gave a short history of the discussion of the intermaxillary, a discussion he considerably expanded in the third edition (1795) of his *De generis*.

edition of his *De generis* (1795), he expanded his discussion of the bone from one line to three pages, giving the history of the controversy over it (without mentioning Goethe by name) and detailing his own further research on the subject. He concluded that while the human embryo did show the *os intermaxillaris* as separate, the adult human skull exhibited only vague outlines but no true sutures. He also pointed out, however, that in several ape skulls he investigated, the sutures were also absent. There was then really very little difference between Blumenbach's position and Goethe's. Both granted that the *os intermaxillaris* did not serve as a natural sign of human superiority.[141]

Blumenbach and Soemmerring were not the only professionals to concede something to Goethe. In 1779 Félix Vicq-d'Azyr (1748–1794), a young comparative anatomist and physician to the queen, had noted "traces [of the intermaxillary] in the os maxillarius superiorius of the human fetus."[142] He published, though, a more decisive account in his *Discours sur l'anatomie* (1786), where he remarked that the bone was "separated from the os maxillarius by a small suture, which is quite distinct in the fetus but hardly visible in the adult."[143] Soemmerring assumed that Vicq-d'Azyr's elaboration of the claim was the result of having gotten wind of Goethe's work.[144] Whatever might have been the stimulus for the Frenchman's pronouncement, it is quite clear that in the wake of Goethe's essay, the question of this small feature of the human countenance had taken on large proportions.

Goethe was quite proud of his discovery and became rather irritated when Camper and Soemmerring—and even his friend Merck—seemed to ignore it.[145] But the true significance of Goethe's work lay not in the simple presence or absence of the bone—about which there was actually little dif-

---

141. Blumenbach, however, complained to Soemmerring that Herder, as wonderful as his *Ideen* was, yet had made some serious errors. He thought Herder had misinterpreted what he had said in the *De generis humani varietate nativa*—he "misunderstood completely many things and too quickly composed his book, e.g., apropos the *os intermaxillaris* in my treatise." See Blumenbach to Samuel Thomas Soemmerring (3 May 1785), in Rudolph Wagner, *Samuel Thomas von Soemmerrings Leben und Verkehr mit seinen Zeitgenossen*, 2 vols. (Stuttgart: Gustav Fischer Verlag, [1844] 1986), 1:307.

142. Vicq-d'Azyr cites his earlier observation in his *Discours sur l'anatomie*, vol. 4 of *Oeuvres de Vicq-D'Azyr*, ed. J. L. Moreau, 6 vols. (Paris: L. Duprat-Duverger, 1805), 4:159.

143. Ibid., 4:26–27.

144. In a letter to Johann Merck (11 November 1786), Soemmerring remarks: "As I hear, Vicq-d'Azyr has taken up Goethe's so-called discovery in his work. I have not yet seen the book." The letter is in *Soemmerring's Leben und Verkehr mit seinen Zeitgenossen*, 1:293.

145. This polite dismissal has been repeated by some later historians, particularly those migrating from biology. Nordenskiöld, for instance, remarks that Goethe "imagined he had 'discovered' the intermaxillary bone in man." Nordenskiöld regarded all the "facts" of the matter having been already established by Vesalius. See Erik Nordenskiöld, *The History of Biology* (New York: Tudor, 1936), p. 280.

ference of opinion—but in the manner in which he pursued his research and in the meaning of his findings. He examined the issue of the human jawbones developmentally and in wide-ranging comparison, whereas the authorities took a more static, restricted, and implicitly theological view. He regarded human worth not to depend on a simple part but on the harmony of the whole arrangement. He made this more generous consideration explicit in a letter to his friend, the court administrator Karl Ludwig von Knebel (1744–1834):

> I have received a response—Herder in his *Ideen* remarks that one cannot find any differences between man and animal in any particulars. Moreover, man is very closely related to the animals. Unity of the whole makes every creature into that which it is. Man is man as well through the form and nature of his upper jaw as he is man through the form and nature of the tip of his little toe. And thus is every creature only one tone, one hue of a great harmony, which one must thus study in the whole and at large, lest every particular become a dead letter. I have written this small treatise from that point of view, and that is really the interest that lies hidden therein.[146]

Though the professional scientists did not embrace Goethe's findings with quite the enthusiasm he desired, his appetite for scientific research was not stanched. He next began to investigate the relationship between plants and animals. His microscopic observations, carefully made during April and May 1786, convinced him that minute animals metamorphosed from bits of plants.[147] This interest in the transmutation of species was undoubtedly stimulated by Herder's quasi-evolutionary ideas. At this time he also began reading Linnaeus, whose books he would take on his travels to mining areas.[148] His study of the Linnaean system was abetted by a new acquaintance, the botanist August Johann Batsch, for whom he secured a university position. Comparable to his assumption of a unity of plan in the animal kingdom, Goethe began to think about a unity of organization in

146. Goethe to Karl Ludwig Knebel (17 November 1784), in *Goethes Briefe*, 1:459.

147. Goethe's notes on the generation of infusorial animals are contained in *Sämtliche Werke*, 2.2:563–80. He speculated that various seedlike organisms if exposed to light became plants, but if kept in the dark became animalculae. He soaked various kinds of plant matter (bits of beans, potatoes, mold, and the like) in clear water in a container to see if infusoria could be generated. He watched over several days and indeed got various transformations. Thus on 12 April: "very small spherical animals. The grains of mold seem to become transparent and to be transformed into infusorial animals" (ibid., 2.2:566).

148. See, for instance, his mention of this to von Stein (9 November 1785), in *Goethes Briefe*, 1:489. In the first volume of his *Zur Morphologie* (1817), Goethe recalled that during this Weimar period, "Linnaeus's *Philosophie der Botanik* was my daily study." See Goethe, *Zur Morphologie*, in *Sämtliche Werke*, 12:22.

the plant kingdom; at least this is what is suggested by a letter to von Stein in summer of 1786:

> I am especially delighted about those creatures of the plant world. . . .
> The enormous realm [of plants] simplifies itself in my soul, so that I can
> dispatch the most difficult task quickly. If only I were able to convey the
> insight and the joy to someone, but it is not possible. And it is no dream,
> no fantasy. It is recognition of the essential form [*Gewahrwerden der
> wesentlichen Form*] with which nature always plays, and in such playing
> brings forth the variety of life.[149]

Goethe's botanical studies would intensify in Italy, to which he would shortly escape. His letter to von Stein indicates that he had already formulated the notion of an *Urpflanze*, the model of the essential plant, whose living embodiment he would attempt to discover in Italy. Goethe's notion of an ideal form, an archetype, by which one could understand the variety of organisms, stemmed from his discussions with Herder and his serious study of the philosopher who exerted so much power over the two friends, Baruch de Spinoza (1632–1677).

The Impact of Spinoza

Heinrich Heine called Goethe "the Spinoza of poetry."[150] He could have added "the Spinoza of science" as well. Spinoza's philosophical work, commonly regarded by the orthodox as deterministic and atheistic, confirmed Goethe's own inclinations and gave them a more profound articulation. For Goethe, Spinoza constituted the rejected stone of the builder that became the cornerstone of a new conception and approach to science. Several aspects of Spinoza's thought touched Goethe at his core, and they became his own: first, that God and nature were fundamentally identical, *Deus sive natura* in Spinoza's terms; second, and closely related, that individual objects have both a material and a spiritual or ideational side—the monistic assumption; third, that one might grasp the spiritual side of things, their essential idea or archetype, in an intuition (something at which Goethe hinted in his letter to von Stein); fourth, that individual parts of an object, or of nature, could be understood only in relation to the whole of which

---

149. Goethe to von Stein (9–10 July 1786), in *Goethes Briefe*, 1:514.

150. Heinrich Heine, *Zur Geschichte der Religion und Philosophie in Deutschland*, in *Sämtliche Schriften*, ed. Klaus Briegleb, 3rd ed., 6 vols. (Munich: Deutscher Taschenbuch Verlag, 1997), 3:618. Heine suggested that Goethe's pantheism, the Spinozistic element, spread throughout his poetry.

they were parts; and fifth, that imagination could easily mislead the careful scientist. Spinoza's disparagement of imagination, typical of the rationalist's evaluation of that faculty, came to characterize Goethe's attitude as well.[151] His view, though, would change in the mid-1790s, after he read Kant's *Kritik der Urteilskraft* (Critique of judgment, 1790), which made imagination an essential faculty for both art and science. Beyond these particular notions, Spinoza's conception of the *amor Dei intellectualis*, that deeply intuitive relation of the individual mind to God-Nature, became emblematic of Goethe's own love and pursuit of nature. The identification of the self with nature meant, as Goethe came to believe, that deep within the soul pathways could be found to hidden aspects of nature and that discoveries in the one would lead to revelations in the other.[152] These features of Spinoza's thought would come also to abide with the Romantics, especially with Schelling and Friedrich Schlegel, who would appreciate this Jewish hyperrationalist in Goethe's embracing light.

Goethe first read Spinoza's *Ethica* in spring and summer of 1773, perhaps to escape the bondage of the passions that constricted him during the wedding of Lotte Buff. Likely his first discussions with Herder and Merck at that time encouraged him to "to see how far I can follow the fellow in the shafts and tunnels of his thought."[153] It would appear that his initial crawl through the more profound passages of the book left a residue, an attitude that there was something quite significant in this Jewish thinker. He undertook serious study of the philosopher, however, only in Weimar, during the winter of 1784–85, when he and von Stein together read the "holy" Spinoza. He took up the book again, it would seem, because of the famous athe-

151. For instance, in an important essay of 1792, which remained unpublished till 1822, Goethe cautions against drawing hasty conclusions from experiments, for along the way one might be ambushed by enemies from within: "Here at this pass, at the transfer from experience to judgment, from knowledge to application is where all of man's internal enemies lie in wait: imagination [*Einbildungskraft*], which lifts him to the heavens on its wings, when he believes he is still on solid ground, impatience, haste, self-satisfaction. . . ." See Goethe, "Der Versuch als Vermittler von Objekt und Subjekt" (The experiment as mediator between object and subject), in *Sämtliche Werke*, 4.2:326. I will discuss this essay in the next chapter.

152. See Goethe, "[Reine Begriffe]", in ibid., 4.2:332–33. In this short unpublished essay of 1792, Goethe argues that the harmony of our internal nature with external nature provides a means to uncover features of external nature that might otherwise remain hidden from us: "Just as our eyes have been harmoniously constructed in respect to visible objects and our ears with respect to the movement of vibrating bodies, so our spirit [*Geist*] stands in harmony with those simpler powers that lie deep within nature; and it is able to represent them to itself just as purely as the objects of the visible world are formed in a clear eye" (4.2:332).

153. Goethe to Ludwig Julius Höpfner (7 April 1773), in *Goethes Briefe*, 1:148. See also Martin Bollacher's discussion of Goethe's relation to Spinoza in his "Baruch de Spinoza," *Goethe Handbuch*, 4.2:999–1003.

ism controversy that broke out at this time. The controversy was instigated by Goethe's old friend Friedrich Heinrich Jacobi (1743–1819).[154]

After Lessing's death in 1781, the philosopher Moses Mendelssohn (1729–1786) decided to write a memoir of the poet. Before he could complete this biography of his friend, however, Jacobi wrote him a long letter describing verbatim a conversation he supposedly had with Lessing, in which the latter revealed his adherence to the doctrines of Spinoza.[155] In the letter he mentioned that he showed Lessing a poem, as yet unpublished, called "Prometheus." He did not tell Lessing the poem was by Goethe. In the poem, Prometheus rails against Zeus and the other gods, protesting that it is only superstition that keeps them nourished. Jacobi related that Lessing not only liked the poem but agreed with its unorthodox sentiments and admitted his further accord with Spinoza. When Mendelssohn read Jacobi's letter, he became quite irritated—they had been at odds with one another for some time. He maintained that Lessing's relationship to Spinoza was more benign than Jacobi was suggesting.[156] Mendelssohn indicated that he would convey to the public his own moderate view of Lessing's Spinozism and that he would respond also to Jacobi's accusations. In some haste, Jacobi tried to anticipate Mendelssohn with his own book, which would contain his correspondence with Mendelssohn, his account of his conversation with Lessing, and his cautionary observation that, as Lessing exemplified, a reason unguided by faith would inevitably lead to Spinozistic atheism.[157] Jacobi's book would be published, in 1785, under the title *Über die Lehre des Spinoza in Briefen an den Herrn Moses Mendelssohn* (On the doctrine of Spinoza in letters to Mr. Moses Mendelssohn).[158] But before he brought it out, Jacobi sent his manuscript to Goethe and Herder for their reactions.

154. Jacobi first met Goethe in 1774. He wrote in his diary for 24 July 1774: "Mr. Goethe had insulted me in print, but he also has written the drama *Götz von Berlichingen*! We shook hands. I looked upon the most extraordinary man, full of high genius, brilliant imagination, deep sensitivity, quick moods, whose strong and sometimes tremendous spirit completely makes its own way." See *Goethe in vertraulichen Briefen seiner Zeitgenossen*, 1:61. Goethe described his friend to Eckermann this way: "Jacobi was really a born diplomat, a good-looking man, a slender figure, with a refined and excellent nature, who as an ambassador would have been entirely in the right place. As poet and philosopher, something was lacking for him to be both." Goethe to Eckermann (11 April 1827), in *Gespräche mit Goethe*, p. 209.

155. Friedrich Heinrich Jacobi to Moses Mendelssohn (4 November 1783), in *Friedrich Heinrich Jacobi Briefwechsel*, ed. Michael Brüggen, Heinz Gockel, and Peter-Paul Schneider, 3 vols. (Stuttgart: Friedrich Frommann Verlag, 1987), 3:227–46.

156. Mendelssohn to Jacobi (1 August 1784), in ibid., 3:339–47.

157. For an extended discussion of the controversy between Jacobi and Mendelssohn, see Frederick Beiser, *The Fate of Reason: German Philosophy from Kant to Fichte* (Cambridge: Harvard University Press, 1987), pp. 44–91.

158. Friedrich Heinrich Jacobi, *Über die Lehre des Spinoza in Briefen an den Herrn Moses Mendelssohn*, ed. Marion Lauschke (Hamburg: Felix Meiner Verlag, 2000).

Jacobi visited Goethe and Herder in September 1784. It was their discussion of the controversy that brought Goethe to undertake his own study
of Spinoza. In November Goethe wrote Knebel: "I'm reading the *Ethics* of
Spinoza with Frau von Stein. I feel very near to him, though his spirit is
much deeper and purer than mine."[159] Herder took part in these readings
and as a Christmas present gave Goethe and von Stein a Latin edition of
the *Ethica*. It carried the inscription: "Let Spinoza be always for you the
holy Christ."[160] Goethe kept Jacobi informed about his study: "I practice
my Spinoza, and I read and reread him again, and wait with anticipation
until the battle over his body will break out."[161] This Spinoza practice led
Goethe to compose a short essay on the philosopher, which he dictated to
von Stein sometime in early 1785.[162] The essay adapted features of Spinoza's system to Goethe's own particular approach to nature. He endorsed
the basic Spinozistic thesis that God and nature were one, so that the world
must be both divine and natural simultaneously. Each individual, Goethe
held, has its raison d'être within itself. An adequate idea would indicate
why the individual had to exist; yet, as he maintained, every individual is
linked to every other, so that the conditions of the whole required the existence of each individual. Hence, all entities must also be conceived in relation to the whole. We cannot, however, grasp the infinite but only form
limited conceptions of the world.[163]

Goethe cultivated his Spinozism in company with Herder, with whom
he fundamentally agreed; both understood that philosophy to have bearing
on the comprehension of nature. For Goethe, the Spinozistic approach to
anatomy meant that one had to examine the range of animal skeletons in
comparative fashion in order to come to an adequate idea, or archetype, of
the animal skeleton. Having achieved such an idea would then indicate
how each of the skeletal parts related internally. If the human skeleton, for
instance, exhibited a pattern comparable to other vertebrates, then one

---

159. Goethe to Knebel (11 November 1784), in *Goethes Briefe*, 1:459.

160. See the commentary by John Neubauer, in Goethe, *Sämtliche Werke*, 2.2:875.

161. Goethe to Jacobi (12 January 1785), *Goethes Briefe*, 1:470.

162. Goethe, "Studie nach Spinoza," in *Sämtliche Werke*, 2.2:479–82.

163. Goethe's interpretation of Spinoza is fairly sound. Spinoza maintained that every object
has its own idea, which forms its essence. To understand an object, then, is to entertain this idea in
an articulate way. But all physical objects and their concomitant ideas form two parallel, infinite
systems within the attributes of extension and thought. These attributes, along with an infinity of
others, constitute the substance that is God or nature. To have an adequate idea of an object, then,
requires extensive experience with bodies, so that we come to understand the necessity of the idea
and its logical relation to other ideas. The total, infinite system of such ideas constitutes God's
knowledge of himself, and insofar as we can begin to comprehend that system, we identify with
God. This is the state of intuitive knowledge (*scientia intuitiva*)and constitutes the intellectual love
of God (*amor Dei intellectualis*).

would expect that every kind of bone would be represented in the human frame, even if at an earlier developmental stage. Hence the expectation that human beings also exhibited an *os intermaxillaris*. Goethe confirmed to Jacobi the significance of Spinoza for his own science:

> When you say man can believe only in God, I say to you that I hold much with seeing. And when Spinoza speaks of *scientia intuitiva*, and says: "Hoc cognoscendi genus procedit ab adequata idea essentiae formalis quorundam Dei attributorum ad adaequatam cognitionem essentiae rerum" [This mode of knowing proceeds from an adequate idea of the formal essence of certain attributes of God to an adequate knowledge of the essence of things]—these few words give me courage to dedicate my whole life to the consideration of things that I touch and of whose formal essence I can hope to form an adequate idea, without worrying how far I will come and what is denied me.[164]

Jacobi had objected to the Enlightenment emphasis on the self-sufficiency of human reason, unaided by religious faith. Unbridled reason, he believed, led inevitably to atheism, as the works of Spinoza and Lessing had shown. Goethe objected vigorously to Jacobi's condemnation of Spinoza as an atheist; he thought the Jewish philosopher to be even, as it were, a "Christianist."[165] But just as vigorously did Goethe reject the assumption that reason needed the guidance of any revealed doctrines. Rather, he thought reason could of itself achieve a kind of divine transcendence when the individual, through the intellectual love of God, recognized the identity of self and nature, that is, self and God. This metaphysical equation would later be further established by members of the Romantic circle, with whom Goethe would be in accord on this question.

By summer of 1786, Goethe had become exhausted and frustrated. His court and administrative duties occupied enormous amounts of time. His love for Charlotte von Stein had settled into a comfortable companionship, with no prospect of his passion being requited; he undoubtedly feared his virginal state might become permanent. His many literary efforts during this Weimar period fell scattered into various unfinished piles, an accumulated experience of *intellectus interruptus*; he thus came to regard his artistic life as nigh on virginal as well. Perhaps as a means to convince himself that his talent had not completely withered, he had contracted, in summer of 1786, to have a small edition of his collected works brought out, including

164. Goethe to Jacobi (3 May 1786), in *Goethes Briefe*, 1:508–9. For the quoted passage, see Baruch de Spinoza, *Ethica*, II, 40, ii, in *Opera*, ed. J. Van Vloten and J. Land, 3 vols. (Hague: Martinus Nijhoff, 1890), 1:104.

165. Goethe to Jacobi (9 June 1785), in *Goethes Briefe*, 1:476.

rough drafts and occasional pieces.[166] Much later he told Eckermann that "in the first ten years of my Weimar ministerial and court life, I accomplished virtually nothing."[167] His scientific and philosophic excitement flickered and threatened also to die away. And he was getting older, approaching his thirty-seventh birthday. He felt he had become time's fool. There was, Goethe believed, but one way out. He had cultivated in his heart since childhood images of a land his father once visited, where disappointment and Teutonic gloom might vanish under a sun-drenched sky; it was a land becoming enshrined in German consciousness as a sensuous retreat where the inhibitions of the north could be shed as easily as a heavy woolen cloak on a warm summer's day. In his unfinished *Wilhelm Meisters theatralische Sendung*, over which he labored at this time, the enigmatic Mignon, an Italian adolescent of mysterious origin, sings a beautiful song of longing:

> Do you know the land where the lemon trees flower,
> Where in verdant groves the golden oranges tower?
> There a softer breeze from the deep blue heaven blows,
> The myrtle still and the lovely bay in repose.
> Do you know it?
> There! There!
> Would I go with you, O my master fair.[168]

Goethe formed secret plans to escape to that land where the lemon trees flowered—to Italy. On 24 July 1786 he set out for Carlsbad, the social and

166. Goethe also had more pragmatic reasons for signing on with the new publisher Georg Joachim Gröschen to bring out a small, eight-volume collection of his works. He needed funds for his planned trip to Italy; and he wished, as well, to forestall the pirated editions of his uncompleted works, especially the poems and plays that had been performed at court and published in the court newsletter. The contract called for an advance of two thousand Reichs dollars, a goodly sum. He completed the manuscripts of the first four volumes just before his departure, leaving it to his friends to read proofs as they came from the press the next year. The final volumes were published in 1790. Goethe had his secretary, Philippe Seidel, do most of the necessary copying of previously published material (*Werther, Götz, Geschwister*, etc.). For the latter he, rather amazingly, had Seidel use a pirated edition, which he had attempted to correct; many changes introduced by the pirate, nonetheless, crept through to the Göschen edition. For a fascinating account of Goethe's relations with his several publishers, see Siegfried Unseld, *Goethe and His Publishers*, trans. Kenneth Northcott (Chicago: University of Chicago Press, 1996).

167. Goethe to Eckermann (3 May 1827), in Eckermann, *Gespräche mit Goethe*, p. 539. He adds that "despair drove me to Italy." Only in light of Goethe's massive output, however, would the first ten years in Weimar seem to have been unproductive.

168. Goethe, *Wilhelm Meisters theatralische Sendung*, in *Sämtliche Werke*, 2.2:170: "Kennest du das Land, wo die Zitronen blühn, / Im grünen Laub die Gold Orangen glühn, / Ein sanfter Wind vom blauen Himmel weht, / Die Myrte still und froh der Lorbeer steht, / Kennst du es wohl? / Dahin! Dahin! / Mögt ich mit dir O mein Gebieter ziehn." In the published version, *Wilhelm Meisters Lehrjahre*, "master" (*Gebieter*) becomes "lover" (*Geliebter*).

therapeutic resort to which the court—including von Stein and Herder—had already decamped. He remained there for the celebration of his birthday on 28 August. But then at three o'clock on the morning of 3 September, after posting notes to the duke and his immediate friends, he slipped away from the company and boarded a coach that would take him to that southern country, where he hoped for a rebirth.

Goethe's Italian Journey: Art, Nature, and the Female

The rebirth that Goethe sought in Italy would begin with his own artistic conception of the journey: he traveled with pen in hand, recording his impressions, thoughts, and reactions in diaries and letters that he prepared for von Stein and his circle of friends in Weimar. He had planned to use this written material to compose a volume, which materialized, however, only during his later years.[169] The *Italienische Reise* re-creates that period that gave his life a new beginning and a new meaning. This rebirth and the artistry of its telling are signaled by the motto Goethe chose for his book: "Auch ich in Arcadien" [I, too, am in Arcadia]. He appears to have taken this resonant epigram from a painting by Guercino (Giovanni Francesco Barbieri, 1591–1666), whose works he viewed in Cento and in Rome. The picture in question portrays two shepherds in an Arcadian setting. They are gazing at a skull atop a rock slab, perhaps a tomb, which is inscribed with the legend "Et in Arcadia ego"—death, too, can be found in paradise. In no less dramatic terms, Goethe would portray his entrance into the Arcadia of Italy, where he would experience a death, that of his former self, and a rebirth.[170] At the beginning of his trip, he wrote von Stein that

169. When he returned to Weimar in June 1788, Goethe began to lose heart for publishing his travel writings; he became, as he told Herder, revolted at this "pudenda," which is "a very dumb matter that now reeks." See Goethe to Herder (July–August 1788), in *Goethes Werke* (Weimar Ausgabe), 133 vols. (Weimar: Hermann Böhlau, 1888), IV.9:9. After completing the third part of his *Dichtung und Wahrheit*, in 1814, he renewed his intention to publish material from the period that so changed his life. The first two parts of his *Italienische Reise* came out in 1816 and 1817 and the less redacted third part not until 1829. He wished the book to be regarded as the continuation of his autobiography.

170. Nicolas Poussin (1594–1665) used the epigram "Et in Arcadia ego" in a painting similar to Guercino's. Poussin's composition, though, suggests that the legend marks the grave of another shepherd, who was "also once in Arcadia." In a famous essay on the epigram, Erwin Panofsky maintained that Goethe did not know the provenance of the phrase and that in his use of the motto "the idea of death has been entirely eliminated." See Erwin Panofsky, *"Et in Arcadia ego*: Poussin and the Elegiac Tradition," in his *Meaning in the Visual Arts* (Chicago: University of Chicago Press, [1955] 1982), p. 319. I believe Panofsky to be wrong on both counts. Goethe would use a similar epigram for the third installment of his autobiography, *Campagne in Frankreich* (see chapter 2), the story of his part in the German attempt to quash the French Revolution. The motto of that latter work—"Auch ich in Campagne"—suggests he was there amidst a great deal of death. The epigram stood on the first two installments of the *Italienische Reise* (1816, 1817), and conveys, perhaps, not only

the rebirth that is transforming me from the inside to the outside continues apace. I thought I would, indeed, learn something here, but that I
would have to return to primary school, that I would have to unlearn so
much—well, I didn't count on that.[171]

And on the denouement of the journey, he explained to Carl August the rationale for this reeducation and its scope:

> The chief reason for my journey was to heal myself from the physical-
> moral illness from which I suffered in Germany and which made me use-
> less; and, as well, so that I might still the burning thirst I had for true
> art. . . . When I first arrived in Rome, I soon realized that I really under-
> stood nothing of art and that I had admired and enjoyed only the pedes-
> trian view of nature in works of art. Here, however, another nature, a
> wider field of art opened itself to me.[172]

He might have added, as he did in other letters to his patron, that a wider
field of the female also opened before him.[173] The *Italienische Reise* and
other material from the time thus speak insistently about the major features
of his rebirth and, indeed, the great themes of his life: women, art, and nature. What must be argued, though, is the intimately affective and causally
transforming relations instantiating these features. The changes, the *Bildung* he experienced, though retrospectively poured by Goethe into the
two years of his journey, spilled over into the next decade and a half of his
life, when his new self developed in the cultural milieu of his friendship
with Schiller and his interactions with the Romantic group at Jena.[174] But
the changes do have their source in his travels through Italy.

Goethe's Itinerary and the Roman Community

Goethe crossed the Alps at the Brenner Pass into Italy on 9 September
1786 and moved south, reaching Verona on 16 September and Venice on
28 September. He remained in Venice for two weeks, enjoying its wonderful natural and human environment, and departed on 14 October. He

---

the idea of Goethe's death and rebirth, but also the wistful recognition that a long-departed self had
once visited paradise.

171. Goethe to Charlotte von Stein (20 December 1786), in *Goethes Briefe*, 2:33.

172. Goethe to Carl August (25 January 1788), in ibid., 2:78.

173. I will discuss this subject below.

174. The fifteen-year period after Goethe's return from Italy has traditionally been designated
"Weimar Classicism," a name that reveals more the distaste of later scholars for the Romantic
movement, as they understand it.

reached Florence on 23 October, but only stayed for two days. Rome was his destination, and he entered the eternal city on 1 November 1786. Goethe remained in Rome for four months. He then moved south, arriving in Naples on 25 February 1787. He loved this beautiful city, though finally, after a month, yielded to the pull of the southern sun. He sailed to Sicily, reaching Palermo on 2 April. He traveled first to the western part of the island, along the coast, and then turned east, moving across country and finally north to Messina, reaching the city on 10 May. He lingered in Messina for three days and then sailed back to Naples, suffering a near shipwreck along the way. He docked on 14 May and remained in the city for three weeks, before the lure of Rome drew him on. Goethe would remain in Rome this second time for almost a year. In late April he reluctantly packed up his belongings and headed back home. He entered Weimar on 18 June 1788, almost two years after he had originally departed.

The initial goal and stabilizing anchor of Goethe's travels in Italy was Rome, the city his father embellished in so many tales during Goethe's childhood. Shortly after reaching his destination, on 1 November 1786, he made the acquaintance of the German community of artists, with whom he quickly dropped his protective pseudonym of Herr Möller. The community, though initially in awe of the great poet amongst them, adopted him as a friend and provided an intimate circle, which he always remembered with great affection.[175] The company included the art critic Johann Heinrich Meyer (1760–1832), a student of Johann Joachim Winckelmann (1717–1768) and someone with whom Goethe would have a lasting relationship; the painter Angelika Kauffmann (1741–1807), whom Herder would call "perhaps the most cultivated woman in Europe"[176]—her portrait of Goethe, for whom she had taken a fancy, rendered him quite dashing; Friedrich Bury (1763–1823) and Johann Heinrich Lips (1758–1817), both of whom would later paint portraits of an older Goethe; and Johann Friedrich Reiffenstein (1719–1793), another pupil of Winckelmann and, by reason of age and wisdom, the padrone of the group. The two members of the community that most influenced Goethe were the painter Johann Heinrich Wilhelm

---

175. Often enough others complained that he kept fairly exclusive company with the German community of artists, as did Cardinal Count Herzan to Prince Kaunitz (24 March 1787), in *Goethes Gespräche*, 1:415.

176. Johann Gottfried Herder to Karoline Herder (28 October 1788), in Herder, *Italienische Reise: Briefe und Tagebuchaufzeichnungen, 1788–1789*, ed. Albert Meier and Heide Hollmer (Munich: Deutscher Taschenbuch Verlag, 1988), p. 401. On his own trip to Italy, Herder became quite entranced with this gifted painter. Goethe also enjoyed many an afternoon strolling through Roman galleries with her.

Tischbein (1751–1829), who would leave memorable portraits of his new friend (especially *Goethe in der Campagna*), and the psychological writer Carl Philipp Moritz (1757–1793). Tischbein later recalled their first meeting in Rome, when the poet, in green coat and warming himself by the fire at an inn near St. Peters, got up and went to him saying: "I am Goethe."[177] The artist would be Goethe's roommate and guide in Italy, and the one who most educated his eye for painterly styles and techniques. Tischbein's considerable abilities seem to have eventually convinced Goethe, by painful comparison, of his own lack of real talent in the medium. Shortly after Goethe met Moritz, this melancholic fellow suffered a badly broken arm. With great solicitude, Goethe visited his new friend, sometimes twice daily, during the six weeks of Moritz's convalescence.[178] They perhaps spoke about the newly published first sections of Moritz's powerful autobiographical novel *Anton Reiser* (1785–90), the eponymous hero of which is transformed by reading *Werther*. Goethe's personal kindness transformed the author himself, who seems never to have before experienced such unselfish care; he came virtually to worship his benefactor. As he wrote to a friend: "To associate with him fulfills the most beautiful dreams of my youth, and his appearance, like a beneficent genius, in this realm of art is for me, as for many others, an unhoped-for joy."[179] Goethe, for his part, regarded the man, as he wrote von Stein, like a "younger brother," someone who "stood like a mirror before him," since Moritz also had escaped to Italy because of unrequited love.[180] This sad writer was a constant companion to the poet during their time in Italy, and Goethe came to so respect Moritz's critical sense that he included an essay on aesthetics by his friend in the *Italienische Reise*.

177. Tischbein to Goethe (1821), quoted by Petra Maisak, "Wir Passen Zusammen als Hätten Wir Zusammen Gelebt," in *J. H. W. Tischbein, Goethes Maler und Freund*, ed. Hermann Mildenberger (Neumünster: Karl Wachholtz Verlag, 1986), p. 24.

178. Carl Phillip Moritz to J. H. Campe (20 January 1787), *Goethe in vertraulichen Briefen seiner Zeitgenossen*, 1:326.

179. Moritz to Campe (20 November 1786), in *Goethes Gespräche*, 1:406. Goethe-worship was something that deeply irritated Herder. When Moritz later came to Weimar to visit Goethe, Herder wrote his wife: "His talk [Moritz's] . . . exhibits neither clarity nor sprightliness, and he is fundamentally a depressed, sick individual. His considerations mean nothing to me. . . . With Goethe it's otherwise, since Moritz is a fool over him and has aimed his entire philosophy to deifying him as the totality of humanity. This is all due to the history of their Roman existence, when Moritz was very depressed and Goethe must have seemed like a god to him." See Johann Gottfried Herder to Karoline Herder (10 February 1789), in *Goethe in vertraulichen Briefen seiner Zeitgenossen*, 1:383.

180. Goethe to Charlotte von Stein (14 December 1786 and 20 January 1787), in *Goethes Briefe*, 2:28–29, 45.

Figure 10.9
Goethe in the apartment
he shared with J. H. W.
Tischbein on the Via del
Corso in Rome. Aquarel,
chalk, and pencil sketch
(1787) by J. H. W.
Tischbein. (Courtesy the
Goethe Museum,
Frankfurt.)

"Without love . . . Rome would not be Rome"[181]

As Goethe entered the city of Rome, he expected the very stones to cry out the genius of their past. But he also listened most attentively for the whisper of a beckoning female voice. Charlotte von Stein had not responded physically to his overtures, and so he abandoned the northern chill for a warmer climate that promised more. Hardly had he stepped on Italian soil, when he began to be on the watch. In Vincenza he noticed some "quite pretty creatures, especially the sort with black curly hair"; and in Venice he observed "some beautiful faces and figures."[182] During his second stay in Rome, he fell for a striking Milanese girl, whom he tutored in English; but

181. Goethe, *Römische Elegien*, I, in *Sämtliche Werke*, 3.2:39.
182. Goethe, *Italienische Reise* (22 and 29 September 1786), in *Sämtliche Werke*, 15:67, 80.

he underwent a Werther-like moment when he learned she was already betrothed.[183] And in Naples he would become entranced by Emma Lyon, the mistress of the English ambassador, Sir William Hamilton. Hamilton was fifty-seven at the time and she twenty-two, "very beautiful and well-built." She entertained Hamilton's guests by striking a series of "attitudes"—appearing draped in a shawl or in Greek costume, now in this pose, now in that. Her performances seemed to Goethe like a dream, one that Tischbein attempted to capture on canvas.[184] From Goethe's descriptions one can well understand how this woman could entice Hamilton, who loved pretty objects, into marrying her and later seduce Lord Horatio Nelson into a scandalous affair.[185]

And then, of course, there were the artists' models. They were generally "lovely and happy to be seen and to be enjoyed." Yet, as he wrote Carl August, "the French disease made this paradise uncertain."[186] By the middle of his second Roman stay, Goethe had grown quite disconsolate. He wrote the duke another letter, lamenting the condition in which the god Priapus had left him:

> The sweet, small god has relegated me to a difficult corner of the world. The public girls of pleasure are unsafe, as everywhere. The *zitellen* [the unmarried women] are more chaste than anywhere—they won't let themselves be touched and ask immediately, if one does something of that sort with them: *e che concluderemo* [and what is the understanding]? Then either one must marry them or have them married, and when they get a man, then the mass is sung. Indeed, one can almost say that all the married women stand available for the one who will take care of their families. These, then, are the wretched conditions; and one can sample only those who are as unsafe as the public creatures. What concerns the heart doesn't belong to the terminology of the present chancellery of love.[187]

183. Ibid. (October 1787), 15:506–11.

184. Ibid. (16 March 1787), 15:258.

185. Goethe was also interested in Hamilton's other kind of erotica. The English ambassador had excavated statuary dedicated to Priapus in Isneria, about fifty miles outside of Naples. Hamilton's friend Richard Payne Knight described this collection of ancient erotica in his *Account of the Remains of the Worship of Priapus* (London: Spilsbury, 1786). Goethe was familiar with Knight's work and likely examined the collection itself when he viewed Hamilton's "secret art treasury." See Goethe, *Italienische Reise* (27 May 1787), in *Sämtliche Werke*, 15:400. Goethe's poem cycle *Römische Elegien* (see below) originally had as introduction and envoi two poems in which the garden god Priapus brandishes his tool.

186. Goethe to Carl August (3 February 1787), in *Goethes Briefe*, 2:48.

187. Goethe to Carl August (29 December 1787), in ibid., 2:75. See two notes above for reference to the "sweet, small god." The letter continues with a passage that has led to some gross exaggerations concerning Goethe's supposed homoerotic tendencies. This is the passage: "After this

Figure 10.10
Emma Lyon
(1765–1815), mistress
and later wife of Lord
William Hamilton and
mistress of Lord Horatio
Nelson. Portrait (1788)
by Angelika Kauffmann.
Tischbein, who also used
her as a model for a
number of classical
depictions, thought she
had "a beauty that one
seldom sees, and the only
one I have seen in my
life." (Courtesy Archiv
für Kunst und
Geschichte.)

Goethe no doubt showed proper caution, even against the swelling tide of powerful desire. This was not the sort of caution that Carl August himself practiced, however, and he strongly admonished his friend to follow suit.[188] Remarkably, Goethe appears to have taken the advice to heart, or at least fortuitous occasion accomplished what friendly council had prescribed. On 16 February 1788, he wrote Carl August:

---

contribution to the statistical account of this land, you will understand how poor our conditions must be and will be able to grasp a peculiar phenomenon, which I've nowhere seen as strongly present as here. This is the love of men for one another. Presumably it is seldom driven to the highest level of sensuality, rather lingering in the middle region of desire and passion. So I can say that I have seen with my own eyes the most beautiful examples thereof, comparable to what we have from the Greek heritage (see Herder's *Ideen*, volume III, p. 171); and as a watchful researcher, I have been able to observe the physical and moral features of the phenomenon. This is a matter about which one can hardly speak, let alone write, so let it be reserved for future conversation." Later (1805), in his homage to Winckelmann, Goethe would remark: "Thus we find Winckelmann so often in the company of beautiful young men, and he never seemed as alive and lovable as when with them, if often only for a fleeting moment." See Goethe, *Winkelmann und sein Jarhundert*, in *Sämtliche Werke*, 6.2:356. Robert Tobin strikes an appropriate balance of recognizing Goethe's interest in homosexuality, while setting this interest within the thoroughly heterosexual inclinations and practices of this female-obsessed poet. See Robert Tobin, "In and Against Nature: Goethe on Homosexuality and Heterotextuality," in *Outing Goethe and His Age*, ed. Alice Kuzniar (Stanford: Stanford University Press, 1996), pp. 94–110.

188. Wilhelm Bode, *Goethes Leben: Rom und Weimar, 1787–1790* (Berlin: Mittler & Sohn, 1923), p. 95. Carl August's actual letter has not survived, but its content is easily surmised from Goethe's response.

Your good advice, transmitted 22 January directly to Rome, seems to have worked, for I can already mention several delightful excursions. It is certain that you, as a doctor *longe experientissimus*, are perfectly correct, that an appropriate movement of this kind refreshes the mind and provides a wonderful equilibrium for the body. I've had this experience more than once in my life, and also have felt discomfort when I have deviated from the broad road to the narrow path of abstinence and safety.[189]

The adolescent boast of the last line undoubtedly suggested to the duke, as it does to us, that Goethe had only recently acquired more than theoretical knowledge of that delightful motion.

One of the greater mysteries of Goethean scholarship has been the identity of his Italian mistress—assuming, of course, that his representation to Carl August had substance. In the *Römische Elegien*, the poem cycle that he began either during his last weeks in Italy or shortly after his return to Weimar, the poet sings of his affair with a young Italian woman named Faustine.[190]

Then one day she appeared to me, a brown-skinned girl,
Whose dark, luxuriant hair tumbled down from her brow.[191]

Blessedly, she knows nothing of Lotte and Werther. They nightly honor the god Amor "devilishly, vigorously, and seriously."[192] The object of this affection has been not been identified with any certainty; though, from the evidence of the poems, she seems to have been a young widow of modest circumstance with a small child, an individual presumably safe enough

189. Goethe to Carl August (16 February 1788), in *Briefwechsel des Herzogs-Grossherzogs Carl August mit Goethe*, ed. Hans Wahl, 2 vols. (Berlin: Ernst Siefgried Mittler und Sohn, 1915), 2:117.

190. The *Römische Elegien* were originally published as a twenty-poem song cycle in Schiller's *Horen* in 1795, though completed, it seems, by 1790. Goethe had initially composed a cycle of some twenty-two poems, bearing a draft title of *Erotica Romana*. Schiller deemed two of the poems (numbers 2 and 16 of the original design) to be too sexually explicit, and so Goethe withheld them from publication (e.g., from the second poem: "We are delighted by the joy of the true, naked Amor / As well as by the lovely creaking music of the rocking bed"). Two other elegies, apparently meant for preface and envoy to the cycle, have been found in manuscript. Goethe himself seems to have had second, prudential thoughts about these latter, so called "Priapus Elegies." Aside from matters of public taste, their rough, vulgar tone jars with the gently erotic feeling of the other poems—e.g., the poet threatens any bluenose that should he protest "the fruits of pure nature, he would be stuck from behind with the red rod that sprouts from Priapus's hips." The elegies and their variations are printed in Goethe, *Sämtliche Werke*, 3.2:38–82. Eva Bernhardt discusses the literary character of the poems, especially their structure as a cycle, in *Goethe's Römische Elegien: The Lover and the Poet* (Frankfurt a. M.: Peter Lang, 1990). I will return to the poems below.

191. Goethe, *Römische Elegien* (no. 4), in *Sämtliche Werke*, 3.2:44: "Einst erschien sie auch mir, ein bräunliches Mädchen, die Haare / Filen dunkel und reich über die Stirne herab."

192. Ibid., 3.2:45: "Schalkhaft, munter und ernst begehen wir heimliche Feste. . . ."

from the French disease.[193] Yet the persona of the poems seems to mask a second woman as well, Christiane Vulpius, whom he met back in Weimar shortly after his return and who quickly became his mistress and later officially his wife.[194] During the period of his reworking of the poems, he rejoiced in love and sexual pleasure with this new mistress, and her scent lingers over their final composition. In this way, Goethe's art transformed the artist himself, that is, he became in Weimar the lover the poems describe; and to further this metamorphosis, he refused to staunch speculation about the authenticity of the events depicted in the poems. The cycle, though, celebrates Rome and the experiences of the poet in that city—and it is, to that extent, biographically revealing, even if the original experiences have been transmuted through a resexualized imagination.[195] More importantly, though, these poems also express Goethe's deeply felt convictions about the relationship of the female ideal to art and science—a complex matter I will discuss in greater detail below.

The Artist in Rome

Goethe traveled to Rome with a satchel full of manuscripts, undoubtedly expecting the new milieu to liberate an inspiration that had been stunted in colder climes. He also had the task of making these manuscripts publishable for the edition of his collected works. Before leaving Weimar, he had a decent draft of a prose version of his play *Iphigenie auf Tauris;* but Italy warmed him to the more classical format of a verse rendition, which he completed in the middle of December 1786. The next year he finished another play previously set aside, his *Egmont,* a tale that celebrates moments

---

193. Boyle (*Goethe,* 1:506) tentatively identified her as Faustiṇa Antonini, an innkeeper's daughter; but this has been disputed by Roberto Zapperi in his *Das Inkognito: Goethes ganz andere Existenz in Rom,* trans. Ingeborg Walter (Munich: Beck, 1999), p. 206. After considerable detective work (pp. 201–38), Zapperi is yet quite convinced of the real existence of Goethe's Roman affair.

194. See Bernhardt, *Römische Elegien: The Lover and the Poet,* p. 21. Given the concrete descriptions of the individual (a young, brown-skinned widow with a child, as elegies 4 and 6 indicate) and the specific Roman surroundings, the poet must have had Faustine in mind as much as Christiane. This is further confirmed by the title he initially had planned for the published poems—*Elegien. Rom, 1788;* moreover, he described his poems to his publisher Göschen (4 July 1791) as "the little book I wrote in Rome" (*Goethes Werke,* IV.9:276). The poems finally appeared in *Die Horen* only with the title *Elegien.* The title *Römische Elegien* was attached to the 1806 edition. See the introduction to the elegies by Hans-Georg Dewitz, in *Sämtliche Werke,* 3.2:444–53.

195. I will simply reiterate here an observation made earlier, namely, that while seeking biographical clues in works of art is risky, there is no reason not to make prudential judgments. As R. G. Collingwood urged against the practices of pedestrian historians, the epistemic value of any document—whether it be letter, book, or recorded conversation—must be judged on the basis of evidence. Imaginative literature should be no different. Poetry, especially Goethe's, can be as revealing, sometimes more revealing, than, say, polite letters that disguise deeper sentiments.

of happiness even in the darkening shadow of mortality. Italian musical heritage, as well as popular songs heard on every street corner, provided the atmosphere, if not the needed genius, for rewriting two operettas he had earlier composed, *Erwin und Elmira* and *Claudine von Villa Bella*.[196] He worked on his early draft of *Faust* and completed a few more scenes that gave the drama a new orientation; but his vision did not quite embrace the whole, and *Faust* remained a fragment. During the last months of his sojourn, however, he all but finished the play *Torquato Tasso*. Italy obviously sharpened a pen that had become dull with producing official documents and frustrated love letters.

When he came to Italy, Goethe also brought his sketchpads and always attempted to quicken his eye to the art that could be found in museums and even beneath his feet in that ancient land. Tischbein, Meyer, and later his new artist friend Christoph Heinrich Kniep (1748–1825) provided constant instruction not only in art history but in the practice of sketching and painting. To this end, he was especially keen to pursue again the study of anatomy, which he had begun with Loder in Weimar. Soon after settling in Rome, he undertook a detailed investigation of the human skeleton, with all its attached ligaments; and he continued this kind of study after his return for the second visit to the city in June 1787.[197] Toward the end of his journey, Goethe reluctantly came to the conclusion that he had only a modest talent for painting. He simply was not able to accomplish what Tischbein and Kniep could in a few deft strokes. Yet his efforts at acquiring artistic skill revealed to him how such practical experience proved a necessary condition for the intuition of ideal structures that might lie more deeply below surface appearances.[198]

Goethe studied the paintings and statuary of the ancients with Winckelmann's *Geschichte der Kunst des Altertums* (History of ancient art, 1764) as his *vade mecum*. Winckelmann disdained those critics who constructed their histories and judgments not from the meticulous examination of the works themselves but from journals and books. This insistence on immediate experience of objects appealed to Goethe, though Winckelmann's considerations of what constituted ideal beauty and how it should be realized— namely, through ideal proportions of different bodily parts, representations of dress, and other marks of a distinctively Greek culture—left Goethe

---

196. In private notebooks, he remarks on the various songs circulating through Rome. See *Auszüge aus einem Reise-Journal*, in *Sämtliche Werke*, 3.2:198–214.

197. Goethe, *Italienische Reise*, in *Sämtliche Werke*, 15:194.

198. Werner Bush investigates, with considerable sensitivity, the transforming experience Goethe's study of art had on his mode of thought. See his "Die 'grosse, simple Linie' und die 'allgemeine Harmonie' der Farben," *Goethe-Jahrbuch* 105 (1988):144–64.

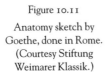

Figure 10.11
Anatomy sketch by
Goethe, done in Rome.
(Courtesy Stiftung
Weimarer Klassik.)

rather confused and uncertain.[199] As a result he became initially convinced of a sharp boundary between the realm of art and the realm of nature. Just after receiving a new translation of Winckelmann's book in early December, he wrote Duchess Luise (wife of Carl August) to explain his abrupt departure from Weimar. In the course of his account of the artistic riches of Rome, he observed

> that nature is more congenial and easier to view and evaluate than art. The lowliest product of nature has a sphere of perfection within itself, and I need only have eyes to see in order to discover these relationships; I am convinced that a whole, true existence is enclosed within this small circle. A work of art, on the other hand, has its perfection outside of itself. The "best" of it is in the idea of the artist, which he seldom or never achieves, and follows from certain assumed laws, which are derived from the character of art and of craft. But these latter are not so easy to understand and decipher as the laws of living nature. In works of art, there is

199. See Johann Joachim Winckelmann, *Geschichte der Kunst des Altertums* (Darmstadt: Wissenschaftliche Buchgesellschaft, [1764] 1972) , pp. 168–88. Much later, Peter Eckermann mentioned to Goethe that he thought Winckelmann's book not particularly clear in its argument. Goethe responded: "You're quite right; one finds him usually groping about. Yet this is what is so great about him: his groping always points to something. He is like Columbus before he discovers the New World—he has presentiments. You *learn* nothing when you read him, but you *become* something." See Eckermann, *Gespräche mit Goethe* (16 February 1827), p. 208.

tradition; the works of nature, however, are always like words immediately spoken by God.[200]

Winckelmann had insisted that artistic genius harbored an ineffable ideal of beauty that could never be given adequate expression. A given work of art might only dimly reflect that beauty, refracting it through historically determined styles of realization. Greek art—sculpture in particular—did, however, achieve the most luminous instantiations of ideal beauty. And this is why ancient art offered models for the cultivation of taste and aesthetic judgment. The great critic contended that only after coming to appreciate ideal beauty as exemplified in the work of the Greeks could one begin aesthetically to judge nature according to the proper standard.[201] Goethe endorsed these Winckelmannian notions, though his letter to Luise suggests they were not his spontaneous convictions.[202] He continued to immerse himself in the art of the ancients, attempting to discover therein the ideals of beauty that his guide so evocatively described. And as he confronted the art of different centuries—from the ancient period to the Renaissance and modern times—he came quickly to appreciate what Winckelmann taught about the historical character of artistic expression. "No judgment in this field is possible," Goethe wrote in January 1787, "unless one has made it in light of historical development."[203] Thus the relativity of artistic styles had to be distinguished from the essentiality of nature, whose eternal structures could be read from the surface of their appearances—at least this was Goethe's opinion at the beginning of his journey. Only after he had reflected more deeply on the character of nature, especially as the result of his experiences in Sicilian parks and Roman arms,

200. Goethe to Duchess Luise (12–23 December 1786), in *Goethes Briefe*, 2:31. Though Goethe undoubtedly felt the beauty of nature more directly than the beauty of art, his views perfectly reflected also those of Winckelmann, who remarked: "The differences of opinion [about what constitutes beauty] show themselves more in the judgments over constructed beauty in art rather than over the beauty of nature itself." See Winckelmann, *Geschichte der Kunst des Altertums*, p. 140.

201. Johann Joachim Winckelmann, *Gedanken zur Nachahmung der griechischen Werke in der Mahlerey und Bildhauerkunst*, 2nd ed. (Stuttgart: Reclam, [1756] 1995), p. 13: "Does it not follow that it is easier to discover the beauty of Greek statuary than the beauty of nature, and that, indeed, the former is calmer, not so fragmented, but more unified than the latter? The study of nature must, therefore, be at least a longer and more difficult path to knowledge of perfect beauty than is the study of the ancients."

202. Bush ("Die 'grosse, simple Linie'") argues persuasively that Goethe rather submissively acknowledged certain conventions of landscape painting then regnant (many adopted by Moritz), whereas his own spontaneous descriptions of nature in his book have a considerably more intuitive feel and break through such conventions. Winckelmann's fundamental aesthetic principle—that an ideal beauty lay deep in the artist's mind—would, however, be confirmed by Kant's philosophical views, which Goethe would later embrace.

203. Goethe, *Italienische Reise* (28 January 1787), in *Sämtliche Werke*, 15:200.

would his notions of nature and art undergo yet further development. By the time he returned to Weimar, Goethe would forcefully reverse Winckelmann's priority of art over nature and would come to see as well the essential in art and the historical in nature—and all of this as confirmed by an Italian girl.

The Morphological Conception of Nature and Art

On Ash Wednesday, 21 February 1787, just after the madness of the Roman Carnival had subsided, Goethe wrote Frau von Stein one of his typically hesitating letters, professing both love ("all the fibers of my being depend upon you") and enervation ("today I am confused and rather weak").[204] He was getting ready to depart for Naples and Sicily, and his preparations reminded him of what he had left in Germany. He was drawn irresistibly south, where grew the lemon trees whose remembered fragrance had produced great longing in Mignon. Those lands promised adventure as well—he had learned that ancient Vesuvius was spewing fire.

He arrived in Naples on 25 February and a week later set out to climb the fabled volcano—as his great admirer Haeckel, in conscious imitation, would three-quarters of a century later. In his first attempt to reach the new crater, the sulfurous fumes drove him back. He tried again four days afterward, this time with a reluctant Tischbein in tow. A small eruption showered them with ashes and rocks, and one large stone came close enough to produce that frisson of imminent death that Goethe on this and other occasions found exhilarating. He would return once more two weeks later to watch the lava flow from the new site.[205]

The dramatic experience of wildly beautiful and sublimely dangerous nature subtly suggested to Goethe the poverty of the merely artistic imagination, Winckelmann's assertions notwithstanding. The calmer scenes of beauty had no less a striking impact. A few days after his last experience with Vesuvius, he was returning to Naples in a small coach driven by a young lad. They came upon an elevated view of the city, in all its magnificence, with the harbor in the distance. The boy let out a yell, which startled Goethe, who turned in irritation to the rough Neapolitan youngster. But the boy, pointing his finger toward the city, exclaimed, "Sir, forgive me. This is my native land!" Goethe confessed, "Tears came to the eyes of this poor northerner."[206]

---

204. Goethe to Charlotte von Stein (21 February 1787), in *Goethes Briefe*, 2:50.
205. See Goethe, *Italienische Reise* (2, 6, 20 March 1787), in *Sämtliche Werke*, 15:229–30, 235–40, 265–67.
206. Ibid. (23 March 1787), 15:275.

Goethe and his new friend, the painter Kniep, left Naples by boat for Palermo on 29 March 1787. The sailing took five days, for most of which Goethe lay seasick and delirious in his cabin.[207] In the early morning of 1 April, a gale tossed the ship about, creating not a little fear in this man of German soil. However, by dawn the sky cleared, and at noon the coast of Sicily came into view. They landed in Palermo on 2 April, and would remain for two weeks in the city, with ventures into the countryside. Just before departing, a set of events occurred that gave impetus to Goethe's developing understanding of the relations holding among art, nature, and the female.

On 16 April 1787, Goethe went to the public gardens in the city to relax with a copy of the *Odyssey*. He was reading Homer as preparation for the composition of a play, tentatively titled *Nausikaa*. In Homer, Odysseus, naked and filthy, washes up on the shore of an island and immediately falls asleep. He is awakened by the play of a most beautiful girl, Nausicaa, daughter of King Alcinous. Once cleaned and properly dressed, the now magnificent Odysseus awakens keen desire in Nausicaa. She takes him to the garden of Alcinous, while she goes forth to prepare the court, where he will sing of his many adventures.[208] Goethe, who regarded the garden of Palermo more lovely than the imagined garden of Alcinous, returned the next day to continue his meditation on his proposed play, which ultimately never got beyond a few lines. But as he sat down to ruminate, before he knew it, as he recalled, "another spirit seized me, which had already been tailing me during these last few days." He gazed around the garden, and inquired of himself:

> Whether I might not find the *Urpflanze* within this mass of plants? Something like that must exist! How else would I recognize that this structure or that was a plant, if they were not all formed according to a model.[209]

Goethe thus moved in imagination from contemplating the lovely Nausicaa and the glorious Odysseus in the garden of Alcinous to the striking notion that in the real garden of Palermo he might discover a comparably beautiful and magnificent form.

Though Goethe believed he might actually find the *Urpflanze* in Sicily, his conception of this entity remained labile, as indicated by a hypothesis he formed at this time. With respect to the organic parts constituting the

---

207. A story later told to Zacharias Werners by Kniep, from Werner's *Tagebuch* (7 May 1810), in *Goethes Gespräche*, 1:421.

208. Homer, *Odysseus*, 6.

209. Goethe, *Italienische Reise* (17 April 1787), in *Sämtliche Werke*, 15:327.

Urpflanze, he supposed: "All is leaf, and through this simplicity the greatest multiplicity is possible." The leaf, he thus conjectured, might be transformable into all the other parts. In this sense, "a leaf that only sucks fluid under the earth we call the root; a leaf that spreads out from those fluids we call a bulb, an onion, for instance; a leaf that stretches out we call the stem," and so on for all the other parts of the plant.[210] But the leaf, as Goethe here mused, had to be understood symbolically: it represented a unitary dynamic force beneath the multiple transformations to which it gave rise—indeed, a short time later he would refer to "the leaf in its most transcendental sense."[211] Just so, the *Urpflanze* itself might be conceived as the dynamic idea that lay behind the variety of all plants. Much later, when he published the *Italienische Reise* in 1816, he wrote Christian Nees von Esenbeck that his friend would be bemused at the diary of his journey because "I sought at that time the *Urpflanze*, unaware that I sought the idea, the concept whereby we could develop it for ourselves."[212] But that awareness had already reached the penumbra of Goethe's consciousness, as his remarks on the leaf suggest. Spinoza had set him on the path that would lead to Kant and then to the Romantics.

A theme of this chapter has been the deep emotional and aesthetic connection between Goethe's experience of female forms—in literature and life—and his ideal biological structures. The eternal feminine and the eternal plant were for Goethe both ideals of beauty and models for the comprehension of their many empirical instantiations—illustrated in the former instance by the many women Goethe conjured up in his autobiographical writings and sung of in his poetry. Another tale from the *Italienische Reise* strongly evinces this conjunction. On 25 March 1787, just a few days before Goethe left for Palermo, Kniep invited him to the roof of his apartment to enjoy the magnificent view of the bay of Naples and the coast of Sorrento. As they were engrossed in the scene, all of a sudden, up from the trapdoor leading to the roof, a beautiful head emerged, that of Kniep's mistress. To Goethe she seemed like the angel of the Annunciation: "But this angel really had a beautiful figure, a pretty little face, and a fine natural comportment. I was delighted to see my new friend so happy under this wonderful

---

210. These lines come from brief notes Goethe made just after his time in the gardens of Palermo. See Goethe, *Zur Morphologie: Von den Anfängen bis 1795, Ergänzungen und Erläuterungen,* in *Goethe: Die Schriften zur Naturwissenschaft,* ed. Dorothea Kuhn, 21 vols. (Weimar: Hermann Böhlaus Nachfolger, 1977), 2.9a:58.

211. The phrase comes from unpublished notes that stem from late 1788. See "Einleitung," in *Sämtliche Werke,* 3.2:306.

212. Goethe to Christian Nees von Esenbeck (middle of August 1816), in *Goethes Werke,* IV.27:144.

sky and in view of the most beautiful region of the world."[213] Immediately after he narrated this story, Goethe went on to say that he took a walk along that magnificent seacoast, where he had a flash of insight concerning his botanical ideas. The passage reads: "I have come to terms with the *Urpflanze*; only I fear that no one will wish to recognize the rest of the plant world therein."[214] Beautiful nature, a beautiful woman, and the primal plant.

I have drawn these conjunctions out of Goethe's *Italienische Reise*, a book he composed of letters and diaries over a quarter of a century after the events spoken of took place. When he cast a melancholic gaze back over this time in Italy, he recalled it as that period in his life when he discovered "what it really was to be a man [*Mensch*]"; and in reference to his life in Rome, he judged that he had "never again been happy."[215] Whether the connections I have followed, then, were the actual fusions of ideas and motivational associations made during the period or whether they were later imposed by a reminiscent imagination is, of course, impossible to know, since few of the original letters and diaries have survived. Yet other, more contemporary sources are available, namely, new scenes for his *Faust* and his poem cycle *Römische Elegien*. He began both of these during his last weeks in Italy and worked further on them shortly after his return to Germany, while memory and longing were still green.

"I have softly beat out the measure of hexameters,
fingering along her spine"

A decade and a half prior to his Italian journey, Goethe had composed a draft of *Faust*, which he had only desultorily reexamined in those intervening years. But now, in Italy, he was determined to finish it, since he wanted to include it in the Göschen edition of his collected works. He added several scenes, yet he simply could not bring the work to completion, at least to his own satisfaction. It would be published with Göschen as *Faust, a Fragment*. The most important of the added scenes speaks to the theme of this chapter. In "A Witch's Kitchen," Mephistopheles brings Faust to a witch who will concoct a magic elixir to make him young again.[216] While in the kitchen, Faust gazes into an enchanted mirror and sees

213. Goethe, *Italienische Reise* (25 March 1787), in *Sämtliche Werke*, 15:276.
214. Ibid. (25 March 1787), 15:277.
215. Goethe to Eckermann (9 October 1828), in *Gespräche mit Goethe*, pp. 248–49.
216. The inserted scene remains, even in the later-completed play, incongruous, since at the beginning of the drama, Faust is a young man and hardly requires rejuvenation—it was the forty-year-old Goethe who, rereading his old play, felt the need for the elixir that Italy provided.

The image of a woman so very fair!
Can a woman be that exquisite?
Should I see in this body lying there
The essence of a heavenly visit?
Is there anything on earth to compare?[217]

After Faust quaffs the magic potion and wishes once more to look at the enticing image, Mephistopheles exclaims,

No! No! You shall soon enough see the model rise
Of every woman bodily before your eyes.[218]

And then adds as an aside:

With this drink you will see in flesh and nail
Soon lovely Helen in every female.[219]

In the Italian redraft of *Faust*, the force of the story begins to shift from that of the seduction of a girl—Gretchen—to the quest for an ideal of beauty and love. Goethe suggests that experience must drive down to the ideal, to the active force that lies in the depths of reality, a force that must be comprehended in order to construe its empirical appearances. Helen—the very form of beauty itself—is present in every female, something that Faust must comprehend, must come to see, even if too late in the case of Gretchen. The fundamental model, the force that furnishes the rationale for Faust's quest, is that of beauty, of Helen, of the very form of the female.[220]

Goethe's *Römische Elegien*, the tales of the poet's encounter with the Roman girl Faustine, express a similar theme, though with greater lyrical intensity. Of the twenty published poems of the cycle, the fifth achieves an elegance and beauty not otherwise matched. In this particular poem,

217. Goethe, *Faust, ein Fragment*, in *Sämtliche Werke*, 3.1:548: "Das schönste Bild von einem Weibe! / Ist's möglich, ist das Weib so schön? / Muss ich an diesem hingestreckten Leibe / Den Inbegriff von allen Himmeln sehn? / So etwas findet sich auf Erden?"

218. Ibid., 3.1:553–54: "Nein! Nein! Du sollst das Muster aller Frauen / Nun bald leibhaftig vor dir seh'n."

219. Ibid., 3.1:554: "Du siehst, mit diesem Trank im Leibe, / Bald Helenen in jedem Weibe."

220. In the late 1790s, when Goethe again took up his *Faust*, he added more material that modified the theme of the search for ideal beauty. In one addition (scene 7), Faust wagers that Mephistopheles will never produce a situation to cause him to say "linger awhile, you are so beautiful." Faust now feels driven to expanded his experience and to continue striving (*streben*), perhaps without ever realizing the ideal. The Romantic conceit that life is an eternal striving after the absolute—man as always becoming—left a deep impress on Goethe's drama. In the next chapter, I will indicate how Goethe himself interpreted his drama as in the Romantic mode. While conceding that the drama has many literary features in common with Romantic writing, Stuart Atkins denies it advances any Romantic view of life. I think Atkins shows a rather narrow understanding of that Romantic view. But see his "The Evaluation of Romanticism in Goethe's *Faust*," *Journal of English and Germanic Philology* 54 (1955): 9–38.

Goethe blends images, feelings, and ideas within alternating hexameter and pentameter lines that evoke the poetry of the Roman elegists. Yet the poem is distinctively Goethe's own: it erotically synthesizes experiences of classical sculpture and of a real woman, showing them to be constituted by the same eternal form.

> Happy, I feel myself inspired now in this classical setting;
> The ancient world and the present speak so clearly and evocatively
>   to me.
> Here I follow the advice to page through the works of the ancients,
> Daily with busy hands and with renewed joy.
> Ah, but throughout the nights, Amor occupies me with other matters.
> And if I wind up only half a scholar, I am yet doubly happy.
> But do I not provide my own instruction, when I inspect the form
> Of her lovely breasts and guide my hands down her thighs?
> Then I understand the marble aright for the first time: I think
>   and compare,
> And see with feeling eye, and feel with seeing hand.
> Though my beloved steals from me a few hours of the day,
> She grants me in recompense hours of the night.
> We don't spend all the time kissing, but have intelligent conversation;
> When sleep overcomes her, I lie by her side and think over many things.
> Often I have composed poetry while in her arms, and have softly
>   beat out
> The measure of hexameters, fingering along her spine.
> In her lovely slumber, she breathes out, and I inspire
> Her warm breath, which penetrates deep into my heart.
> Amor trims the lamp and remembers the time
> When he performed the same service for his three poets.[221]

---

221. Goethe, *Römische Elegien*, in *Sämtliche Werke*, 3.2:47: "Froh empfind' ich mich nun auf klassischem Boden begeistert, / Lauter und reizender spricht Vorwelt und Mitwelt zu mir. / Ich befolge den Rat, durblättre die Werke der Alten / Mit geschäftiger Hand täglich mit neuem Genuss. / Aber die Nächte hindurch hält Amor mich anders beschäftigt, / Werd ich auch halb nur gelehrt, bin ich doch doppelt vergnügt, / Und belehr ich mich nicht? Wenn ich des lieblichen Busens / Formen spähe, die Hand leite die Hüften hinab. / Dann versteh ich erst recht den Marmor, ich denk' und vergleiche, / Sehe mit fühlendem Aug', fühle mit sehender Hand. / Raubt die Liebste denn gleich mir einige Stunden des Tages; / Gibt sie Stunden der Nacht mir zur Entschädigung hin. / Wird doch nicht immer geküsst, es wird vernünftig gesprochen, / Überfällt sie der Schlaf, lieg ich und denke mir viel. / Oftmals hab' ich auch schon in ihren Armen gedichtet / Und des Hexameters Mass, leise, mit fingernder Hand, / Ihr auf den Rücken gezählt, sie atmet in lieblichem Schlummer / Und es durchglüht ihr Hauch mir bis ins tiefste die Brust. / Amor schüret indes die Lampe und denket der Zeiten, / Da er den nämlichen Dienst seinen Triumvirn getan." The poets mentioned in the last line are the Roman elegists Propertius, Catullus, and Tibullus. One must exercise

Even in translation, perhaps some impression of the whole has been preserved. A conceptual rendition of the poem further strips it of its aesthetic meaning. While recognizing that crucial liability, I will attempt one reading that coheres with much else that I have concluded thus far about Goethe's aesthetics and metaphysics. In the first part of the poem, the poet claims to understand great classical art only when he can embrace its living embodiment. The white, hard marble of the statue speaks to him only after he has experienced the brown, pliant flesh of the girl ("I inspect the form / Of her lovely breasts, and guide my hands down her thighs"). As a result, the visual aspects of each have taken on a new measure of tactility and, reciprocally, haptic awareness has been imbued with visual sensibility ("see with feeling eye, and feel with seeing hand")—thus he might "page through the works of the ancients" with eyes manually instructed. The first part of the poem, then, suggests that the sculptured marble and the living girl embody the same *Urtypus* of the female, a form necessary for both the actual creation of each—by the artist and by nature—and for the aesthetic comprehension of each. This attitude may be regarded as a kind of gloss on Winckelmann's interpretation of the Laocoon group: namely, "that the artist must feel in himself the strength of spirit that he impresses on his marble."[222] The second part of the poem relates how this newly understood dynamic form transmutes into another artistic instantiation, this time in poetry. In an unforgettable image, the poet lies in the arms of his lover, inspiring her living spirit and beating out the hexameters of a poem he is composing—the very poem we are reading?—by counting along her vertebrae. The poet is actually following the natural form of his lover in order to impress that same form on his words. The erotic power of nature thus transmutes into an artistic force that realizes beauty in another medium. Touch and vision, reason and sense combine in love to produce an aesthetic and intuitive understanding of the unity grounding nature and art. Well, there are obvious dangers in conceptually disassembling what exists only in poetic fusion.

---

considerable imagination to accept the suggestion of Sander Gilman that the "Roman Elegies" really have as their object a boy and that they detail in disguised form homosexual encounters. See his "Goethe's Touch: Touching, Seeing, and Sexuality," in his collection *Inscribing the Other* (Lincoln: University of Nebraska Press, 1991). Gilman thought that while in Rome, Goethe would have turned to boys to avoid venereal disease. Goethe's mistress—being a recent widow with young child—would, it seems, have been the better solution.

222. Winckelmann, *Gedanken zur Nachahmung der griechischen Werke*, p. 20.

Journey's End

The conclusion of the Italian journey marked a new beginning for Goethe as he returned to Weimar. The new beginning had several interrelated components. With Winckelmann, he maintained that the artist worked with an ineffable conception of beauty, whose expression would be conditioned by historical circumstance. Classical artists—whether Phidias or Homer—had achieved a realization of the archetype of beauty in a more objective way than modern artists. Goethe made this point in a letter to Herder upon returning from Sicily:

> Let me express my thoughts briefly. They [ancient writers] represent real existence, we usually represent its effects. They portray what is frightful, we portray frightfully; they the pleasant, we pleasantly, and so on. As a result, everything we produce is overdone, mannered, without real grace, a mess.[223]

Goethe then immediately confessed to his friend that he perhaps did not initially appreciate what Homer had wrought; but now, on his return from Sicily—his head reeling with images drawn from immediate experience of rocky coasts and sand-strewn bays—"the Odyssey for the first time has become for me a living word."[224] Just so the marbles of Polycleitos and Myron required the real experience of the lover's caress to appreciate what they had achieved. Only under the erotic authority of the living female could forms buried in marble come alive.

In the *Italienische Reise*, Goethe juxtaposes his remarks to Herder on the ancients' objectifying ability with a passage from a letter to von Stein and the Weimar group. The letter, actually written two weeks after the one to Herder, relates what he had learned about the *Urpflanze*:

> Tell Herder that I am very near to the secret of the generation of plants and their organization and that it is the simplest thing conceivable. Under these skies, one can make the most beautiful observations. Tell him that I have very clearly and doubtlessly uncovered the principal point where the germ [*Keim*] is located and that I am in sight, on the whole, of everything else and that only a few points must yet be determined. The *Urpflanze* will be the most wonderful creation on the earth; nature herself will envy me. With this model and its key, one can, as a consequence, dis-

---

223. Goethe, *Italienische Reise* (17 May 1787), in *Sämtliche Werke*, 15:393. This letter in the book is addressed to Herder; the original does not survive.

224. Ibid.

cover an infinity of plants—that is, even those that do not yet exist, because they could exist. It will not be some sketchy or fictive shadow or appearance, but will have an inner truth and necessity. The same law
[*Gesetz*] will be applicable to all other living things.[225]

In his *Italienische Reise*, Goethe juxtaposed his observations about the objective idea used by superior artists with this passage, in which he described
the objective idea—or law—used by nature. The connection between the
artist's idea and nature's idea is, via this conjunction, only implicit. But in
another letter to von Stein and his Weimar circle, written not long after the
one quoted above, he made the connection between the artist and nature
quite explicit:

> These great works of art are comparable to the great works of nature; they
> have been created by men according to true and natural laws. Everything
> arbitrary, imaginary collapses. Here is necessity, here is God.[226]

By the end of his journey, Goethe had begun to reverse the priority Winckelmann had suggested, namely, that artistic beauty was more fundamental
for aesthetic understanding than natural beauty. Goethe now proposed that
the objective idea employed by nature in the creation of beauty is that very
idea used by the artistic genius.

The notion that the artist operates according to the same laws as nature
received comparable expression in an essay that Goethe jotted down in his
travel diary at this time (and published shortly after his return). In "Ein-
fache Nachahmung der Natur, Manier, Styl" (Simple imitation of nature,
manner, style, 1789), he distinguished artists of modest ability, who would
imitate nature very precisely in their simple compositions, from those of
greater talent, who would discover within themselves a language, a manner,
by which to express more complex subjects. Artists of insight would, however, penetrate to the essential structures of their subjects, casting away
what was superficial. But the truly great artist would dialectically incorporate all of these stages and move beyond them. He or she would study the
varying phenomenal aspects of natural objects, penetrate to their essential
features, and bring the particular and universal together in a distinctive
manner. Artists manifesting this greater talent would be able to create in an
artistic medium what nature creates in her own sphere.[227] Goethe's asser-

225. Goethe to Charlotte von Stein (8 June 1787), in *Goethes Briefe*, 2:60; also partly in *Italienische Reise*, in *Sämtliche Werke*, 15:394.
226. Goethe, *Italienische Reise* (6 September 1787), in *Sämtliche Werke*, 15:478. The original of
this letter has not survived.
227. Goethe, "Einfache Nachahmung der Natur, Manier, Styl," in *Sämtliche Werke*, 3.2:186–
91: "Style rests on the deepest foundation of knowledge [*Erkenntnis*], on the essential nature of

tions in this essay again reversed the priority that Winckelmann had given to the imitation of ancient art. The great critic had proclaimed that immersion in ancient works would develop the kind of taste necessary to appreciate beauty in nature.[228] Goethe proposed that authority flowed in the opposite direction: namely, that perception of nature revealed a kind of beauty and reality that would illuminate the work of ancient authors. Despite this alternative emphasis, the stamp of Winckelmann on Goethe's thought is unmistakable. Goethe reached his conclusion about the artist working according to the same laws as nature through his reading of Winckelmann, but a Winckelmann sifted through his own experiences in Italy— and added to this mix a tincture of Spinoza. This conviction would leaven Goethe's biological science during the rest of his life. He would find a comparable conception, somewhat differently formulated, in Kant's third *Critique*, which he would take up not long after his return to Germany. And later, under Schiller's tutelage, he would come to admit that nature, as experienced, has its humanly constructed features, but also to recognize with the Romantics that there was a deep aspect of human nature that connected it with external nature, making the interchange between man and the world harmonious. As he epitomized this notion in his posthumously published maxims: "There is an unknown, lawlike something in the object that corresponds to an unknown, lawlike something in the subject."[229]

In this chapter I have been arguing that nature manifest in the form of the female had a command over Goethe, an erotic authority that directed him to a deeper understanding of that larger nature, especially as it opened

---

things [*Wesen der Dinge*], insofar as it is given to us to know that in visible and tactile forms" (3.2:188). Boyle (*Goethe*, 1:561–62) gives this essay an odd reading, seeing in it the view that "art, like life, was all matter, and externality" and reflected nothing of the internal life of the artist. He further maintains that Goethe believed that the aim of the artist was to create "a thing that gives pleasure to his patron." But the man who was at the time composing the *Römische Elegien* on the vertebrae of a young woman had not sacrificed an inner life; rather he found a language in himself that could formulate the laws that joined self to nature. The very concept of "style," as Goethe used it, meant an individual manner of expression, which in the great artist nonetheless revealed ideal beauty.

228. In his *Gedanken zur Nachahmung der griechischen Werke*, Winckelmann had argued that the surest way to develop taste would be in imitation of the art of the ancients rather than the imitation of nature. The former would ultimately lead not to mere imitation of ancient models but to the formation of an ideal of beauty, which one could then apply to nature (p. 15). Though Goethe reversed the priority in his essay on "Simple Imitation," he would have lent a ready ear to Winckelmann's environmentalism, according to which the warm sunny air of Greece invigorated naked youth, who by custom and inclination were given over to the kind of exercise that produced beautifully formed bodies. These well-muscled boys and healthy girls served as models for sculptors, stimulating in them the kind of sensitivity required to produce great art (p. 8).

229. Goethe, *Maximen und Reflexionen* (no. 1344), in *Sämtliche Werke*, 17:942.

to the complementary approaches of art and science. In rendering this argument, I have violated the *Römische Elegien* by stripping the poems of their aesthetic beauty. Perhaps, though, W. H. Auden offered the only kind of comment suitable to the medium—another poem. In "Good-bye to the Mezzogiorno," he recognized Goethe (despite their different sexual tastes) as the very emblem of the poet:

> Goethe,
> Tapping homeric hexameters
> On the shoulder-blade of a roman girl, is
> (I wish it were someone else) the figure
> Of all our stamp.

Two years after Goethe had returned to Weimar, he attempted a second journey to Italy; he felt the need to recapture that originating experience. After about six weeks, however, he admitted that he could not make the past come alive again. He reluctantly headed back home in melancholy disappointment but never thereafter ceased to recall the time of his rebirth. Auden ends his poem with a remark about Italy that might reflect Goethe's own feeling about the land that so changed his life:

> Though one cannot always
> Remember exactly why one has been happy,
> There is no forgetting that one was.[230]

Conclusion

In a late reminiscence, Goethe recalled that during his close association with Schiller, he was constantly defending "the rights of nature" against his friend's "gospel of freedom."[231] Goethe's characterization of his own view was artfully ironic, alluding as it did to the French Revolution's proclamation of the "Rights of Man."[232] His remark implied that values lay within nature, values that had authority comparable to those ascribed to human

230. W. H. Auden, "Good-bye to the Mezzogiorno," in *Selected Poems*, ed. Edward Mendelson (New York: Vintage Books, 1979), pp. 241–42.

231. Goethe, "Einwirkung der neueren Philosophie," *Zur Morphologie*, 1.2, in *Sämtliche Werke*, 12:97.

232. The phrase "rights of nature" may originally have come from Schiller's monograph *Über Anmut und Würde*. See *Schiller Werke und Briefe*, ed. Otto Dan et al., 12 vols. (Frankfurt, a. M.: Deutscher Klassiker Verlag, 1988–), 8:344. See also the next chapter. I owe to Lorraine Daston the suggestion that Goethe's use of the phrase "rights of nature" likely also alludes to the "Rights of Man."

beings by the architects of the Revolution. During the time Goethe made this defense, he also faced another revolution, one in which Schiller was a partisan—that of Kant in the intellectual sphere. Both upheavals had undermined the autonomy of nature, replacing her authority with that of human will and understanding.

But these two revolutions concluded a long history in the shifting fortunes of nature that brought her to this stage of jeopardy. In the very early classical period, nature as a unified whole had not yet arisen. Animate and inanimate objects had natures—characteristics modes of action—but there was as yet no articulate concept of nature as a whole standing over against human beings.[233] During the late Hellenic and medieval periods, nature came to be personified in the form of a didactic female, a figure imaginatively based, it would seem, upon the system of natural philosophy (*philosophia naturalis*) that stood in contrast to the revealed wisdom of God. Nature in this guise yet derived her authority and nurturing capacity from that higher divine power.[234] During the seventeenth century, writers like Bernard de Mandeville began to suspect that nature might be a chimera, a fictive creature that disguised humanity's own hidden desires and inclinations.[235] These doubts grew during the next century—with the likes of David Hume accelerating the skepticism—until finally, in the two revolutions that so troubled Goethe, nature was stripped of her authority.

Goethe had become confirmed in his defense of the rights of nature during his travels to Italy. After he returned to Germany in summer of 1788, he set out to develop a science that would recognize nature's autonomy and authority. In this reconfiguration, though, nature would come to exhibit features distinctively altered from those of her earlier incarnations. Goethean morphology would not reinstate nature as emissary of an aloof, divine power. Nature would no longer stand apart from human beings, designed for their instruction, but would encompass the authority of both divinity and humanity: fecund, creative nature would replace God; and man would find himself an intrinsic part of nature and able to exercise, in the role of the artist, her same creative power. But as in the ancient period, the symbol for the generative power of self and nature would be the female. This transition

---

233. Laura Slatkin discusses the very early Greek concepts of nature in "Measuring Authority, Authoritative Measures: Hesiod's Works and Days," in *The Moral Authority of Nature*, ed. Lorraine Daston and Fernando Vidal (Chicago: University of Chicago Press, forthcoming).

234. Katherine Park has investigated the personification of nature in her "Nature in Person: Medieval and Renaissance Allegories and Emblems," in ibid.

235. See Danielle Allen, "Burning the Fable of the Bees: The Incendiary Authority of Nature," in ibid.

in the conception of nature, the eroticization of nature, would ultimately lead to the kind of evolutionary theory that Goethe himself would introduce and Darwin would later cultivate. The hinge of this great transition was, for Goethe, the experience of his southern travels, when so much depended on an Italian girl.

Chapter 11

GOETHE'S
SCIENTIFIC REVOLUTION

That the French Revolution was also a revolution for me, you can
well imagine.

—Goethe to Jacobi

As a poet, Goethe cultivated desire in a disciplined way; it was a desire that
encompassed nature and nature's emblem, the female form. Just before his
Italian journey, that desire began to reach out to nature through a complementary mode of engagement—through scientific discipline. The first fruit
of this new mode to achieve public notice was an inauspicious little book,
Die Metamorphose der Pflanzen (The metamorphosis of plants, 1790), which
he produced shortly after his return to Weimar. The book marked a pivot
point in his intellectual life; and through the development of its ideas, he
seeded a revolution in thought that would transform biological science during the nineteenth century. Helmholtz rightly judged that Goethe's theories of morphology became the established mode of biology in the first half
of the century and that they cleared the way for the triumph of Darwinian
science in the second half.[1] Indeed, one might even say, without distortion,
that evolutionary theory was Goethean morphology running on geological
time. And Goethe himself initially calibrated the trajectory.

This understanding of Goethe's scientific accomplishment deviates from
recent evaluations. His efforts have been judged sometimes as failed science
but more often as not even science. There is a sense—not a very considered
one—in which all past science is failed science, since older theories have
been replaced by more recently accepted ones. In that restricted view,
Goethe might be judged a failure; he would in that case, though, share the
judgment with some quite distinguished company, from Harvey and Newton to Darwin and Maxwell. But if we estimate his efforts in light of the criteria for scientific practice of the period, we should not, I think, regard him

1. See the introduction to chapter 10 for a discussion of Helmholtz's views on Goethe.

a failure. Moreover, if we rely on the evaluations of his contemporaries (Blumenbach and Soemmerring, for instance) and those of the immediately succeeding generation (Whewell, Helmholtz, and others), then there can be little doubt of Goethe's success in biological science. The refusal to regard him as a scientist suggests that the criteria employed suit some mythical ideal and not the actual standards set by significant practitioners in the nineteenth century.[2]

In this chapter I will argue that Goethe's understanding of scientific procedure marked him not simply a good scientist for the time, but a good scientist for all time. His conceptions in morphology, the science virtually of his own creation, had a solid empirical footing and provided purchase for the emergence of evolutionary theory in Germany and England. The controlling idea for this science—that of the *Urtypus*, or archetype—itself went through an evolution, from its first visibility just prior to his Italian journey, through the years of the journey and those immediately following, to its final form during the first three decades of the nineteenth century. During this period, the epistemological forces on his considerations kept shifting. His scientific conceptions initially took on the ballast of realism— deriving from his inveterate attitude and further weighted by devotion to Spinoza and experiences in Italy. Those conceptions then came under the pressure of the new Kantian philosophy, which his friend Schiller urged him to adopt. During the late 1790s, the vectors dramatically altered again as he entered the circle of the Jena Romantics and came under the magnetic influence of the young Friedrich Schelling. Accordingly, his loosely bound epistemological and metaphysical beliefs moved from a rationalistic realism to a critical idealism and finally to an absolute ideal-realism. These forces on his thought did not have the airy lightness of disembodied ideas

2. Beddow represented Goethe as a failed scientist in his review of the second volume of Nicholas Boyle's biography *Goethe: The Poet and the Age*, 2 vols. (Oxford: Oxford University Press, 1991–2000). See Michael Beddow, "Don't Fell the Walnut Trees," *Times Literary Supplement*, no. 5054 (11 February 2000), pp. 3–4. Gray has challenged this description. He argues that the poet was not even a scientist. Gray believes Goethe should not be accorded the accolade because he formulated his notion of the primal plant in light of alchemical ideas and because there was no way to test his botanical conception. See Ronald Gray's letter to the *Times Literary Supplement*, no. 5056 (25 February 2000): 17; see also Gray's *Goethe the Alchemist* (Cambridge: Cambridge University Press, 1952), especially pp. 71–100. Gray embosses the judgment earlier rendered by Nordenskiöld, who dismissed Goethe's *Metamorphosis of Plants* with the remark that "it is romantic philosophy from beginning to end; it bears no resemblance whatever to modern natural research." See Erik Nordenskiöld, *The History of Biology* (New York: Tudor, 1936), p. 282. Gillispie, I believe, reflects the still-standard evaluation of Goethe's scientific accomplishments. He urged that one simply had to recognize the poet's "intrusion as profoundly hostile to science, hostile to physics and misleading, even if stimulating, to biology." See Charles Coulston Gillispie, *The Edge of Objectivity* (Princeton: Princeton University Press, 1960), p. 197.

but were energized by relations to particular individuals to whom he became deeply attached. Yet he hardly moved through these waters like a light skiff, having no resistance to the powers operating on him. Rather, he felt the forces of the surrounding culture, as well as the impact of immediate experience, but charted a direction, as Schiller remarked, "in his own style and way."[3] If we follow Goethe along this course, a picture of him will emerge that displays all the colorings of a Romantic biologist and, as well, of a good and fruitful scientist—or so will be the historical portrait I render.

Homecoming

Goethe returned from Rome to Weimar on 18 June 1788 in a subdued state of mind. His closest intellectual friend, Herder, was just at that time preparing for his own Italian journey, finally departing on 15 August. But Herder's Roman experience soured, especially as he found his friend's scent everywhere. Sensing too keenly the poet's presence even amid the clouds of Catholic incense, this Lutheran clergyman was happy, after an unpleasant year, to rid himself of the eternal city; he came back to Weimar on 9 July 1789. Goethe, meanwhile, felt keenly Herder's absence; and in memory yet green, he would frequently travel back to the land that had so changed his own life. He perhaps too readily revealed to his immediate circle "how much I have suffered by my departure from Rome, how painful it was for me to leave that beautiful country."[4] He returned, though, according to his agreement with Carl August, to diminished administrative duties and the promise of freedom for intellectual and artistic work. He continued to supervise the mining commission, which had to deal with flooding of the mines at Ilmenau, and he remained a potent influence in the privy council. And as always, he stood on the watch to improve the cultural life of the duchy and his own life as well. Soon after his arrival back in Weimar, Goethe read, with some distaste, Friedrich Schiller's play *Die Räuber* (The robbers). Its impetuosity and moral paradoxes reminded him of his own youthful period; still, he rather liked the younger man's more recent poetry, particularly "Die Götter Griechenlandes" (The gods of Greece). They met for the first time at a country party in early September, but the *Geheimrat* kept a friendly distance. Goethe obviously perceived in Schiller an individual of significant talent, for when the latter's history of the revolt of the

3. Friedrich Schiller to Christian Gottfried Körner (1 November 1790), in *Schillers Werke* (Nationalausgabe), ed. Julius Petersen et al., 55 vols. (Weimar: Böhlaus Nachfolger, 1943–), 26:54–55.
4. Goethe to Johann Heinrich Meyer (19 September 1788), in *Goethes Briefe* (Hamburger Ausgabe), ed. Karl Robert Mandelkow, 4th ed., 4 vols. (Munich: C. H. Beck, 1988), 2:102.

Netherlands appeared a short time later in *Der Teutsche Merkur*, Goethe arranged for his appointment to the university at Jena, though without regular salary.[5] Their close friendship, however, would only begin some six years hence.

While Herder was in Italy and tending to the Dowager Duchess Anna Amalia, whose entourage had also decamped for those southern skies, he received a handsome offer from the university at Göttingen. Goethe did not wish to lose his friend and particularly his friend's wife, Karoline, with whom he had become quite close. He arranged with Carl August for an increase in Herder's salary and a decrease in his duties, which convinced the family to stay in Weimar. While Goethe cultivated Karoline's friendship, he irrevocably lost that of Charlotte von Stein—because he had discovered a new, native love.

About a month after Goethe returned from Italy, in the first part of July 1788—perhaps as he strolled along the river Ilm, in the park not far from his garden house—a young woman approached him with a petition from her father who sought a suitable situation for his son, Christian August Vulpius (1762–1827), a fledgling author in financial difficulties, who would in time achieve a modest commercial success.[6] Goethe had known of the boy prior to Italy, but now he found himself agreeing to secure what aid he could—more, it seems, because of the messenger than the message. Goethe immediately fell for the woman carrying the plea, Christiane Vulpius (1765–1816). Within a few weeks of their meeting, she became his mistress.

Christiane had long brown curls, a broad face, a strong nose, and was just a bit zaftig. She had little formal education, just enough to read and write. Much later, in 1807, Goethe himself remarked that "my wife has not read one line of any of my works. The realm of the mind has no existence for her; she is made only for the household."[7] But as a young woman of twenty-three, she seemed to Goethe from head to foot made for love. Perhaps she

---

5. Schiller's *Geschichte des Abfalls der vereinigten Niederlande von der spanischen Regierung* (The history of the revolt of the united Netherlands from the Spanish regime, 1788) was a bit of a patch job, which he cobbled together from other sources. Yet it had many narrative qualities and artistic turns that Goethe liked. Schiller wrote his fiancée Charlotte von Lengefeld (23 December 1788) that he had visited Goethe, who "was busy with this affair" of his appointment. See the letter in *Goethes Gespräche*, ed. Flodoard von Biedermann, expanded by Wolfgang Herwig, 5 vols. (Munich: Deutscher Taschenbuch Verlag, 1998), 1:460–61.

6. Friedrich Wilhelm Riemer, secretary to Goethe and tutor to his son August, related that the poet met Christiane "on a walk in the park," with the young woman carrying the abovementioned request. See Friedrich Wilhelm Riemer, *Mitteilungen über Goethe*, ed. Arthur Pollmer (Leipzig: Insel Verlag, [1841] 1921), p. 164.

7. Christine von Reinhard to her mother (5 July 1807), in *Goethes Gespräche*, 2:235. Von Reinhard here recounts to her mother a conversation Goethe had with her husband.

reminded him of the woman he left in Italy; she was obviously a suitable enough target for his liberated sexual desires. He arranged for her to make discreet visits to his lodgings. For several months she was his unabated passion: "Reverence casts me at her feet, / But joy throws me on her breast."[8] They had, it seems, all the pleasures proved; they also tested his bed, which was in constant need of repair.[9] She was to the forty-year-old poet "this child whom I have chosen, / in the happiness of a most beautiful marriage, / which lacks only the blessing of the priest."[10] Against the chilly German night, she opened up a vista of sunny pleasure that warmed them both against a disapproving world. Schiller thought him "a bit daft" and believed that in Christiane "the love of women [*Weiberliebe*] that plagues him seems to have taken its revenge."[11] Von Stein was devastated and turned her back on her onetime companion.

When von Stein explained to Karoline Herder her cool behavior toward Goethe, her high-mindedness must have sounded like breaking glass. Karoline wrote her husband:

> I now have the secret from Stein herself, why she and Goethe no longer get along. He has taken the young Vulpius as his Clärchen and has her come to him often. She dismisses him completely for this. Since he is so estimable a man and already forty years old, he simply should not do anything to reduce his worth to the common run—what do you think about that? All of this *sub rosa*.[12]

By mid-spring Karoline feared that von Stein would be driven mad with suffering over Goethe. And by June Goethe himself could not find even a polite way to be in her company. He wrote her what amounted to the parting letter:

> I have to say that I cannot stand the way you have treated me up to now. When I have tried to engage in conversation, you have shut my lips; when I have tried to be communicative, you have accused me of indiffer-

8. Goethe, "Genuss" (Pleasure), in *Sämtliche Werke nach Epochen seines Schaffens* (Munich Ausgabe), ed. Karl Richter et al., 21 vols. (Munich: Carl Hanser Verlag, 1985–98), 3.2:27. The poem was written at the time of his first meeting with Christiane.

9. Sigrid Damm found Goethe's bills for several repairs of broken bed slats. In this and other ways, she has tried to defend Christiane against a still-disapproving world. See Sigrid Damm, *Christiane und Goethe: Eine Recherche* (Frankfurt a. M.: Insel Verlag, 1999), p. 118.

10. Goethe, "Genuss," in *Sämtliche Werke*, 3.2:26.

11. Schiller to Christian Gottfried Körner (1 November 1790), in *Schillers Werke*, 26:55.

12. Karoline Herder to Johann Gottfried Herder (8 March 1789), in Johann Gottfried Herder, *Italienische Reise: Briefe und Tagebuchaufzeichnungen, 1788–1789*, ed. Albert Meier and Heide Hollmer (Munich: Deutsche Taschenbuch Verlag, 1988), p. 373. In Goethe's play *Egmont*, Clärchen is the lover of the Dutch hero.

Figure 11.1 Christiane Vulpius (1765–1816), with a copy of Tischbein's *Goethe in the Campagna* over the table. Sketch (1791) by J. H. Lips. (Courtesy Stiftung Weimarer Klassik.)

ence; when I have helped friends, you have complained of my coldness and neglect. You have checked my demeanor, you have criticized my movements and my way of being, and have always put me *mal a mon aise* [ill at ease]. How can trust and openness thrive, when you intentionally shove me away. I would like still to add many things, were I not afraid that they would wound your composure rather than provide any comfort.[13]

Goethe's future domestic life became set along a definite path when he learned, sometime in summer of 1789, that Christiane was pregnant. On Christmas day she gave birth to a son, the first of five children and the only one to survive infancy. He was baptized August Walther Vulpius.[14] Eventually Goethe came to settle into a modestly comfortable life with Christiane and his son. But passion for his mistress had subsided into companionability. A year later he told Schiller that "if he married the girl, it would be for the sake of the child."[15] Goethe did finally marry her, but by that time, in

13. Goethe to Charlotte von Stein (1 June 1789), in *Goethes Briefe*, 2:116.
14. Boyle, *Goethe: The Poet and the Age*, 1:598.
15. Schiller to Christian Gottfried Körner (1 November 1790), in *Schillers Werke*, 26:55.

1806, the child was fifteen years old. And von Stein? After several years she and Goethe renewed their relationship, though at a much lower pitch of intensity, meeting at teas and informal gatherings. Though Goethe's sexual passions were banked, they would flare up in his later years when a bright young woman would make his acquaintance.

The Foundations of Morphology

The Primal Plant

In the period between 1788 and 1790, Goethe not only completed the volumes for his collected works and composed his beautiful *Römische Elegien*, but set out on two scientific trajectories that would carry him through the rest of his life and fix his scientific reputation for succeeding generations: his investigations of optics and morphology. I will not discuss his theories of light and color in any depth, but I will explore, below, the methodological and epistemological considerations prompted by his optical work. Here I will focus on the studies that led to the publication, in 1790, of his *Metamorphosis of Plants*.

Prior to his Italian journey, Goethe had developed a taste for horticulture. He laid out for himself garden plots at his house just outside the city and would often accompany his letters to von Stein with some asparagus or other treats. He also had begun to speculate on the nature of plants and had formulated certain ideas, greatly influenced by Spinoza, concerning how we could know something to be a plant. Several individuals living in Weimar abetted Goethe's interest in botanical questions, for instance: Friedrich Gottlieb Dietrich (1768–1850), a local boy who had a natural gift for identifying plants—Goethe helped him acquire a university education at Jena; and Christian Wilhelm Büttner (1716–1801), a retired professor from Göttingen and mentor to Blumenbach. Büttner had given his library to Jena in exchange for a pension from Carl August. He and Goethe spent considerable time in the 1780s arguing Linnaean systematics. And in 1786 Goethe arranged for the support in Jena of August Johann Georg Carl Batsch (1761–1802), the naturalist with whom he would avidly discuss botanical matters after his return from Italy. During his journey Goethe's concern with plants deepened, as did many of his other passions. Back in Weimar he had opportunity to cultivate his interests in a more systematic fashion at his garden house. There he could observe the flora surrounding the property, and he lovingly tended his own plots of flowers and vegetables.[16] Goethe's

16. See Dorothee Ahrendt, "Gärten," in *Goethe Handbuch*, ed. Bernd Witte et al., 4 vols. (Stuttgart: J. B. Metzler, 1996–98), 4.1:334–36.

Figure 11.2 Goethe's house on Der Frauenplan. He rented rooms here from 1782 to 1789. Carl August purchased the house and presented it to Goethe, who occupied the entire structure from 1792 until his death in 1832. The house was reconstructed after the Second World War. (Photography by the author.)

botanical writings during this period make it clear that he based his theories on careful observation of the growth and development of a large variety of flowering plants and ferns.

Two fluctuating approaches to observation guided his study: in the first phase, he sought to discover, through comparative analysis, a law or laws governing the formation of plants; but then, in a second phase of research, he attended to an exact description of plant growth. These descriptions took strength from the hypothesis that the variously developing parts of plants could be understood as modifications of a single part, in his terms "the leaf in its most transcendental sense" [*das Blatt in seinem transzenden-tellsten Sinne*].[17]

Goethe initially tried, through empirical observation and comparison, to specify a conception of the *Urpflanze*, that is, the archetype of all plants. In notes that probably come from late 1788, he wrote:

> It is very difficult generally to establish the type of a whole class so that it is appropriate for each sex and every species; for nature can produce only

17. Goethe, in *Sämtliche Werke*, 3.2:306.

Figure 11.3 Gardens at the rear of Goethe's house on Der Frauenplan. (Photography by the author.)

through her genera and species, while the type, which is prescribed for her by universal necessity, is such a Proteus that it escapes the sharpest comparative sense and can hardly even be partly grasped, and then only, as it were, in contrast.[18]

In these notes Goethe began the project that would unite his efforts with those of natural researchers since the scientific revolution, namely, to bring the diverse phenomena of nature under unifying law. This effort had powerfully succeeded in the physical sciences, but doubt remained about its possibility in the life sciences—indeed, Kant, in his *Kritik der Urteilskraft*, would show why there could be no "Newton of the grass blade." Goethe certainly would not aspire to be a Newton—the Englishman, he would shortly argue, was an impatient observer and poor experimenter. He did, however, remain confident throughout his life that the underlying structural laws of living nature might be revealed to the researcher who cultivated those scientific virtues that Newton lacked.

The notes that Goethe composed at this time indicate that for him the complex of laws governing plant development would be expressed in the archetype, the *adequate idea* of the Spinozistic system. But the *Urtypus* would

18. Ibid., 3.2:303.

be more than this: he began to think of it not as a static form but as an active power—a Proteus, as he here described it—that would give rise to endless varieties of plants. (In the second part of *Faust*, Proteus, in the form of
a dolphin, carries on his back the Homunculus, who will be transmuted
into higher living forms.)[19] At this time, in 1788, Goethe thought a law
governing the structures of all plants had to exist, for otherwise we could
not recognize something *as* a plant—a quasi-Platonic principle that carried
enormous weight with him.[20] He tried formulating a series of such laws (for
instance, "every node of a plant has the power to develop, advance, and reproduce another plant node"), but he must finally have concluded that he
was merely describing the stages of plant development.[21]

In these notes two tangential features of his immediate concerns appear
that would later take on larger proportions. The first has to do with a long-
standing controversy over ontogenetic development, that is, the dispute
between the epigeneticists and the evolutionists (i.e., preformationists).
Goethe believed that his theory could show the merits of and thereby sub-
sume both points of view. Plants did not merely unfold, or "evolve," from a
miniature plant in the seed; rather, close observation revealed that something came temporally to be that did not before exist. Yet—and this was
the preformed aspect—organic development had to be governed by an un-
changing law or power that was realized in empirically variable phenomena. The leaf, the "transcendental leaf," did exist already in the seed. The
other consideration, which seems first to have appeared in these notes, is
Goethe's assumption of an economic balance within organic life—excessive development of one part of the plant required a reduction in another.
As he put it in these notes: "One part thus cannot be added to unless something is taken from another."[22] This law of compensation, as it would later
be dubbed, would appear at various junctures in Goethe's subsequent considerations of plant and animal development.

In November 1789 Goethe began to compose the little book that signaled a turning point in his scientific life, his *Versuch die Metamorphose der
Pflanzen zu erklären* (Experimental observations to elucidate the metamor-

19. Goethe, *Faust: Die Tragödie Zweiter Teil*, in *Sämtliche Werke*, 18.1:227 (8321–26).

20. Goethe, in *Sämtliche Werke*, 3.2:307: "The greatest difficulty in laying out this system consists in this, that one must treat as immobile and fixed that which in nature is always in movement,
that one must reduce to a simple, visible, and graspable law what eternally changes and under our
very eyes conceals itself now under this form, now under that. If we were not able to convince ourselves, a priori as it were, that such laws must exist, it would be a brash act to wish to seek and uncover such. But this must not prevent us from going forward. The most unpracticed sensibility can
distinguish a plant from other objects of nature."

21. Ibid., 3.2:312; he formulated three such laws in ibid., 3.2:312–13.

22. Ibid., 3.2313.

phosis of plants).[23] After the book was published the following Easter, he sent his friend Knebel a copy and proclaimed: "With this I begin a new course, to which I will apply myself, not without considerable difficulty. My inclinations drive me now more than ever to natural science [*Naturwissenschaft*], and I am only surprised that in this prosaic German land a small cloud of poetry still remains floating over my skull."[24] To Jacobi, he made the same proclamation, though expressed more dramatically and with sardonic intent: "That the French Revolution was also a revolution for me, you can well imagine."[25]

The title of the work, *Metamorphose*, already indicates the leading idea of the treatise: one part of the plant in the course of development transmutes into something apparently quite different, just as the caterpillar metamorphoses into a butterfly. His tract, then, follows the growth of the plant through its various transitional stages. The first organs to appear aboveground are the seed leaves, or cotyledons. Whereas later foliage leaves usually spiral alternately up the stem, the cotyledons initially form an opposing pair, perched at the end of the small stem and attached at the node. Soon from the node a minute plumule, or bud, can be seen, which slowly expands into foliage leaves and further extension of the stem. Depending on the species of plant, successively appearing leaves gradually change patterns as they move up the stem. Another distinct stage of growth is achieved when the leaves at the top of the stem, called sepals, begin to form a cup, or calyx. The calyx contracts and therein colored petals (collectively called the corolla) gather to produce the flower, which expands into a bright display. During the process of flower production, several of the petals contract into the stamens (the male organs, each consisting of a rodlike filament topped by an anther, which produces the pollen) and the pistil (the female organ, composed of a tubular style capped by its stigma, to which pollen adheres). With pollination, the fruit and the seed it contains form within a capsule, whose leaflike features are prominent. One finds in "the seed, the highest level of contraction and internal formation."[26]

Goethe portrayed the various stages of plant development as expressive of the universal forces of expansion and contraction—or, as one found them in the nonanimate sphere, forces of repulsion and attraction. In plants these forces operated on the same organ, the leaf, which became

---

23. See below for a discussion of the term *Versuch*.

24. Goethe to Karl Ludwig von Knebel (9 July 1790), in *Goethes Briefe*, 2:128. With his botanical tract, Goethe also sent his friend a copy of his unfinished *Faust*. It is thus unclear whether Goethe meant his remark about poetry to refer to his tract on plants or his play—perhaps both.

25. Goethe to Friedrich Heinrich Jacobi (3 March 1790), in ibid., 2:121.

26. Goethe, *Versuch die Metamorphose der Pflanzen zu erklären*, in *Sämtliche Werke*, 3.2:350.

transmuted into the various parts of the plant in sequential fashion. Yet, as he reminded his reader, he had only adopted a common word, "leaf" [Blatt], to designate the organ that metamorphosed into the variety of forms assumed by different parts of the plant. "One could just as well say," he cautioned, "that the petal is an extended stamen as that the stamen is a contracted petal."[27]

In notes that come from May 1787, which he jotted down shortly after his experience in the gardens of Palermo, Goethe had already formed the notion that a *Keim*, or germ, served as the generator of all the parts of plants.[28] He identified this germ with the leaf—"all is leaf, and through this simplicity the greatest multiplicity is possible." In these Italian notes, he sketched the sort of transformations that would later be minutely described in his book: "A leaf that only sucks fluid under the earth we call the root; a leaf that spreads out from those fluids, we call a bulb, an onion for instance; a leaf that stretches out, we call the stem," and so for all the other parts.[29] Goethe appears to have chosen the leaf as the symbolic organ for two principal reasons. First, because in the many varieties of flowering plants he investigated, he noticed that in some he could fairly well see alterations occurring between one stage and another as leaflike transformations. For example, in the Paduan public gardens, he saw a fan palm that still retained leaves in several states of transition.[30] Later he observed in some species of tulip a leaf along the stem that was partly green and partly the color of the petals, and that reached up to form a section of the corolla.[31] He also noticed that in many species of flower the sepals of the calyx would often display partly the green color of the foliage leaves and partly the bright color of the petals. The leaf, then, could be detected in various transitional states. The second reason for Goethe's choice of the leaf as the elemental organ is perhaps because of its veined structure. He found a comparable network of veins in the sexual organs of plants. Raw juices, he thought, flowed through the veins of the foliage leaves and stem, producing growth; and as those juices passed through the finer network of vessels in the sexual organs, their

27. Ibid., 3.2:366 (para. 120).

28. He wrote to von Stein (8 June 1787) that he had found the secret of plant generation in the "germ" (see the previous chapter for the passages).

29. Goethe, *Die Schriften zur Naturwissenschaft*, ed. Dorothea Kuhn, 21 vols. (Weimar: Hermann Böhlaus Nachfolger, 1977), 2.9a:58.

30. Goethe recounted prominently the example of the palm in his reminiscent essay from 1831 "Der Verfasser teilt die Geschichte seiner botanischen Studien mit" (The author shares the history of his botanical studies), in *Die Schriften zur Naturwissenschaft*, 1.10:333. Goethe was so taken with these unusual specimens that he dried and preserved them on pasteboard and retained them through his last years.

31. Goethe, *Metamorphose der Pflanzen*, in *Sämtliche Werk*, 3.2:336.

powers became more spiritualized (*geistigere Kräfte*)—that is, more refined—and yielded the reproductive sap.[32] Thus the upper organs of the plant displayed the same physiology as the lower, namely, that of the leaf. Undoubtedly at the beginning of his considerations of the primal plant, Goethe did understand the leaf as the kind of organ that might undergo the various transformations he described. But by the time he composed his book, the leaf had become only symbolic—a *transcendental* leaf; it represented the underlying forces that in their various instantiations formed the transmuted empirical structures of the plant.

Second Italian Journey

During early 1790, after he had finished his little book on plants and arranged for its publication, Goethe had a recurrence of the malaise that initially drove him to Italy. Again he conjured up warm, healing breezes wafting through groves of lemon trees and pleasures of the sort provided by Palermo's gardens—and the back streets of Rome. On 19 March, after an official mission—attendance at the funeral in Augsburg of the Holy Roman Emperor Joseph II—Goethe slipped southward. He wrote Herder just before he left to inform him, with apologies, that he told Christiane she could turn to the Herders should something occur she could not handle.[33] He arrived in Verona on 25 March and reached Venice shortly thereafter, during Holy Week. He joined the party of Duchess Anna Amalia, and traveled with them from Venice to Padua, Vicenza, Verona, and Mantua. He wrote Carl August in early April that already the trip "has restored me and will do both my body and my soul good."[34] As the duchess's party prepared for its return to Germany, Goethe had to decide whether to accompany them or to continue on to his once-beloved city of Rome. But the *Amor* of Roma had faded, washed over by another longing—that for Christiane and his son. He wrote the Herders from Mantua: "I confess that I passionately love the girl. How close I am joined to her, I first felt on this trip. I deeply long to

32. Ibid., 3.2:364. Gray sees in Goethe's description of the purification of the sap the influence of the Hermetic tradition. Terms such as "spiritual power," he argues, are "not scientific terms." By such anachronistic judgment, Gray seeks to indict Goethe's work as not being science at all. A comparable charge might be leveled at Newton, I suppose, especially when he identifies space and time with God's sensorium. While Goethe's or Newton's terms would not appear in contemporary science (at least not often), this simply means that our standards for scientific discourse were not those of an earlier time. But see Gray, *Goethe the Alchemist*, p. 77. In any case, Goethe used the term *geistiger* metaphorically and only meant thereby "more refined." This usage is preserved in the phrase "spirits of wine," the highly distilled liquor used to preserve organic specimens in the nineteenth century.

33. Goethe to Johann Gottfried Herder (12 March 1790), in *Goethes Briefe*, 2:122.

34. Goethe to Carl August (3 April 1790), in ibid., 2:124.

be home. I have completely moved out of the ambit of Italian life."[35] Goethe's passion for Christiane, though, seems to have been the kind cultivated best in absence. His desire for settled domesticity must have been equally keen. He turned back north to Weimar.

Though brief and melancholic, Goethe's time in Italy was productive. He began the composition of a series of pungent Martial-like epigrams that sparkle with wit and irreverence. Through the cracks in the verse, though, the sulfur of deep disappointment rises to sting the eyes, as when he wrote:

> O, Venice, if you had girls that spread out like your canals,
> Girls with cunts like your narrow alleys
> You would be the most wonderful city.[36]

The magic of Venice evanesced into the desolation of its dark passages, the romance of the past into the vulgarity of the present. In the epigrams, however, he did console himself with another kind of passion, one that abided:

> Have you given yourself over to botany? to optics? What are you doing?
> Is it not a lovelier prize to move a tender heart?
> But oh, the tender hearts! A bungler is able to move them;
> Let my only happiness be to touch you, nature![37]

Two years before he had been enraptured by Faustine and a dew-fresh Christiane, who guided him to nature's hidden forces; but that path now seemed blocked. He then turned directly to his never-failing companion, nature, and would attempt more immediately to move her to disclose her secret parts. And chance favored this jaded lover.

While in Venice, on 22 April, Goethe and his amanuensis Paul Götze went strolling down the Lido, along a stretch of sand between the sea and a Jewish cemetery. For a lark, Götze tossed him a broken sheep's skull, pretending it was a Jewish head. When Goethe looked at it, he immediately saw something striking. He wrote Karoline Herder to say that the skull enabled him "to take a great step in explaining the formation of animals."[38] Over a quarter of a century later, he more explicitly described the event: he

---

35. Goethe to the Herders (28 May 1790), in ibid., 2:127–28.

36. Goethe, *Venezianische Epigramme*, in *Sämtliche Werke*, 3.2:86: "Ach ein weiter Canal tat sich dem Forschenden auf! / Hättest du Mädchen wie deine Canäle Venedig und Fotzen / Wie die Gässchen in dir wärst du die herrlichste Stadt."

37. Ibid., 3.2:142: "Mit Botanik gibst du dich ab? Mit Optik? Was tust du? / Ist es nicht schönrer Gewinn, rühren ein zärliches Herz? / Ach, die zärtliches Herzen! Ein Pfuscher vermag sie zu rühren; / Sei es mein einziges Glück dich zu berühren, Natur!"

38. Goethe to Karoline Herder (4 May 1790), *Goethes Briefe*, 2:126. A few days before, he had indicated his findings, in much the same terms, to a woman who attracted his momentary interest. See Goethe to Charlotte von Kalb (30 April 1790), in ibid., 3.2:125.

suddenly saw confirmed that the vertebrate skull "had arisen from trans-formed vertebrae."[39] Just as the parts of the plant consisted of transformed leaves, so the skull was composed of metamorphosed vertebrae—or at least that was how he retrospectively represented his insight. Lorenz Oken, who claimed in 1806 also to have discovered the vertebral character of the skull, accused Goethe of stealing his originality with this post hoc gloss on the Venetian episode. The events following Goethe's discovery do seem a bit strange. After he had returned from Italy, with a flush of enthusiasm, he be-gan outlining a theory of the *Urtypus* of the vertebrate skeleton, compara-ble to the *Urtypus* of the plant.[40] Yet he made no mention of the vertebral composition of the skull, which he would later represent as one of his most important discoveries. The fact of his silence makes the question nagging: To what, exactly, was Goethe referring in his letter to Karoline Herder? I will try to answer that question below. In any case, Goethe abruptly shelved his essay on the *Urtypus* of the animal, since he thought his considerations at the time unripe. The finer development of his morphology, its compre-hensive extension and articulation, would await the stimulus and guide of the Kantian philosophy, whose leading advocate among Goethe's acquain-tances was Friedrich Schiller.

Friendship with Schiller and Induction into the Kantian Philosophy

A Most Fortunate Encounter

The individual who exercised the greatest impact on Goethe's intellectual and artistic life was undoubtedly Friedrich Schiller. Their intense friend-ship, which began in 1794 and ended only with Schiller's death in 1805, encompassed many dimensions of their lives. At a deeply personal level, they found great satisfaction in one another's company. Goethe dined often at Schiller's house; he had known Charlotte von Schiller—née von Lenge-feld (1766–1826)—since she had been a girl and a frequent visitor at the home of Charlotte von Stein. The two poets read each other's manuscripts, Schiller offering commentary on *Wilhelm Meister*, the *Römische Elegien*—which he thought "lubricious and not very decent, but among the best

---

39. Goethe, *Tag- und Jahres-Hefte* (1790), in *Sämtliche Werke*, 14:17. Goethe composed this di-ary retrospectively for the edition of his works that came out with Cotta during the last part of his life. That Goethe discovered something in Venice is confirmed by his letters. Exactly what it was that he discovered is not clear from contemporary evidence. Only much later did he detail his claims, which Lorenz Oken thought outright lies.

40. Goethe, "Versuch über die Gestalt der Tiere," in *Sämtliche Werke*, 4.2:134–44. This essay is discussed below.

things he has done"[41]—*Venezianische Epigramme*, and *Faust*, while Schiller received Goethe's views on *Über die ästhetische Erziehung des Menschen*, *Über naive und sentimentalische Dichtung*, and *Wallenstein*. Individual poems sailed in both directions. Goethe contributed to Schiller's journal *Die Horen*, and they composed together a series of mordant epigrams, the *Xenien*, which brought obsequious complaint to them both. Schiller edited Goethe's play *Egmont* for production in the theater they had worked to create in Weimar. And it was Schiller, through conversation and particularly through his monograph *On Naive and Sentimental Poetry*, who brought Goethe more completely into the Kantian fold. Goethe thought his relationship with Schiller "the most significant that fortune provided me in my later years."[42]

These two poetic geniuses and friends yet differed in temperament and intellectual attitudes, approaching common issues from quite distant poles. Schiller, Goethe later recalled, "preached the gospel of freedom; I certainly did not want to abridge the rights of nature."[43] This simple but trenchant characterization crystallized several facets of their intellectual differences: Schiller displayed a kind of religious fervor, Goethe a cooler, almost legal demeanor; Schiller emphasized the creative freedom of the artist, Goethe the constraints imposed by nature; Schiller looked inward, Goethe outward; Schiller was a Kantian idealist, Goethe—initially at least—a Spinozistic realist. But as their friendship matured, their ideas and attitudes began to migrate toward more common ground.

Initially, there seemed only distance between them. They had known of each other for some time, but Goethe was wary of the man who wrote the paradox-mongering play *Die Räuber*. When they first met on 7 September 1788 at a party given by the Lengefelds, Schiller's future in-laws, Goethe treated the younger poet in a friendly fashion but did not long linger with him. He seems mostly to have spent his time entertaining the other guests with stories of his Italian adventures. A few days after the meeting, Schiller's rather critical review of *Egmont* appeared in the *Allgemeine Literatur-Zeitung*.[44] No name was attached to the review, but soon enough acquaintances discovered its author, though Goethe seems not to have been immediately ap-

41. Friedrich Schiller to Charlotte Schiller (20 September 1794), in *Schillers Werke*, 27:49.

42. Goethe, "Glückliches Ereignis" (A fortunate event), *Zur Morphologie*, in *Sämtliche Werke*, 12:86.

43. Goethe, "Einwirkung der neueren Philosophie" (Impact of recent philosophy), in ibid., 12:97. Goethe's language here comes virtually directly from Schiller's monograph *Über Anmut und Würde*. See the discussion immediately below.

44. The review is reprinted in *Schiller Werke und Briefe*, ed. Otto Dann et al., 12 vols. (Frankfurt, a. M.: Deutscher Klassiker Verlag, 1988–), 8:926–37.

prised; for in late fall of that year, he helped arrange a teaching position for Schiller at Jena. Schiller's feelings about Goethe became more public in 1793, when his monograph *Über Anmut und Würde* (On grace and dignity) appeared. The essay, read at one level, only attempted to rework some Kantian distinctions in aesthetics; at another level, it loosed several shafts, which sank deeply into Goethe's sense of artistic self-worth.

In the essay Schiller suggested that some individuals might only passively respond to natural beauty and feel obliged simply to protect the "rights of nature." Other individuals, among whom he placed himself, strove to re-create nature with a moral spine. Schiller implied that the artist of natural genius, whose description certainly sounded like Goethe's, would express beauty almost automatically, as virtually a necessary reaction, while the graceful artist struggled to transform nature morally through the free play of imagination. He described the former disposition as indicating "talent" but the latter as revealing "personal merit."[45] During the next few years, he would further develop these categories of the natural poet and the graceful poet, retaining Goethe and himself as their type specimens. In this later deployment, however, natural genius received a less deprecating account.

Goethe was vexed with Schiller because of the essay. Yet when the young poet approached him the next spring with a proposal to become a participant, along with Fichte and Wilhelm von Humboldt, in his new journal *Die Horen,* Goethe graciously accepted.[46] On 20 July the *Geheimrat* attended a meeting of the Natural History Society in Jena, conducted by the botanist Carl Batsch, and then two days later discussed with Schiller and his coworkers the operations of *Die Horen.* Probably not by chance, Schiller also was present at Batsch's meeting. The occasion was fateful for the relationship of the two men.

As they left the meeting and Goethe could not avoid exchanging a few pleasantries, they began to discuss Batsch's lecture on plants, which both agreed had a dusty, empirical quality that declined into disconnected particulars. Goethe suggested another, more interesting approach that began

45. Friedrich Schiller, *Über Anmut und Würde*, in *Werke und Briefe*, 8:330–94. "Architectonic beauty . . . is that part of human beauty that is produced not only through a natural power (which is true of every appearance) but which is determined through natural power alone." This is to be distinguished from grace: "Grace is beauty of form under the influence of freedom, the beauty of that appearance which the person determines. Architectonic beauty gives honor to the author of nature; grace and gracefulness give their possessor honor. The former is talent, the later is personal merit" (ibid., 8:335, 344). The phrase "rights of nature" occurs on p. 344.

46. Schiller to Goethe (13 June 1794) and Goethe to Schiller (24 June 1794), in *Sämtliche Werke*, 8.1:11–12.

with the "living whole" in order to get a sense of the function and character of the parts—especially as comparable parts could be found in different species. Schiller used the event to invite Goethe back to his house for further conversation. After they arrived, Goethe began illustrating what he meant by drawing a "symbolic plant"—the Urpflanze that he had sought in Italy. Schiller listened politely, shook his head, and exclaimed: "that's no observation, that's an idea." Goethe recalled being struck by this remark and not a little irritated. He replied: "Well, I am quite happy that I have ideas without knowing of them and that I can even see them."[47]

Goethe's account of his conversation with Schiller appeared a quarter of a century after the event. The recollection obviously profited from certain epistemological convictions he had already assumed prior to his meeting with Schiller and that he would more assiduously develop after the meeting—the encounter thus served more as a memorial focal point for the summation of a long-term development. And it was a deliciously ambiguous reconstruction at that. Certainly before the meeting, Goethe had already become convinced that surface phenomena could not be equivalent to the archetypal ideas lying behind the appearances: Spinoza had taught him that adequate ideas, which embodied the essence of natural objects, had to be discovered through considerable experience and research. The transcendental leaf, so discovered, constituted an ideal structure that could be realized now as seed leaves, now as stem, now as petals or seed, but never identified with any one state. Such ideal structures, as Goethe would come explicitly to maintain, could not be represented by particular, empirical objects; they could not be seen with the physical eye but only with the inward eye. Yet under Schiller's tutelage, he would come to adopt the quasi-Kantian conception of the artist as sharing in nature's own intellectus archetypus; the artist of requisite genius might thus employ the kind of ideal structures that nature herself would use in the production of objects—so paradoxically, the artist would indeed be able to draw a plant that embodied the ideal, though the result would be a piece of art that expressed an ideal detectable only by the inward eye. Goethe's recollection of his "fortunate encounter" captured a reality, but one that far outstripped the immediate event.[48] This much about their meeting, however, seems factually clear:

47. Goethe recalled this interchange for his Zur Morphologie; see "Glückliches Ereignis," in Sämtliche Werke, 12:88–89.

48. A few years before the publication of his Zur Morphologie, Goethe mentioned his interchange with Schiller to Sulpiz Boisserée, a young art critic and collector who had been a private student of Friedrich Schlegel in Paris and Cologne (1803–6). Boisserée became quite close to Goethe in his later years. He recorded in his diary the poet's recollections, in 1815, of his meeting with Schiller: "As he afterward [after his Italian journey] saw Schiller in Jena and related to him his

Schiller wanted to win him for *Die Horen,* so, as Goethe recalled, the younger poet continued the conversation tactfully but with the insistence of a well-versed Kantian. Thus began the great friendship that so altered the lives of these two men.

At the end of the next month, Schiller sent him a remarkable letter di-agnosing their intellectual and artistic differences. The analysis flattered Goethe for his genius, but yet did not unduly dim Schiller's estimate of his own virtues. The letter employed the same basic categories as *Grace and Dignity,* but the tone now was more ameliorative. It suggested how intel-lects of such diverse carriage might yet reach a common point. He wrote:

> What is difficult for you to realize (since genius is always a great mystery to itself) is the wonderful agreement of your philosophical instincts with the purest results of speculating reason. Initially it seems that there could not be a greater opposition than that between the speculative mind [*den spekulativen Geist*], which begins with unity, and the intuitive [*den intu-itiven*] which starts from the manifold [of sense]. If the first seeks experi-ence with a chaste and true sense, and the second seeks the law with a self-active and free power of thought [*Denkkraft*], then they cannot fail to meet each other halfway. To be sure, the intuitive mind is only concerned with the individual, the speculative only with the kind [*Gattung*]. But if the intuitive has genius and seeks in the empirical realm the character of the necessary, it will always produce the individual, but with the charac-ter of the kind; and if the speculative mind has genius and does not lose sight of experience—which that sort of mind rises above—then it will always produce the kind but animated with the possibility of life and with a fundamental relationship to real objects.[49]

Schiller further suggested that Goethe had a southern, virtually Greek tem-perament, which could only have realized its potential after coming into contact with original, ancient sources. The lived reality allowed his imagi-nation, "in a rational fashion, to give birth internally to a Grecian land."[50] Yet after this rational re-creation occurred, according to Schiller, it had to

---

view of the matter [his notion of metamorphosis], Schiller immediately cried: 'But that is an idea.' Goethe with his naive sensibility said quickly: 'I don't know what an idea is; I see it actually in all plants.'" The gist of the recollection is obviously the same in both accounts, though, undoubtedly warmed by the intervening years. See Sulpiz Boisserée, *Tagebücher,* ed. Hans-J. Weitz, 5 vols. (Darm-stadt: Eduard Roether Verlag, 1978–95), (3 October 1815), 1:278.

49. Schiller to Goethe (23 August 1794), in *Sämtliche Werke,* 8.1:15.

50. Ibid., 8.1:14.

be turned back into intuitions and feelings, which would then guide artistic production.[51]

Schiller's own intuitive, even poetic description of Goethe's mind convinced the older man to accept the classification—an acceptance that demonstrates, at least, the power of flattery, since the classification was hardly different from that which had so nettled Goethe when it appeared in *Grace and Dignity*. Schiller derived inspiration for his diagnosis, in the monograph and in the letter, from a quite elevated and abstract source (as would be the wont of the speculative mind), namely, Kant's third *Critique*, a book that both he and Goethe (quite independently) had been reading since its appearance four years earlier. According to Kant, the artist creates a mental realm in which concrete images (intuitions) are shaped and reshaped in pursuit of an ineffable idea, a nonarticulable conception born out of the free play of imagination—a conception of purposiveness without specifiable and conscious purpose. But the artist only feels this purposiveness of forms; it is this feeling that guides the aesthetic instantiation of these forms in some medium, that is, in a novel, poem, or statue. Before his contact with Schiller, Goethe had inchoately assumed that the beauty of nature simply rushed into his eyes and gushed out of his pen. Schiller convinced him that rational concepts intervened, that his aesthetic appreciation of nature required the creative potency of ideas, even if those ideas lay deeply below the limen of immediate consciousness. The lesson was more effective than Schiller bargained for, since Goethe would soon entertain the view that nature was just an idea, as he fell under the spell cast by Schelling.

In response to this letter of 23 August, Goethe sent Schiller a brief essay: "In wiefern die Idee: Schönheit sei Vollkommenheit mit Freiheit, auf organische Naturen angewendent werden könne" (The extent to which the idea of beauty as perfection with freedom can be applied to organic nature).[52] The essay recognizes that every animal, insofar as it can exercise its normal functions, should be regarded as perfect. The so-called higher animals, though, exhibit behavior that is beyond any functions needed for simple existence—for example, a horse in happy gallop. In such instances

---

51. Ibid., 8.1:14–15: "This logical direction, which your mind was required to take in a reflective mode, comports ill with the aesthetic direction it takes when it creates. So you had one more task: just as you went from intuition to abstraction, so now you had to move in the reverse direction and turn concepts into intuitions and thoughts into feelings, since only in this way can genius produce anything."

52. Goethe, "In wiefern die Idee: Schönheit sei Vollkommenheit mit Freiheit, auf organische Naturen angewendet werden könne," in *Sämtliche Werke*, 4.2:185–88. This essay was only discovered in the 1950s among Schiller's papers.

where the animal "seems to act and effectively exercise a free will," we will think it "splendid." We will call it "beautiful," however, when we see that it *could* use its power, while yet remaining in repose. Such a beautiful scene, according to Goethe, produces in us additional feelings of "trust and hope." Schiller immediately responded to the essay with complete accord—not surprising, since it echoed some of his own conceptions in *Über Anmut und Würde*.[53] Beyond that, their agreement certainly had roots in a common ground, namely, Kant's third *Critique*. Kant maintained that we could validly call a person or a horse beautiful if we focused solely on its form, assessing the grace and harmony of the form and not its purpose (i.e., not how closely the form of this horse matched some ideal of the animal). Kant called these judgments about natural organisms judgments of "free beauty."[54]

There is no sense, of course, in which the movements of a horse or any other animal could be *morally* free; such movements would only *appear* to be free. Schiller and, with less reflection, Goethe were ready to anchor judgments of beauty not simply in a harmony of forms but in the apparent *free* harmony of forms—for instance, when the graceful movements of a horse provoked comparison with the free movements of a rational, moral agent. Schiller would recognize this as an imposition on experience of a moral idea, namely, that of freedom. Kant himself would not have grounded our judgments of beauty quite in this way, though his language in some instances came close to suggesting that. In any case, agreement about aesthetic matters sealed the friendship of the two poets.

Goethe's Kantianism

In mid-February 1789 Christoph Martin Wieland (1733–1813), acerbic poet and older mentor of Goethe, wrote his son-in-law Karl Leonhard Reinhold (1758–1823), the principal supporter of Kant in the philosophy faculty at Jena, about his friend's new preoccupation: "For a while now, Goethe has been studying Kant's *Critique* with great application and plans

53. Schiller defined "grace" as the "beauty of form in motion by reason of freedom." See Schiller, *Über Anmut und Würde*, in *Werke und Briefe*, 8:345. A bit later he adjusted his notion of beauty so that it itself included the idea of freedom. He wrote his friend Körner that if we observed a natural phenomenon as if it exhibited a pure will—that is, freedom—we would think of it as beautiful. "Beauty," he said, "is nothing but freedom in appearance." See Schiller to Körner (8 February 1793), in *Schillers Werke*, 26:183. To ground the experience of beauty in a concept of freedom, if only analogously applied, moved a step beyond Kant and a step closer to making aesthetics serve morality, which Schiller would later propose in his *Letters on the Aesthetic Education of Mankind*.

54. Kant, *Kritik der Urteilskraft* (§16), in *Immanuel Kant Werke*, ed. Wilhelm Weischedel, 6 vols. (Wiesbaden: Insel Verlag, 1957), 5:310–13.

to hold a lengthy meeting about it with you in Jena."[55] Goethe read the first *Critique* rather thoroughly, as his pencil scorings throughout the book indicate; and he did take instruction with Reinhold.[56] By his own admission, however, he seems not to have penetrated very far into the Kantian doctrine—or perhaps, simply did not believe it: "Sometimes my poetical abilities hindered me, sometimes my mundane understanding, and I felt I had not gotten very far."[57] Herder, who by this time had become an implacable foe of his former teacher, erected only critical barriers: "I could not agree with Herder, but I also could not follow Kant."[58]

Another obstacle to following Kant was likely the construction that Reinhold placed on the doctrine of the first *Critique*. Just at the time Goethe sought guidance from this representative of the Kantian philosophy at Jena, Reinhold had produced a study of the critical philosophy that aimed to make it more rigorous and systematic than its author originally had. In *Versuch einer neuen Theorie des menschlichen Vorstellungsvermögens* (Essay on a new theory of human representational ability, 1789), Reinhold maintained that the critical analysis of reason had to begin with the fundamental, indisputable fact of consciousness—with the ability to represent something, whether that representation be a perception, thought, imagination, or knowledge claim. He articulated the fundamental structure of representation, namely, that "to every representation there belongs a representing subject and a represented object, both of which must be distinguished from the representation to which they belong."[59] From this analysis, Reinhold concluded that the object represented must have a form and content different from that thing-in-itself existing outside of consciousness.[60] Whereas Kant stressed that things outside the mind appeared

55. Christoph Martin Wieland to Karl Leonhard Reinhold (18 February 1789), in *Goethes Gespräche*, 1:470.

56. Karl Vorländer describes Goethe's marginalia. See his *Kant-Schiller-Goethe*, 2nd ed. (Aalen: Scientia Verlag [1923], 1984), especially pp. 283–91. Goethe mentioned to Sulpiz Boisserée that he had a seminar with Reinhold on Kant. See Boisserée, *Tagebücher* (3 October 1815), 1:278.

57. Goethe, "Einwirkung der neueren Philosophie" (1820) in *Sämtliche Werke*, 12:95. Vorländer, in the fundamental treatment of Goethe's Kantianism, judged that the few notes the poet made directly from the *Critique of Pure Reason* indicate that he had not overestimated his penetration of the work. See Vorländer, *Kant-Schiller-Goethe*, pp. 140–44.

58. Goethe, "Einwirkung der neueren Philosophie," in *Sämtliche Werke*, 12:95.

59. Karl Leonhard Reinhold, *Versuch einer neuen Theorie des menschlichen Vorstellungsvermögens* (Jena: C. Widtmann and I. M. Mauke, 1789), p. 200.

60. Ibid., pp. 244–50. Reinhold's conception of the Kantian project would prove crucial for the later development of Fichte's and Schelling's idealism. When these two philosophers rejected so vehemently Kant's *Ding an sich*, they were reacting mostly to Reinhold's version of that Kantian doctrine. The blunt formulations of Reinhold undoubtedly made Goethe's acceptance of the Kantian epistemology difficult, and allowed the poet to be more receptive to Schelling's later advance upon Kant.

in experience, Reinhold made evident that Kant had no grounds for allowing external reality to creep into conscious experience.[61] Goethe would
quite simply conclude that Kantianism was hostile to any immediate contact with nature. The critical philosophy thus left a bitter taste in the
mouth of the poet.

The poet, nonetheless, continued his Kantian studies. Goethe also read
during this same period Kant's *Metaphysische Anfangsgründe der Naturwissenschaft* (Metaphysical foundations of natural science). In that tract Kant
reduced the concept of matter ultimately to that of the powers of attraction
and repulsion. Goethe later remarked that this Kantian principle inspired
him to "develop the concept of the fundamental polarity of all beings, a polarity that penetrates and animates the infinite manifold of appearance."[62]
Kant, then, is perhaps the source for Goethe's conception of plant stages as
alterations of expansion and contraction of the elemental organ.

Goethe's hesitating difficulties with Kant became blanketed in a cloud of
enthusiasm when he took up the newly published *Critique of Judgment*
(1790). In October 1790 Körner wrote Schiller that he had spoken with
Goethe in Dresden: "Where we found the points in common to speak
about—well, you'll never guess. Where else, in Kant! He has found nourishment for his philosophy in the critique of teleological judgment."[63]
Schiller responded with his own tale of discussing Kant with Goethe, on
the occasion of their first meeting: "Yesterday he was with us [at the Lengefelds'] and our conversation soon came to Kant. It's interesting how he
dresses everything up in his own style and manner and renders rather surprisingly what he has read. But I didn't want to argue with him about matters very close to my own interest."[64] Goethe read and reread Kant's
Critique of Judgment, jotted notes in it, and made it his own. He thought the
period during which he took up the book "a wonderful time of my life."[65]
Indeed, so impressed was he with Kant's treatment of biological judgment
that in the cruel autumn of 1792, during the wretched and humiliating retreat of the German armies from France—with soldiers dying from disease
and Goethe himself wracked with cold and hunger (see chapter 2)—he

---

61. Ameriks argues that Reinhold distorted the Kantian theory of the thing-in-itself. He contends that Kant did not mean to cut consciousness epistemologically off so radically from things
outside the mind. See Karl Ameriks, *Kant and the Fate of Autonomy: Problems in the Appropriation of
the Critical Philosophy* (Cambridge: Cambridge University Press, 2000).

62. Goethe, *Campagne in Frankreich, 1792* (November), in *Sämtliche Werke*, 14:468. Schelling
made comparable use of that Kantian principle; see chapter 2.

63. Christian Gottfried Körner to Friedrich Schiller (6 October 1790), in *Goethes Gespräche*,
1:497.

64. Schiller to Körner (1 November 1790), in *Schillers Werke*, 26:54–55.

65. Goethe, "Einwirkung der neueren Philosophie," in *Sämtliche Werke*, 12:96.

could yet engage a young schoolmaster met along the way in a discussion of the third *Critique!*[66]

Goethe resonated to several features of Kant's conception. The third *Critique*, first of all, demonstrated that his two favorite occupations, art and biological science, had intimate connections; they were, in Goethe's terms, both "supported by the same faculty of judgment." Moreover, both art and nature, according to the Kantian view, worked "from the inside out," that is, they both exhibited an internal structure and dynamic that got expressed in external action and impact. More fundamentally, both existed in their own right and for themselves. Organisms might be shaped by the external environment, but their internal structure was neither explained nor justified by that environment nor by any other *external* cause, human or divine. Art as well had been freed, Goethe thought, from the oppression of final causes. These ideas, then, constituted the fruits of Goethe's first study of Kant's third *Critique*.[67] They would play an important role in his developing conceptions of morphology and its methods, which I will take up in a moment. Goethe would return to Kant's book in 1817, when preparing his series *Zur Morphologie*. At that time, certain other aspects of the Kantian approach to nature would spring to life for him as a result of an injection of Schelling's idealism at the turn of the century. But in the mid-1790s, the vehicle that most impressed Goethe and drove the Kantian message home was Schiller's great monograph *Über naive und sentimentalische Dichtung*.

Schiller's *On Naive and Sentimental Poetry*

Goethe credits Schiller's monograph with introducing into aesthetic discourse the categories of the *classical* and the *Romantic*, when these were synonyms for the real and the ideal.[68] The Schlegel brothers, he maintained, almost immediately took up these categories and gave them widespread currency. Goethe also asserted that this work of Schiller's brought him to see that he himself was a Romantic. He later explained all of this to Eckermann:

> The concept of classical and Romantic poetry, which now has spread over the entire world and has caused so much strife and division . . . orig-

66. Goethe, *Campagne in Frankreich* (25 October), in *Sämtliche Werke*, 14:439. Goethe epitomized for the teacher what he thought to be the message of Kant's argument: "A work of art should be treated as a work of nature, a work of nature as a work of art; and the value of each should be developed out of the intrinsic character of each, and each considered for itself alone."

67. Goethe sketched the ideas that most struck him in "Einwirkung der neueren Philosophie," in *Sämtliche Werke*, 12:96–97.

68. Ibid., 12:97.

inated with me and Schiller. For poetry I had the maxim of objective experience and only wanted this to obtain. Schiller, however, argued entirely for the subjective, took his sort to be the only kind, and in order to defend against me, wrote the essay on naive and sentimental poetry. He demonstrated to me that I myself, against my will, was a Romantic. . . . The Schlegels seized upon the idea and took it further, so that it has now spread over the whole world and now everyone talks about classicism and Romanticism, whereas fifty years ago no one had even thought of them.[69]

The Schlegels, in defining Romantic literature, had Goethe specifically in mind (see chapter 2). So by reason of theory and definition, Goethe became, even against his will, a Romantic (see below for a further discussion of Goethe's Romanticism). Let me explore the theoretical side of the question as Schiller developed it in his monograph. For Goethe, though, the full meaning of Schiller's essay would come only after a further theoretical contribution from Schelling.

Schiller's tract classifies and analyzes two different kinds of genius, and assumes the Kantian interpretation of that faculty. In the third *Critique*, Kant describes genius in this way:

> Genius is the talent (natural gift) that gives the rule to art [*Kunst*]. Now since talent is an inborn, productive ability of the artist, it belongs to nature. So we might also express it this way: Genius is the inborn mental trait (*ingenium*) through which nature gives the rule to art.[70]

The definition suggests—at least it did to Schiller—that ineffable rules for the creation of beauty arise from the artist's nature, which is of a piece with nature writ large. Those rules cannot be made fully conscious, though Schiller does suggest that the sentimental poet—that is, the reflective poet—struggles to bring them to light, if ultimately unsuccessfully. These rules, mutely sounding in the free play of imagination, express themselves in particular aesthetic feelings, in a distinctive kind of pleasure. The naive artist, according to Schiller, follows these rules of beauty immediately and unreflectively, his pen or brush guided by the sheer sense of aesthetic rightness. The difference between the naive poet and the sentimental poet, as Schiller reconstructs their activities, rests not, therefore, in the use of ideas—both employ ideas that at a deep level join their natures with external nature. It is, rather, that the naive poet does not reflectively struggle

---

69. Johann Peter Eckermann, *Gespräche mit Goethe in den letzten Jahren seines Lebens*, 3rd ed. (Berlin: Aufbau-Verlag, 1987), (21 March 1830), p. 350.

70. Kant, *Kritik der Urteilskraft* (§46), in *Werke*, 5:405–6.

with the ideas in the manner of the sentimental poet. With a penetrating eye, the naive poet sees through the artificiality of technique and society to an authentic ground in nature, and he represents this nature directly and sensuously. Homer, for instance, depicted nature more simply, immediately, and objectively than the modern, sentimental poet. The latter, by contrast, perceives the absence of nature in human affairs, but seeks it as an ideal to be realized. His commerce is with the truth of ideals, with recollections of a nature lost after the passing of the ancient period. His is the poetry of longing (Sehnsucht) for a once and future integration into the natural world.

Schiller's categories of the naive and the sentimental, while they have an aesthetic backbone, are yet fleshed out with moral interest. They are meant to express a mode of artistic engagement with nature that goes beyond the merely aesthetic: both the naive poet and the sentimental have an interest in nature, a moral commitment to the existence of natural creatures, and their poetry expresses that commitment.[71] Most individuals, according to Schiller, take an interest in nature simply because it is nature—that is, the existence of things on their own, according to their own laws. The satisfaction one takes in such experiences is a moral one because, as he views it, we actually love and wish to preserve not so much nature herself but a universal idea represented in natural creatures and natural scenes. "We love in them," he contends, "the quietly creative life, the calm autonomous activity, the existence according to their own laws, the inner necessity, the eternal unity with themselves." Young children become for Schiller a model for the relation to nature that the artist must achieve. Children react spontaneously to their surroundings, in innocence, without partition between what they believe and how they act. "They are," he pronounced, "what we were; they are what we again should become."[72] Yet the poet can become as a child only by the adult becoming civilized, that is, by attaining culture; the reason and freedom that culture brings allows the artist once again to approach nature in a childlike way. The two kinds of poets, however, will relate to nature differently. The poet of naive temperament—like his Greek forebears—will respond to a circumscribed nature intuitively, spontaneously, and without artifice; his poetry itself will virtually be an ob-

71. Kant argued that authentic judgments of beauty could not be based on any moral evaluation of the object or on any other kind of extrinsic interest we might have in it. He did not deny, however, that aesthetic judgment might be linked to moral judgment. Kant rather thought those of a good moral disposition would naturally take an interest in preserving beauty, especially beauty in nature. See ibid., pp. 395–400 (A 163–71, B 165–73). Kant concluded that "if an individual has a direct interest in the beauty of nature, there is reason to suppose he has the disposition to have a good moral attitude" (ibid., p. 398 [A 167, B 169–70]).

72. Schiller, Über naive und sentimentalische Dichtung, in Schiller Werke und Briefe, 8:707, 708.

ject of natural beauty. The sentimental poet will emphasize the disjunction between man's present state of disconnectedness with nature and an ideal of natural unity that should be reestablished; his poetry will struggle to realize an infinite ideal in a finite form. As Schiller epigrammatically expressed his analysis, the poet "either is nature or he will seek nature."[73] Further, each kind of poet will achieve a moral perfection commensurate with his respective genius. The naive individual can achieve virtually an absolute perfection, but of a limited ideal, while the sentimental type strives after the infinite, ultimately fails, but still advances to a relative perfection, to a partial realization of an infinite ideal. This latter kind of temperament embodies the progressive spirit of civilization:

> The path of the newer poets is the same as that which mankind generally, in the individual and in the whole, must follow. Nature makes him one with himself, while art separates and divides him; but through the ideal he returns to unity. Since the ideal is infinite, he can never reach it; just so the cultivated man, in his mode, can never achieve the perfection that the natural man is able to achieve in his mode. . . . The one acquires his worth through the absolute achievement of a finite greatness, the other through the approximation to an infinite greatness. . . . Insofar as the final goal of humanity can be achieved only through progress, and it is through cultivation that one returns to nature, there is no question which of the two has the greater advantage in respect of the final goal.[74]

It is this note of becoming, of striving after the infinite, that Friedrich Schlegel embraced as the essence of Romantic poetry and of man as a progressive being (see chapter 2).

One would not wish to press Schiller's analysis too far, lest the interesting categories, which produced such a revolution in aesthetics, collapse with the release of their somewhat misty and opaque air. After all, both types of poets that Schiller discriminated employ ideas of ineffable beauty; and it would be difficult to say what exactly would constitute a limited idea of beauty and what an infinite idea. Moreover, the ideas of beauty that artists of both types express are supposed to be, in the Kantian view, unconscious, not conceptually articulable. In what sense, then, does the sentimental artist "struggle" with these ideas? Schiller described such poets as Homer and Shakespeare simply *as* nature—flattering company, no doubt, for Goethe. Yet when Friedrich Schlegel, after reading Schiller's essay, launched the comparable categories of the classical and the Romantic, he

73. Ibid., 8:732.
74. Ibid., 8:735.

also placed Goethe alongside Shakespeare but regarded them both as modern writers, as opposed to classical. So it is obvious the categories lacked a
certain analytic clarity. Schiller did not wish to crucify his graceful concepts on the cross of logic. His categories do, though, mark off two distinctive temperaments.

While Schiller held that both types of temperament were requisite for
the complete ideal of humanity, there is little doubt as to which of the two
he thought the moral advantage accrued. Goethe's overwhelmingly positive response to the essay and his recognition that he himself should be
classified as a Romantic seem, though, to suggest that he had come to regard
his kind of poetry at least as much sentimental as naive—that is, a poetry
that strove to realize infinite ideals, as well as being directly responsive to
nature. Schiller made this complex assessment even easier for his friend by
suggesting, in passing, that the same poet might embody both naive and
sentimental modes, and might even express them in the same work, "as for
example, in *Werthers Leiden*."[75]

The Science of Morphology

From the period during which he composed his *Metamorphosis of Plants*
roughly to the end of the 1790s, Goethe worked assiduously, if intermit-
tently, on constructing a science of morphology, laying down its general
definitions and its methods, and empirically determining its content. His
reading of Kant's *Kritik der Urteilskraft* had a measured effect on his conception of the general character of morphology, and Schiller's essay endorsed
Goethe's own mixed rational-empirical approach to the subject. His empirical work in morphology extended from comparative studies in plants and
in animals, to human osteology, to dissectional descriptions of frogs and
snails, and to careful observation of and experiment on the metamorphosis
of butterflies. The extent and depth of Goethe's close empirical work
should erase all doubts about his capacities as a scientist. This does not
mean, of course, that his theoretical conclusions might today be accept-
able, but, then, neither are Newton's. Yet like Newton's, they were immediately adopted by his contemporaries. Moreover, they pumped swiftly
moving ideas into a major tributary of biological thought during the nineteenth century, something amply attested to by several authorities of that
period. William Whewell, for example, added his voice to that of
Helmholtz in recognition of Goethe's formulation of the "laws" of plant organization, which "have been generally accepted and followed up." In the

75. Ibid., 8:734–35n.

area of animal morphology, Goethe again, according to Whewell, was the first clearly to develop the notion of the type, a concept that then spread to such anatomists as Oken, Meckel, Spix, and Geoffroy Saint-Hilaire, with Cuvier employing it to greatest effect.[76] Finally, as Helmholtz observed at the end of the century, Goethe's morphology made the way smooth for the quick acceptance of Darwin's evolutionary theory in Germany. Helmholtz did not realize, however, the various ways in which Goethe's ideas were instrumental in the very construction of Darwin's theory. In Darwin's early essays, for instance, he made Goethean metamorphosis central to the notion of unity of type, of which his own theory would give account.[77] The great poet's remarkable accomplishments, in the estimation of principal authorities of the period, thus led to the most vital developments in nineteenth-century biology.

The Methods of Science

During the decade of the 1790s, Goethe worked on methodological questions, especially as they pertained not only to morphology but also to his work in optics; and, of course, he assiduously pursued the empirical content of these sciences. Often the methodological and epistemological issues became intertwined with the empirical work, though one essay in particular stands out as more indicative of his general methods of conducting experimental and experiential science: "Der Versuch als Vermittler von Objekt und Subjekt" (The experiment as mediator between object and subject, 1792).[78]

76. William Whewell, *History of the Inductive Sciences*, 3rd ed., 3 vols. (London: Parker and Son, 1857), 3:359–76; citations from pp. 360, 370.

77. In the epilogue I will count the ways in which Darwin was indebted to the Romantics in general and Goethe in particular. In the case of unity of type, in early essays Darwin made unmistakable reference to Goethe's fundamental proposals. He wrote: "There is another allied or rather almost identical class of facts [he had initially mentioned Owen's theory of limb homologies] admitted under the name of Morphology. These facts show that in an individual organic being, several of its organs consist of some other organ metamorphosed: thus the sepals, petals, stamens, pistils, &c. of every plant can be shown to be metamorphosed leaves. . . . The skulls, again, of the Vertebrata are composed of three metamorphosed vertebrae, and thus we can see a meaning in the number and strange complication of the bony case of the brain." See Charles Darwin, *The Foundations of the Origin of Species: Two Essays Written in 1842 and 1844 by Charles Darwin*, ed. Francis Darwin (Cambridge: Cambridge University Press, 1909), p. 215. The comparable reference occurs in Darwin's section on morphology in the *Origin of Species* (London: Murray, 1859), pp. 434–39. See also my *Meaning of Evolution: The Morphological Construction and Ideological Reconstruction of Darwin's Theory* (Chicago: University of Chicago Press, 1992), chaps. 3, 4.

78. Goethe, "Der Versuch als Vermittler von Objekt und Subjekt," in *Sämtliche Werke*, 4.2: 321–32. The essay is dated 28 April 1792, but did not receive its title till publication in 1823.

Goethe thought this essay on experimentation, which he wrote in connection with his optical studies, a quite important statement of his scientific procedure. When he composed it in 1792, Kantian considerations had
only begun to take hold. The reigning spirits of the essay are rather Spinoza
on the positive side and Newton on the negative. Schiller, to whom he sent
a copy in 1798, marked up the essay, urging the more idealistic perspective,
and to this Goethe acceded in a modest revision of the tract for his series
Zur Naturwissenschaft überhaupt, which appeared a quarter of a century
later.

The essay provided a sustained analysis of the subjective snares that
might catch up the unwary researcher and of the objective measures that
must be taken to avoid these dangers. To comprehend exactly what was at
stake for Goethe in his analysis, one first needs to consider what he meant
by "experiment." In the original, the word is *Versuch,* a term for which "essay" would today be a ready translation. In English "essay" not only means a
genre of writing but also an "attempt," a "trial," which comes closer to
Goethe's usage. Several of Goethe's tracts on morphology would have the
term as the first word in their titles, and these instances are indeed most
naturally translated into English by "essay," as for example, "Versuch über
die Gestalt der Tiere" (Essay [Experiment] on the form of animals, 1790).
Goethe, though, explicitly gave the term a meaning that neither "experiment" nor "essay" completely captures:

> When we intentionally, either by ourselves or in company with others,
> repeat experiences [*Erfahrungen*] that have occurred before, and we again
> present the phenomena that arise either [naturally] by chance or by arti-
> ficial means, we call this an experiment [*Versuch*]. The worth of an exper-
> iment resides chiefly in this, that the experiment, whether simple or
> compound, which occurs under determinate conditions of a known ap-
> paratus and technical expertise, can be repeated at any time, provided
> those conditions again obtain.[79]

Goethe included under the rubric of *Versuch,* then, what we would think of
as repeatable observations of natural phenomena (e.g., the developmental
pattern exhibited by plants or animals) as well as repeatable observations of
artificially induced phenomena (e.g., the spectrum of colors produced by a
prism). His various essays on morphology would thus describe his *Versuche,*
his repeatable observations concerning plant and animal forms. In
Goethe's science, then, what counts is the experimental intention, the aim

to make observations that can be repeated under specified natural or artificial conditions.[80]

This last criterion suggested to Goethe the critical difference between art and science. What was virtue in artistic practice—the use of imagination, the desire to entertain, the solitary effort, the unique product, and the need to present only a completed work to an audience—became, he argued, vices in scientific practice.[81] To protect against these liabilities, which he believed led Newton and his followers quite astray in optics, one must: (1) work cooperatively with other observers, sharing information; (2) publish empirical evidence as it becomes available; and (3) repeat experiments and conduct many of them to ascertain those relationships actually characteristic of nature. Goethe thought Newton sinned particularly against this last requirement. The Englishman conducted only a few prism experiments and thus precipitously established his theories on shifting and shallow ground. Goethe suspected that Newton had already convinced himself of his theory and had selectively employed a few experiments to prove what he had already assumed. He lacked sufficient reflection to recognize that his own desires, imagination, and self-satisfaction hastily propped up what a more thoroughly objective researcher might easily topple over. A more careful investigator would, as the history of science evinced, conduct many experiments, compare them with one another, and order them in sequence prior to any formulations of a more abstract sort. Goethe believed this sequential ordering of phenomena essential so as "to capture connections that are true to nature."[82] Had he himself examined only, say, the stem, flower, and seeds of a plant, without carefully following in sequence all the intermediate stages of development—and then repeating such observations for a number of different plants—he would not have concluded that all the organs of a plant were modifications of a fundamental element. So when empirical evidence has been gathered and arranged in the way prescribed, the experimenter might then be able to derive certain propositions of a higher level that would indeed connect disparate phenomena, as Goethe thought himself able to do both in optics and in biology. And if successful at this second level, he might, perhaps, discover a yet more comprehensive law. Goethe summed up his analysis of the proper use of experiment in this way:

80. Even today the critical feature of empirical science is controlled (thus, repeatable) observation, whether that observation is yielded by controlling an apparatus (for instance, a cyclotron) or by controlling a conceptual situation (for instance, a plan of observation in astrophysics).

81. Though Goethe mentioned the need to present a completed work as characteristic of art, he himself often published uncompleted poems and plays, as if they were *Versuche*.

82. Ibid., 4.2:326.

My intention is to collect all experience in this area, to perform all experiments [*Versuche*] myself, and to do so throughout all their varieties,
and thereby make it easy to repeat them and thus allow other researchers
the chance to observe them. In this way, I hope to establish from a higher
standpoint the propositions that express these observations and to determine whether such propositions can be brought under a yet higher principle.[83]

The epistemological and methodological proposals contained in this essay suggest that by 1792 Goethe had not yet advanced beyond the Kant of
the first *Critique*. He would later, for instance, soften to indistinction the essay's sharp division between art and science. Kant had indeed maintained
that the artistic endeavor was grounded in a subjective feeling, albeit of
universal extension, and that proper, mechanistic science (as optics was
thought to be) had anchors in necessary, objective nature. After the tutelage of Schiller and his friendship with Schelling, Goethe would find the
message of Kant's third *Critique*—or at least the Romantic interpretation of
that book—more compelling: namely, that art and science had deep foundations within a nature that encompassed both the subjective and the objective. He would find in imagination not the betrayer of truth but the
faculty of creative possibility. And he would be more circumspect concerning scientific theory and the way it might guide one to sound observation.[84]
Indeed, after a more studied examination of Kant and the constant urgings

83. Ibid., 4.2:332. In 1798 Goethe sent his essay to Schiller, who commented on it (12 January
1798), characterizing the older poet's procedure as a "rational empiricism" [*rationelle Empirie*].
Schiller affirmed the need to follow phenomena carefully, on the one hand, but on the other, also
"to allow the imaginative powers and the faculties of combination the freedom to explore according to their pleasure, with this restriction, that the imaginative power should stay within its own
world and not attempt to constitute anything in the world of facts." Goethe answered the next day
(13 January 1798): "I have moved, by sound step and careful attention, from a stiff realism and
plodding objectivity to where I can endorse your letter as my own confession of faith." See the exchange of letters in *Sämtliche Werke*, 8.1:491–95. Goethe then sent to Schiller another brief essay
that distilled his method into a three-stage procedure: one began with ordinary "empirical phenomena"; this would then be turned into "scientific phenomena" through experiment and ordering
in a sequence; and, finally, from these experiments and observations, a "pure phenomenon" would
result. This latter would, he contended, not be isolated, but perceived in the "continuous series of
appearances." What he seems to have meant by this last note was that the pure phenomenon—the
Urtypus of his morphological essays—would not be an entity abstracted from individual experience, but one that existed in experience, that organized experience, and that simultaneously organized the objects of experience. See Goethe, ["Das reine Phänomen"], in ibid., 6.2:820–21.

84. In the published version of the essay, Goethe dropped this line: "Imagination, which lifts
him to the skies on its wings, when he believes his feet to still be on the ground." He also dropped:
"that one should not attempt to demonstrate something immediately from experiments [*Versuche*],
nor confirm some theory by experiments." See Goethe, "Der Versuch als Vermittler von Objekt
und Subjekt," in ibid., 4.2:326.

of Schiller, Goethe would come to hold that observation itself was theory-laden. One simply had to be aware of that fact and become reflectively cautious.[85] In the earlier years, though, he had a more limited conceptual horizon but a greater irritation. He was driven by what he thought the defects of the Newtonian approach: insufficient experiment and hasty generalization, which violated "the rights of nature." And in the essay, the powerful influence of Spinoza can still be felt insofar as Goethe understood the trajectory of scientific work to move from a careful, sequential ordering of experimental results to higher principles of connection. Such an ordering would allow the kind of cognition that might constitute an adequate idea and ultimately an intuition into the whole (*scientia intuitiva*, properly speaking).[86]

This more Spinozistic reading of Goethe's "Versuch als Vermittler" seems confirmed by a very brief essay he composed a little after this time. It concerns the nature of the "pure concept" of natural forces that lie beyond observable phenomena. He wrote:

> Since the simpler powers of nature are often hidden from our senses, we must seek, through the powers of our mind [*die Kräfte unseres Geistes*], to reach out to them and represent their nature in ourselves . . . [for] our mind stands in harmony with the deeper lying, simpler powers of nature and so can represent them in a pure way, just as we can perceive the objects of the visible world with a clear eye.[87]

A Spinozistic monism, in which natural processes have a mental counterpart that can be recognized through careful experimental procedure and that can be captured at a higher level of cognition—these assumptions

85. In the foreword to his *Farbenlehre* (1810), Goethe warned against the naive assumption that our experience of nature is free of theory: "The most amazing demand is sometimes made, though the individuals who make it cannot themselves meet the demand, namely: that one should undertake experiential observations [*Erfahrungen*] without any theoretical assumptions and that it should be left to the reader or student to form his own conclusions. But we cannot make any progress through mere observation of something. Every observation becomes a consideration, every consideration becomes a reflection [*Sinnen*], every reflection becomes an inference [*Verknüpfen*]—so one can say that we theorize every time we attend to something in the world. But we should be conscious of this, we should be self-aware of it, and undertake it freely—and, to use a bold expression—with irony. Such a caution is necessary if the abstraction, about which we are wary, is to be rendered innocuous if the experiential result, for which we hope, will be really vivid and useful." See Goethe, "Vorwort," *Zur Farbenlehre*, in *Sämtliche Werke*, 10:11.

86. Spinoza distinguished three stages of cognition: at the first level was opinion and imagination; at the second, reason—which produced adequate ideas; and finally *scientia intuitiva*, which "proceeds from an adequate idea of the formal essence of some attribute of God to an adequate knowledge of the essence of a thing." See Spinoza, *Ethics*, paras. 2, prop. 40, schol. 2, in *Benedicti de Spinoza Opera*, ed. J. Van Vloten and J. Land, 3 vols. (Hague: Martinus Nijhoff, 1890), 1:104–5.

87. Goethe, ["Reine Begriffe"], in *Sämtliche Werke*, 4.2:332.

seem to lie behind Goethe's methodological pronouncements. The "simpler powers of nature," if thus resident in an external world, would remain inaccessible to a Kantian mind. Herein would lie the attractiveness of Schelling's idealistic Spinozism, which took its departure from Kant, yet which spoke of a nature that did not lie irrevocably hidden in a noumenal world but was continuous with our very selves.

Essays toward a Theory of the Animal Archetype[88]

After he saw into publication his *Metamorphosis of Plants*, Goethe turned to consider animals. He wished to provide structural and developmental descriptions comparable to those he had offered for plants. In the first instance, these descriptions would be based on the conception of a dynamic type—which would later be dubbed an "archetype." The archetype would furnish a model by which to understand the structural and developmental features of all animals. But the archetype, as he gradually came to conceive it, would be more than a simple pattern useful for comparative zoology: it would be a dynamic force actually resident in nature, under whose power creatures would come to exist and develop. From 1790 through 1797, Goethe composed five essays that advanced and refined his conception of that object for which he was groping.

Goethe's first effort in October 1790, his "Versuch über die Gestalt der Tiere" (Essay on the form of animals), may well have been inspired by the revelation that occurred on the Lido in Venice, when he saw something in that battered sheep skull. But he could not shape this essay to his satisfaction and abruptly ended his study after three weeks.[89] He thought his work both too abstract, needing more empirical content, and too raw, needing more theoretical articulation. Disappointment also cooled his enthusiasm for such work, since gloomy clouds of negative reaction trailed his recently published *Metamorphosis of Plants*. Even his friends seemed to think Bonnet's encapsulation theory more plausible than the transformational pro-

---

88. For reasons of distinctiveness, I will generally refer to Goethe's theory of the *Urtypus* (the fundamental type) as his theory of the "archetype," a Kantian term he would introduce in the 1820s. As I will discuss, his theory of the archetype differs considerably from that more familiar conception advanced by the British, especially Richard Owen.

89. Goethe to Karl Ludwig von Knebel (1 January 1791), in *Goethes Briefe*, 2:133: "Hardly had I returned home than I undertook to write an essay on the form of animals. I was encouraged to take this on especially because of a collection of animal skeletons that I found in Dresden. I had spent about three weeks thinking and dictating, but finally I could not get on with this abstract material and had to set it aside." Around this same time, he began another essay on plants, "Metamorphose der Pflanzen, zweiter Versuch," but it ended about as abruptly as his essay on animals. See *Sämtliche Werke*, 3.2:367–70.

posal of the *Metamorphosis*. "No one could understand," he felt, "the serious passion with which I pursued this business; no one perceived how it came from the depths of my inmost being."[90]

Goethe found greater immediate promise in his optical study, which he obsessively pursued from early 1790, when he first looked into a prism, to mid-1794, when he conducted experiments on the physiology of color perception. So great were his hopes for this line of work that on the campaign against the French in fall of 1792 and during the siege of Mainz the following spring (see chapter 2), he carried his optical manuscripts with him and continued to perform experiments on light and color—often in the most extraordinary and trying circumstances. He believed that as a result of the multitude and variety of experiments he was conducting that Newton's hypothesis would "collapse like an old wall as soon as I will have undermined its foundation."[91] From these research efforts, he published, in spring of 1791 and in spring of 1792, two long essays in a series for which he had mounting expectations, his *Beiträge zur Optik* (Contributions to optics). The reaction of the learned public was, however, mixed.[92] With blighted hopes and after the devastating experience in France, Goethe returned to literature to expiate his depressed cynicism about human folly.[93] From early 1793 through fall of that year, he composed an epic poem in twelve parts, a retelling of an animal fable that commented obliquely on the sclerotic ancien régime of France and the chaotic excess that succeeded it. Herder thought his *Reineke Fuchs* (Reynard the fox) "the first and greatest epic of the German nation, indeed of any nation since Homer."[94] Through the rest of 1793, Goethe occupied himself as well with several plays and some renewed effort with the much-lamented *Wilhelm Meister*.

90. Goethe, *Campagne in Frankreich, 1792* (Pempelfort, November 1792), in *Sämtliche Werke*, 14:467–68. In these passages, Goethe relates that his friends at Pempelfort (just outside of Düsseldorf) found his botanical work unconvincing but that they at least found novelty in some experiments with color that he performed for them.

91. Goethe to Johann Friedrich Reichardt (17 November 1791), in *Goethes Briefe*, 2:142.

92. Goethe was quite sensitive to what professionals would say about his work in science; but, like many of us, he often gave disproportionate weight to the negative reactions. Lichtenberg, certainly one of the eminent men of physics in his day, was slow to respond to the *Beiträge zur Optik*, which irritated Goethe considerably. But Lichtenberg had been quite ill, something he explained in a long letter to Goethe. In the letter he praised the poet for the work he had done on the color of surrounding shadows and wished himself to investigate the matter. It was a supportive letter and not, as far as one can tell, written out of the desire to simply humor the great man—the letter was one of substance. He tactfully omitted mention, however, of other aspects of Goethe's theory. See Georg Christoph Lichtenberg to Goethe (7 October 1793), in *Briefe an Goethe* (Hamburger Ausgabe), ed. Karl Robert Mandelkow, 3rd ed., 2 vols. (Munich: C. H. Beck, 1988), 1:135–40.

93. After a short refractory period, Goethe returned to experiments and reading in optics, all of which eventuated in his *Zur Farbenlehre* (1810).

94. Herder to Johann Wilhelm Gleim (12 April 1793), in *Goethes Gespräche*, 1:540.

Gradually his thoughts drifted back to biological work. In 1794 he took up again the problem of the archetype, this time with a charnel house of bones stacked all around him. In "Versuch einer allgemeinen Knochenlehre" (Essay on a general theory of bones), he began an exact empirical investigation of the vertebrate skeleton, which yielded rather detailed descriptions of the head and backbone. During this same year, he composed a methodological-epistemological tract, "Versuch einer allgemeinen Vergleichungslehre" (Essay on a general theory of comparison). These two incomplete works, along with his earlier incomplete "Essay on the Form of Animals," became the basis for a more extensive study in January 1795, his *Erster Entwurf einer allgemeinen Einleitung in die vergleichende Anatomie, ausgehend von der Osteologie* (First sketch of a general introduction to comparative anatomy, based on osteology). About a year after completion of this latter essay, Goethe began a redraft that seemed to have been intended for a more comprehensive study. In later publication, the redraft carried the inelegant title *Vorträge über die drei ersten Kapitel des Entwurfs einer allgemeinen Einleitung in die vergleichende Anatomie, ausgehend von der Osteologie, 1796* (Lectures on the first three chapters of the sketch of a general introduction to comparative anatomy, based on osteology, 1796). The *Erster Entwurf* and the *Vorträge* made it into his *Zur Morphologie* a quarter of a century later, whence their influence radiated to several generations of biologists. Let me offer brief analysis of each of these five essays, which sought to establish the archetype. I will then say a few words about other scattered essays composed from 1796 to 1797 that began to formalize the discipline of morphology and to give it a name.

"Essay on the Form of Animals." Goethe planned this study immediately after the publication of his *Metamorphosis of Plants* and finally began it in early October 1790, some six months after his second Italian journey and his experience with the sheep skull on the Lido. He quite specifically directed his attention not to all animals or to all parts of animals—for example, he excluded the soft parts, hoping later to do something with them—but only to the vertebrate skeleton. He never got beyond some prefatory considerations, however. He seems to have thought the comparative resources were, though growing through the efforts of numerous researchers, yet insufficient to construct a "fundamental image" [*Urbild*], or "common type" [*allgemeiner Typus*], of the animal in general. He recognized the need for wide-ranging comparisons in order that one not be misled by examining only, say, the features of the human adult skeleton. The tracing of a single bone through animals of different species and at different ages would prevent us, he believed, from erring "when this or that part in one class or

genus at a certain age, under certain circumstances, escapes our senses." Extensive comparison would keep all the parts of vertebrates, despite aberrations, "visible to our understanding."[95] Goethe's approach differed considerably from that of other anatomists of the period, who tended to focus their studies on particular vertebrate species (most often the human) and paid scant attention to elucidating a common form that might unite these various groups.[96] His efforts would also differ from those later anatomists, like Richard Owen, who would pursue a general archetypal pattern but one that illustrated the least common denominator of the vertebrate class, describing the vertebrate archetype as essentially a string of vertebrae. By contrast, Goethe conceived the archetype as an *inclusive form*, a pattern that would contain all of the parts really exhibited by the range of different vertebrate species.[97] Since corresponding parts of various groups would vary considerably from one another (for instance, the limb of a horse and that of a human being), the archetypal form would not be representable to the external eye but only to the inward eye. Moreover, even at this early juncture, Goethe hoped to be able to reconstruct the bones of the vertebrate "in the manner that nature has distinguished, constructed, and determined them"—a kind of Spinozistic wish to identify with the constructive power of nature.[98] In 1790, however, these goals faded in the brighter light of his optical studies.

"Essay on a General Theory of the Bones." Some five years later, after the disappointing reception of his optical work, Goethe began another investigation of the vertebrate skeleton. Now he was supplied with sufficient osteological material, as well as with Samuel Thomas Soemmerring's great treatise on the human skeleton.[99] In his essay Goethe described in considerable detail the bones of the head, beginning with the intermaxillary (also called the incisor bone, since it held the incisor teeth). This bone, of course, meant much to Goethe's own research history. He began with its de-

---

95. Goethe, "Versuch über die Gestalt der Tiere," in *Sämtliche Werke*, 4.2:144.

96. Typical of the kind of study taking the more narrow view was Blumenbach's description of the human skeleton, published a few years before Goethe's effort. See Johann Friedrich Blumenbach, *Geschichte und Beschreibung der Knochen des menschlichen Körpers* (Göttingen: Johann Christian Dieterich, 1786).

97. Ryan Boynton drew my attention to this fundamental feature of Goethe's concept of the archetype.

98. Goethe, "Versuch über die Gestalt der Tiere," in *Sämtliche Werke*, 4.2:138.

99. In spring 1791 Soemmerring sent Goethe the first volume of his *Vom Baue des menschlichen Körpers* (On the construction of the human body), which was on the bones. Goethe wrote him in late winter of 1794: "During the past few days, I have read your anatomical guide book through completely, and have taken great pleasure in the content as well as in the method." See Goethe to Soemmerring (January–February 1794), in *Goethes Briefe*, 2:526.

scription, he said, because of its significance for understanding the life of the entire animal. Since a creature used this bone initially to grab food, its particular form mirrored external relationships, "for it determined whether an animal peacefully pulled up grass and grazed, or gnawed on firmer bodies, or forcefully held living creatures." And like all other bones, the intermaxillary also reflected the entire structure of the animal, since it functioned "in general harmony with all the [other] parts of the living creature."[100] Cuvier would later elevate these two relational aspects of animal structure into axioms of morphological understanding. He referred to them as the "conditions of existence" and the "subordination of traits."[101]

From the intermaxillary, Goethe went on to describe the other structures of the head: the upper jaw, zygomatic arch, the lachrymal bone, the palate, the frontal bone, the sphenoid, the postsphenoid, the squamosum, the petrosum, and the tympanicum. In a brief extension of this effort, he trailed down the vertebrae, gave a short description of the sternum, and then turned back to the lower jaw and teeth.[102] After this his stamina seemed to flag for such exacting empirical analyses. He nonetheless felt confident enough to begin generalizing his research experience in a modestly extensive essay on comparative anatomy.

"*Essay on a General Theory of Comparison.*" This small treatise of 1794 was directed specifically to morphological questions, and it shows the impact of Kant's third *Critique* and discussions with Schiller. In the essay Goethe highlighted a particular aspect of Kant's proposal concerning teleological judgment, namely, that organisms, while they displayed an internal teleology, should not be regarded as elements of an external teleology—final causes in a more cosmological sense. In this Kantian light, he urged that the

---

100. Goethe, "Versuch einer allgemeinen Knochenlehre," in *Sämtliche Werke*, 4.2:147.

101. Georges Cuvier, *Le Règne animal distribué d'aprè son organisation*, 2nd ed. 5 vols.; vols. 4 and 5, ed. P. A. Latreille (Paris: Déterville, 1829–30), 1:5–9: "Natural history, moreover, depends on a rational principle, peculiar to itself, which it employs with advantage on numerous occasions, namely, the *conditions of existence*, more commonly known as *final causes*. As nothing can exist that does not bring together those conditions which render its existence possible, the different parts of each being must be coordinated in a way that renders the whole being possible, not only in respect to itself but also in respect to the relations of its environment. The analysis of these conditions often leads to general laws that are as demonstrable as those which derive from mathematics or experimental observation [*l'expérience*]. . . . One can employ an assiduous comparison of a being, if one is guided by the principle of the *subordination of traits*, which is itself derived from the principle of the conditions of existence. Since the parts of a being all arise from a mutual adaptation, there are some traits which, by reason of their character, exclude others, while some require others. When one recognizes this or that trait in a being, one can calculate those others that coexist with them or those that are incompatible with them." Cuvier seems not to have owed anything to Goethe for these principles.

102. Goethe, "Versuch einer allgemeinen Knochenlehre," in *Sämtliche Werke*, 4.2:146–79.

anatomist not conceive the structures of animals and plants as designed for human usage or divine messages, rather that the researcher should understand those structures as having their raison d'être in the functional organization of the entire creature.[103] To adapt Voltaire's piquant example, we should not marvel at a superior wisdom that supposedly designed the bark of the cork tree so fine wines could be preserved; rather, we should try to understand how that part functioned within the organization of the cork tree itself and how it was affected by its geographical circumstances. The relationship of organisms to their environment, according to Goethe, had to be regarded as nonintentional: the environment had an impact on creatures, changing their outward shape to conform to particular requirements and thus giving an organism "its purposiveness in respect to that external environment" [*seine Zweckmässigkeit nach aussen*]. But that was only part of the story, for the internal structures of plants and animals showed another force at work. There was also an "inner kernel" [*innerer Kern*] that provided a general structural pattern for an organism. Extrinsic forces might particularize this internal pattern in different ways: the seal, for instance, had a body formed by its aquatic environment, but its skeleton displayed the same general pattern as that of land mammals.[104]

Living organisms thus derived their structures from two forces: an intrinsic one, which determined the general pattern—and Goethe ventured in his essay that there was a single pattern (*einziges Muster*) for all animals; and an extrinsic force, which shaped an organism to its particular circumstances. He conceived this latter, environmental force much as Lamarck and the young Darwin would, namely, as a direct effect on the organism that adapted it to particular circumstances ("purposiveness in respect to that external environment"). Goethe had replaced divine teleology with natural causality, though a causality that retained a telic feature. Some time later Cuvier and Owen would reach comparable conclusions, though they still detected the ultimate intentions of the Creator expressed in such proximate causes.

After composing these essays in 1794, Goethe again abandoned his morphological study. He had found in Schiller a vital companion who might respond in positively critical ways to his literary work; but he had not found a comparable mind for his scientific work, that is, not until Alexander von Humboldt visited Jena.

103. Goethe had plucked out of Kant's third *Critique* this antiteleological message as early as 1790. At some point late that year, he started a second essay on the metamorphosis of plants but dropped it after only a few pages. He did caution that Kant had provided a cure for these overreaching teleological tendencies that come so easy to the human mind. See "Metamorphose der Pflanzen, zweiter Versuch," in *Sämtliche Werke*, 3.2:367–70.

104. Goethe, "Versuch einer allgemeinen Vergleichungslehre," in ibid., 4.2:182.

First Sketch of a General Introduction to Comparative Anatomy. During the Christmas season of 1794–95, the Humboldt brothers, Wilhelm and Alexander, arrived in Jena.[105] Goethe could not refrain from talking over his anatomical ideas with them, especially with Alexander, whom he thought amazingly learned in science and a sympathetic, if young critic—he was twenty-six years old at the time. Goethe would later say of him that "he has a knowledge and vital wisdom whose like we will not see again."[106] Discussions with the brothers generated the needed enthusiasm and confidence for Goethe to continue his morphological studies. With his previous essays at hand and after further osteological investigation, he dictated to Max Jacobi (Friedrich's son, who was a medical student in Jena) his *First Sketch of a General Introduction to Comparative Anatomy.*[107]

Goethe reiterated his position that the archetype must be empirically derived through extensive comparison over the range of animal species. But he now emphasized that the very idea of the archetype—that there existed a common pattern uniting all animals—was itself a hypothesis that comparison might test. Simultaneously this conception would function as the rational idea in light of which the common *Urform* might be extracted from further empirical data.[108] Since this archetype had to serve as the standard by which many different kinds of animals, including man, would be related to one another, "no particular animal can be used as the canon of comparison."[109] The pattern itself had to be an idea genetically formed, thus incorporating the range of possibilities across species and across developmental stages. As a consequence, the idea would only be visible to the mind's eye.[110] Schiller's argument took hold—Goethe's genius also required the intervention of ideas, ideas he would come to see in their instantiations.

Kant's third *Critique* reverberated through Goethe's essay. He described "the individual animal as a small world that exists for itself and through itself"—that is, he thought of the structures of animals as displaying an in-

---

105. Alexander von Humboldt stayed in Jena from 14 December to 19 December 1794. And he returned the next year for the month of April.

106. Goethe to Eckermann (11 December 1826), in *Gespräche mit Goethe*, p. 161.

107. Goethe, *Tag- und Jahres-Hefte* (1795), in *Sämtliche Werke*, 14:36.

108. Goethe, *Erster Entwurf einer allgemeinen Einleitung in die vergleichende Anatomie, ausgehend von der Osteologie*, in *Sämtliche Werke*, 12:122: "Initially, experience must teach us about those parts that all animals have in common, and wherein these parts differ. The idea must govern the whole and must work to abstract, in a genetical fashion, the general image [*das allgemeine Bild abziehen*]. Even if such a type is proposed only as an experiment [*Versuch*], we can still use the customary comparative method to test it out."

109. Ibid.

110. Ibid., 12:140.

ternal perfection without external purpose, much as Kant had suggested. Goethe seems to have supposed, however, that the perfection flowed not because of the form of this or that particular species but because the general pattern itself was perfect. The actual form a particular species displayed was, as he had before maintained, determined by a balance of internal and external forces. But in this 1795 essay, he introduced a concept—the *Bildungstrieb*—to describe the power of the archetype to cause a development that was "continually renewed through the circle of life."[111] He seems to have thought talk of a *Bildungstrieb* comparable to that of an *innere Kern*—both represented the dynamic force resident in natural objects that caused metamorphic development. The *Bildungstrieb* worked against the external forces of the environment to exact, he believed, a kind of economic or compensatory redress. In the actual expression of the type in a species, if one part was exaggerated, then another had to be diminished in compensation: so, for example, "in the giraffe, the head and extremities are developed at the cost of the main body, while in the mole the reverse is the case." Nature, like a good Weimar administrator, "can never be in debt or certainly not become bankrupt."[112]

The term *Bildungstrieb* was originally coined by Blumenbach; and in several treatises from 1780 through 1789, he explored different aspects of this dynamic force that explained distinctively organic processes (see chapter 5). Goethe was certainly aware of the Göttingen zoologist's theory of the *Bildungstrieb*. But Blumenbach may not have been the proximate source of Goethe's explicit introduction of the concept. In the same year of 1795, Schiller published in *Die Horen* an essay by Alexander von Humboldt entitled "Die Lebenskraft oder der rhodische Genius, eine Erzählung" (The force of life or the Rhodian genius, a tale).[113] The young naturalist used a poetical device to describe the character of the force that worked to unify life processes and to hold in check the tendency of matter so organized to disintegrate, to decompose into elemental parts. Humboldt had met Goethe for the first time in March 1794, shortly after his brother Wilhelm had moved to Jena, and he returned in December of that year for about a week. The next April he stayed in the city to conduct experiments on ani-

111. Ibid., 12:126.

112. Ibid., 12:125–26. Jackson draws some interesting comparisons between Goethe's work as an administrator and his employment of budgeting metaphors in his morphological essays. See Myles Jackson, "Natural and Artificial Budgets: Accounting for Goethe's Economy of Nature," *Science in Context* 7 (1994): 409–31.

113. Alexander von Humboldt, "Die Lebenskraft oder der rhodische Genius, eine Erzählung," *Die Horen*, part 5 (1795): 90–96. Humboldt reprinted his fable in the second and third editions of his *Ansichten der Natur*. See Alexander von Humboldt, *Ansichten der Natur*, 3rd ed. (Stuttgart: Cotta'schen Buchhandlung, 1849), pp. 315–24.

mal electricity.[114] The juxtaposition of Humboldt's visits with Goethe's *Erster Entwurf* suggests that the polymathic researcher was the likely source of the older man's inspiration for the *Bildungstrieb*.[115] Moreover, Goethe simply came to think of Humboldt himself as a veritable life force, so knowledgeable and energetic was he.

In his *First Sketch*, except for the explicit use of the idea of a *Bildungstrieb*, Goethe reiterated notions of the year before. What he mostly added to this essay was a systematic detailing of the different ways in which conditions might alter the expression of the archetype across species and developmental stages. So, for instance, he cautioned that the comparative morphologist had to examine the fetuses of animals, since bones separated in the young might be fused in the adult. Further, the same bone might have a different size in different organisms, a different shape (thus the jaw of deer and lion), a different location (for instance, in the ox the upper jaw is separated from the nose bone by the lachrymal, but no such division occurs in most other animals), a different number (so, for instance, the digits in the ape and horse), or be absent (e.g., many animals have no clavicle). Goethe even recognized that some animals might have parts calcified where others would have cartilage (the baculum or "penis bone" in some bats).[116] Despite these phenomenal differences, Goethe insisted that the same archetypal force operated to produce the same pattern in all animals.

Under the rubric of "animal," Goethe wished to include insects, at least the more perfect ones. These latter could be divided into a fore section, midsection, and rear section, which were also the major divisions of the vertebrate skeleton. The location of insects within the animal archetype conformed to Goethe's basic notion that this *adequate idea*, as Spinoza would have termed it, represented the potential of all animals; it was an inclusive idea as opposed to a minimalist idea. At this stage of his own considerations, though, he was content to focus on one aspect of the archetype, the bony structures of higher animals.

The empirical part of Goethe's essay had an introductory enumeration of the bones of the vertebrate archetype, with scattered and modest descriptions. His brief listings make it clear that the methodological and epistemo-

---

114. See Kurt-R. Biermann et al., *Alexander von Humboldt: Chronologische Übersicht über wichtige Daten seines Lebens* (Berlin: Akademie-Verlag, 1968), pp. 12, 14.

115. Later in his series *Zur Morphologie*, Goethe would declare his preference for the idea of *Trieb* (drive) as representing the developmental dynamic of living nature, since *Kraft* (force) indicated a material principle that could not explain organization. See Goethe, "Bildungstrieb," *Zur Morphologie*, in *Sämtliche Werke*, 12:100–2. In the 1790s Goethe seems to have displayed no comparable linguistic abstemiousness.

116. Ibid., 12:136–44.

logical aspects of archetype theory were of more interest to him than were precise empirical analyses. He undoubtedly believed he could make a greater contribution to biological science at this more abstract level. The essay did not push his work much beyond the efforts of the previous year; and at the end of January 1795, he turned his energies elsewhere, particularly to the composition of his novel *Wilhelm Meisters Lehrjahre*, which he had begun a few months prior.

Lectures on the Sketch of a General Introduction. Sometime in 1796, perhaps in the fall of that year, Goethe went through another spurt of work in pursuit of the archetype, which yielded his *Lectures on the Sketch of a General Introduction.* These lectures consisted of a redraft of the initial two chapters of his *First Sketch of a General Introduction*, which advanced his considerations modestly, and a third section, which incorporated some newly emerging ideas. After three sections, however, he left off. He seems never actually to have delivered this essay as a lecture, but he might have been inspired to consider doing so by Alexander von Humboldt, who lectured on his own research into animal electricity in Jena during late winter of 1797.[117] Goethe attended the young scientist's lectures, conferred with him frequently, and witnessed some of his experiments on electrical transmission in the nerves.[118] Humboldt seems constantly to have prodded Goethe into more work on anatomy.

The first section of Goethe's essay gracefully argues the advantage of understanding the structure of animals so as to learn about the more intricate and complex arrangements of the human form. Were researchers merely to settle for an examination of outward phenomena, they would easily be led astray by the different manifest features of animals and by the assumption of divine purpose in their forms. One had to explore something deeper. The second section introduced the concept of the type, which could only be discovered through anatomical investigation of different animal groups. Goethe remarked that the common type had been obscured by the practice of anatomists, who at best undertook pair-by-pair comparisons (say, a wolf with a lion) but did not extend those paired comparisons to other animals. But once a common pattern had been empirically derived, we could use it to investigate the structures of a multitude of animals without having to refer each new structure to all of the others, only to the common standard.

117. The time of his lectures is given in Biermann, *Alexander von Humboldt*, p. 18. I conjecture that Goethe might later have attached the title "lecture" to his essays because of the inspiration provided by Humboldt.

118. See Goethe, *Tagebücher* (1–16 March 1797), in *Goethes Werke* (Weimar Ausgabe), 133 vols. (Weimar: Hermann Böhlau, 1888), III.2:58–61.

Such a standard—that is, the archetype—could not, therefore, be identified with a particular animal group, since it had to serve for all groups.

Up to this point, Goethe had only reissued his earlier notions. In the second section, however, he modestly extended his considerations by emphasizing features of archetype theory that had more pronounced Kantian contours, but a Kant to Goethe's own liking. He asserted that we could be sure this talk of a fundamental type did not rest on mere hypothesis, since it followed from "the concept of a living, determined, fully differentiated, and spontaneously effective natural being."[119] Such a being would have its parts mutually dependent upon one another and comprehensible only in relation to the whole. Now, of course, Kant urged much the same, though he specified that such a teleological concept, while characteristic of the human mode of thought, was only regulative, not constitutive of external nature. In one sense, then, Kant's conception of a living being was precisely hypothetical and could not function in authentic science—since legitimate science could only be mechanistic.[120] Goethe used the Kantian framework to picture another conception, which he put in this way:

We are thus assured [by reason of our concept] of the unity, variety, purposiveness, and lawfulness of our object. If we are thoughtful and forceful enough to approach our object and to consider and treat it with a simple, though comprehensive mode of representation [Vorstellungsart], one that is lawfully free [gesetzmässig-frei], lively, yet regular—if we are in a position, employing the mental powers that one usually calls genius (which often produces rather dubious effects), to penetrate to the certain and unambiguous genius of productive nature—then we should be able to apply this meaning of unity in multiplicity to this tremendous object. If we do so, then something must arise with which we as men ought to be delighted.[121]

119. Goethe, Vorträge über die drei ersten Kapitel des Entwurfs einer allgemeinen Einleitung in die vergleichende Anatomie, ausgehend von der Osteologie, 1796, in Sämtliche Werke, 12:203.

120. See chapter 5 for a discussion of Kant's conception of biology as an imperfect science.

121. Ibid. The text is ambiguous as to whether the concept "lawfully free" (gesetzmässig-frei) and the others in that string of adjectives should be applied to the mode of representation or to the object represented (i.e., the type). Kant cultivated the notion that our moral and aesthetic representations were the result of being oneself freely under law. Goethe conceived of the archetype as embracing a lawlike pattern, but one that permitted the freedom of empirical expression. Insofar as Goethe attributed the same concept to the genius of nature as well as to that of man, perhaps he intended the ambiguity. More generally, Goethe liked the notion that constraint imposed a kind of freedom. Fichte, whose work Goethe knew quite well, had elevated the notion of representation (Vorstellung) to a prominent place in the epistemology of science, and he, too, insisted on the freedom of the act of representing.

In this turbulently flowing passage, several eddies of Kantian meaning form around the bedrock of Goethe's realism. First, he suggests here that the necessary concept we have of living beings assures us of their fundamental unity of type, a unity that nonetheless permits great variety in realization. Second, he thinks of this concept as being lawfully free (*gesetzmässig-frei*), a phrase and conception that seems to come straight from Kant's description of aesthetic judgment as stemming from "the free lawfulness of the understanding" [*die freie Gesetzmässigkeit des Verstandes*].[122] But Kant held aesthetic judgments, as well as judgments of intrinsic teleology, not to be determinative but reflective; and such reflective judgments in the case of organisms (i.e., teleological judgments) are regulative (i.e., virtually hypothetical). Yet the archetype for Goethe, as he here thinks of it, is not merely a regulative or hypothetical concept, since it is the same one that the productive genius of nature employs—in Kant's terms, it would be determinative. Goethe here obviously plays off Kant's definition of genius as "nature giving the rules to art," though the art in this instance is the science of morphology. Goethe's residual Spinozism, according to which productive ideas reside in nature, gives a decided turn to his newly adopted Kantianism, a twist that most Kantians would find simply confusing if not confused. After all, why should we assume that our mode of conceiving nature has purchase on nature herself? Schelling will have an answer to this question.

In the third section of the *Lecture*, Goethe introduced a new proposal that would distinguish archetypes according to the kinds of development they specified. His proposal rested on the differences he now recognized in the organizational structure of mineral bodies, plants, insects, and higher animals. The first important divide was between the inorganic and the organic. Any change of a constituent of an inanimate body, he observed, had an immediate effect on the form of the whole. Further, mineral bodies could be separated into their parts and then reconstituted from those parts or combined with other elements according to the elective affinities of the components. Organic bodies, by contrast, had a subordination of parts; and should their unity be destroyed, one could not reconstitute them from the detritus. Though Goethe did not use precisely the same language, these differences marked for Kant the distinction between a *mechanical* system and an *organic* system. In Kant's terms, the parts determined the whole in me-

---

122. Kant, *Kritik der Urteilskraft* (§22), in *Werke*, 5:325 (A 69, B 70). Goethe frequently urged the idea that freedom was only found in restraint. His poem "Natur und Kunst" ends with the line "And only the law can give us freedom" [*Und das Gesetz nur kann uns Freiheit geben*]." See "Natur und Kunst," in *Werke*, 1:245.

chanical systems, whereas the whole determined the parts in organic systems.[123]

Goethe, however, went beyond Kant to make some important distinctions between the kinds of metamorphosis undergone by plants and by animals. Generally speaking, the parts of plants arose through transformations of the same organ—the leaf—through *successive* stages. The parts of animals, however, were not formed from the same organ; moreover, they arose not successively but *simultaneously*.[124] This difference meant that the parts of a plant (for instance, cuttings from a plant) could generate a new, whole organism. Obviously this could not be done with animals; for in the animal economy, one part could not recapitulate the whole. Plants, as Goethe now came to understand them, were not like individuals, as animals were, but like an assembly (*ein Vielfaches*).[125] Further, though the parts of plants had a successive ordering—for example, from calyx to corolla—some reversal of order was possible—thus, from corolla to calyx. But even the imperfect animals, which displayed both simultaneous and successive development, could not retrace their stages: nature only moved from egg to caterpillar to butterfly, but did not reverse. In the higher animals, simultaneous metamorphosis dominated development. So, for instance, though the vertebrate backbone was composed of the same type of bone—the vertebra—yet as the string of vertebrae simultaneously appeared in the fetus, they developed through reciprocity with other parts of the organism, so that finally the vertebrae in the neck would be shaped rather differently from those in the tail. This example of the same part (the type of the vertebra) being modified extensively through "the harmony of the organic whole" indicated that, in the world, organisms as different as giraffe and elephant could nonetheless be modeled on the same plan. Goethe concluded by thus observing two different kinds of metamorphosis in the higher animals:

> First, that which we have already seen in the backbone, when identical parts, according to a certain schema, are differentially modified in the most continuous way by the formative force [*die bildende Kraft*], whereby the type in general is made possible; and second, the metamorphosis according to which the particular parts identified in the type are continually changed throughout all the animal races and species.[126]

123. Kant, *Kritik der Urteilskraft* (§77), in *Werke*, 5:526 (A 347, B 351).

124. Goethe means this to be the general pattern distinguishing plants from animals. Certainly in plant development some organs will arise simultaneously and in animals there will be a sequential appearance of some parts.

125. Goethe, *Vorträge über die drei ersten Kapitel des Entwurfs*, in *Sämtliche Werke*, 12:207.

126. Ibid., 12:211.

We must note two crucial aspects of Goethe's analysis here. First, he does not claim that a single part, namely, the vertebra, underlies the plan of the whole vertebrate skeleton, but only that identical parts in the skeleton— for instance, digits in front and rear limbs—have become differentially modified at the same time, just as have the vertebrae in the neck and in the tail. This was not the view of most morphologists following in the Goethean tradition. Lorenz Oken and Carl Gustav Carus, for example, would argue that all the bones of the skeleton could be understood as modifications of the vertebra—skull, limbs, and ribs would all be interpreted as transformed vertebrae. For them, the archetype of the vertebrate skeleton would become only a string of vertebrae, or, rather, a single vertebra, just as the plant archetype was only a single element, the leaf. They thus collapsed the differences between the plant and the animal archetypes, whereas Goethe kept them structurally distinct. The British inherited Oken and Carus's version of archetype theory. The second important feature of Goethe's conception is that he thought of the unified archetypal plan as including all of its possibilities—again, a very different idea from the minimalist interpretation of his immediate successors and the British. Their archetype represented the lowest common denominator, the vertebra, shared by all members of the group.

Morphology as a Science

Goethe was not the first to bring the term *Morphologie* to print. That credit goes to Karl Friedrich Burdach, who, in 1800, used the word to describe the general form of the human being.[127] Goethe, however, coined the term at least four years earlier. In November 1796 he wrote Schiller about his recent geological studies and remarked that without geological considerations, "the famous morphology would not be complete."[128] His conception of that "famous" science underwent rapid development, which can be followed through several scattered notes and brief essays that he composed roughly between 1796 and 1798. The fundamental conception, though, is contained in a brief definition, which comes from the time he wrote to Schiller:

> Morphology rests on the conviction that everything that exists must express and indicate itself. The first physical and chemical elements to the spiritual manifestation of the human being serve to confirm this funda-

127. See Karl Friedrich Burdach, *Propädeutik zum Studium der gesammten Heilkunst* (Leipzig: Dyt'schen Buchhandlung, 1800), p. 62.
128. Goethe to Schiller (12 November 1796), in *Sämtliche Werke*, 8.1:268.

mental principle. We turn immediately to what has form. The inorganic, the vegetative, the animal, the human—each expresses itself, each appears as that which it is to our outer sense and to our inner sense. Form is a moving, a becoming, a passing thing. The doctrine of forms is the doctrine of transformation. The doctrine of metamorphosis is the key to all signs of nature.[129]

Goethe's theory of the *Urtypus* was thus central to his conception of the science of morphology. While he initially included geological formations as part of this science, the Kantian distinction between the inorganic and the organic seems to have convinced him that rock forms should be decisively distinguished from organic forms. The principal difference was that geological structures lacked dominion over their parts; they arose passively, merely as a consequence of the features of their component minerals. In the organic realm, form had both epistemological and ontological priority, the latter most conspicuously displayed by the ability of plants and animals to reproduce their forms in offspring, even while component parts of progeny might differ from the parent.[130]

In the above definition, Goethe firmly reiterated his fundamental epistemological assumption that our senses met an endlessly various world but that they could lead us to the deeper structures of that world, which the inward eye might gaze upon. With the mental eye, we would see form as dynamic, as containing its infinite variety of transformations within a unity, hence his claim that "the doctrine of forms is the doctrine of transformation." Goethe never abandoned the assumption that nature existed as a reality quite apart from our subjective experience and that through effort we might gain access to her secrets. He did, however, have to find a way of squaring these convictions with his grudging recognition of the soundness of the Kantian epistemology.

Though the above definition makes morphology one of the very fundamental sciences, Goethe located it within a hierarchy of the other sciences. In two sets of notes from 1796–97, he tried out a couple of different ways of situating morphology.[131] The slight differences between the notes are not as important as what they further reveal about the deeper nature of Goethe's reflections. His initial ordering of the sciences and his description of morphology's relation to them are not remarkable. *Natural history* considers

---

129. Goethe, "Morphologie," in ibid., 4.2:188.
130. Goethe, ["Betrachtung über Morphologie"], in ibid., 4.2:197.
131. These notes have been given the title "Betrachtung über Morphologie" (Observations on morphology) and "Betrachtung über Morphologie überhaupt" (General observations on morphology), in *Sämtliche Werke*, 4.2:197–204.

the variety of different species and genera and arranges them systematically. Anatomy examines the internal structures of those different species. *Morphology*, unlike natural history, finds the unity within the diversity of species; and unlike anatomy, it begins with the whole organizational structure, using that structure to comprehend the various parts. In addition to natural history and anatomy, Goethe distinguishes two other sciences in regard to which morphology may be related, and therewith another feature of his conception emerges. These are the sciences of *Zoonomie* and *Physiologie*. By *Zoonomie*, a term apparently taken from Erasmus Darwin, he means a "study of the whole insofar as it lives, with this life supported by a particular physical force."[132] *Physiologie* is a "study of the whole insofar as it lives and acts, with this being supported by a mental force [*eine geistige Kraft*]." *Morphologie*, he proposes, has a relation to both of these disciplines, according to two specifications he provides: Morphology is, first, a "study of the form both in the parts and in the whole, in its commonality [over different species] and its deviations, without any other considerations." From this perspective, morphology also examines the physical force that form represents. Second, morphology is a "study of the organic whole through a representation of all these considerations [of its unity and diversity] and their connections through the power of the mind [*die Kraft des Geistes*]."[133]

Goethe did not publish his notes relating morphology to the other sciences, and they were left in an inchoate and tentative condition after his death. They yet suggest a definite direction of his thought. He conceived morphology as the science of form, the whole and its parts, including the deviations or possibilities contained in the form. He also thought of morphology as a study of the forces of development. Such forces had been discussed by Blumenbach under the rubric of the *Bildungstrieb*, which he likened to the physical force of gravity (see chapter 5). Goethe understood that force to have two sides, an empirical side visible in its particularity to the physical eye and a nonphysical side visible in its generality to the mind's eye. But not only could we observe the empirical and the nonphysical aspects of the life force in other individuals, as it were from the outside, Goethe believed we could also experience them in ourselves from the inside: the healthy individual could become conscious of that feeling of well-

---

132. Darwin's book *Zoonomia, or the Laws of Organic Life* (1794–96) was translated into German, in several parts, almost immediately. See Erasmus Darwin, *Zoonomie, oder Gesetze des organischen Lebens*, trans. J. D. Brandis, 3 vols. in 5 (Hannover: Gebrüder Hahn, 1795–99).

133. Goethe, ["Betrachtung über Morphologie"], in *Sämtliche Werke*, 4.2:200. In the companion notes ["Betrachtung über Morphologie überhaupt"], he distinguishes *Zoonomie* into two parts, one that examines the physical force underlying life and one the spiritual force (ibid., 4.2:201).

being, the feeling of life, that resides not in a particular part but in the whole.[134]

Summary of Goethe's Morphological Considerations

In 1786, during Goethe's last months in Weimar before his Italian journey, we have the first glimmers of his musing on the concept of a *Urpflanze*. In Italy the idea began to flower, and through the next decade, he continued to cultivate that notion, developing and generalizing what became the idea of the *Urtypus*—or archetype—in botany and zoology. By 1797 he had formulated these theoretical proposals, supported as they were by considerable empirical research, into the science of morphology. The results of this research, along with some refining considerations, finally became public in his series *Zur Morphologie*, which appeared from 1817 through 1824 (discussed below).

Morphology was to be a science of organic forms, finding its place within a collection of disciplines, like natural history and anatomy, and serving as an instrumental science (*Hülfswissenschaft*) for them all. Archetypes would be established for the plant and animal kingdoms by thoroughgoing comparative research. Once fixed they might serve as models for investigating the whole organic realm. But they were more than ideational plans. Goethe understood them also as the dynamic causes of development in those realms. As a standard that included all possible forms, the archetype could serve for comparative research into the variety of species; and as a dynamic force, it could explain the proliferation of those species forms. The archetypal force would express itself according to certain (if unknown) principles of economic compensation: any expansion of one part would be balanced by the contraction of another. The plant archetype would represent that single organ (call it the "leaf"), which could be transmuted into the various parts of a plant. Plant development under its aegis would, generally speaking, occur successively, consisting in the sequential appearance of the one organ in various guises. The animal archetype would encompass all of the animal parts (most conspicuously the bones), and it would control development of those parts appearing simultaneously (for instance, the vertebrae from neck to tail). The form of actually living plants and animals would re-

---

134. Goethe, ["Betrachtung über Morphologie überhaupt"], in ibid., 4.2:202: "We ourselves are conscious of such a unity: for we are conscious of being in a perfect state of health when we sense the whole and not its parts. This whole can exist only insofar as natures are organized, and they are organized and act only through the condition we call life." Johann Christian Reil would make a comparable observation a few years later, calling that experience of well-being *Gemeingefühl*. For a discussion of Reil's theory see chapter 7.

sult from an interaction of the archetype (now as *Bildungstrieb*) and the external forces of the environment.

Was Goethe's science good science? Good science, I believe, is not essentially characterized by the discovery of unalterable truths but by fruitful ideas that have the authority of empirical confirmation. To believe otherwise would be to deny that any scientist prior to the late twentieth century practiced good science—and we may expect that our contemporary science will not be able to apply the philosopher's stone to turn all of its theories to lasting gold. Though Goethe often insisted that he did not employ speculative hypotheses—the Newtonian warning echoed down the generations— he obviously did assume some hypotheses. Yet his ideas were empirically grounded in two ways: they were suggested, in part, by his observational studies, and they were confirmed by the extension of those studies into new areas. And the ideas were fruitful, as generations of later scientists would attest (Whewell and Owen conspicuously among them). These ideas invaded not only botany, establishing therein the science of plant morphology, but they also reached over to zoology, grounding there a comparable science of animal morphology, which itself became the backbone of evolutionary theory. Though botany and zoology today focus more on the microstructures of organisms, especially cellular phenomena and genetics, the broader morphological considerations that Goethe initiated continue to be an important part of these sciences. Perhaps had Goethe come to his senses and confined himself to science instead of exercising a polymathic genius in works of extraordinary imagination and sensitivity, the harsh rejections by many contemporary critics might well have evanesced before they took substance. His poetic gifts, however, seem only to have invited suspicion of his science.

Putting aside the philosophical question of the "goodness" of Goethe's science, a lingering historical question, which I have mentioned several times already, remains: Since Goethe accepted the basic Kantian epistemology, how could he assume that a researcher might have in consciousness the same archetypal ideas actually used by nature in her constructions? I will offer an answer to this question in the next section.

The Romantic Circle and Schelling

The Meaning of Romanticism for Goethe

In the last decades of his life, Goethe occasionally uttered disparaging remarks about Romanticism in general and several Romantics in particular. In 1808, during a conversation with Friedrich Wilhelm Riemer (1774–

1845), his secretary and the tutor of his son August, he compared ancient tragedy with contemporary Romantic drama:

> Ancient tragedy is humanly played. With the Romantic there is nothing natural, original, but something contrived, labored, overblown, over-done, and bizarre, descending into the grotesque and into caricature. . . . So-called Romantic poetry attracts particularly our young people, since it is arbitrary, sensuous, tending to disconnectedness—in short, it flatters the inclinations of youth.[135]

And he famously remarked to Eckermann in 1829 that

> a new expression has occurred to me . . . that rather well captures the re-lationship [between the classical and the Romantic]. The classical I call healthy and the Romantic sick. . . . Most of the new poetry is not Ro-mantic because it is new, but because it is weak, sickly, and ill, and the old is not classical because it is old, but because it is strong, fresh, cheerful, and healthy.[136]

These *obiter dicta*, frequently cited, seem utterly to condemn Romanticism in the poet's eyes.[137] What is usually overlooked, however, is that in both of these instances, Goethe was reacting to Romantic literature in France after the turn of the century, not to the literature of the early Romantic move-ment in Germany. Were the above remarks meant to be universal indict-ments, he would not likely have admitted so sanguinely to Eckermann that Schiller "demonstrated to me that I myself, against my will, was a Roman-tic."[138] His more studied considerations of Romanticism provide further counterbalance. In an essay entitled "Klassiker und Romantiker in Italien, sich heftig bekämpfend" (Italian defenders of the classic and romantic in mighty battle, 1820), he judged that the war then being waged between classical and Romantic writers in Italy was already being resolved (*zu ver-ständigen anfangen*) in Germany. Because of the often-abstruse character of Romanticism, many critics, he observed, simply gave up and used the word to refer to whatever was dark, foolish, and incomprehensible. Indeed, "even

135. Goethe in conversation with Riemer (28 August 1808), in Friedrich Wilhelm Riemer, *Mitteilung über Goethe*, ed. Arthur Pollmer (Leipzig: Insel Verlag, [1841] 1921), pp. 295–96.
136. Goethe to Eckermann (2 April 1829), in Eckermann, *Gespräche mit Goethe*, p. 286.
137. The conventional view that Goethe defended the classical mode of thought against the Romantic is argued in several articles by Stuart Atkins. See, for example, his "*Italienische Reise* and Goethean Classicism," in *Aspekte der Goethezeit*, ed. Stanley Corngold, Michael Curschmann, and Theodore Ziolkowski (Göttingen: Vandenhoeck & Ruprecht, 1977), pp. 81–96. Atkins cites many other authors who make the same assumption.
138. Goethe to Eckermann (21 March 1830), in Eckermann, *Gespräche mit Goethe*, p. 350. This passage is discussed above.

in Germany the most noble title of *Naturphilosoph* has become debased in a cheeky way as a byword and an insulting term." But the Romantic movement in Germany, he believed, had already passed through its religious phases, which, indeed, had at times sunk into the morass of viscous obscurity. Goethe concluded his essay by simply remarking that "a word, through consequences of its usage, can take on a completely opposite meaning—since in our tradition nothing lies closer to the Romantic than the Greek and Roman."[139] Since notions of the classical and Romantic arose out of both Schiller's reflections on the naive and sentimental and the Schlegel brothers' studies of Greek and Roman poetry, the categories, indeed, had deeply rooted connections.

In the last years of his life, as he composed the enigmatic second part of *Faust*, Goethe believed his own poetry to be effecting that reconciliation between the classical and Romantic about which he had remarked in his essay (and both Schiller and Friedrich Schlegel, in years past, had observed even in his earlier works).[140] He recognized that passages in his play would undoubtedly provoke critics to use the description "Romantic" as an epithet. In justification of those obscure places both in his *Faust* and in his earlier poetry, Goethe offered what must be regarded as a defense of the Romantic mode in literature:

> Thus because of other darker places in my earlier and later poetry, I would like to offer the following for consideration: Many things in our experience do not allow for full expression and direct communication; so I have, for a long while now, chosen to reveal to the attentive the more hidden meaning through opposition and multiply reflecting images.[141]

And this, of course, is the mode not only of Romantic poetry but of all good literature, as Friedrich Schlegel would have observed.

139. Goethe, "Klassiker und Romantiker in Italien, sich heftig bekämpfend," in *Sämtliche Werke*, 11.2:258–62.

140. Certainly a major theme of *Faust* has a Romantic construction in part 1 and a classical-Romantic resolution in part 2. In a scene that Goethe composed in the years 1797–1801—during the pinnacle of the Romantic rush in Jena—he introduced the wager Faust strikes with Mephistopheles, namely, that the spirit cannot produce a situation which would cause Faust to say: "Tarry awhile, you are so beautiful" [*Verweile doch! Du bist so schön*] (l. 1700). This is because, as the Romantics urged, man is a creature of eternal becoming, of striving (*Streben*), not a classical being of Greek perfection and stasis (ll. 1675–77). In part 2, written in the last months of his life, Goethe has Faust undertake a plan with whose realization he could indeed rest content. With that plan he both loses his bet with Mephistopheles, because he utters the formula, and wins, because the foreseen event lies in the distant future and transcends Faust himself (ll. 11580–86). And with the resolution of the wager—and after an angelic battle—choirs of angels carry his soul up to heaven (ll. 11612–843) . See *Faust I*, in *Sämtliche Werke*, 6.1:581, 580; and *Faust II*, in ibid., 18.1:335, 337–43.

141. Goethe to Carl Jacob Ludwig Iken (27 September 1827), in *Goethes Werke*, IV.43:83.

Goethe directed his animadversions on Romanticism principally to the excesses of French poets and to the later, religious turn of Romantic writing. He reprehended the Schlegel brothers in particular; because after their conversions to orthodoxy in the new century, they prized only those artistic works that advanced a religious purpose.[142] But in the late eighteenth century, he expressed a quite different attitude about the spring flowers of the Romantic movement and their works. I have described the figures of the early movement in previous chapters, but let me quickly recapitulate Goethe's relationship to them. I will then focus on his interactions with Schelling, which had significant consequence for them both.

The Gathering of the Romantics

Goethe realized that the university at Jena suffered a tremendous loss in 1794 when Karl Leonhard Reinhold accepted a position at Kiel. The *Geheimrat* found, however, opportunity rapidly to make good the departure with the man whom Kant himself initially admired, Johann Gottlieb Fichte. The philosopher's first weeks at the university produced tremendous excitement. He and Schiller befriended one another immediately, and Goethe took to reading the new arrival's works. Goethe's grasp of the philosophy impressed Fichte: "He [Goethe] has represented my system so succinctly and clearly that I myself could not have done it more clearly."[143] During his five years at Jena, Fichte raised the philosophical stakes of discussion; and he made Goethe aware—as much through force of a powerful personality as through his difficult ideas—of the liabilities of crucial aspects of the Kantian philosophy, especially the notion of a *Ding an sich*. By 1799 Fichte's relationships with other faculty members at the university, many jealous of his standing and irritated by his imperious attitudes, became so frayed as to leave him dangerously dangling before less sophisticated detractors. A devoted cadre of students stayed loyal, but the fraternities continued to give him a hard time for his grousing about their rowdiness. Fichte finally stumbled in his relationship to Carl August when a charge of athe-

142. What really irritated Goethe about the Schlegels was not so much their later sharp turn toward religion (Friedrich converted to Catholicism in 1808), but the intrusion of religion into aesthetic evaluations. Friedrich thought the modern painter had to have deep religious convictions to be great. The year before he died, Goethe rendered a final judgment on the brothers: "During their entire lives, the Schlegel brothers were and are, despite so many wonderful abilities, unhappy men. They would imagine more than nature granted them and wanted to have more of an impact than they were able; accordingly they have worked considerable mischief in art and literature." See Goethe to Carl Friedrich Zelter (20 October 1831), in *Goethes Werke*, IV.49:118.

143. Fichte was quoted by Wilhelm von Humboldt to Schiller (22 September 1794), in *Goethes Gespräche*, 1:564.

ism was brought against him. On this occasion, Goethe (who had some sympathy for Fichte's religious position)—even Goethe could not save him from himself. Fichte was dismissed from the university in June 1799.

Two years after Fichte's arrival in Jena, Wilhelm Schlegel moved to the city, carrying with him many literary ambitions, along with his new wife Caroline Michaelis Böhmer Schlegel and her daughter Auguste. He had been recruited to collaborate with Schiller and Goethe on the new journal *Die Horen*. In 1798, after a falling out with Schiller, he became professor at the university. Goethe immediately discovered in Schlegel one whose aesthetic judgment coincided with his own, although he was initially wary of his new friend's "democratic tendencies."[144] They conferred about Schlegel's translations of Shakespeare, and Goethe found in this genial critic, translator, and poet a literary confidant second only to Schiller. Schlegel, for his part, came, in the words of his brother, to "worship" Goethe.[145] About a month after Wilhelm and Caroline Schlegel had settled in Jena, Friedrich Schlegel—more volatile, brash, and passion-ridden than his older brother—set up house not far from the family. Goethe had read with great interest the younger Schlegel's collection of essays on Greek and Roman poetry, with its laudatory preface about Schiller's *Über naive und sentimentalische Dichtung*.[146] He soon invited this exuberant historian of ancient literature to accompany him on afternoon walks. Though Friedrich became an extreme admirer of Goethe, his relationship to Schiller went quickly sour when he published some political tracts in the republican journal *Deutschland;* he then added personal insult to these political injuries with essays critical of Schiller's poetry and judgment. The suspicion that Caroline gave succor to these attitudes was hardly amiss; and Schiller, in his turn, cultivated a hearty disdain for "Madam Lucifer." In June 1797, as his social circumstances became quite uncomfortable in Jena, Friedrich left for Berlin, where he met and embraced the friendship of Friedrich Daniel Schleiermacher, who had just embarked on that philosophical-religious course that would show the way for Protestant thought in Germany. When Friedrich returned to Jena in 1799, shortly followed by Dorothea Mendelssohn Veit, the woman he would live with in imitation of Goethe's arrangement with Christiane, the friendship between the older man and his younger admirer again flowered. In January 1800 Friedrich began serializing

144. Goethe to Johann Heinrich Meyer (20–22 May 1796), in *Goethes Briefe*, 2:221–22.

145. Friedrich Schlegel to Wilhelm Schlegel (1799), in *Friedrich Schlegels Briefe an seinen Bruder August Wilhelm*, ed. Oskar Walzel (Berlin: Speyer & Peters, 1890), p. 410.

146. See Goethe, *Tagebücher* (20 March 1797), in *Goethes Werke*, III.2:62. Friedrich Schlegel's *Die Griechen und Römer* appeared in 1797 (though some of the essays had been printed earlier), bearing the preface that celebrated Schiller's essay. See chapter 2.

his *Gespräch über die Poesie* in the brothers' journal *Athenaeum*. In that monograph he maintained that only Goethe could stand with such "Romantic" poets as Cervantes and Shakespeare. Indeed *Wilhelm Meister* had achieved that ideal of beauty, a fusion of classical and Romantic styles so as to make it, in Friedrich's estimation, a tendency of the age, along with Fichte's *Wissenschaftslehre* and the French Revolution. Shortly after Goethe's death, Heinrich Heine judged, with some hyperbole, that the poet "had owed the greatest part of his fame to the Schlegels."[147] And so, by virtue of being the type-specimen, Goethe's work became, nolens volens, definitional of the Romantic ideal.

The Romantic circle achieved philosophical completion when Friedrich Schelling arrived in Jena to take up the post that the combined forces of Fichte, Schiller, and Niethammer strove to make possible. Goethe was at first hesitant because he suspected him a thinker too strongly enticed by the idealistic unreality conjured by Fichte. But when he met Schelling at a party thrown by Schiller, he came away singularly impressed with this very young philosopher, who showed a knowledge of real natural science, particularly Goethe's own works, and who seemed untainted by the kind of Jacobin inclinations displayed by Fichte. Goethe wrote Privy Councillor Voigt to extol this new star on the philosophical horizon:

> Schelling's short visit was a real joy for me. For both him and us, it would be a wish realized were he to be brought here. For him, so that he might enter an active and energetic company—since he has had a rather isolated life in Leipzig—a company in which he might be guided in experience and experiments, and prosecute an enthusiastic study of nature, so that his beautiful mental talents might be applied with appropriate purpose. For us, the presence of so estimable a member would help us tremendously, and . . . my own work would be considerably advanced with his aid.[148]

Goethe and Schelling grew quite close during the six years of the philosopher's stay in Jena. Caroline regarded the poet's solicitude as quite paternal. So when Schelling declined into a depressive melancholy after the death of Caroline's daughter Auguste, the grieving mother felt she could ask the great man to care for her lover during the Christmastide of

147. Heinrich Heine, *Die Romantische Schule*, in *Sämtliche Schriften*, ed. Klaus Briegleb, 3rd ed., 6 vols. (Munich: Deutscher Taschenbuch Verlag, 1997), 3:388.
148. Goethe to Christian Gottlob Voigt (21 June 1797), in *Goethes Briefe*, 2:615.

1800–1801. Goethe also facilitated the divorce between Caroline and Wilhelm Schlegel, so that she and Schelling could marry. The deep personal relationship between Goethe and Schelling inevitably affected their intellectual lives, and the reciprocal impact was striking.

Goethe and Schelling: The Adjustments of Realism and Idealism

Goethe is usually portrayed as utterly rejecting the scientific and metaphysical aspirations of the Romantics.[149] Historians who make this judgment do so by considering under the rubric of "Romantic writer" such diverse individuals as Schelling, Henrik Steffens, Gotthilf Heinrich von Schubert, and Lorenz Oken. Goethe certainly thought the latter three often indulged in *Schwärmerei* and obscurity.[150] But as I have attempted to show in preceding chapters, those individuals who might carry (or have pinned on them) the banner of Romanticism expressed distinctive philosophical views and dispositions. Goethe reacted to each differently, and his feelings for each altered over time. When the Schlegels expressed more orthodoxly religious sentiments after the turn of the century, Goethe became disappointed, suspicious, and irritated with them—especially as those sentiments infected their aesthetic principles. For Schelling, though, he harbored quite warm affection, which hardly abated over the years. And his enthusiasm for the young philosopher's ideas continued to grow from their first meeting until the time when Schelling left Jena in 1803. Thereafter he kept up with Schelling's changing philosophical interests, always indicating positive re-

---

149. Dietrich von Engelhardt has offered the most sensitive consideration of Goethe's relation to the science and philosophy of the Romantics. He argues: "Though Goethe's research into nature doubtlessly is close to metaphysical *Naturphilosophie* and Romantic research into nature, it nonetheless is quite clearly distinct from these positions in its specific connections with aesthetics, philosophy, science, and biography." I rather believe that Goethe shows the most intimate connections with *Naturphilosophie* precisely in these last mentioned areas. See Dietrich von Engelhardt, "Natur und Geist, Evolution und Geschichte: Goethe in seiner Beziehung zur romantischen Naturforschung und metaphysischen Naturphilosophie," in *Goethe und die Verzeitlichung der Natur*, ed. Peter Matussek (Munich: C. H. Beck, 1998), p. 18.

150. I will discuss Goethe's dispute with Oken below. After reading Henrik Steffens's *Grundzüge der philosophischen Wissenschaften* (1806), Goethe observed to his friend Friedrich August Wolf: "The preface to the little book has indeed a mellifluous edge, but we amateurs choke mightily on the contents." See Goethe to Wolf (31 August 1806), in *Goethes Werke*, IV.19:187. Goethe complained to Sulpiz Boisserée about Schubert's Protestant "mysticism": "Thus the pitiable Schubert, with his attractive talent, attractive remarks, etc.—he plays now with death and seeks his healing in decay—and indeed, he is already half decayed himself, that is, he quite literally has consumption." See Boisserée, *Tagebücher* (4 August 1815), 1:232.

gard, if not complete acceptance.[151] In the near term, though, the young philosopher secured the lines of Goethe's drifting metaphysical views, providing many of his instinctive attitudes hard rational demonstrations, while shifting others into more dangerous currents. And the reciprocal pull on Schelling's own philosophy was hardly less dramatic.[152]

During Schelling's years in Jena, he and Goethe met frequently to discuss philosophical, scientific, and artistic matters. Goethe—the poet, scientist, and Weimar genius—like a whirlpool of creative energy, carried the young philosopher into the center of his interests and flooded him with reorienting conceptions. His diverting power had its effect almost immediately. In the winter term 1798–99, Schelling began lecturing at Jena on *Naturphilosophie*, lectures which by Easter would yield his *Erster Entwurf eines Systems der Naturphilosophie* (First sketch of a system of nature philosophy). In November he and Goethe met to discuss the nature of *Naturphilosophie* and particularly the problems of organic metamorphosis.[153] After the publication of his lectures, Schelling, under the influence of Goethe, felt the need to clarify and develop an aspect of *Naturphilosophie* that he had neglected, namely, the role of experiment and observation. During a particularly intense period, from the middle of September to the middle of October 1799, the two met almost daily to discuss this problem, and together they spent a week going over Schelling's *Einleitung zu dem Entwurf eines Systems der Naturphilosophie* (Introduction to the sketch of a system of nature philosophy).[154] Schelling remarked to a friend that the conversations had produced a great "fluorescence of ideas" for him.[155] The *Einleitung*, as we have seen (chapter 3), stated unequivocally the necessity

151. For instance, even after Schelling had turned more to investigate religious phenomena, Goethe read his works with avid curiosity, as he indicated to Jacobi: "I owe [my attitude] to the more elevated standpoint to which philosophy has raised me. I have learned to value the Idealists. . . . Schelling's lecture has given me great joy. It sails in those regions in which we both like to tarry." See Goethe to Friedrich Heinrich Jacobi (11 January 1808), in *Goethes Werke*, IV.20:5.

152. My view of the relationship between Goethe and Schelling differs sharply from that of Boyle, who, in the second volume of his monumental Goethe biography, attempts to show that Goethe, though initially friendly toward Schelling's *Naturphilosophie*, simply rejected the idealism of the young philosopher. See, for instance, Boyle, *Goethe: The Poet and the Age*, 2:667–68, 675–77.

153. Goethe, *Tagebücher* (12–13, 16 November 1798), in *Goethes Werke*, III.2:222–23.

154. Goethe read Schelling's *Einleitung* on 23 September and talked with him about it; and then from the 2 to 5 October, they read through the work together. See Goethe's *Tagebücher*, in *Goethes Werke*, III.2:261–63.

155. Schelling to F. A. Carus (9 November 1799), in F. W. J. Schelling, *Briefe und Dokumente*, ed. Horst Fuhrmans, 3 vols. to date (Bonn: Bouvier Verlag, 1962–), 1:176–77: "A short while back he [Goethe] spent several weeks here. I was with him for a long time every day, and had to read aloud my work on *Naturphilosophie* and explained it to him. What a fluorescence of ideas these conversations have produced for me, you can well imagine."

of experiment in discovering the laws of nature. And, indeed, Schelling—the knight-errant of idealism—proclaimed that "all of our knowledge is constituted by experience."[156] It is hard to doubt that Goethe did anything but stimulate, promote, and encourage this appeal to experience as the true Excalibur of natural science. The *Einleitung* clearly marked the deviant path of Schelling's idealism, which led him within two years to develop the kind of Spinozistic objectivism that Fichte scorned. Though I have suggested that many diverse pressures on Schelling shaped this trajectory, there can be little doubt that powerful Goethean forces pulled him sharply toward that ideal-realism he would finally espouse.

Goethe, as well, shifted orientation. He began to rethink the relationship between art and science, especially their underlying connection in an identity between mind and nature, a move inspired, as we have seen, by his reading of Kant and the insistent ideas of his dear friend Schiller. Already in 1796 he had pricked the ears of Jacobi with the remark that his friend would "not find me any more the rigid realist."[157] Schelling, however, accelerated Goethe's move toward idealism. It began with their meeting in late May 1798 to conduct optical experiments together.[158] Then in early June, Goethe took up Schelling's book *Die Weltseele*. Like Jacob with the angel, Goethe had to struggle mightily with the tome.[159] He later recalled this time when "I found much to thinking about, to examine, and to do in natural science. Schelling's *Weltseele* required our utmost mental attention. We saw it everywhere incorporated into the eternal metamorphosis of the external world."[160] This angel, having been brought to submission, inspired in Goethe a poem, itself entitled "Weltseele." It sang of that dream of the gods, the world soul, which

Grasps quickly after the unformed earth
And with creative youth, does not cease

156. Schelling, *Einleitung zu dem Entwurf eines Systems der Naturphilosophie*, in *Schellings Werke*, ed. Manfred Schröter, 3rd ed., 12 vols. (Munich: C. H. Beck, 1927–59), 2:278 [III:278]. Wetzels, as most other historians of the period, thinks Goethe rejected the presumed deductively a priori scientific methods of Schelling: "Goethe refused to embrace the approach advocated and practiced by Schelling, namely, to proceed from preconceived general hypotheses about nature, and to deduce from them the properties of reality." See Walter Wetzels, "Organicism and Goethe's Aesthetics," in *Approaches to Organic Form*, ed. Frederick Burwick (Dortrecht: Reidel, 1987), pp. 71–85; quotation from p. 74. Of course, Goethe did reject such a-priorism, but then, so did Schelling.

157. Goethe to Jacobi (17 October 1796), in *Goethes Briefe*, 2:240.

158. Goethe's diary indicates they met several times during Schelling's first year (1798) in Jena, and at the end of May, they performed those optical experiments mentioned above. See Goethe, *Tagebücher* (28–30 May 1798), in *Goethes Werke*, III.2:109.

159. Ibid. (7–8 June 1798), III.2:110–11.

160. Goethe, *Tag- und Jahres-Hefte* 1798, in *Sämtliche Werke*, 14:58.

To animate and to bring to birth
Ever more life in measured increase.

So that finally "each mote of dust lives."[161] These few lines capture in poetic form the kind of *dynamische Evolution* that Schelling himself portrayed in his tract and that Goethe would endorse.

Goethe's poem not only indicates a locus of interest that Schelling's work had for him; it also signals a transformation in his attitude about the relationship between art and science. We have seen that Goethe had, in the early 1790s—especially in his essay "The Experiment as Mediator between Object and Subject"—drawn sharp methodological distinctions between art and science—though, of course, breaching these distinctions often enough in practice. Schelling, by contrast, began developing his philosophy precisely along the lines prescribed in the Romantic mandate of Friedrich Schlegel: "All art should become science and all science art; poetry and philosophy should be made one."[162] Certainly the Romantics were as much influenced in this regard by Goethe, the poet-scientist, as he was by them. But during his interactions with Schelling, he came to understand the ways in which poetry and science could and must come together.

A few days after struggling with Schelling's *Weltseele*, Goethe began discussing with Schiller "the possibility of a presentation of natural science [*Naturlehre*] through a poem."[163] His initial inspiration for this project may have come from reading, the previous January, Erasmus Darwin's long didactic poem *The Botanic Garden*. While he rather liked some of the ideas of Darwin's book *Zoonomia*, he thought, as Coleridge, that Darwin's verses settled on the senses as did the cold mists that sometimes gather at the foot of Parnassus. The Englishman, in his pedantic, measured way, simply lacked poetic feeling.[164] The more proximate source for Goethe's plan for a great scientific poem, however, seems to have been the metaphorical presentation of Schelling's own philosophy, draped as it was in the ancient myth of a world soul. Just a few days after reading Schelling, Goethe began to enact his project, or at least he initiated a small part of it, with his poem "Metamorphose der Pflanzen."

---

161. Goethe, "Weltseele," in *Sämtliche Werke*, 6.1:53–54: "Ihr greifet rasch nach ungeformten Erden / Und wirket schöpfrisch jung / Dass sie belebt und stets belebter werden, / Im abgemess'nen Schwung." This quasi-evolutionary perspective also echoed the first part of Herder's *Ideen* (see the previous chapter). The exact date of the poem is uncertain, but it seems to have come from sometime in 1798–99.

162. Friedrich Schlegel, *Kritische Fragmenta* (115), in *Kritische Friedrich-Schlegel-Ausgabe*, ed. Ernst Behler et al., 35 vols. to date (Paderborn: Verlag Ferdinand Schöningh, 1958–), 2:161.

163. See Goethe, *Tagebücher* (18 June 1798), in *Goethes Werke*, III.2:212.

164. Goethe made these observations to Schiller in a letter of 26–27 January 1798, in *Sämtliche Werke*, 8.1:506–9.

The verse is an elegy done in hexameter lines and is addressed to the poet's beloved. The poet desires to reveal to her, amid the confusion of plants, the "secret law" that underlies them all. He describes the common transformations of the organs of all plants, transformations ultimately yielding "the most delicate forms, striving together to climax in union." These metamorphoses become as a mirror reflecting the development of a "friendship, which powerfully unfolds from within, into the flowers and fruits engendered by Amor."[165] Embraced by these sensually vivid images are rather more strained lines carrying the details of plant development. Some time later—the date is not exactly certain—Goethe turned to animal morphology, rendering that science also in hexameters, measures that would, one suspects, not play very well down a girl's spine. The "Metamorphose der Tiere" (Metamorphosis of animals) is stuffed with morphological ideas, from the notion "every animal is an end in itself" to "the burden of compensation destroys all beauty of form." The central message is that "the rarest form preserves in secret the archetype [*Urbild*]."[166] Even to Goethe himself, these lines probably sounded a bit too ploddingly like Darwin's verse. He finally decided a long, continuous poem ill suited to his aspirations.[167]

Despite the doubtful merits of this morphology-ridden poetry, it yet makes clear, I think, that Goethe and the Romantic circle were forcefully drawn together. As we have seen (chapter 3), with the help of Caroline Schlegel and Friedrich Schlegel, Schelling himself began to work seriously during winter of 1799–1800 on his own large philosophical poem. Like Goethe's project, though, Schelling's did not get much beyond a few stanzas. Yet for both, the ideal drew them on. In Schelling's case, it greatly colored his treatise *System des transscendentalen Idealismus*, lectures he gave in winter term 1799–1800 and published at Easter. He sent Goethe a copy, and the poet immediately responded that as far as he had quickly read, he thought he understood his young friend's argument. He was sure that "in this kind of presentation, there would be great advantage to anyone who was inclined to practice art and observe nature."[168] Goethe could rarely be moved to dispense patronizing flattery, and this certainly was not a case. He expressed his admiration for Schelling's ideas quite openly, as Friedrich

---

165. Goethe, "Metamorphose der Pflanzen," in *Sämtliche Werke*, 6.1:16. The poem appeared in Schiller's *Musenalmanach für das Jahr 1799*.

166. Goethe, "Metamorphose der Tiere," in *Sämtliche Werke*, 6.1:17–19.

167. In discussing his long nature-poem with Boisserée, he indicated that he changed his mind, since a number of smaller poems collected together would yield a more satisfactory result. See Boisserée, *Tagebücher* (3 October 1815), 1:277–78.

168. Goethe to Schelling (19 April 1800), in *Goethes Briefe*, 2:405.

Schlegel mentioned, with some pique, to his brother: Goethe, he wrote, "talks of Schelling's *Naturphilosophie* constantly with particular fondness."[169]

Two weeks after sending a copy of his *System* to Goethe, Schelling left Jena with Caroline Schlegel and Auguste for Bamberg, where, in midsummer, tragedy befell them (see chapter 3). In Schelling's absence, Goethe decided to continue study of this new brand of idealism, different as it was from Fichte's. He wrote the young philosopher, whom he feared might never return to Jena, that he looked forward to "complete harmony" with the new conception:

> Since I have shaken off the usual sort of natural research and have withdrawn into myself like a monad and must hover over the mental regions of science, I have only occasionally felt a tug this way or that; but I am decisively inclined toward your doctrine. I wish for complete harmony, which I hope to have effected sooner or later through the study of your writings or, preferably, from your personal presence and, as well, through the elevation of my particular notions to the universal. This harmony must accordingly become more pure, since the more slowly I absorb this, the necessarily truer I am compelled to remain to my own way of thinking.[170]

Goethe mentioned in his letter that he had been taking instruction in the new idealism with Niethammer (who was in the philosophy faculty at Jena). They met almost daily for a month, from early September to early October 1800.[171] After Schelling returned in the fall to Weimar, his philosophical élan slowly faded, replaced by a growing depression over the death of Auguste. By Christmas he was in such a state that Caroline believed he might commit suicide, and so she arranged for Goethe to take in her despairing lover. Schelling spent the Christmas holiday with Goethe, who apparently wrought the right kind of psychological cure. Schelling recovered, and his gratitude for Goethe's personal solicitude mixed sweetly with admiration for the poet's genius. When he lectured on the philosophy of art shortly thereafter, he did not hesitate to proclaim the poet's *Faust* "nothing else than the most intrinsic, purest essence of our age."[172] While Schelling remained at Weimar, he and Goethe continued to meet; they would discuss

---

169. Friedrich Schlegel to Wilhelm Schlegel (26 July 1800), in *Friedrich Schlegels Briefe an seinen Bruder August Wilhelm*, p. 431.

170. Goethe to Schelling (27 September 1800), in *Goethes Briefe*, 2:408.

171. Goethe, *Tagebücher* (5 September–3 October 1800), *Goethes Werke*, III.2:304–8.

172. Schelling, *Philosophie der Kunst*, in *Schellings Werke*, 3:466 [V:446].

the philosopher's new projects, projects such as the *Bruno*, which gave forceful and fairly accessible expression to Schelling's Spinozistic identity theory, a theory in which reality, fractal-like, is reflected both in idea and in empirical existence. Indeed, Schelling's assertion of the hyper-reality of ideas found echo in Goethe's astounding play *Die natürliche Tochter* (The natural daughter, 1803). The vivacious heroine of the play, whom the father believes dead, yet lives, not in a dream or memory but in the absolute idea. In a father's words:

> You are no dream figment, as I see you;
> You were, you are. The divinity has you
> Perfectly, once you are thought and represented.
> So you are part of the infinite,
> Of eternity, and you are eternally mine.[173]

I believe there should be little doubt of Goethe's admiration for Schelling or his enthusiasm for the new philosophy. Goethe explained this engagement to Schiller: "Since one cannot escape considerations of nature and art, it is of the greatest urgency that I come to know this dominating and powerful mode of thought."[174] This "dominating and powerful mode of thought" solved for Goethe several deep problems concerning nature and art about which he worried continually. Let me indicate specifically just how Schelling's philosophy accomplished this.

First and most important, Schelling's philosophical view, especially as developed in his *System des transscendentalen Idealismus*, demonstrated that scientific understanding and artistic intuition did not play out in opposition to one another, as Goethe once thought, but that they reflected complementary modes of penetrating to nature's underlying laws. For Goethe this liberated his sense of the intimate connection between the scientific and the artistic approaches to nature, which he consequently expressed, as was his wont, in a poem—"Natur und Kunst" (Nature and art)—that he composed at this time:

> Nature and art seem each other to flee,
> Yet, each finds the other before one can tell;

---

173. Goethe, *Die natürliche Tochter* (iii.4), in *Sämtliche Werke*, 6.1:290–91: "Du bist kein Traumbild, wie ich dich erblicke; / Du warst, du bist. Die Gottheit hatte dich / Vollendet einst gedacht und dargestellt. / So bist du teilhaft des Unendlichen, / Des Ewigen, und bist auf ewig mein." Boyle sees in these lines the shadow of Schelling's *Bruno*—another fantasy of the deluded father. I suspect, though, Goethe's poignant lines reveal greater sympathy for his characters than Boyle allows. See Boyle, *Goethe: The Poet and the Age*, 2:785.

174. Goethe to Schiller (16 September 1800), in *Sämtliche Werke*, 8.1:814.

The antagonism has left me as well,
And now they both attract me equally.[175]

If, as Schelling maintained, "the world is the original, yet unconscious poetry of the mind [Geist],"[176] then the poet might construct beautiful representations by employing those same principles that went into the original creation of the natural world. So, again, there would be philosophical justification for assuming complementarity of scientific and aesthetic judgment: the poet, through creative genius, could compose those works that would have the authority of nature herself—an authority that became a lived reality for Goethe while he was in Italy, but one he could not philosophically justify. And reciprocally, the aesthetic might lead us to nature's concealed laws, to those archetypes according to which nature creatively expressed herself. As Goethe epigrammatically formulated it: "The beautiful is a manifestation of secret laws of nature, which without its appearance would have remained forever hidden."[177] This Goethean conception ran counter to the deep separation that Kant constructed between determinate judgments of nature and regulative judgments of art. It was Schelling, though, who demonstrated how aesthetic judgment opened the heart of nature for scientific examination.

Schelling had also shown in the *System* that Kant should not have restricted genius to the aesthetic realm. Genius, he argued, could also be found in science. As Goethe met resistance from professionals to his work in optics and morphology, he would, undoubtedly, rest more comfortably in the knowledge, thanks to his young friend's analysis, that his genius in science need not adhere to conventional scientific wisdom. His aesthetic intuitions might probe more deeply, might lead more surely to new discoveries in science than could the plodding, tradition-bound studies of his critics. Moreover, Schelling had argued that the laws of nature, which the poet-scientist might comprehend, would be also laws of free creativity. Though Goethe probably did not inquire too deeply into Schelling's argument that the free creativity of mind conformed exactly to the fixed laws of nature, the argument, nonetheless, confirmed his settled belief, which extended from ethics and politics to aesthetics, that true freedom was only realized in limitation. As he expressed it in the concluding lines of his "Natur und Kunst":

175. Goethe, "Natur und Kunst," in *Werke*, 1:245: "Natur und Kunst, sie scheinen sich zu fliehen, / Und haben sich, eh man es denkt, gefunden; / Der Widerwille ist auch mir verschwunden, / Und beide scheinen gleich mich anzuziehen."
176. Schelling, *System des transscendentalen Idealismus*, in *Schellings Werke*, 2:349 [III:349].
177. Goethe, *Maximen und Reflexionen*, in *Sämtliche Werke*, 17:749.

He who would be great must act with fine aplomb;
In constraint he first shows himself the master,
And only the law can give us full freedom.[178]

Schelling's philosophical principles also resolved for Goethe that deeper conundrum plaguing all who became persuaded, as he did, of the Kantian epistemology: namely, how might we have authentic understanding of external nature, if we were shielded by our own ideas from reality? If our mental constructions erected only a faux nature? The resolution, from Schelling's perspective, was simply that mind may indeed construct nature, but that there was no *Ding an sich* standing behind the construction. Nature really was as she appeared to be. So the bright colors and forms that dazzled the eye were not meretricious and superficial traits—they inhered in nature. In Schelling's view, the true idealism was the most authentic realism. Moreover, as Schelling drove his philosophy to an absolute ideal-realism, his position merged with that of Spinoza: the ideas that constituted nature's creations were not captives of individual minds, but stood beyond self and nature, though were realized in both. Hence the solution to the puzzle of Goethe's epigram that "an unknown, lawlike something in the object corresponds to an unknown, lawlike something in the subject."[179] The connection between object and subject occurred through the organic activity of absolute mind and its ideas, ideas that functioned as the archetypal concepts at the foundations of morphology and as the creative forces in nature.[180] Schelling's impact on Goethe reverberated through the years, and again became particularly manifest during the time he worked on *Zur Morphologie*.

Zur Morphologie

Schiller's Death and Napoléon's Invasion

The Weimar court celebrated the beginning of the new nineteenth century with a great masquerade party on 2 January 1801. Henrik Steffens recalled

178. Goethe, "Natur und Kunst," in *Werke*, 1:245: "Wer Grosses will, muss sich zusammenraffen; / In der Beschränkung zeigt sich erst der Meister, / Und das Gesetz nur kann uns Freiheit geben."

179. Goethe, *Maximen und Reflexionem* (no. 1344), in *Sämtliche Werke*, 17:942.

180. My analysis of Goethe's debt to Schelling differs sharply from those who only see disparity between Goethe's and Schelling's views. Thus, for instance, R. H. Stephenson: "Schelling's position and, perhaps, Goethe suspected, all abstract thought in which an identity is posited between the Idea at work in human thinking and the Idea at work in Nature, is fundamentally at odds with Goethe's." See R. H. Stephenson, *Goethe's Conception of Knowledge and Science* (Edinburgh: Edinburgh University Press, 1995), p. 29.

the event fondly, especially a particularly lively discussion on aesthetics en-gaged in by Goethe, Schiller, and Schelling.[181] The sparkling conversa-tion, he remembered, sailed along on strong currents of champagne. Goethe, who was in a particularly good humor, jabbed with delight at the more earnest doctrines being advanced by Schiller. Schelling, though, kept his composure during this mock battle, still apparently preoccupied with the death of Auguste. The exertion of the evening, however, was too much for Goethe; and during the next several days he went into collapse. Steffens marked this as the point after which the immortal poet began to "suffer from thoughts of impending death."[182] That ultimate event would not oc-cur for another three decades; but at the turn of the century, Goethe was over fifty, had grown rather fat, his kidneys troubled him, and his teeth were none too good. He heard the low voice of mortality whispering in his ear.

Early in February 1805, Goethe again contracted a grave illness, this time with a lung infection that his friends feared might be fatal. After three days, however, his fever broke; and he could say to Johann Heinrich Voss (1779–1822), a young visitor who stood watch at his bed: "My dear boy, I'm still with you. You must not cry any more."[183] But soon Goethe's own tears would begin to flow without restraint. During the time of his illness, from early February through the end of April, his friend Schiller, who had visited him often during this period, had himself become unwell, a flare-up of his chronic lung disorder. On 1 May he stopped by Goethe's house in the evening for a short visit before attending the theater. During the performance he be-came acutely ill and five days later went into a coma. Goethe was beside him-self. Voss, who visited him during this time, described his state:

> During Schiller's last illness Goethe was quite crushed. I found him once in his garden crying. But those were merely some tears brimming from his eyes. His spirit cried, not his eyes. And in his look I saw that he felt some-thing enormous, something unearthly, something infinite. I told him sev-eral things about Schiller, to which he listened with indescribable composure. "Fate is inexorable and man little." That was all he said.[184]

Schiller died on 9 May 1805 at age forty-five. No one, though, had the courage to tell Goethe. Christiane took him to bed and watched over him during the night. When he awoke the next morning, as Voss tells the story, he asked her:

181. Henrik Steffens, *Was Ich Erlebte*, 10 vols. (Breslau: Josef Max, 1840–44), 4:407–13.
182. Ibid., 4:412.
183. Johann Heinrich Voss, in *Goethes Gespräche*, 1:986.
184. Ibid., 1:999.

"Schiller was *very* ill last night, wasn't he?" The emphasis that he placed on "very" struck her so forcefully that she could no longer contain herself. Instead of answering him, she began to sob loudly. "Is he dead?" Goethe asked with firmness. "You have said it yourself," she answered. "He is dead," Goethe repeated, and turned away, covered his eyes with his hands and cried, without uttering a syllable.[185]

Goethe slowly returned to work after Schiller's death, but his own health visibly suffered. Nephritis gave him trouble, and he had bouts of melancholy as he recalled all that Schiller meant to him—and he meant so much. Goethe's preoccupation with the death of his friend rendered his personal diary virtually empty for the entire year. But his life did brighten a bit at the end of May, when the great classicist and Homeric scholar from Halle, Friedrich August Wolf, visited. He was a man of such enlarged and acute learning about ancient lands, peoples, and authors, that he could "conjure up the past vividly to the mind's eye."[186] Goethe had previously consulted with Wolf over a small book that he had just published on the history of art and Winckelmann's contributions to its evaluation and understanding.[187] Wolf himself supplied an essay for the volume (anonymously included) on Winckelmann as philologist. Goethe meant to defend Winckelmannian concepts of beauty, which were rooted in Greek ideals, against those critics, like Friedrich Schlegel, who came to believe only a deeply religious sentiment could produce great art.[188] The book's disparagement of Christianity's unhappy influence on aesthetic ideals caused something of a stir— Friedrich Schlegel thought the tract simply a confession of Goethe's paganism.[189] Goethe wanted to discuss the issues further with Wolf, and they

185. Ibid., 1:1000.

186. Goethe, *Tag- und Jahres-Hefte* (1805), in *Sämtliche Werke*, 14:132.

187. Goethe's *Winckelmann und sein Jahrhundert* was published in summer of 1805 and is reprinted in *Sämtliche Werke*, 6.1:195–401.

188. Carl Schüddekopf and Oskar Walzel discuss Goethe's opposition to the religious aesthetics developed by the Schlegels after the turn of the century. See their introduction to *Goethe und die Romantik: Briefe mit Erläuterungen*, vol. 13 of *Schriften der Goethe-Gesellschaft* (1898): L–LI.

189. Friedrich Schlegel to Wilhelm Schlegel (15 July 1805): "You should try to get a copy of Goethe's book on Winckelmann. It is rather funny and in its own way remarkable. The old rascal has officially proclaimed therein his paganism; he has never given himself over so to his restlessness. He advances rather poorly armored; the old Duchess [Anna Amalia, to whom it is dedicated] in front and Wolf in the rear. . . . Someone recently wrote me that he is rather sick and has weak nerves. The book will cause quite a stir; Schelling will now, doubtlessly, again confess paganism and construe it as it deserves. German literature increasingly leaves a bad taste in your mouth." Schlegel's letter is quoted by Victor Lange in his commentary to *Winckelmann und sein Jahrhundert*, in *Sämtliche Werke*, 6.2:1054.

spent considerable time doing so. The visit had the added grace of Wolf's young daughter, who quickened the sad poet to "the charm of a fresh youth that competes with the spring."[190] Wolf reciprocated the poet's hospitality in July, when Goethe passed through Halle on his return from the summer theater he helped create at Bad Lauchstädt (near Merseburg, just south of Halle). During Goethe's visit, Franz Joseph Gall, on his tour through Germany, stopped at Halle to give lectures and demonstrations of his phrenological doctrine (see chapter 7). Goethe attended the public performances and even received private seminars in his rooms during that time when he had to take to his bed with an attack of kidney problems. Gall demonstrated fair knowledge of osteology and brain anatomy, but really no more than Goethe himself possessed. Though tolerant of the notion that organs of the brain could dispose people to certain kinds of action (from criminality to charity) and bemused by the flattery that the itinerant phrenologist shoveled his way, Goethe thought the series of lectures just a bit pedantic and not very revelatory.[191] While in Halle, Goethe sought a consultation with Johann Christian Reil, hoping the famed doctor might be able to alleviate his kidney problems. He thought very highly of Reil, but little seems to have been done (or perhaps could have been done) for his malady.

During the first part of 1806, Goethe, his health much improved, worked steadily on optical experiments and completed part 1 of his *Faust*, which he gave to his publisher Cotta in May. During February his quiescent thoughts about morphology sprung to life again because of the receipt of several books that returned him to the subject.[192] As a result of this reading, he decided to republish his *Metamorphose der Pflanzen* and to include several of his accumulated essays on morphology. But his plans for publication came to naught at this time, as war and rumors of war began to spread throughout the region.

In August 1805 Britain, Russia, and Austria formed a coalition to restrain the grasping hand of the French, who, under the command of Emperor Napoléon, had already taken large parts of Europe. On the sea, the allies triumphed with Nelson's defeat of the French navel fleet at Trafalgar on 21 October; but virtually at the same moment, Napoléon's land forces of

190. Goethe, *Tag- und Jahres-Hefte* (1805), in *Sämtliche Werke*, 14:132.

191. Goethe described his encounters with Gall in ibid., 14:136–38.

192. The books that he received in February that brought him back to his morphological work were Samuel Thomas Soemmerring, *Abbildung des menschlichen Hörorgrans* (Frankfurt a. M.: Varrentrapp und Wenner, 1806); Alexander von Humboldt, *Ideen zu einer Physiognomik der Gewächse* (Tübingen: Cotta, 1806); and Heinrich Cotta, *Naturbeobachtungen über die Bewegung und Funktion des Saftes in den Gewächsen* (Weimar: Hoffmann, 1806). The last, in particular, inspired Goethe, because of its fine woodcuts. See Goethe, *Tag- und Jahres-Hefte* (1806), in *Sämtliche Werke*, 14:171–72.

over 200,000 men crushed the Austrian army of some 70,000 at the city of Ulm (in Bavaria) before Russian reinforcements could arrive. The emperor then quickly turned his armies toward the remaining allied forces and in November pushed them out of Vienna and into Moravia (now the Czech Republic). The Austrians and Russians took a stand just outside the city of Austerlitz. Through the blundering of the Russians and the strategic brilliance of the French, the allied forces suffered a great defeat, causing the Hapsburg Francis I to sue for peace in early December 1805. Because the French armies were stretched out and perhaps too complacent, Frederick William III of Prussia saw what he thought might be an opportunity. He signed a secret alliance with the Russians in July 1806, and they set about planning their attack. As the Prussian forces, aided by the Saxons (including Carl August, now in arms with 750 men), began building and with the Russians moving back into the Germanies, Napoléon rushed into Thuringia in early October to meet the Germans, who were camped in two armies, one near Jena and the other near Auerstädt (twelve miles north of Jena). The split German forces were outmaneuvered by clever French tacticians and utterly defeated. Russia and Prussia were forced into signing a costly peace treaty the following year (the Treaty of Tilsit, 7 and 9 July), and therewith Napoléon effectively controlled the Prussian and Austrian lands. In the immediate aftermath of the Battle of Jena, French troops began ransacking the surrounding areas, including Weimar.

On Tuesday morning, 14 October 1806, in the garden at his large house on Der Frauenplan, Goethe and his friends could hear the cannonade from the battle at Jena.[193] By noon the firing in the distance ceased. But hardly had his company sat down to their three o'clock meal, when they heard shots fired nearby and saw hordes of retreating Prussian hussars flowing past; and these retreating troops were quickly joined by those billeted in Goethe's house. Shortly thereafter, the advancing French began to enter the town. A French cavalry officer set out to find Goethe, a one Baron von Türckheim—the son of Lili Schönemann, Goethe's former love. The poet and the soldier spent some time talking, and one can guess that Goethe was not only relieved to see a friendly French face, but quite touched to learn of one whose past was so intimately intertwined with his.

Goethe had to quarter some sixteen French officers in his house and reserve room for the erratic but charismatic Marshal Michel Ney (1769–1815). After consuming vast quantities of beer and wine, the billeted soldiers passed out in various areas of the house. Goethe and Christiane retired

---

193. The story of the French occupation of Weimar and of Goethe's house is told by Friedrich Riemer, who was on the scene. See Riemer's *Mitteilungen über Goethe*, pp. 167–73.

to their rooms. Friedrich Riemer, the house tutor and Goethe's secretary, described how he remained at the bolted door awaiting Ney's arrival while attempting to dissuade those seeking entrance with the warning that a field marshal was about to occupy the house. But he could not persuade a couple of drunken solders, who threatened to burn the place down unless admitted. Riemer relented. As the two louts became ever more drunk, Goethe came down in his night robes, looking like an Old Testament prophet, to admonish them. His imposing mien quieted them for the moment. The two later made their way to the room reserved for the marshal, and only by Riemer's rousing another officer could they be driven out. What Riemer only found out the next day, however, was that the two subsequently had forced their way into Goethe's rooms and threatened him. Christiane escaped to call for help, and one of the billeted officers quickly ran the two drunks out of the rooms and from the house. Ney and his company arrived that morning.

A good deal of Weimar was burned or looted by the French soldiers. Goethe gave thanks that his own house and papers remained intact—especially the manuscript for his *Farbenlehre*, part of which was already in press. The events of the week, nonetheless, had a profound effect on him. On Sunday, 19 October 1806, a few days after the initial occupation of Weimar, Goethe took Christiane to the Schloss church, and there in the sacristy, with his son August, his secretary Riemer, and a few other close friends attending, he married his longtime companion. Riemer believed that Goethe took this step in gratitude to Christiane for saving his life.[194]

Goethe's Evolutionary Theory

Napoléon exacted no reprisals against Carl August or other Saxon nobles, and the region fell into a period of relative quiet: Carl August came back to Weimar; the theater, which had been damaged during the invasion, reopened; and Goethe's life returned to normal. He again thought about republishing the *Metamorphose der Pflanzen*. In an introduction he prepared at this time, he remarked that the dangers of recent days had counseled him to include in his planned volume many of his fugitive essays on biological form.[195] He wrote his publisher Cotta (who was in the process of bringing out his collected works) that "having worked through my ideas on organic formation, especially on the osteological type, I wish now to have these

194. Ibid., pp. 173–74.
195. Goethe, "Das Unternehmen wird entschuldigt" (An apologia for the enterprise), in *Sämtliche Werke*, 12:12. The essay is dated "Jena, 1807."

published this winter."[196] A few sheets of the planned volume did go through the presses, and Cotta's catalog for the following Easter (1807) announced the title *Goethes Ideen über organische Bildung*.[197] The book, however, never appeared. Goethe seems to have become fully occupied with another project he had begun over a decade and a half before, his *Farbenlehre*. From January 1806, when the first part of this huge work went to the printer, till the appearance of the finished volume in 1810, his concern to complete the research and writing, while simultaneously correcting proofs, rose in a crescendo of effort; and his diary entries for the period are replete with records of that activity.

Goethe did, however, compose two introductory essays for his intended volume on morphology. In these essays he reiterated two ideas, now more saliently than before.[198] The first derived directly from his theory of plant metamorphosis: namely, that what we take to be individual organisms are composed, as it were, of other individuals; and that only among the more perfect animals is there a subordination and greater differentiation of parts. The second idea really grounded his entire notion of the archetype and its instantiations: the science of morphology deals with objects that are constantly being transformed; nowhere is there stasis, no fixed *Gestalten*, forms, rather only *Bildung*, progressive formation.

In urging this second idea, Goethe hinted at just how radical his understanding of *Bildung* and metamorphosis really might be. There was the whisper of a theory that harkened back to the 1780s, to his experiments on infusoria and his discussions with Herder about the development of life on the planet, a theory that, in its further development, he shared with Schelling. Undoubtedly Goethe thought the evidence for this conception insufficient, yet he seems nonetheless to have been convinced of its truth. It was a theory of the metamorphosis of species. In his essay he wrote:

> If one observes plants and animals in their imperfect state, one sees they are hardly distinguishable. A point of life, rigid, moving or halfway between can hardly be observed through our senses. Whether these first beginnings are determinable and lead over on one side to plants through [the effects of] light or on the other to animals through [the effects of] darkness, we ourselves do not have confidence to decide, although opinions and analogies seem to suggest this. But we can say this much: that

196. Goethe to Johann Friedrich Cotta (24 October 1806), in *Goethes Werke*, IV.19:218.

197. See Goethe, *Tagebücher* (1807), in *Goethes Werke*, III.3:190 (28 January 1807); and his *Tag- und Jahres-Hefte* (1807), in *Sämtliche Werke*, 14:184.

198. Goethe, "Die Absicht eingeleitet" (The intention introduced), in *Sämtliche Werke*, 12: 12–17.

out of a relationship that can hardly be analyzed as to what is plant and what animal, creatures gradually emerge in two opposite directions toward perfection, with the plant finally reaching glory in the tree, perduring and rigid, and the animal in human beings, the epitome of mobility and freedom.[199]

Goethe's mention here of the transmutation of the minute point of life sounds rather like the comparable reference made by Wagner (*Faust II*), who, before the eyes of Mephistopheles, chemically concocted a homunculus in a retort. This protoman developed "from the tender point from which life sprang" and began its metamorphosis through a myriad of forms.[200]

Was the transformation suggested only symbolic, or did Goethe perhaps have in mind an actual transformation of species in time—evolution in our sense? Darwin, in the *Origin of Species*, listed Goethe as one of his predecessors.[201] More recent historical scholarship has rejected Darwin's secondhand contention. Goethe, it is maintained, did not endorse anything like the transmutation and descent of species over time.[202] I rather believe that Darwin had it right. Goethe did not think the empirical evidence for transmutation reached a level that would compel unequivocal assent; but he was convinced of transmutation—or so I will argue.

Let me say first what I think the features of Goethe's evolutionary theory were. They can be summarized in five principles: (1) As Herder and Blumenbach believed (see chapter 5), in times past, out of the organic earth, less perfect plants and animals arose under the influence of a *Bildungstrieb*; (2) this force operated according to eternal archetypes of the organic—in-

199. Ibid., 12:15–16.

200. Goethe, *Faust II*, in *Sämtliche Werke*, 18.1:177–78 (ll. 6836–60).

201. Charles Darwin, "Historical Sketch," *On the Origin of Species by Charles Darwin: A Variorum Text*, ed. Morse Peckham (Philadelphia: University of Pennsylvania Press, 1959), p. 61. The "Historical Sketch" was introduced in the 3rd edition (1861).

202. In commenting on the passage just quoted on infusoria, Becker warns about "drawing the impermissible conclusion that he [Goethe] is going along with Darwinian thought." See Hans Becker, "Kommentar zu *Zur Morphologie*," in *Sämtliche Werke*, 12:924–25. Wenzel simply says that "an evolutionism . . . establishing an historical transformation in the world of biological phenomena over generations lay far beyond Goethe's horizon." See Manfred Wenzel, "Naturwissenschaften," in *Goethe Handbuch*, ed. by Witte, 2:781–96; quotation from p. 784. Mayr concurs, remarking that "these ideas [Goethe's] had nothing to do with evolution." See Ernst Mayr, *The Growth of Biological Thought* (Cambridge: Harvard University Press, 1982), p. 458. Wells, who has given the most thorough consideration to the question of Goethe as an evolutionist, concludes: "He [Goethe] was unable to accept the possibility of large-scale evolution." Though Wells does concede that Goethe made favorable comments about those who held Lamarckian views and even speculated on the descent of the giant Megatherium from a sea creature, he still thinks Goethe understood biological transformations only in an "ideal sense, not as a real process whereby one form gave rise to another." See George Wells, *Goethe and the Development of Science, 1750–1900* (Alphen aan den Rijn: Sijthoff & Noordhoff, 1978), pp. 45, 46.

deed, these archetypes constituted the creative force; (3) these archetypes became realized in the empirical world over time, such that less perfect creatures were followed by more perfect creatures; (4) the external environment modified the surface aspects of living beings and shaped their basic archetypal forms—though the essential internal structures remained constant; and these modifications were passed to descendants; and (5), as Schelling urged, forms tended to become ever more individualized, and after considerable alteration over time would perdure in a fairly stable way. (This last principle indicates a residual teleological assumption that perfection and beauty had an endpoint.) I will discuss these principles in context below.

It should be mentioned, if only in passing, that prior to about 1850, Charles Darwin believed versions of these principles: through spontaneous generation, several independent archetypal lines of descent arose, such that less perfect forms gave rise to more perfect, under the aegis of a *Bildungstrieb*, namely, natural selection. Darwin, as well, constructed his theory from principles of ontogenetic development.[203] I note this only to point out that the kind of evolutionary theory I have just outlined was contemplated by Darwin as well as by Goethe. The point of demonstrating that Goethe in fact had an evolutionary theory of the sort I have described is twofold: it shows the depth of his commitment to metamorphosis and indicates one of the sources of evolutionary ideas in Germany during the first part of the nineteenth century.

Recall that in the 1780s, Goethe had been speculating along with Herder on the history of life on earth, according to Charlotte von Stein. Goethe, she maintained, had been entertaining the possibility that we were "once plants and animals" (see chapter 10). Goethe referred to these speculations in 1816, when he prepared for publication his *Metamorphose der Pflanzen* and other essays on morphology. At that time he recalled his conversations with Herder: "Our daily discussions concerned themselves," he wrote, "with the origins of the water-covered earth and its organic creatures, which have developed [*entwickelt*] on it from very ancient times."[204] These remarks certainly seem to indicate that Goethe and his friends considered a transition in species forms. Johannes Daniel Falk (1768–1826), a satirical writer and casual acquaintance in Weimar, reported that he had directly put to Goethe the question of species change. Falk asked

whether it did not seem likely to him that all the many different animals have arisen from one another through a metamorphosis similar to that by

203. Darwin's early theory is described in my *Meaning of Evolution*, pp. 91–126.
204. Goethe, "Der Inhalt Bevorwortet" (The content prefaced), in *Sämtliche Werke*, 12:16.

which the butterfly has arisen from the grub. Goethe, however, answered that nothing certain can be firmly established on that question and that he would prefer to refrain from an opinion. Yet it is as clear as day that the whole realm of appearance is an idea and a thought. . . . "We are now awfully close to the chemistry of the whole thing, yet we all now choose terminology so that nothing in life is transformed. Out of all your bits of filings, you can't make a single mote of steel. I have given in my *Metamorphosis of Plants* the law whereby everything in nature is built up (this is through polarity, through generation [*Geschlechter*]). According to this law, things move into ever more splendidly and progressively higher syntheses [*immer prächtiger in gesteigerten höheren Verbindungen*].[205]

Falk was obviously befuddled by several of Goethe's remarks; he exclaimed that the poet often engaged in "oracular speech" [*Orakelsprüche*].[206] But Goethe's response was more coherent than Falk realized.[207]

Goethe would likely have been circumspect in answering Falk's query, since the writer was a notorious satirist who loved the exaggerated bon mot—hence Goethe's reluctance to offer an opinion. But then, obviously, he did offer one: our customary language prevented us, he claimed, from perceiving that organisms were being transformed, moving "into ever more splendidly and progressively higher syntheses." According to Falk's report, the two friends had just been talking about Schelling before the question concerning transmutation arose. Goethe's answer echoed the letter Schelling sent to the poet sometime before. In that letter (quoted at length in chapter 3), the young philosopher explained that his mentor's work on metamorphosis had brought him "very near to the inner identity of all organized beings among themselves and with the earth, which is their common source." He then gave a brief indication of the theory of "dynamic evolution" that he had developed in the *Erster Entwurf*, according to which the earth, already an organic entity, "can become plants and animals" through chemical processes.[208] Further, Goethe's remark that "the realm of appearance is an idea and a thought" certainly sounds like an oblique reference to the idealist position that he shared with Schelling.

205. Johannes Daniel Falk (15 March 1813), in *Goethes Gespräche*, 2:786. Falk saw Goethe frequently in Weimar during this period.

206. Ibid.

207. Falk's memoir has to be used with caution. Riemer, who lived in the Goethe household for many years, thought Falk quite unreliable in his views. He pointed out several instances of very unlikely remarks attributed to Goethe. Falk also had a dislike of the Romantics. See Riemer, *Mitteilungen über Goethe*, pp. 38–46.

208. Schelling to Goethe (26 January 1801), in Schelling, *Briefe und Dokumente*, 1:243. See chapter 8 for a fuller quotation of the letter and for a discussion of Schelling's theory of "dynamic evolution."

Some historians have argued that it is a mistake to attribute to thinkers like Schelling and Goethe any notion of an actual transmutation of species since the metaphysics of idealism simply could not sustain such a theory. But only a brief consideration of the problem ought to shelve that objection: the ideal structures of nature have to be realized in time and space (no matter how these modalities are construed), and they might well do so gradually and transitionally. Such a transmutation might be already contained, as it were, in the creative archetype of an organism, but this does not prevent phenomenal transmutations, as Goethe's own work on plants and insects perfectly well showed. All vertebrates might have a common identity in the idea while their increasingly perfect forms gradually become instantiated through time—which is what Goethe's cryptic remarks suggest, I believe.

This invites a further consideration. In his morphological doctrine of animal development, Goethe developed the idea that internal structures were causally modified by the external environment: for example, he supposed that the aquatic environment of fish and seals modified their common vertebral structure, rendering it suitable to that environment. In holding this position, he had to have asserted one of three things. The first is that, perhaps, he really did not mean what he said and held the causality of the environment to be only ideal—but his language and argument move forcefully in the opposite direction, toward an explicitly real, efficiently causal conception of formation. Or he thought that water, for instance, acted to modify seals ontogenetically, that is, only during the lifetime of each animal, and that these modifications were not inherited—but Goethe investigated the embryology of such creatures and knew that species-specific structural patterns developed already in the embryonic state before the environment could have an effect. Last, he might well have thought the changes wrought on animals were heritable, as Erasmus Darwin and Lamarck believed—and this final possibility seems the most reasonable of the three. Indeed, Goethe read Darwin's *Zoonomia*, which contains an elaboration of the Englishman's transmutational theory, and said he liked the book.[209] Under my construction, Goethe and Schelling's theory, though, was not exactly Erasmus Darwin's, nor certainly that of his grandson. But through the nineteenth century, several different concep-

---

209. Goethe read the German translation of Erasmus Darwin's *Zoonomia* in the late 1790s. Schiller informed him of the translation (29 December 1795), which came out in parts from 1795 through 1798. Goethe expressed his sympathy with Darwin, who was, as Goethe himself, both poet and zoologist (30 December 1795) and indicated he was favorably disposed (*günstig*) to Darwin's book (26 January 1798). See the letters in *Sämtliche Werke*, 8.1:145, 146, 509.

tions of species descent appeared, each dependent on rather different modes of species alteration. The theory of Goethe and Schelling, I believe, was one of these.

I have argued that only if Goethe actually believed in transmutation, can we make sense of his various other pronouncements. Yet we need not remain solely dependent on an appeal to the general coherence of his thought. He was more explicit about his transmutation speculations in *Zur Morphologie*.

Goethe eventually did complete his planned volume on morphology. In gathering and redacting his morphological essays, he added others of a more general scientific interest. They appeared in periodical form, from 1817 to 1824, in six numbers, which were gathered into two volumes carrying the general title *Zur Naturwissenschaft überhaupt, besonders zur Morphologie*. The first number reprinted, at long last, his *Metamorphose der Pflanzen*, along with a series of essays explaining its purpose and history. In subsequent numbers, he published several of his essays on animal morphology written in the 1790s. But he also included new essays, book reviews, and apposite poetry. I will talk a bit more about the structure of his work below. Of particular interest, however, for the question of Goethe's transmutational ideas are two essays in the fourth number of the first volume, which appeared in 1822.

In one essay Goethe reported on the comparative analyses offered by a Stuttgart physician, Georg Friedrich Jäger (1785–1867), of a giant fossil ox and a modern animal. Jäger explained the large eye sockets, ear openings, and more vaulted skull of the modern animal as the result of a "mighty animal instinct for preservation and nutrition" that led to "ever greater impressions of the sensible world absorbed" by the primitive animal over the course of thousands of years. Goethe not only thought Jäger's Lamarckian analyses "beautiful observations"; he undertook his own comparative study of a similar fossil ox that had been brought to Weimar. He particularly focused on the inward and inelegant turn of the horns on the fossil. He argued that herdsmen, presumably over long stretches of time, changed the shape of the horns for reasons of "utility," which also happened to produce a more beautiful formation. The alteration wrought by the herdsmen consequently became heritable.[210] We might call this an example of artificial improvement, in which a more ancient and primitive organism had been transformed. Goethe's essay here, while not directly endorsing a general evolutionary conception, is quite consistent with one. His second essay, however, is perfectly explicit.

210. Goethe, "Fossiler Stier," *Zur Morphologie*, in *Sämtliche Werke*, 12:252–59.

Goethe reviewed the first two parts of a multimonographic work by Christian Pander and Eduard d'Alton, *Die vergleichende Osteologie*.[211] The initial volume (1821) was a comparative study of the fossil Megatherium, or extinct giant sloth.[212] Starting in 1818, Pander undertook with d'Alton, his illustrator, a round of visits to natural history museums throughout Europe to do comparative studies of mammals and birds, including fossil representations. Their first trip took them to Madrid, where the remains of a prehistoric monster, found around Buenos Aires in 1789, were on display. Cuvier had earlier given a description of the giant, which was about the size of a rhinoceros; he called it a Megatherium and remarked on its close affinity to the modern sloth. Pander and d'Alton agreed that the fossil beast bore resemblance to living species of sloth—but, unlike Cuvier, they argued that the similarity was founded on descent (*Abstammung*).

Pander and d'Alton, in the introduction to their volume, expressed their transformationist theory as an application of Goethe's morphological ideas:

> The differences in formation of fossil bones in comparison with those of still-living animals are greater the older the rock formations in which they are found (with the fossil remains of the most recent formations quite similar to the structures of living animals). This common observation supports the assumption of an unbroken train of descent [*ununterbrochenen Folge der Abstammung*] as well of the progressive transformation of animals in relation to different external relationships. The observation that animals during the last millennium have reproduced with specific similarity in no way contradicts the theory of a general metamorphosis; rather such an observation only demonstrates that during this time no significant alteration in the external conditions of development has occurred.[213]

This last remark was obviously directed to Cuvier's objection to Lamarckian theory, namely, that mummified animals recovered from ancient tombs in Egypt showed no deviation from extant forms.

211. Goethe, "Die Faultiere und die Dickhäutigen," *Zur Morphologie*, in *Sämtliche Werke*, 12: 244–50.

212. Christian Pander and Eduard d'Alton, *Das Riesen-Faultier Bradypus Giganteus* (Bonn: Weber, 1821). Wells gives a description of this work in his *Goethe and the Development of Science*, pp. 33–38; he has, though, been misled into thinking this part was by d'Alton alone. Goethe himself only refers to d'Alton as the author, which seems to have been due to a curious circumstance: d'Alton reissued a couple of the parts of *Die vergleichende Osteologie* with his name only, including this part. The original edition, however, carries the names of both Pander and d'Alton; and the introduction, which Goethe cites, employs the first person plural ("*wir*") throughout and refers to a work of Pander using the same locution.

213. Pander and d'Alton, "Einleitung," in *Riesen-Faultier*, [p. 6].

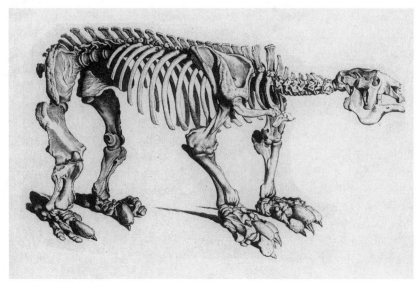

Figure 11.4 Megatherium drawn by D'Alton in Pander and D'Alton's *Das Riesen-Faulthier* (1821).

Through comparative analysis, Pander and d'Alton attempted to demonstrate that the living modern sloth descended from the extinct giant through a metamorphosis of the sort they specifically identified with Goethe's conception.[214] In particular they argued that the internal structures of living sloths closely resembled those of the extinct creature (especially the skull), while their rather different external forms must have been due to a changing environment in ancient times: if food supplies had been reduced, there would have been a comparable reduction in the size of the primitive animal; and a slower moving creature would have allowed for an expansion of the limbs, thus producing the awkward modern sloth.[215] These sketchy notions about how transmutation might have occurred comport with ideas of Erasmus Darwin, Lamarck, Geoffroy Saint-Hilaire, and, of course, Goethe. The authors, however, only mentioned Goethe's *Metamorphose der Pflanzen* as a theoretical source of their considerations; but they were obviously aware of Lamarck's work as well—at least as negatively represented by Cuvier.

In his review of the monograph on the giant sloth, Goethe declared that "we are in perfect agreement with the author as concerns the introduction." He went on to say:

214. Ibid.
215. Ibid., pp. 15–17.

We share with the author the conviction of a common type, as well as of the advantages of an empirical [*sinnig*] representation of a sequence of forms; we also believe in the eternal modifiability of all forms in appearance.[216]

Goethe, like Schelling (and Darwin), thought the process of transformation went from a more generalized form to one more specified and individualized. He thus maintained that "certain forms, having become generic, specified, and individualized [*generisiert, spezifiziert, individualisiert sind*], remain stable through many generations and over long periods of time, and, indeed, always essentially continue the same through the greatest deviations."[217] Goethe meant this last remark to be exemplified by the case of the giant sloth, the basic pattern of which remained constant through the greatly deviating forms of the two modern species (i.e., the aï and the unau).

Goethe not only endorsed the authors' evolutionary analyses, he even offered a "poetic" sketch of how the descent of these creatures might have occurred. He supposed the giant sloth first existed as a kind of whale that got trapped along a swampy, sandy beach. To bear its great weight on land, it would have had to develop large limbs, which would then be passed to descendants. Subsequent generations would then further adapt to the land, achieving their modern ungainly structure.[218] Well, the story, as he remarked, is "poetic" and many theoretical details are wanting; it has the piquant flavor of the tale Darwin himself told in the *Origin of Species* of how a bear, swimming with its mouth open to catch insects, might have gradually been transformed into a whale.[219]

Goethe developed his ideas about metamorphosis in relation to the ontogeny of individual organisms. But their application more generally to species could be and was easily made. His developing theory received conceptional support both from his early work with Herder on the formation of the earth and, later, from Schelling's ideas about *dynamische Evolution*. Goethe quite naturally also extended his theory of metamorphosis to

216. Goethe, "Die Faultiere und die Dickhäutigen," *Zur Morphologie*, in *Sämtliche Werke*, 12: 245.

217. Ibid. Goethe may have been aware that Karl Ernst von Baer had argued that during ontogeny the creature developed from a more generic state to a more individualized state—that is, he held that the human embryo went through stages of a generalized vertebrate, a generalized mammal, to the more specific human structure and then the particular form of the individual. Darwin would characterize the phylogenetic process of transformation in the same way. See my *Meaning of Evolution*, chap. 3, for a discussion of von Baer's and Darwin's use of the principle.

218. Ibid., 12:246–47.

219. Charles Darwin, *On the Origin of Species* (London: Murray, 1859), p. 184. Darwin dropped this fanciful story from the third and subsequent editions of the *Origin*.

species in order to accommodate mounting paleontological evidence, with which he was quite familiar. He was certainly encouraged in this direction by the transformational use made of his morphological ideas by the likes of Schelling, Pander, and d'Alton. From our perspective, many loose ends yet dangle from his evolutionary notions—for instance, whether the several lines of animals arose independently or were produced out of one another. It would be historiographically quite satisfying to be able to weave these disparate threads neatly back into the whole. Theoretical urgency, without much concern for conceptual tidiness, however, led Goethe to the evolutionary hypothesis. We cannot expect that issues that have become poignant for us only after Darwin's accomplishment should have beseechingly entreated scientists several decades before that accomplishment. Yet there can be little doubt that Goethe believed in a gradual transmutation of species over long periods of time and that he thought such change occurred by thoroughly natural processes. These ideas lay inextricably mixed with his general morphological doctrines and thus surreptitiously enjoined those adopting his theories also to incline to the evolutionary hypothesis, thinkers such as d'Alton and Pander at the beginning of the nineteenth century and Haeckel at the end.

The Creation of a Scientific Life: *Zur Morphologie*

The long gestation of Goethe's *Farbenlehre* came to term in May 1810. Immediately several critics went after the new babe with lethally sharpened pens. In July an anonymous review in the *Neue oberdeutsche Allgemeine Literaturzeitung* of Munich declared:

> Several individuals, who were familiar with the activities of the author, were of the opinion, even before the book appeared, that science would gain nothing with its publication. One knew that he understood nothing of mathematics and was not a practicing physicist. One knew that he belonged to that school that did not exactly shine through the clarity and precision of their principles. . . . In this light it was almost predetermined that this new theory of colors would be Romantic, poetic, and certainly not plainspoken, and that we would not be able to expect anything more than the explanation of well-known natural phenomena travestied in the artificial language of transcendentalism.[220]

---

220. [Anonymous], "Rezension der Farbenlehre," in *Sämtliche Werke*, 10:1034–39; quotation on p. 1034.

Goethe had his supporters, of course. His friend Johannes Daniel Falk gave it a laudatory review, not being at all hampered by ignorance of optics.[221] But Goethe could only be embarrassed by the likes of Falk. He sought the approval of experts like Jakob Friedrich Fries (1773-1843), the Kantian philosopher and mathematician at Heidelberg, who praised the poet for his artful presentation and condemned him for his science.[222] Helmholtz's reevaluation would undoubtedly have salved some of the deeper wounds, but his consoling words sounded only at the end of the century. Goethe had devoted almost two decades to the research and composition of his color theory, and his contemporaries rejected his efforts, rejected that part of his genius, that part of his life. This devastating judgment came during the period when Schiller's loss was still agonizingly felt, when his own health had declined, and when the French invaders threatened everything.

In 1808 part 1 of *Faust* appeared. The dedicatory poem, composed during the time that Schiller urged Goethe to renewed work on the drama (1797), yet carried a sentiment fit for these later troubled times. The poet envisioned "wavering forms" that arose out of the past, that brought him "images of happier days" of love and friendship; but now his song would fall on the ears of an unknown company. The four stanzas end with the poignant lament of memory:

What I possess, I see as if in a far locality,
And what has vanished now becomes for me the reality.[223]

Goethe penned a comparable sentiment to a friend during this period of intensified focus on the loss of Schiller and on his own mortality: "I have turned to thinking vividly about the past, and rather passionately feel the duty we have to preserve in memory what seems to have vanished forever."[224]

These various forces of rejection, loss, and mortal fear drove Goethe to contemplate his own life and to consider how such a life developed like a poetic line, partly from contingency, partly from necessity—but a line that could be intelligibly followed and aesthetically appreciated. In fall 1809 he began collecting materials—letters, reminiscences of friends, diaries, literature of an earlier day—to compose that life. *Dichtung und Wahrheit* appeared in four volumes, in 1811, 1812, 1814, and posthumously in 1833.

---

221. Johannes Daniel Falk, "Erstes sendschreiben über die Goethesche Farbenlehre," in ibid., 10:1040–46.

222. Jakob Friedrich Fries, "Rezension," in ibid., 10:1047–48.

223. Goethe, "Zueignung," in *Sämtliche Werke*, 6.1:535–36: "Was ich besitze seh' ich wie im weiten, / Und was verschwand wird mir zu Wirklichkeiten."

224. Goethe to Philipp Hackert (4 April 1806), in *Goethes Briefe*, 3:20.

The story would be continued in *Italienische Reise*, published in 1816, 1817, and 1829. These autobiographical constructions led up to and embraced his work *Zur Morphologie*, which appeared in six parts—1817, 1820 (two numbers), 1822, 1823, and 1824. These he also regarded not merely as objective studies of biological and other scientific subjects but as exercises in aesthetic construction (several poems are included) and reminiscent consciousness. In one of the essays for this collection, he made the self-revelatory intent of his work clear:

> In the present volume as in the earlier, I have followed this path: to express how I view nature, simultaneously to reveal, to a certain extent, myself, my interior, my way of being, insofar as that would be possible. . . . In this I admit that I have always been suspicious of the great and so portentous sounding phrase "Know thyself." . . . A person can know himself only insofar as he knows the world, which he perceives only in himself, and himself only in it. Every new object, well investigated, opens up a new organ in one's self.[225]

A severe dualism would have separated the interior subjective realm from the external objective realm (a gulf that Kant himself could not bridge). But Goethe had, with the idealists, rejected this bifurcation of reality. And so, the many essays previously written (along with the dates of their initial composition), poetry, reviews, and new essays on various scientific topics constitute a work in which the self becomes expressed by immersion in the world and the world takes on the colors of the self. I do not wish to suggest that twenty years after his intense relationship with Schelling, he retained a firm belief in his friend's abstruse philosophy of ideal-realism; but he did assume, in a more casual way, the identity of self and world, and the instantiation of ideas in the natural realm.

This kind of idealism was clearly expressed in one of the new essays for *Zur Morphologie*, "Anschauende Urteilskraft" (Intuitive judgment, 1820). The essay returns to Kant's third *Critique*, as Goethe himself did at this time, to consider the philosopher's distinction between reflective and determinative judgment. It will be recalled that Kant classified judgments of beauty and judgments appropriate to biology (ends-means assessments) as *reflective*. Such judgments arose in attempting to understand the relationship of parts to whole, either in a work of art or a work of nature. In our appreciation of an artistic work, our understanding considers its various parts,

---

225. Goethe, "Bedeutende Fördernis durch ein einziges geistreiches Wort" (A significant advance through a particularly meaningful expression, 1823), *Zur Morphologie*, in *Sämtliche Werke*, 12:306. He goes on to say (as William James later would) that our self-knowledge is increased by the reflection of ourselves in the minds of others.

allowing the free play of imagination, to get a sense of the harmony of forms; as imagination re-creates those forms and compares their arrangements with the requirements of understanding, a feeling of purposiveness arises, an aesthetic feeling that we generically call a feeling of beauty. Kant's analysis assumes that this feeling will be similar to the one that guides the artist and that emerges from his very nature, a nature that bubbles over with inarticulable but harmonious forms. The situation of the biologist is comparable. When he assesses the traits of an organism, the same reflective procedure occurs: through an initial exploration of the parts, he formulates an idea of the whole—though now a conscious and articulate conception of the whole, an archetype—and thereby understands the organism's particular traits in relation to the whole. Indeed, the student of nature must, according to Kant, judge the structures investigated *as if* they came to exist by reason of the archetype, as if nature herself (or nature's God) artistically created organisms according to such an ideal. But in this instance, the biologist makes only a heuristic assessment, and does not—cannot—presume the idea at which he arrives to have actually caused the structure. Our intellect, according to Kant, can only make *determinative* attributions of mechanical causes, not of ideational causes, to explain natural phenomena.

In our scientific understanding of nature, according to Kant, we apply categories like causality and substance determinatively to create, as it were, the phenomenal realm of mechanistically interacting natural objects. But in considering biological organisms, we must initially analyze the parts in reflective search of that organizing idea that might illuminate their relationships. Kant suggested that we could yet conceive of another kind of intellect, one other than ours, which might move from the intuition of the whole to that of the constituents, instead of following our path from parts to whole. This would then be an *intellectus archetypus*, whose very ideas would be creative.[226]

Concerning this Kantian notion of an *intellectus archetypus*, Goethe made a trenchant and many-layered observation:

> The author seems here, indeed, to refer to a divine understanding. Yet, if in the moral realm we are supposed to rise to a higher region and approach the primary being through belief in God, virtue, and immortality, then it also should be the same in the intellectual realm. We ought to be worthy, through the intuition of a continuously creative nature, of mental participation in its productivity. I myself had incessantly pushed, initially unconsciously and from an inner drive, to the primal image [*das*

226. Kant, *Kritik der Urteilskraft*, in *Werke*, 5:526 (A 346–47, B 350–51).

Urbildliche] and the type [das Typische]. Fortune smiled on this effort and I was able to construct a representation in a natural way; so now nothing more can prevent me from boldly undertaking that "adventure of reason," as the grand old man from Königsberg himself has called it.[227]

Goethe here attempts to muscle into philosophical acceptance a thesis similar to one of Schelling: namely, if moral experience requires us to postulate God to make sense of that experience,[228] then our experience of organisms should also require us to postulate an intellectual intuition to make sense of such experience. Goethe immediately concluded that such intellectual intuition, if it indeed allows us to understand organisms, ought to be our own as well as nature's. That is, he maintains, as did Schelling, that if archetypal ideas are necessary for our experience of organic nature, then they must be constituents of that experience—mentally creative of that experience. And there is the further implication of this analysis, namely, that in such mental creations, we share in nature's own generative power. Goethe thus reaffirms a Schellingian Spinozism: God, nature, and intellect are one.

Goethe's final remark in the quotation above draws out the ultimate consequence of this ideal-realism. In the third Critique, Kant recognized that the variety of organic forms yet displayed "a common archetype" [ein gemeinschaftliches Urbild], and thus might be produced, as he put it, by "a common primal mother." This might lead to undertaking "a daring adventure of reason," namely, that the earth be thought of as having given birth to less purposive forms and these to more purposive, until the array of currently existing organisms should have appeared. Kant thought this transformational hypothesis would be logically possible if we assumed the initiating cause of the series was itself organic. He yet rejected this evolutionary hypothesis because he did not think we had any empirical evidence of the generation of a more organized form from a less organized one.[229] Schelling's theory of dynamic evolution, which Goethe accepted, postulated an organic foundation for a transformational series (namely, in absolute

227. Goethe, "Anschauende Urteilskraft," Zur Morphologie, 1.2, in Sämtliche Werke, 12:98–99.
228. Though a bit vaguely stated, this was Kant's position in the second Critique and reiterated in the third Critique. But Kant thought the postulate of God necessary as a practical matter—rather like a heuristic or regulative idea. The argument from moral experience could not be used as a proof of the existence of a transcendent entity. In the third Critique, Kant made this point explicit in a footnote to his moral argument: "This moral argument should not be taken to provide an objectively valid demonstration of God's existence, not an argument that might prove to the skeptic that there is a God—rather that if he wishes to think consistently about morals, he must assume this proposition [that God exists] among the maxims of his practical reason." See Kant, Kritik der Urteilskraft, in Werke, 5:577 (B 424–25).
229. Ibid., 5:538–39 (A 364–65; B 368–69). See also chapter 5.

mind), and Goethe's own theory of the archetype augmented the Schellingian conception; moreover, by the time of *Zur Morphologie*, researchers had accumulated fossil evidence of such transformations. Goethe was thus ready, as he concluded, boldly to undertake that adventure of reason rejected by Kant, namely, that of evolution.

The evolutionary hypothesis as applied to nature reflected Goethe's own mental development: *Zur Morphologie* tracked the gradual ascent of his morphological ideas, and those ideas gave rise to the transformational hypothesis he rather daringly embraced in the book. The metaphysical foundation of these two evolutionary series—of self and of nature—rested ultimately on the kind of ideal-realism for which Schelling had argued and which Goethe embraced. But the historiographic enterprise, which *Zur Morphologie* represented—along with his other, explicitly autobiographical writings on which he worked at this time—was a constructive effort, ultimately a fabrication by the self from the resources of memory. This construction certainly had ample documentary evidence to which Goethe could refer: his voluminous correspondence, his many manuscript essays, and numerous published works. Nonetheless it was a new creation, but not the only poem that could have been sung of those times. Goethe's bitter dispute with Lorenz Oken demonstrates this with some poignancy.

The Vertebral Theory of the Skull: Goethe's Dispute with Oken and the Truth of Memory

In the second number of the first volume of *Zur Morphologie* (1820), Goethe appended a long extension to his 1786 essay on the intermaxillary bone. He stated that this extension was written with a certain question in mind: "Whether one can really derive the bones of the skull from the vertebrae and whether one is able to and should recognize their initial form despite their great and considerable change?" He said he "happily admitted that he had been convinced of this secret relationship for thirty years and had always advanced this conviction."[230] In the first number, second volume of *Zur Morphologie* (1823), he gave a more exact description of his discovery, tracing it to his second Italian journey at the beginning of the 1790s:

I had already recognized the three hindmost bones of the skull [as vertebral]. But it was only in the year 1791 [*sic*, 1790], as I lifted a battered sheep's skull from the dune-like sands of the Jewish cemetery in Venice,

230. Goethe, "[Nachträge]," in *Sämtliche Werke*, 12:188.

that I immediately perceived the facial bones were likewise to be traced to the vertebrae.[231]

Goethe here clearly claimed that he discovered the vertebral origin of the skull even before his trip to Italy and that he confirmed and extended the theory during his stay in Venice, actually in 1790, not 1791, as he falteringly recalled. Well, the event in that cemetery occurred over thirty years before, so some discrepancies in recollection would not be surprising. Moreover, his letter from Venice to Karoline Herder, quoted earlier in this chapter, certainly indicates he discovered something in 1790.

In a subsequent essay in the same number of *Zur Morphologie*, Goethe returned to the vertebral theory of the skull. In his "Das Schädelgerüst aus sechs Wirbelknochen auferbaut" (The frame of the skull is constructed out of six vertebrae), he provided a detailed description of the vertebral bones of the skull and then added with pique that "in the year 1807 this theory sprang chaotically and defectively into the public arena" and that an "unripe mode of presentation" had distorted the idea.[232] This condemnation, directed to an unnamed lecture that took place over a decade and a half earlier, must have struck Goethe's readers as odd; though if they followed him through the three essays I have cited, they would have clearly perceived his assertion of priority for the discovery. The object of his obloquy was Lorenz Oken, whose inaugural lecture in Jena, *Über die Bedeutung der Schädelknochen* (On the significance of the skull bones), occurred in July 1807.[233] Oken also claimed to have discovered the vertebral origin of the skull.

Oken's Early Career

Lorenz Okenfuss (1779–1851) received his medical degree from Freiburg and passed his medical exams in 1804, and at that time he shortened his rather comical name of Oxenfoot. He then enrolled at Würzburg, where he attended the lectures of Ignaz Döllinger on physiology and probably got a whiff of the Schellingian *Naturphilosophie* that had begun to leaven Döllinger's work. At this time, too, he became friends with Schelling's disciple Henrik Steffens. Very quickly he published a small monograph, *Die Zeugung* (1805), that attracted some attention. In the preface to that work, he allied himself with the epistemological views of Fichte and Schelling. He main-

---

231. Goethe, "Bedeutende Fördernis durch ein einziges geistriches Wort," in ibid., 12:308. I have discussed in the first part of this chapter Goethe's assertion that he discovered the skull to be composed of vertebrae.

232. Goethe, "Das Schädelgerüst aus sechs Wirbelknochen auferbaut," in ibid., 12:359.

233. Lorenz Oken, *Über die Bedeutung der Schädelknochen: Ein Programm beim Antritt der Professur an der Gesammt-Universität zu Jena* (Jena: Göpferdt, 1807).

tained that a priori principles allowed us to recognize the fundamental nature of reproduction (*Zeugung*) but that those principles could also be confirmed through empirical examination. Reason alone indicated that the organic could not arise out of the inorganic; and empirical study confirmed this: simple infusoria, he observed, did not arise spontaneously but were the constituents of more complex organisms—plants or animals—that had decayed.[234] This meant, looking in the other direction, that the higher animals arose in a kind of synthetic process, out of the agglomeration of these elemental animals. Oken conceived this to occur as the male semen (full of infusoria) was received into and given form by the female.[235] In the production of the human being, according to Oken, the fetus went though stages of recapitulation, in which it took on sequentially ever-higher forms, analogous to the hierarchy of animals, until the perfection of the human form was achieved.[236]

In spring 1805 Oken traveled from Würzburg to Göttingen to hear lectures by Blumenbach, who, the young doctor discovered, had passed his prime. That winter (1805–6) Oken taught as *Privatdozent* while conducting an extensive embryological study of higher organisms—pigs, dogs, and the like.[237] He undertook the investigation under the supervision of Blumenbach and in collaboration with his friend Dietrich Georg Kieser. They wished to determine the relationship of the umbilical cord to the intestines. In this work Oken also crystallized several ideas that became part of his more general conception, especially the idea that "man is the unification of all animal characters, while animals are only the particular expressions of these particular characters."[238] The lower animals, he maintained, tended to have less differentiated parts than the higher and were governed by one central organ—thus, worms, according to Oken, were animals chiefly of an "epidermal system," birds of a "bony system," fish of a "liver system," and so on. This occurred because, in a given class of animals, one organ system received most of the nourishment, while the others were greatly diminished because of lack of nutrition.[239] Yet all animals exhibited a common "funda-

234. Lorenz Oken, *Die Zeugung* (Bamberg: Goebhardt, 1805), p. 19: "The origin of infusoria is . . . a release [*Freiwerden*] from the bonds of larger animals, a dissolution of the animal into its constituent animals. Those later are not the result of a synthesis of crude, dead stuff, which could never produce an organic life no matter what form that stuff might take. Out of absolute death no life can be born!"

235. Ibid., p. 103.

236. Ibid., pp. 146–47.

237. Lorenz Oken and Dietrich Georg Kieser, *Beiträge zur vergleichenden Zoologie, Anatomie und Physiologie*, 2 vols. (Bamberg: Goebhardt, 1806–7).

238. Ibid., 1:103. This section was written by Oken.

239. Ibid., 1:103–4.

Figure 11.5
Lorenz Oken
(1779–1851). An etching
done for his friends.
(Courtesy Wellcome
Institute Library.)

mental type" [*Grundtypus*], and their differences resulted from the nutritional expansion of one set of organs at the expense of others. Human beings, however, enjoyed the greatest perfection of their individual organs and thus recapitulated the animal kingdom in its various classes.[240]

Oken's embryological work, if not exactly his more general recapitulational ideas, won the praise of the redoubtable Karl Ernst von Baer. Von Baer thought the work flawed, yet "it obviously belongs to the most exact we have on the mammals." He believed that "Oken's investigations had been the turning point for the more correct recognition of the egg of mammals," a discovery that von Baer himself had made in 1827.[241]

During his time in Göttingen, Oken began corresponding frequently with Schelling, informing the philosopher of his own work, discussing the fortunes of *Naturphilosophie,* and in some plaintive terms describing his wretched circumstances.[242] Those circumstances finally drove him to ask Schelling for a small loan, which the philosopher provided. But Oken's for-

240. Ibid., 1:103–4, 109, 114.

241. Karl Ernst von Baer, *Über entwickelungsgeschichte der Thiere,* 2 vols. (Königsberg: Bornträger, 1828–37), 1:xvii–xviii.

242. Oken's correspondence with Schelling is contained in Alexander Ecker, *Lorenz Oken: Eine biographische Skizze* (Stuttgart: Schweizerbart'sche Verlagshandlung, 1880), pp. 177–211.

tunes changed dramatically when, in 1807, because of a budding reputation, he received a call to Jena, where he took up duties in July of that year.
His inaugural lecture, however, became the ember that ignited a bitter priority dispute with Goethe that flared some years thereafter.

Oken's Lecture on the Vertebral Construction of the Skull

In his lecture Oken began with a statement of a different kind of recapitulational idea than he had voiced before:

> A vesicle becomes calcified, and that is a vertebra. A vesicle becomes
> elongated into a tube, becomes articulated, calcified, and you have a
> spine. The tube produces (according to laws) dead-end side branches,
> they become calcified, and you have the skeletal trunk. This skeleton is
> repeated at both poles, each pole repeating itself in the other, and you
> have head and pelvis. The skeleton is only a fully grown, articulated,
> repetitive vertebra, and the vertebra is the preformed germ [Keim] of the
> skeleton. The entire human being is only a vertebra.[243]

Oken's idea here is very close to that of Goethe in the Metamorphose der
Pflanzen: one part of an organism gives rise to all the others through various
transformations of a basic type. Goethe might reasonably have seen his own
general theory of metamorphosis reflected in Oken's conception. But his
objection, at least as later expressed, went to a more specific thesis advanced in Oken's lecture. The young naturalist maintained that just as "the
brain is the spinal marrow that has greatly developed into a more powerful
organ, so the brain case is the more greatly developed backbone."[244] Consequently, the bones of the skull ought to be regarded as further developed
vertebrae. The empirical evidence, Oken urged, could be found in the skull
bones of a young sheep—that is, the frontal, parietal, ethmoidal, and temporal bones of its head. These, he assured his audience, formed a kind of
bony column, whose similarity to the vertebral column, when viewed in
this way, would be obvious to an experienced anatomist.[245] Oken's representation did not make it precisely certain how many vertebrae he believed
went to form the bones of the skull—he explicitly stated three, but suggested more.[246] He also proposed that since the vertebrae had expanded to
form hands, feet, thorax, and so on, one could compare the various skull

243. Oken, Über die Bedeutung der Schädelknochen, p. 5.
244. Ibid.
245. Ibid.
246. Oken later made it clear he thought the skull to consist of three modified vertebrae. See
his [Review of Über die Bedeutung der Schädelknochen], Isis 1 (1817):1204–8.

bones with these other parts as well. Analogy thus permitted one to think of the upper jaw as comparable to the hands and to describe the nasal and ethmoidal bones as forming the "thorax of the head."[247] Comparisons of this kind—certainly exaggerated from our point of view—would later be termed by Richard Owen "serial homologies," and despite the weirdness of some of the particular comparisons, they have a sound-enough rationale. Oken attempted to uncover the recapitulation of relationships governing the bones of the body in those governing the bones of the skull. A more modest effort might simply have observed the homologies holding, say, between the vertebrae themselves or between front and rear limbs. Even in the estimation of many contemporaries, Oken stretched these comparisons into comical and even grotesque proportions.

One particular comparison of the serial homology type would certainly have caught Goethe's attention, namely, that concerning the intermaxillary bone. Oken claimed that he had investigated "dozens of children's skulls," which convinced him that the intermaxillary bone was also to be found in human beings and that this bone could be compared to the thumb (insofar as the upper jaw might be a repetition of the hand)![248] Oken's subtle suggestion that he had at least confirmed the existence of the intermaxillary in man—in addition to his wild comparison of that bone with the thumb—might well have provoked Goethe. After all, he might have learned of Goethe's views about the intermaxillary from Blumenbach or from the various publications (those of Soemmerring, for instance) that had Goethe's name attached to the idea.[249] But Oken made no mention of Goethe's work.

In any case, shortly after Oken sent Goethe a copy of this inaugural lecture, the *Geheimrat* invited him to several social gatherings.[250] The young anatomist was quite apprehensive about the meetings—or so he indicated to Schelling. He simply felt out of place in such august company.[251] Schelling cautioned him about his awkward manner and exaggerated expressions—such behavior, the philosopher warned, might endanger his relationship to Goethe, and only he would be the loser for that.[252] Prior to 1820, when Goethe first publicly insinuated that Oken had dealt nefari-

247. Oken, *Über die Bedeutung der Schädelknochen*, pp. 12, 14.
248. Ibid., p. 14.
249. See the previous chapter for a discussion of the ways in which Goethe's name got publicly attached to the idea of the human intermaxillary.
250. Goethe's *Tagebuch* for 1807 indicates they met on 13 and 29 November and 4 December. Goethe also invited Oken to a social gathering a couple of times in 1808 (27 and 29 October). See Goethe, *Tagebücher*, in *Goethes Werke*, III.3:295, 302, 304, 395.
251. Oken to Schelling (3 November 1807), in Ecker, *Lorenz Oken*, p. 201.
252. Schelling to Oken (26 November 1808), in ibid., p. 119.

ously with claims to priority, they both met often enough on various occasions to confer on scientific matters or for social gatherings.[253] And Oken
himself thought they were on good terms, or at least that is what he told
Schelling.[254]

The Priority Dispute: Who Discovered the Vertebral
Construction of the Skull?

In the 1820s Goethe begin to make public protest about the usurpation of
his vertebral theory of the skull. Even earlier, apparently with allusion to
Oken, he complained to Steffens concerning his osteological theory that
"some naturalists had abused his trust and announced discoveries without
referring to him as if they were the ones who introduced them."[255] But did
Goethe originate the notion that the skull consisted of transformed vertebrae—something he said he discovered while playing with a battered sheep
skull in Venice in 1790? Scholarly consensus takes Goethe at his word and
often enough charges Oken with intellectual theft—suggesting that the
young anatomist somehow learned of Goethe's idea and appropriated it for
his own.[256]

A few years after Oken had declared the vertebrate skull to be composed
of transformed vertebrae, several other anatomists sanctioned and re-
affirmed the claim. In 1811, for instance, Johann Friedrich Meckel (1781–
1833) at Halle suggested that "the head is only an enlarged vertebra, or perhaps a synthesis of several vertebrae."[257] He cited Oken as originally
proposing this thesis. In 1815 Johann Baptist von Spix (1781–1826) devoted a spectacularly illustrated book to the further establishment and refinement of the idea, which he also credited to Oken.[258] Attributions of
this sort undoubtedly brought Goethe into high dudgeon.

253. For instance, in 1809 Goethe and Oken met, according to Goethe's *Tagebuch*, some seven
times, either in Weimar or in Jena.

254. Oken to Schelling (25 January 1809), in Ecker, *Lorenz Oken*, p. 205.

255. Steffens, *Was ich erlebte*, 6:252. Steffens recalled this conversation to have occurred about
1809, around the time of the publication of Oken's *Lehrbuch der Naturphilosophie* (1809–11).

256. John Williams, in the recent literature, has simply accepted Goethe's version. See his *The
Life of Goethe: A Critical Biography* (Oxford: Blackwell, 1998), p. 265. Hermann Bräuning-Oktavio
provides the most detailed study of the dispute. See his *Oken und Goethe im Lichte neuer Quellen*
(Weimar: Arion, 1959). See also Helmut Müller-Sievers's perceptive analysis "Skullduggery:
Goethe and Oken, Natural Philosophy and Freedom of the Press," *Modern Language Quarterly* 59
(1998): 231–59. Both of these scholars side with Goethe.

257. Johann Friedrich Meckel, *Beyträge zur vergleichenden Anatomie*, 2 vols. in 3, 2nd no. (Leipzig: Carl Heinrich Reclam, 1812), p. 74.

258. Johann Baptist von Spix, *Cephalogenesis; sive, Capitis ossei structura, formatio et significatio
per omnes animalium classes, familias, genera ac aetates digesta* (Monachii: F. S. Hübschmannii, 1815),
p. 12.

Oken, likely getting wind of Goethe's pique, published in 1818 a short account of his own discovery of the vertebral skull. He said he had become convinced by 1805, from the study of insects, that the jawbones of vertebrates were analogous to hands and feet (presumably because in insects mandibles appear comparable to hands). Then on a trip in August of the next year, he claimed to have made the fateful discovery. He and two students (one he named, the other whose name he had forgotten) were traveling through the Harz Mountains, and during their rambling he came upon a bleached skull of a deer. He said that as he gazed at it, the idea came to him in a flash: "It is a vertebral column!" And thereafter "the skull has been for me a vertebral column."[259] After his 1818 account, Oken maintained, in increasingly strident tones, his own priority for the discovery. Even as late as 1847, when he was sixty-seven years old, he still complained, to his friend Richard Owen in England, that "Goethe had the audacity [Keckheit] to maintain that he had discovered [the vertebral composition of the skull] twenty years before [Oken's 1807 lecture]." Oken added that "one need only read Goethe's other osteological treatises in order to recognize immediately that he had had no idea of it."[260]

Oken was certainly correct. None of Goethe's osteological essays written prior to the publication of his Zur Morphologie give any indication of his belief in the vertebral structure of the skull, even when that idea should naturally have been discussed. What he emphasized in an essay written shortly after returning from Italy, where he made his mysterious discovery, is something quite different. In the summary paragraph of his incomplete "Essay on the Form of Animals," he wrote:

> We must maintain that the structure of the bones of all mammals, to be brief, is formed according to the same pattern and concept, not only on the whole, but also that the particular parts are to be found in every creature—this often escapes our eye because of the form, measure, direction or exact connection with other parts and only remains visible to our understanding. All parts, I repeat, are present in every animal, and we must seek them and discover them by our effort and keen perception—that concept is the thread of Ariadne.[261]

---

259. Lorenz Oken, [Discussion of Über die Bedeutung der Schädelknochen], Isis 1 (1818): 510–11; quotation from p. 511.

260. Oken to Owen (12 January 1847), in the papers of Richard Owen, London Natural History Museum.

261. Goethe, "Versuch über die Gestalt der Tiere," in Sämtliche Werke, 4.2:146.

The notion that every part was present in every vertebrate—insofar as the animal realized the vertebrate archetype—was, I believe, the thread Goethe found wrapped around the battered sheep skull on the Lido in Venice. But what, then, led him later to claim that he discovered at that time the vertebral composition of the skull? The answer lies, I believe, in the work of a protégé of Goethe, Friedrich Siegmund Voigt (1781–1850), physician and botanist at Jena.

Goethe thought the young botanist, as he mentioned to Charlotte von Stein, a "learned and cultivated young man."[262] He and Voigt had been conferring on osteological work during late 1806 and early 1807, just before Oken's inaugural lecture. Goethe lent the young scientist some of his unpublished essays on osteology. These essays stimulated Voigt in his own research on the vertebrate skeleton. On 20 February 1807, he wrote Goethe the following:

> I would like to show you how much your new views and ideas have struck me; in so many ways, they have provided material for further investigations. . . . In describing the vertebra of the backbone from the simplest to the highest vertebra of the head, I have not lost sight of the further development of their branches. I am thus in a position to show all the processes of the vertebrae to be found also in the sphenoidal bone [of the skull], and I can clinch this comparison with a very interesting demonstration of the foramen opticum [i.e., opening for the optic nerve in the sphenoid], which is analogous to the canal of the vertebra. . . . The idea, which I have inchoately entertained for a long time, has become clear through your representation, namely, that the skull consists of three vertebrae. This idea now completely alters the organization of the bones. . . . In my view, the comparison of the facial bones with those of the insects makes the origin of the former clear. I compare the beetle and bee initially with these mammals, whose feet are no longer able to serve for grabbing (the hoofed animals), and who, for that reason, show a greater development of the facial bones. The whole thing rests on the comparison of the upper jaw with the pincers of the insects.[263]

---

262. Goethe to Charlotte von Stein (11 May 1810), in *Goethes Briefe*, 3:125.

263. Friedrich Siegmund Voigt to Goethe (20 February 1807), as quoted by Bräuning-Oktavio, *Oken und Goethe im Lichte neuer Quellen*, pp. 55–56. Bräuning-Oktavio uses this letter to argue that Goethe had the idea prior to Oken. But the straightforward reading of the letter suggests that Voigt himself had the idea that the skull consisted of vertebrae and that Goethe's morphological ideas only inspired him to make the discovery.

This letter seems to show that Voigt himself developed the idea of the vertebral construction of the skull and that the insect comparison first led him to that notion, aided, as he says, by inspiration from Goethe's morphological work. Goethe seems not to have remonstrated with Voigt that he, Goethe, was rather the author of the idea—something one might have expected in light of his later claims to priority.

Voigt seems to have come to the idea about the same time as Oken. Oken, however, had the idea by at least December 1806, two months before Voigt's letter to Goethe. We know this because Oken mentioned his idea in a letter to Schelling on 27 December 1806. "Take a look at the skull (its bones) of a sheep," he suggested to his friend, "and you will find that it consists of several extended vertebrae, and so also does that of the human being. I have nearly completed work on the significance of the system of bones."[264] The letter, posted from a small town on the North Sea (where he stayed until spring), makes plain that Oken had been working on his interpretation of the skull prior to his appointment to Jena and before he could have had any access (if he ever did) to Goethe's unpublished work. He simply could not have stolen the idea from Goethe.

Oken continued in his position as professor of anatomy at Jena for several years. In 1816 he founded the journal *Isis*, which published articles of general cultural interest, especially concerning science. The essays, though, quickly sounded a rather radical tone. Oken ran reviews critical of other governments and university operations, and these brought protests to Weimar. The Weimar administration warned Oken that such reviews were beyond the officially sanctioned mandate of the journal. Then in summer 1817, the *Burschenschaften*—the student fraternities—of Jena held a festival at the Wartburg. They burnt some symbols of what they took to be political oppression, but otherwise behaved rather well. Yet a rumor spread that the young men were conspiring with revolutionary intent. Oken reported on the student meeting in *Isis*, and that report raised howls of protest from the Prussian government. Carl August gave Oken the alternative of ceasing to publish the journal or of giving up his teaching position. Oken, who made most of his living from *Isis*, resigned from Jena and moved to Leipzig, where he continued to publish. He later received a post at the university in Munich, due to Schelling's influence, though they too eventually fell out with one another. Because of his work on *Isis*, Heine judged Oken "one of Germany's greatest citizens," who "roused the enthusiasm of young Germans for the fundamental rights of man, for freedom and equality."[265]

---

264. Oken to Schelling (27 December 1806), in Ecker, *Lorenz Oken*, p. 199.
265. Heine, *Religion und Philosophie in Deutschland*, in *Sämtliche Schriften*, 3:637. Müller-Sievers

In considering the case of Oken right after the Wartburg incident, Carl August asked Goethe to write a report as to what action should be taken. Goethe himself advised complete suppression of *Isis*, advice Carl August did not follow. Goethe's judgment seems, in retrospect, quite harsh; and since Carl August did not follow the advice, he too seems to have believed it was. The issue, however, was complex. Another journal, the *Allgemeine Literatur-Zeitung*, had already secured an exclusive privilege to publish criticism and reviews. The approval for Oken's journal restricted publication in those areas, and he initially confirmed to the Weimar administration that he would stay away from those kinds of articles and certainly not engage in political criticism. When he nonetheless began to run reviews, the editor of the *ALZ*, Abraham Eichstädt, notified the government of Oken's breach of understanding. After some further complaints by other German governments that had received criticism, Oken was warned that further infraction would forfeit the journal. His report on the Wartburg student festival consequently brought the government into action and Carl August's request of advice from Goethe.

The request, however, came just at the time that Goethe began feeling aggrieved—perhaps really for the first time—over Oken's claims to priority. For Oken had published in the spring 1817 issue of *Isis* a full account of his 1807 *Bedeutung der Schädelknochen*.[266] He said he did so because so many illustrations of the idea had arisen in Europe, he needed to correct errors and clarify ambiguities. One ambiguity he certainly clarified was his claim to the discovery of the vertebral construction of the skull. This *Isis* article appeared just before Goethe made his recommendation to put Oken out of business. The article undoubtedly also provoked Goethe to make public his own assertion of priority.

The evidence of this history leads, I believe, to several likely conclusions: first, Goethe did not have a clear notion of the vertebral theory of the skull before he received the letter from Voigt in 1806; second, since Voigt attributed his insight to Goethe, the poet came to feel, after some period of time, that he not only inspired the discovery but that he actually made the discovery himself; and third, since he remembered having discovered something while gazing at that sheep skull in 1790, he amalgamated his later conviction of the vertebral construction of the skull to that earlier feeling of discovery. I do not think Goethe intentionally made a false claim. Just at the time Oken reasserted his priority in 1817, Goethe was seeing

elaborates on this theme in his "Skullduggery: Goethe and Oken, Natural Philosophy and Freedom of the Press."

266. Oken, [Review of *Über die Bedeutung der Schädelknochen*], pp. 1204–8.

Figure 11.6

Bust of Lorenz Oken on
the Fürstengraben in
Jena, erected in 1857.
The inscription reads:
"Lorenz Oken,
1779–1851, great natural
scientist, civil democrat,
1807–1819 professor in
Jena, emigrant, 1833 the
first rector, University of
Zürich." The subscription
for the monument began
in the year of his death,
just after the
revolutionary struggles of
1848. (Photography by
the author.)

through the press the first two parts of his *Italienische Reise* (1816, 1817)—
the narrative of his rebirth in Italy. With *Zur Morphologie*, he recorded another aspect of that rebirth: he peeled back the caul of memory to reveal the
vertebral skull. Trusting to Goethe, many contemporary scholars have simply consumed this afterbirth.

Chapter 12

CONCLUSION:
THE HISTORY OF A LIFE
IN ART AND SCIENCE

We tend to see Goethe as he came to see himself—in memory. His various late publications—*Dichtung und Wahrheit, Italienische Reise, Campagne im Frankreich, Zur Morphologie, Tag- und Jahres-Hefte*—look back on an earlier time, and we look with him. We see him as a rather young man, poetry gushing from his pen, quickly composing plays and taking a leading role in them, traveling to romantic environs—Switzerland and Italy—participating in the great revolutions of the age—the French Revolution and the Kantian—making discoveries in science and challenging encrusted thought. The many loves of his life, affairs of the heart that punctuated his existence and warmed his recollections, these we can never forget because he did not. And though the artifice of memory gave these many events novel-like features, they yet seemed to preserve a truth, both factual and emotional. In a historical effort, however, we must always test that truth, as best we can, against contemporaneous sources. But even here, we are compelled to make a judgment about what captures more of the reality: Goethe's memory or documents composed at the time. Those documents—a flattering letter, an essay that only sketches a partial reality—may be more unreliable than what the author preserved in his heart and later recorded in his books. And there is the added consideration that an event's significance may only be realized at a later time, when memory and judgment have fleshed out their consequences.

We remember the Goethe described by the physician Zimmerman as that "most dangerous man" for the heart of a woman: the man who wooed Charlotte Buff and then transformed that infatuation into the highest art; the man who carried on an intimate if not physical affair with Charlotte von Stein and let that affair bleed out of his pen; the man who found sexual fulfillment in Rome and who lived openly with a mistress for a decade and a half while serving in the court of a German prince. For these faces of Goethe there is substantial evidence. When he returned from Italy, the dangerous man could still be found, though he gazed out at the world through a physical body deeply gouged by time. The young physician David

Johann Veit met Goethe in 1793 and described him in a letter to Rahel
Levin, whose salon in Berlin rivaled that of Henriette Herz. Veit compared
Goethe to his recent portrait (1791), which had been rendered by Johann
Heinrich Lips (fig. 10.1):

> He is rather greater than of usual height and, for this height, relatively
> fat, broad-shouldered. If you know my uncle Salomon Veit, you'll have
> the similarity of figure; but Goethe is rather larger and stronger. His fore-
> head is quite attractive, more so than I have seen; the eyebrows in the
> painting are exactly right, but the very brown eyes incline a bit lower. . . .
> In his eyes there is much spirit, but not the consuming fire that one hears
> about. Under his eyes, he already has wrinkles and considerable bags. He
> certainly appears to be an individual of about forty-four or forty-five. . . .
> [I]t must be true what everyone in Weimar says, that he aged markedly
> during his stay in Italy. His nose is certainly aquiline, though the bump in
> the middle softly fades away. . . . The mouth is very attractive, small, and
> extremely mobile; only when he smiles does it distract from his appear-
> ance, because of his yellow and very crooked teeth. When he is silent, he
> seems very earnest, though really not ill-tempered, and no thought, not a
> hint of being full of himself. His face is full, with somewhat sagging
> cheeks. On the whole, the portrait is well done; but it gives an entirely
> wrong picture of him. You would certainly not recognize him.[1]

Goethe's friend Lips saw through the man sitting before him to the idealized youthful figure yet confined in the declining, middle-aged body. The
Lips portrait is my preferred image of the poet whose fire still burned bright
in the minds of those who knew him well.

We tend to remember Goethe in ways that conform to the traditions of
interpretation that have accumulated about him since he first made his
mark with *Leiden des jungen Werthers*. These traditions bear on our understanding of Goethe as a Romantic writer and as a biological thinker. The
conservative turn that Romanticism took during the first several decades of
the nineteenth century—depicted with Schadenfreude by Heine in his *Romantische Schule*—sequestered Goethe from this movement, despite the
nature poetry of his middle years, the Gothic arabesques of the two parts of
Faust, his many love affairs, and his own final assessment that Schiller
showed him that he was indeed a Romantic. Yet when we come to under-

---

1. D. J. Veit to Rahel Levin (20 March 1793), in *Goethes Gespräche*, ed. Flodoard von Biedermann; expanded by Wolfgang Herwig, 5 vols. (Munich: Deutscher Taschenbuch Verlag, 1998),
1:536–37.

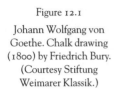

Figure 12.1
Johann Wolfgang von
Goethe. Chalk drawing
(1800) by Friedrich Bury.
(Courtesy Stiftung
Weimarer Klassik.)

stand the character of Romanticism, especially as Friedrich Schlegel and
Friedrich Schiller construed that mode of thought, then it seems perfectly
condign to regard him as an emblem of the type.

Biologists and historians of biology, starting at the beginning of the past
century, have usually denigrated Goethe's scientific accomplishments. On
the one hand, his arguments for the presence of the intermaxillary bone in
man have been dismissed, and his entire theory of the archetype thought a
Platonic shadow that had to fade in light of the new evolutionary ideas just
emerging. On the other hand, uncritical enthusiasts for Goethe have
granted him the palm when faced with an obviously even more unworthy
claimant, as Lorenz Oken was thought to be. Oken was a follower of
Schelling and the doctrines of *Naturphilosophie*—doctrines that the more
sober Goethe, at least as he is viewed by many historians, was thought to
have ridiculed. The preceding two chapters have been an effort to portray
Goethe in a quite different light.

Goethe constructed his science out of many elements: the physical evidence before him, the ideas of his predecessors (like Spinoza) and contemporaries (like Kant and Schelling), his own artistic imagination, and the

features of a complex personality and tremendously active life. I have tried to show how these various elements contributed to his theory of the archetype. Goethe undertook extensive botanical and osteological examinations, which certainly belie the assumption that he was mostly an armchair theorist. Of course, he brought conceptual inclinations to his empirical studies: Spinoza provided the notion that creative ideas were resident in nature, ideas that were both productive and standards for judgment; Schelling reinforced the Spinozistic thesis, but revealed the intimate relations with the self that such ideas trailed in their wake; and Kant indicated the structural connections between judgments of beauty and judgments of art. These conceptual forces, together with his empirical work, moved Goethe to advance the thesis that science and art provided complementary approaches to the same archetypal foundations of nature. The archetypes that underlay the variegated displays of nature would not be the anemic, minimal structures later adopted by the English, who held the vertebrate archetype, for example, to be but a string of vertebrae; rather, they would be the robust conceptions of plenary structures—that is to say, Goethe made the vertebrate archetype, for instance, to encapsulate potentially all the specific articulations of animals from bony fish to bipedal humans. This is why, of course, Goethe insisted that archetypes of plants and animals could only be viewed with the mind's eye, not the physical eye. And these archetypes, as Goethe more cryptically suggested, would be instantiated progressively in time, in an evolutionary development similar to that envisioned by Erasmus Darwin and Lamarck.

Charles Darwin, to be sure, introduced history into biological analysis in a profound and dramatic way, eclipsing the contributions of his grandfather and his French predecessor. But at the end of the eighteenth century in Germany, the application of historical analysis to nature had already begun with the work of Herder and the Romantics. They understood human beings to be constituted historically; man was a creature in a state of becoming, a creature profoundly influenced by the habits and customs of his time and directed toward a more perfect, if infinitely receding, fate. According to the Romantics and Schiller, modern poetry reflected this temporal aspect of humanity. Further, the metaphysics of absolute idealism revealed that a temporally constituted self had as its causally reciprocal correlate a developing nature. Goethe, as the previous chapter has argued, absorbed these features of Romanticism. Certainly during the last three decades of his life, he became obsessed with constructing a historical understanding of himself, as the huge edifice of his autobiographical writing attests. His scientific writing also exemplifies this historical mode of thought. The first volume of his *Farbenlehre* is devoted to a history of optics; and many of the essays of his *Zur Morphologie* carry the dates of their composition—little tags that indicate to the reader the progress of a mind. These essays also include reflections on his own developing ideas about organic phenomena—thus he gives us both text and historical commentary on the text. It would be passing strange, then, if Goethe did not apply this historical mode of analysis directly to nature—and I believe he did. The trajectory of his thought and work led to an embrace of an evolutionary conception of life.

The abstractions of morphology and the historicity of nature would not have had resonance for the poet without their embodiment in his immediate experience and artistic imagination. His Italian adventures, for instance, brought him to the concrete realization that in the passionate embrace of another human being, one could discover the beauty and creativity of nature—it was a realization that breathed life into the more remote possibilities of archetypal theory and marked a distinctive temporal stage in his own understanding. His scientific thought pulsed with the flow of aesthetic fancy, which moved him beyond the more cautious and unimaginative theories of his contemporaries. His initially tentative evolutionary notions, accordingly, became more vivid as they took poetic form in the play of the Faustian Homunculus or in re-creations of the lumbering Megatherium.

Even at the edge of death, Goethe's life and art merged, at least in the memories of his friends. About the middle of March 1832, Goethe began quickly to decline. He suffered severe chest pains and grew very cold, probably the result of one or more heart attacks and collapsing circulation. Early in the morning, while in his dimly lighted sick room, he asked of those who had gathered around what day it was. When told it was March 22, he replied: "Ah, so spring has begun and we can recover more quickly."[2] The remark lifted, momentarily, the spirits of his friends. Shortly thereafter, however, he fell into a fitful delirium, muttering in a rather incomprehensible way; the regions of his mind had become a tangled growth and the end was near. His physician, Carl Vogel, recalled that his last clear words, uttered to his servant, were "more light."[3] Goethe, he said, hated darkness, no matter what its form. The request for light certainly resonates with his art and science, and generations have found this a poetic conclusion to a

2. This remark was recorded by his friend Clemens Coudray; see *Goethes Gespräche*, 3.2:883.

3. Carl Vogel, in his *Die letzte Krankheit Goethes*, recalled these as the famous last words; see *Goethes Gespräche*, 3.2:882.

Figure 12.2

Dedication page of
Alexander von
Humboldt's *Ideen zu einer
Geographie der Pflanzen*
(1806). The artist Apollo
reveals the character of
the nature goddess Isis, at
whose feet lies Goethe's
*Metamorphose der
Pflanzen*.

poetic life. But his valet recalled the event differently. He said Goethe did make a last remark to him. The sublime poet had asked not for more light but for a chamber pot.[4]

4. Goethe's servant, Gottlieb Friedrich Krause, made this correction to the story of Goethe's last moments; see ibid., 3.2:889.

Part Four

Epilogue

◆◆◆

Chapter 13

THE ROMANTIC
CONCEPTION OF LIFE

In his *History of Sexuality*, Michel Foucault stated that the intention of his study was "to learn to what extent the effort to think one's own history can free thought from what it silently thinks, and so enable it to think differently."[1] No doubt Foucault felt that his particular investigation of the construction of sexual self-awareness and modes of desire was personally liberating. His general conviction, though, is one that most historians would themselves confess, at least as a greatly desired consummation, namely, that historical consciousness might make things happen, might change attitudes, and, consequently, actions. Like poetry, history should be both *dulcis et utilis*. Foucault believed previous modes of thought could be lethally confining, sealed caskets that had the smell of death. Minds living those modes, he presumed, usually failed to plunge beneath the surface to conceptual foundations, remained ignorant of antecedents, and neglected to pursue consequents. What he did not argue, though perhaps suggested, was that historical consciousness also allows us to rethink past events, to discover another significance that might lie below the surface of conventional understanding. The positive pursuit of history can thus promote another kind of liberation: cracking thought encrusted by traditional presumption and surface interpretation, so that we might reconceive the meaning of past events to discover in them a new character, a new being. History perpetually renews itself in this way. I have tried to accomplish something like this in the preceding pages: to expose the roots of nineteenth-century biological thought, so that this thought might be reconsidered.

My historical reconstructions have incorporated wider biographical accounts than might be expected in a history of science and philosophy. Most histories of thought have centered their analyses on the detached ideas, theories, and observations of a period. This emphasis must certainly be maintained as a guide to any other approach. However, it is easy to forget

1. Michel Foucault, *The Use of Pleasure*, vol. 2 of *The History of Sexuality*, trans. Robert Hurley (New York: Vintage Books, 1990), p. 9.

that these intellectual achievements arose in lives whose contours had spread out like a costal shelf, with submerged hopes fears, and desires. A history of Romantic biology can, perhaps, more easily expose this underlying terrain, since the individuals involved tended to lead lives whose personal relations and emotional encounters cannot be easily ignored. Such a history can show more perspicuously that abstract ideas or even concrete observations might be sustained by a causal matrix volatile with ardent features. If a history of science attempts to explain the origin of scientific ideas—which must be its raison d'être—then one must surely attend to this larger environment, to the concourse of individual lives.

I am not unmindful that an emphasis on passional concerns and personal entanglements suggests that the science under consideration formulated its notions not necessarily for good reasons but for apparently extrinsic causes, which might be thought to undermine its validity. While this suggestion lingers, I remain convinced that nineteenth-century science—including Romantic biology—had a strong purchase on reality and that its theories grasped part of the truth about natural objects and their interactions. Perhaps the rough edge of this conundrum can be smoothed a bit with the recognition that the very lived experience of these Romantic thinkers might not have dragged their theories off course but rather anchored them in the very nature that they engaged.

I have argued that certain fundamental features of nineteenth-century biology—especially archetype theory and its articulation in morphological and evolutionary thought—came to life in the soil of the German Romantic movement. The Romantic movement, to be sure, has been dismissed as a source of genuine scientific accomplishment during the period. After all, poets and philosophers—and biologists comparably inclined—could have nothing of importance to contribute to that science, especially when it is presumed that these individuals endorsed a mystical, anti-empirical approach to nature. I hope that such presumption has now become less certain for the reader. When the serious biology of the period has been relocated within the Romantic movement, I think two fundamental features will stand out, features that have been either denied or squeezed into the periphery of historical concern: namely, the aesthetic and the moral dimensions of the science of biology. If the conception of nature, as the Romantics at the beginning of the nineteenth century generally contended, had its source in the self, then location of aesthetic and moral values within nature would not seem so surprising. And these values might have yet perdured, though secreted away, in the biological science cultivated by succeeding generations.

This history has set its temporal limits at the age of Goethe, roughly the sixty years between 1770 and 1830. Yet really to sustain my thesis about the centrality of ideas flowing from the Romantic movement, attention should also be given to biological thought developed beyond Goethe's lifetime. In this epilogue, then, I will sketch what I believe to be the further reaches of the history laid down in the previous chapters of this volume. I will focus on the dominant accomplishment of biology in the nineteenth century—Darwin's theory of evolution. I will suggest, in brief compass, the ways in which the heart of Darwin's theory pulsated with ideas drawn from the Romantic movement, to show that in important respects Darwin was a Romantic biologist.

Chapter 14

DARWIN'S
ROMANTIC BIOLOGY

I am at present fit only to read Humboldt; he like another Sun illu-
mines everything I behold.

—Charles Darwin, *Beagle Diary*

Upon learning that Alexander von Humboldt's health had worsened, Darwin wrote his friend Joseph Hooker, then in Paris:

I grieve to hear Humboldt is failing; one cannot help feeling, though un-
rightly, that such an end is humiliating: even when I saw him he talked
beyond all reason.—If you see him again, pray give him my most respect-
ful & kind compliments, & say that I never forget that my whole course
of life is due to having read & reread as a Youth his Personal Narrative.[1]

Alexander von Humboldt—Romantic adventurer, friend of Goethe, and
the very doyen of German science in the first half of the nineteenth century. He preceded Darwin to the Americas on a comparable five-year voyage. He wrote of his explorations in his *Personal Narrative of Travels to the
Equinoctial Regions of the New Continent, during the Years 1799–1804* and
elaborated their findings in his greatly celebrated *Cosmos*. These two multivolume books had a tremendous impact on Darwin and helped forge the
Romantic conception of nature that underlay his theory of evolution. That
Romantic conception reveals itself most clearly in two areas of Darwin's
thought: in the assumptions about mind that grounded his construction of
nature and in the moral evaluation of nature and man that he plaited
through his two principal works, the *Origin of Species* and the *Descent of
Man*.

Darwin's conception of nature and man has, of course, been typically re-

---

1. Charles Darwin to Joseph Hooker (10 February 1845), in *The Correspondence of Charles Darwin*, 12 vols. to date, ed. Frederick Burkhardt et al. (Cambridge: Cambridge University Press,
1985—), 3:140.

garded as the very antithesis of the Romantic.[2] Romantics such as Schelling and Goethe argued that the usual dualism between mind and nature was founded in a faulty metaphysics, especially of the Kantian variety. They were convinced that the naturalism of Spinoza stood closer to the mark— that it was *Deus sive natura*, God and nature were one. More specifically, they discovered mind to be deeply implicated in the organic structure of nature. Darwin, by contrast, is thought to have conceived nature not organically but mechanistically—as if he had to reach back to physics to secure the basic principles of his biology. The Spinozistic formulation would, it appears, be quite foreign to this sober Englishman. The moral theory most often ascribed to him harkens back to Hobbes: ethical propositions are merely flimflam for efforts at selfish aggrandizement.[3] This perception of Darwin's moral theory flows from a further presumption, namely, that he eliminated from nature the kinds of values the Romantics thought secreted therein. As one recent analyst of the *Origin* puts it: "Darwin does not moralize when describing nature. . . . Nature is harsh and strictly amoral. Notions of good and evil do not exist in the natural world."[4] And why is Darwinian nature indifferent to human moral aspirations? Well, because "nature brings forth her products without ever being distracted from her task, untiring, without regard for persons, blind and cruel—in one word [*sic*]: machine-like, mechanical."[5]

This conception of Darwinism and its implications is vintage, but more like strong vinegar than fine wine. At the beginning of the past century, George Bernard Shaw sounded this once and future theme in the preface to his play *Back to Methuselah*, where he disclosed what he thought to be the real meaning of Darwinian nature: "When its whole significance dawns

2. David Kohn provides one of the very few exceptions to this generalization. See his "Aesthetic Construction of Darwin's Theory," in *Aesthetics and Science: The Elusive Synthesis*, ed. Alfred Tauber (Dordrecht: Kluwer, 1996). Kohn analyzes two salient metaphors in the *Origin of Species*, the "entangled bank" trope, which occurs on the penultimate page of the book, and the "wedging" simile, which characterizes the way selection forces organisms into the economy of nature.

3. Michael Ghiselin offers a self-reveling example of this assumption, when he asserts that, in the Darwinian view, "an 'altruistic' act is really a form of ultimate self-interest." See Michael Ghiselin, "Darwin and Evolutionary Psychology," *Science* 179 (1973): 964–68; quotation from p. 967.

4. Ilse Bulhof, *The Language of Science, with a Case Study of Darwin's* The Origin of Species (Leiden: Brill, 1992), p. 95.

5. Ibid., p. 82. Even historians engaged in the study of Romanticism's impact on science usually do not strain against these conventional assessments of Darwin's accomplishment. In an essay that briefly touches (but hardly wrestles with) this subject, the author affirms: "Darwin's great synthesis in the *Origin of Species* was not rooted in Romanticism but in the very different tradition of Paley and Thomas Malthus." See David Knight, "Romanticism and the Sciences," in *Romanticism and the Sciences*, ed. Andrew Cunningham and Nicholas Jardine (Cambridge: Cambridge University Press, 1990), p. 22.

upon you, your heart sinks into a heap of sand within you. There is a hideous fatalism about it, a ghastly and damnable reduction of beauty and intelligence, of strength and purpose, of honor and aspiration." "If it could be proven that the whole universe had been produced by [Natural] Selection," he thought, "only fools and rascals could bear to live."[6]

In this epilogue I will try to demonstrate that the usual interpretation of Darwinian nature is quite mistaken, that Darwin's conception of nature derived, via various channels, in significant measure from the German Romantic movement, and that, consequently, his theory functioned not to suck values out of nature but to recover them for a de-theologized nature. I will indicate how his Romantic assumptions led him to portray nature as organic, as opposed to mechanistic, and to identify God with nature, or at least to reanimate nature with the soul of the recently departed deity. Further, I will try to dig out the obscure roots by which his conception of mind was nourished through the Romantic movement. Finally, I will sketch the ways in which Darwin's Romantic inclinations led him to attribute to human beings a moral conscience that sought not selfish advantage but one that would respond altruistically to the needs of others.

The Romantic Movement

In the preceding chapters, I have portrayed the life and thought of those like-minded individuals—the Schlegels, Novalis, Schelling, Böhmer, Schleiermacher, Humboldt, and Goethe—who between 1770 and 1830 came together in Jena, Weimar, Berlin, and Halle, and through an indefinable magic exerted peculiar forces on the thought of the age. The kind of scientific perspective that emerged in Germany usually bears variously the names *Naturphilosophie* and "Romantic science." I distinguished the different sets of ideas traveling under these designations. *Naturphilosophie* specifically focused on the organic core of nature, its archetypal structure, and its relationship to mind, while Romanticism added aesthetic and moral features to this conception of nature. Thus, as I have used the terms, all Romantic thinkers were *Naturphilosophen*, though not all *Naturphilosophen* were Romantics. Friedrich Schlegel, who coined the term *romantisch*, used it to indicate a specific kind of poetic and morally valued literature. I have used it to distinguish a type of science that retains this aesthetic and moral heritage.

6. George Bernard Shaw, "Preface," *Back to Methuselah* (London: Penguin Books [1921] 1961), pp. 33–34, 44.

Naturphilosophie

Many of the main figures of early-nineteenth-century German biology—
Carl Gustav Carus (1789–1869), Lorenz Oken (1779–1851), Karl Burdach (1776–1847), Ignaz Döllinger (1770–1841), and Karl Ernst von Baer
(1792–1876), for instance—adopted a conception of the unity of natural
types that had its foundations in the philosophy of nature developed by
Kant and especially by Schelling, and in the anatomical studies of Goethe.
"Archetype" theory, as it became further elaborated, regarded living nature
as exhibiting fundamental organic types (*archetypi*, *Urtypen*, *Haupttypen*,
Urbilden, and the like). In France, Étienne Geoffroy Saint-Hilaire (1772–
1844) and Georges Cuvier (1769–1832), both of whom knew the German
literature, also assumed this doctrine of unity of type, though without connections to the idealist metaphysics that underlay the German versions of
the theory. Cuvier distinguished what he took to be the four most basic animal structures—the radiata (e.g., starfish and medusae), articulata (e.g.,
insects and crabs), mollusca (e.g., clams and octopuses), and vertebrata
(e.g., fish and human beings)—and by the 1830s these types had become
canonical. Both Darwin and Ernst Haeckel, for instance, accepted the idea
that nature exhibited such structural patterns; their task, of course, was to
account for them. The archetype of, say, the vertebrata formed the *Bauplan*
for backboned animals. It had ideal reality, according to Schelling, and
formed the metaphysical ground for the infinite varieties it potentially contained. Through application of the ideal type as an investigative standard, a
researcher would be able to understand the intimate physical logic of organic development and would be able to discriminate the unity underlying
the dizzying variability of nature. As Goethe put it: "The creative force produced and developed a more complete organic nature according to a common scheme; through this archetype [*Urbild*], which is perceived only by
the mind and not the senses, we might work out, as by a norm, our descriptions."[7] When archetype theory migrated from Germany to Britain, however, it underwent a reductive process. Goethe thought of the archetype as
containing all its potential variations—hence visible only to the mind's
eye. Richard Owen, who adapted archetype theory to his own needs, regarded the type-pattern as the minimal common feature shared by all organisms of a group—the vertebrate archetype, for instance, consisted only
of a string of vertebrae, according to this British usage. The *Naturphiloso-*

7. Goethe, *Vorträge über die drei ersten Kapitel des Entwurfs einer allgemeinen Einleitung in die vergleichende Anatomie, ausgehend von der Osteologie*, in *Sämtliche Werke nach Epochen seines Schaffens*
(Münchner Ausgabe), ed. Karl Richter et al., 21 vols. (Munich: Carl Hanser Verlag, 1985–98),
12:201–2.

phen usually invoked special causal forces to explain the instantiation of archetypes and their progressive variations, forces that were transformations of physical powers—for example, Schelling's polar forces or Goethe's *Bildungstrieb*. With Darwin that force became natural selection.

The *Naturphilosophen*, as I have portrayed them, commonly thought individual organisms and nature as a whole to be teleologically structured. They rejected Kant's heuristic analysis of biological explanation and located telic structures within nature. The British, by contrast, generally attributed to God the designing power through which nature came to realize distinctive means-ends patterns. When Darwin abandoned the Creator God, he did not, as I will argue, eviscerate living nature of teleological structure; rather his nature had exactly the same Romantic look as that depicted by thinkers like Alexander von Humboldt. And creative power would be transferred to nature in the gradual, evolutionary unfolding of telic purpose.

Romantic Biology

The German scientists who immediately adopted the legacy of the early Romantic writers generally accepted the metaphysical and epistemological propositions of *Naturphilosophie*. But they also incorporated aesthetic and moral considerations into their thinking about nature. They came to find in nature those very values that previously had been thought deposited there by an independent deity. For them, autonomous nature harbored precisely these kinds of value. This meant artistic experience and expression might operate in harmony with scientific experience and expression: in complementary fashion, the laws of nature might also be apprehended and represented by the artist's painting or the poet's metaphor. And with Darwin, as I will attempt to show, the artful use of metaphor gave structure to a conception of nature far different than modern readers suppose.

The chain connecting the Romantic movement and Darwin's own conceptions of nature and her operations was forged through several links—the writings of Richard Owen, for one—but principally through the work of the man about whom Darwin wrote to Hooker, Alexander von Humboldt.

Alexander von Humboldt

In the late 1790s, Alexander von Humboldt, along with his brother Wilhelm, spent considerable time in Jena involved with a circle intersecting that of the Jena Romantics. His more intimate configuration included

Goethe and Schiller, but he knew the Schlegels and Schelling quite well. Later he would write Schelling of his admiration for the philosopher's conception of nature. Humboldt imbibed the particularly heady atmosphere of philosophy, science, and poetry while at Jena, and these helped form his scientific essays and books into both resources for the Jena Romantics and, indeed, representations of the particular kind of science emanating from that small German redoubt.

In 1799 Humboldt began the first leg of a journey for which he had been preparing for several years. He and his friend Amié Bonpland set off for the New World to engage in research that ranged over botany, zoology, anthropology, geology, and meteorology.[8] Like Darwin's own adventure, Humboldt's trip lasted five years, though his itinerary differed somewhat. His travels took him through the northern parts of South America, into Central America, and finally through North America and a visit to Thomas Jefferson in Washington. His voyage yielded several volumes of specialized zoological and botanical studies, a geological monograph (and others on astronomy and anthropology), as well as a large book describing his travels and researches, his *Personal Narrative of Travels to the Equinoctial Regions of the New Continent, during the Years 1799–1804*. The *Personal Narrative*, like the travel book of his friend Georg Forster, takes the reader chronologically on a journey of adventure and science; but with the range of his investigations, Humboldt brought to fruition what were only suggestions in the work of Forster.

Humboldt, like Goethe, thought that plants exhibited different fundamental forms, of which he identified some sixteen. He regarded the different species and subspecies of plants as playing variations on these constant themes. The same types, he thought, would be found in similar environmental conditions (for instance, of temperature, elevation, and moisture) across the globe. For just as common geological and mineral formations displayed comparable patterns in different, far-flung locations, so, too, did plants:

---

8. Susan Faye Cannon provides a still instructive sketch of the character of Humboldt's science in her "Humboldtian Science," in *Science in Culture: The Early Victorian Period* (New York: Science History Publications, 1978), pp. 73–110. Michael Dettelbach details the features of the science Humboldt himself developed in his various writings on "terrestrial physics." This kind of physics had as its aim the discovery and display of laws of natural equilibria that interact in complex but harmonious ways (e.g., isothermal lines showing constant mean temperatures in related geographical regions of the world). See Michael Dettelbach, "Humboldtian Science," in *Cultures of Natural History*, ed. N. Jardine, J. A. Secord, and E. C. Spary (Cambridge: Cambridge University Press, 1996), pp. 287–304.

All formations are, therefore, common to every quarter of the globe and assume the like forms. Everywhere basalt rises in twin mountains and truncated cones; everywhere trap-porphyry presents itself to the eye un-der the form of grotesquely shaped masses of rock, while granite termi-nates in gently rounded summits. Thus, too, similar vegetable forms, as pines and oaks, alike crown the mountain declivities of Sweden and those of the most southern portion of Mexico.[9]

The discrimination of form, according to Humboldt, is an aesthetic task, not a classification done according to the usual criteria of botanical systems, such as that of Linnaeus. Rather, we must be guided by the painterly eye, which highlights the distribution of leaves, the forms of stems and branches, the height and breadth of the entire plant. Moreover, the geo-graphical distribution of plants depends on surrounding organisms; certain types usually occur together in assemblages—palm and banana trees, for example. The assemblages will yet have distinctive characters that give a geographical location its individual feel. "Swiss scenery" or "Italian sky" are produced by distinctive combinations of elements: "The azure of the sky, the effects of light and shade, the haze floating on the distant horizon, the forms of animals, the succulence of plants, the bright glossy surface of the leaves, the outlines of mountains, all combine to produce the elements on which depends the impression of any one region."[10]

Humboldt understood all of nature and nature's laws to unite in a vast complex of interrelationships—"a Cosmos, or harmoniously ordered whole, which, dimly shadowed forth to the human mind in the primitive ages of the world, is now fully revealed to the maturer intellect of mankind as the result of long and laborious observations."[11] This whole of nature displayed herself not only to the scientific instruments that Humboldt carried with him into the jungles along the Orinoco and the Amazon Rivers, but she also revealed her most intimate self, her deepest lawful relations, to his aes-thetic perceptions. "Everywhere," he exclaimed, "the mind is penetrated by the same sense of the grandeur and vast expanse of nature, revealing to the

9. Alexander von Humboldt, *Views of Nature, or Contemplations on the Sublime Phenomena of Creation*, trans. E. C. Otté and Henry Bohn (London: Henry Bohn, 1850), p. 218. There were two English translations of Humboldt's *Ansichten der Natur* (1808), this one just cited and one com-pleted the year before. Darwin read the work in February 1852. See Darwin's "Reading Notebooks," in *Correspondence of Charles Darwin*, 4:487.

10. Humboldt, *Views of Nature*, pp. 217–18.

11. Alexander von Humboldt, *Cosmos: A Sketch of a Physical Description of the Universe*, 5 vols., trans. E. C. Otté (New York: Harper, n.d.), 1:24. Humboldt's *Kosmos* (1845–58), was translated into English three times in the nineteenth century. Darwin disliked the initial pirated translation of the first two volumes. See Darwin's comments, for instance, to J. D. Hooker (28 September 1846), in *Correspondence of Charles Darwin*, 3:342.

soul, by a mysterious inspiration, the existence of laws that regulate the forces of the universe."[12] It was this sort of aesthetic penetration into nature that led Darwin to exclaim about *Cosmos* that it provided "a grand coup d'oeil of the whole universe."[13]

Humboldt had intended that a reader like Darwin should have exactly this kind of complete engagement with his text, that the reader should respond not only to the science conceptually articulated therein but also to the aesthetic display of nature's other face. Humboldt made this intention explicit in *Cosmos*, where he argued that the natural historian had the duty to re-create in the reader—through the use of artful language—aesthetic experiences of the sort the naturalist had himself undergone in his immediate encounters with nature. A worthy naturalist, Humboldt thought, left no means "unemployed by which an animated picture of a distant zone, untraversed by ourselves, may be presented to the mind with all the vividness of truth, enabling us even to enjoy some portion of the pleasure derived from the immediate contact with nature." A poetic natural historian, in this Romantic mold, utilizes appropriate language to make palpable to an audience what has emanated "from the intuitive perception of the connection [in the naturalist's mind] between the sensuous and the intellectual, and of the universality and reciprocal limitation and unity of all the vital forces in nature."[14] As we will see, Darwin's own prose, which vibrated with the poetic appreciation of nature's inner core, had a comparable end: to deliver to the reader an aesthetic assessment that lay beyond the scientifically articulable.[15]

12. Humboldt, *Cosmos*, 1:25.

13. Darwin to Edward Cresy (May 1848), in *Correspondence of Charles Darwin*, 4:135.

14. Humboldt, *Cosmos*, 2:81. Humboldt's aesthetic philosophy here derives from Kant of the third *Critique* and from Schelling. At the beginning of the second volume of *Cosmos* (p. 19), Humboldt contrasts his "objective," physical account of nature with his observations about the way nature impresses the sensations and fancy of the naturalist. In reading this, one might assume that Humboldt's aesthetics of nature were entirely "subjective" (a term, incidentally, that he does not use). If we make a simple distinction between epistemic and ontological claims, on the one hand, and the objective or subjective bases of those claims, on the other—so, for instance, "The wine is tart" would be ontologically subjective but epistemically objective—then one would have to interpret Humboldt's aesthetics of nature as epistemically objective, since it deals with the *universal* reactions of different subjects.

15. Darwin's own prose became so structured by Humboldt's that his sister Caroline began to complain. She wrote him in 1833: " . . . as to your style. I thought in the first part (of this last journal) that you had, probably from reading so much of Humboldt, got his phraseology, & occasionally made use of the kind of flowery french expressions which he uses, instead of your own simple straight forward & far more agreeable style. I have no doubt you have without perceiving it got to embody your ideas in his poetical language & from his being a foreigner it does not sound unnatural in him." See Caroline Darwin to Charles Darwin (28 October 1833), in *Correspondence of Charles Darwin*, 1:345.

Darwin's Romantic Conception of Nature

Few would deny, I think, that Darwin's perception of nature, his feeling of its lived reality, his understanding of the connections between organic and geological phenomena—that this perception achieved critical form during the five years of his *Beagle* voyage. But it would be a mistake to believe that Darwin, like a cheerful Kurtz, went into the jungle and viewed his surroundings with naked and bewildered eyes, simply registering the beauty and mystery of the flora and fauna, and the exotic behavior of its human inhabitants. His letter to Hooker expressed a deep truth, which an examination of his writing from this period confirms: Darwin experienced the South American environment, the interconnectedness of its various aspects, the sublimity of its scenes, and the moral behavior of its peoples—all filtered through a Humboldtian discourse on these very subjects.

While at Cambridge in the years 1828–31, aside from dinner parties, foxhunting, beetle collecting, and the occasional study of Euclid and Paley—certainly preparations suitable enough for the life of the country parson that he planned—Darwin did spend time on more serious, scientific pursuits. In the lectures on botany he heard from John Stevens Henslow (1796–1861), as well as over tea at the professor's house, the young student was introduced to questions stimulated by German biologists concerning the vital forces that distinguished the organic from the inorganic. As Phillip Sloan has meticulously demonstrated, Darwin became intrigued by the controversy over whether vital force, as Henslow thought, came extrinsically to potentially living matter or whether such force, as the Germans contended, was intrinsic to animate organization.[16] The role of vital force became fascinating for the young Darwin as he attempted to puzzle out the nature of pollen granules and their relation to similar substances found in simple invertebrates. These questions, which whetted the sensibilities of the yet green undergraduate, would be pursued while on the *Beagle* voyage and most assiduously thereafter, as he began working out his hypotheses about species change.

But these more minute research topics for the novice played only circumspectly in the background of his concerns. During Darwin's first year at Cambridge, he began reading Humboldt's *Personal Narrative*.[17] This young German adventurer knew how to compose a tale that would entice a med-

---

16. See Phillip Sloan, "Darwin, Vital Matter, and the Transformism of Species," *Journal of the History of Biology* 19 (fall 1986): 369–445.

17. This early date for Darwin's reading Humboldt is suggested by a friend's discussion about the *Personal Narrative* with him. See Sarah Owen to Darwin (18 February 1828), in *Correspondence of Charles Darwin*, 1:51.

ley of readers (Haeckel also among them), but particularly the young and those whose bones ached, like Humboldt's own, for escape from suffocating surroundings. And if a reader, say, one cloistered in the decaying traditions of an ancient university, had yet discovered a taste for nature and huddled against the warming desire to do something in a scientific mode—a desire stoked, perhaps, by the low expectations of a famous father—then could he fail to escape in his imagination with this polymathic Romantic? Even the twentieth-century reader, of requisite disposition, might feel a stir when he or she would read what Darwin did:

> If America occupies no important place in the history of mankind, and of the ancient revolutions which have agitated the human race, it offers an ample field to the labours of the naturalist. On no other part of the Globe is he called upon more powerfully by nature, to raise himself to general ideas on the cause of the phenomena, and their natural connection. I shall not speak of that luxuriance of vegetation, that eternal spring of organic life, those climates varying by stages as we climb the flanks of the Cordilleras, and those majestic rivers which a celebrated writer [Chateaubriand] has described with so much graceful precision. The means which the new world affords for the study of geology and natural philosophy in general are long since acknowledged. Happy the traveler who is conscious, that he has availed himself of the advantages of his position, and that he has added some new facts to the mass of those which were already acquired![18]

So taken with Humboldt's tale was Darwin that he copied out long passages to read aloud to Henslow and other friends.[19] Like Humboldt, Darwin in his last year at Cambridge was of an age, twenty-one, "when life appears an unlimited horizon," an age "when we find an irresistible attraction in the impetuous agitations of the mind, and the image of positive danger."[20] Quickly the young Englishman became thrall to the idea of undertaking a similar trip of adventure, if not all the way to the New World, then at least to the Canary Islands, with its active volcano and ancient and magnificent dragon tree—all of which Humboldt had so richly described.

Darwin, of course, got his chance. He shipped out on the *Beagle* on 27 December 1831, carrying among his supplies a small library, including Hum-

---

18. Alexander von Humboldt and Aimé Bonpland, *Personal Narrative of Travels to the Equinoctial Regions of the New Continent, during the Years 1799–1804*, trans. Helen Williams, 7 vols. (London: Longman, Hurst, Rees, Orme, and Brown, 1818–29), 1:xlv—xlvi. Humboldt's companion Bonpland supplied many observations, but the tale unfolded in the first-person of Humboldt.

19. Charles Darwin, *The Autobiography of Charles Darwin, 1809–1882*, ed. Nora Barlow (New York: W. W. Norton, 1969), pp. 67–68.

20. Humboldt, *Personal Narrative*, 1:3.

boldt's *Personal Narrative*, which had been given him as a parting gift by Henslow.[21] Ironically, the ship had to bypass the Canaries because of quarantine restrictions. The *Beagle* finally reached the Brazilian coast on 28 February, much to Darwin's great relief, since his seasickness in transit had been monumental. During the voyage, when not retching, Darwin read his Humboldt. And just after landing, he quickly wrote his father of the exciting new land and of the proper way to appreciate it: "If you really want to have a notion of tropical countries, study Humboldt.—Skip the scientific parts & commence after leaving Teneriffe.—My feelings amount to admiration the more I read him."[22]

Though Darwin himself would pay close attention to the kinds of scientific studies Humboldt had undertaken—and kept the sort of notes that would enable him later to produce a travel book precisely modeled after Humboldt's narrative—it was undoubtedly the sensuous and imaginative descriptions to which Darwin most immediately responded. Shortly after landing in Bahia, Darwin took to the jungle. He experienced the luxuriant growth under the emerald canopy bathed in a soft Humboldtian light. He wrote in his diary on the day of disembarking:

> I believe from what I have seen Humboldts glorious descriptions are & will for ever be unparalleled: but even he with his dark blue skies & the rare union of poetry with science which he so strongly displays when writing on tropical scenery, with all this falls far short of the truth. The delight on experiences in such times bewilders the mind. . . . The mind is a chaos of delight, out of which a world of future & more quiet pleasure will arise.—I am at present fit only to read Humboldt; he like another Sun illumines everything I behold.[23]

Throughout Darwin's diary, which he kept on his voyage, entry after entry invokes the name of Humboldt on such varied subjects as the passing of time in the tropics; profusion and forms of vegetation; the constellations of the southern sky; volcanic formations; the color of landscape through the softening effects of the atmosphere; sickness in the tropics; the incursions of Christianity; mountaineering; biogeographical relationships; and much

21. Darwin took at least the first two volumes of the narrative with him. He likely added on his own the remaining five volumes. For the trip, he also packed several other books by Humboldt, chiefly geological. See the list of books taken on the voyage, published in *Correspondence of Charles Darwin*, 1:553–66.

22. Charles Darwin to Robert Darwin (8 February to 1 March 1832), in ibid., 1:204.

23. Charles Darwin, *Beagle Diary*, ed. R. D. Keynes (Cambridge: Cambridge University Press, 1988), p. 42 (entry for 28 February 1832).

more.[24] All of these subjects, braced by the authority of his German predecessor, Darwin lightly transformed, in 1839, into the more public record of
his voyage, his *Journal of Researches into the Geology and Natural History of
the Various Countries Visited by H.M.S. Beagle*.[25] But it wasn't simply Humboldt's individual descriptions or scientific calculations that arrested the
young naturalist's attention. His very experience passed through the lens
provided by Humboldt.

In September 1836, during the long voyage home, Darwin had opportunity to page through his diary and reflect on his experiences. He himself
judged that Humboldt provided the mold that shaped his observations, perceptions, and emotional reactions to the many tableaux that had passed before him. He recorded in his diary this judgment and then transferred it
virtually unchanged into the *Journal of Researches:* "As the force of impression frequently depends on preconceived ideas, I may add that all mine
were taken from the vivid descriptions in the Personal Narrative which far
exceed in merit anything I have ever read on the subject."[26]

The nature that Darwin experienced with the aid of Humboldt was not
a machine, a contrivance of fixed parts grinding out its products with dispassionate consequence. The nature that Darwin experienced was a cosmos, in which organic patterns of land, climate, vegetation, animals, and
humans were woven into a vast web pulsating with life. Darwin's aestheticized experience, rendered in Humboldtian terms, delivered to him a vision
of nature and man that would subtly form his later theory, a vision that can
be glimpsed in a concluding passage of his diary:

> Among the scenes which are deeply impressed on my mind, none exceed
> in sublimity the primeval forests, undefaced by the hand of man, whether
> those of Brazil, where the powers of life are predominant, or those of
> Tierra del Fuego, where death & decay prevail. Both are temples filled
> with the varied productions of the God of Nature:—No one can stand
> unmoved in these solitudes, without feeling that there is more in man
> than the mere breath of his body.[27]

In the jungles of South America and the forbidding coasts around the
Horn, Darwin found the God of nature, not the God of Abraham, Isaac,
and Jacob.

24. For these subjects see ibid., pp. 34 (6 February 1832), 42 (28 February 1832), 48 (24–26
March 1832), 54 (9 April 1832), 67 (26 May 1832), 70 (2 June 1832), 72 (4 June 1832), 267 (26
November 1834), 288 (12 February 1835), and 308 (21 March 1835).
25. Charles Darwin, *Journal of Researches into the Geology and Natural History of the Various
Countries Visited by H.M.S. Beagle* (London: Henry Coburn, 1839).
26. Darwin, *Diary*, p. 443 (September 1836); *Journal of Researches*, p. 604.
27. Darwin, *Diary*, p. 444 (September 1836); *Journal of Researches*, p. 604.

After Darwin had published his *Journal of Researches*, he then, with extreme trepidation, sent a copy to Humboldt himself. The great man responded with a generosity and perceptiveness that leaped beyond any of Darwin's most secret hopes. Obviously Humboldt, who had no reason to flatter Darwin—and by this time Humboldt had achieved an eminence in Europe exceeded by no other naturalist—found in Darwin a kindred spirit.[28] Indeed, Darwin's aesthetic sensitivity and expressiveness made an indelible impression on this Nestor of science. In *Kosmos*, published almost a decade after Darwin's *Journal*, Humboldt's admiration had not abated. He likened the Englishman's "aesthetic feelings" and "vividly fresh images" to those of his own great friend Georg Forster, who first opened the way to "a new era of scientific travel and who aimed at a comparative study of peoples and lands."[29]

It might yet be thought that the influence of Humboldtian Romanticism on Darwin was only of slight significance, that his fundamental conception of nature, at least as presented in the *Origin of Species* and in the *Descent of Man*, had stripped off metaphysical assumption and aesthetic decoration, so that only the cold machinery of the living world might lie exposed. Needless to say, I do not think this conception of Darwin's accomplishment to be sound. I believe a fair examination of his principal evolutionary tracts will reveal the deep penetration into Darwin's thought of a Humboldtian Romantic view of nature. Initially, I'll touch on three aspects of his representations: the archetypal unity displayed by nature; her organic and non-machine-like aspect; and, finally, her moral features. In a subsequent section, I will turn to Darwin's depiction of that smaller nature, encapsulated in the human being, to assess the Romantic strains found therein.

Romantic Nature in the *Origin of Species*

Theory of the Archetype

Archetype theory became a central doctrine for German Romantic writers. Schelling, Goethe, Oken, and Carus, along with *Naturphilosophen* such as von Baer, developed the idea that nature exhibited fundamental unities, of

28. See Humboldt to Darwin (18 September 1839), in *Correspondence of Charles Darwin*, 2: 218–22.

29. Alexander von Humboldt, *Kosmos: Entwurf einer physischen Weltbeschreibung*, 5 vols. (Stuttgart: J. G. Cotta'scher Verlag, 1845–58), 2 (1847): 72. See also *Cosmos*, 2:80. Humboldt's enthusiasm for Darwin's aesthetically sensitive descriptions was shared by many of the reviewers of the *Journal of Researches*. Typical was Broderip's judgment that Darwin's account "fill[ed] the mind's eye with brighter pictures than a painter can present." See William Broderip, "Review of *Journal of Researches* by Charles Darwin," *Quarterly Review* 65 (1839): 194–234.

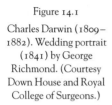

Figure 14.1
Charles Darwin (1809–
1882). Wedding portrait
(1841) by George
Richmond. (Courtesy
Down House and Royal
College of Surgeons.)

which species and individuals played out the variations. In England this ba-sic conception was embraced by Joseph Henry Green (1791–1863), Hunterian lecturer at the Royal College of Surgeons, and most famously by Green's protégé, Richard Owen (1804–1892). Green had studied in Ger-many; and like Schelling and Goethe, both of whom he assiduously read, he argued that nature exhibited essential patterns or rational ideals. These ideals, as existing in the mind of God, served as "archetypes and preexisting models"; but as "acts of the Divine Will manifested in nature," they func-tioned as "laws."[30] Drawing on the "objective idealism of Schelling," Green construed an archetype as

> a causative principle, combining both power and intelligence, contain-ing, predetermining and producing its actual result in all its manifold re-lations, in reference to a final purpose; and realized in a whole of parts, in which the Idea, as the constitutive energy, is evolved and set forth in its unity, totality, finality, and permanent efficiency.[31]

As with Schelling, Kielmeyer, and other German biologists at the begin-ning of the nineteenth century, Green's archetype theory included the no-

---

30. Joseph Henry Green, *Vital Dynamics: The Hunterian Oration before the Royal College of Sur-geons in London, 14th February 1840* (London: William Pickering, 1840), pp. xxv—xxvi. The dy-namical aspect of Green's theory followed the path into "objective Idealism" laid by Schelling (pp. xxix–xxx). He also cited Goethe, Oken, Spix, and Carus as developing the archetype theory that he himself was further elaborating (pp. 57–58).

31. Ibid., pp. xxv.

tion that individual organisms higher in the scale of development recapitulated the same forms achieved during the evolution of those lower in the hierarchy of animals.[32] This doctrine would be embraced by Darwin and made the fulcrum point of Haeckel's evolutionism.

Richard Owen, perhaps the most influential biologist in England during the first half of the century, likewise followed the German lead in constructing his own quite celebrated theory of the archetype. In his *Report on the Archetype* (1847) and *On the Nature of Limbs* (1849), Owen gave his analytical attention to the osteological pattern that lay at the foundation of the vertebrate skeleton.[33] The archetype of the vertebrata, in Owen's construction, was simply a string of vertebrae. According to his theory, different vertebrate skeletons manifested modifications of this basic plan. So, for instance, the bones of the head would be regarded as a development of the several anterior vertebrae, and the ribs, pelvis, and limbs as developments of different processes of more posterior vertebrae. This conception allowed the anatomist to compare the "same" bones in different species or in different higher taxa. Thus the bones in the human hand, in the wing of a bat, and in the paddle of a porpoise would be considered homologous, that is, bones having, as Owen put it in his Germanophilic way, the same "Bedeutung," or meaning.[34] Such bones might be adapted to particular purposes—for instance, tool use, flying, or swimming—but their structure would express their common nature.

Owen drew particularly on the work of Oken and Carus in constructing his theory of the archetype. Similar to these Romantic anatomists and to his mentor Green, Owen thought of the archetype as more than merely a standard by which to conduct comparative zoology. For him, the archetype reared up as a vital force in nature. In the metaphysical considerations occurring at the end of his *Report on the Archetype*, Owen distinguished two

32. Ibid., pp. 39–40. Green did believe in the historical evolution of species, as I have argued in my *Meaning of Evolution: The Morphological Construction and Ideological Reconstruction of Darwin's Theory* (Chicago: University of Chicago Press, 1992), pp. 72–79. His theory was similar to Schelling's "dynamical evolution" [*dynamische Evolution*]. Schelling's theory of dynamic evolution maintained that new species appeared over time, each advancing toward the perfect realization of organism in general, which Green interpreted as the appearance of the human form. See Friedrich Schelling, *Erster Entwurf eines Systems der Naturphilosophie* (1799), in vol. 1 of *Schellings Werke*, ed. Manfred Schröter, 3rd ed., 12 vols. (Munich: C. H. Beck, 1927–59), esp. p. 64. See also chapter 8, above.

33. Richard Owen, *Report on the Archetype and Homologies of the Vertebrate Skeleton*, in *Report of the Sixteenth Meeting of the British Association for the Advancement of Science; held at Southampton in September 1846* (London: Murray, 1847); and *On the Nature of Limbs, a Discourse delivered on Friday, February 9, at an Evening Meeting of the Royal Institution of Great Britain* (London: John Van Voorst, 1849).

34. Owen, *On the Nature of Limbs*, pp. 2–3.

Figure 14.2
Richard Owen
(1804–1892), the
Hunterian Professor.
Engraving after a
daguerreotype. (Courtesy
Wellcome Institute
Library.)

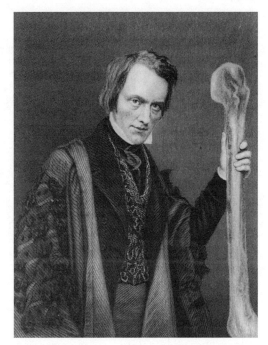

opposing vital forces: "Besides the ἰδέα, organizing principle, vital property, or force, which produces the diversity of form belonging to living bodies of the same materials . . . there appears also to be in counter-operation during the building up of such bodies the polarizing force pervading all space." The Platonic Idea operated to establish the species-specific form impressed on organic matter, while the polarizing force, which he (a bit confusingly) identified with the archetype, constrained activity to produce similarity of form among species (general homology) and repetition of parts within species (serial homology).[35] Perhaps because of the pantheistic implications of two oppositional forces creating living beings, three years later, in his little book *On Limbs*, Owen collapsed the vital forces into one, "answering to the 'idea' of the Archetypal World in the Platonic cosmogony, which archetype or primal pattern is the basis supporting all the modifications of such part [as the limb] for specific powers and actions in all animals possessing it."[36] Construing the archetype as a Platonic Idea presumably made easier its identification with an idea in the mind of the Creator God and thus also easier for Owen to mute any suspicions that he

35. Owen, *Report on the Archetype*, pp. 339–40.
36. Owen, *On the Nature of Limbs*, pp. 2–3.

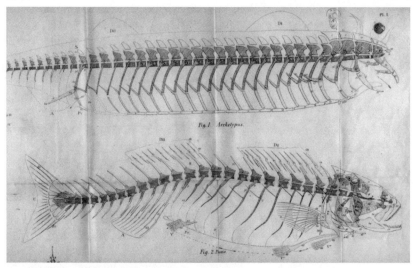

Figure 14.3 Richard Owen's illustration of the vertebrate archetype (*top*), with a fish as the next stage of development; from Richard Owen, *On the Nature of Limbs* (1849).

allowed nature herself to fashion creatures from her own resources.[37] Such suspicions, though, could not be quelled by a subtle alteration of arcane metaphysics, especially when Owen himself invited the idea that he had, in Germanic fashion, attributed to nature herself powers of development. At the conclusion to *On the Nature of Limbs*, he perorated:

> To what natural laws or secondary causes the orderly succession and progression of such organic phaenomena may have been committed we as yet are ignorant. But if, without derogation of the Divine power, we may conceive the existence of such ministers, and personify them by the term "Nature," we learn from the past history of our globe that she has advanced with slow and stately steps, guided by the archetypal light, amidst the wreck of worlds, from the first embodiment of the Vertebrate idea under its old Ichthyic vestment, until it became arrayed in the glorious garb of the Human form.[38]

Orthodox readers of this passage perceived that Owen treaded gingerly along a dangerous pantheistic path laid by German Romantics, who had promulgated the doctrines of naturalistic development and transformation.

37. Nicholaas Rupke discusses Owen's problems over the religious implications of the archetype in his *Richard Owen, Victorian Naturalist* (New Haven: Yale University Press, 1994), pp. 196–204.

38. Owen, *On the Nature of Limbs*, p. 86.

Those guardians of theological rectitude became quite concerned for Owen's welfare and that of naive readers.[39] The less orthodox discovered in Owen's theory of the archetype a comparable message, but felt more sanguine about it.

Owen himself reacted to charges of pantheism with stiff indignation. And though he apparently did harbor a notion of the gradual appearance of new species through a kind of progressive transformation—that is, a transformational theory, but not a genealogical one. Natural powers (directed by God, of course) might introduce new species according to a plan that fossils—those archetypal vestiges—disclosed, but these species would not evolve genealogically out of one another. The genealogical view, Owen thought, was characteristic of German writers of the "transcendental school." In his Hunterian Lecture of 1837, he expressly denied the two genealogical doctrines he saw linked in painful excess: a theory of embryonic recapitulation and a theory of species transformation. He believed anyone holding to the notion that the embryo of more developed organisms went through the morphological stages of those lower in the line of development—such a foolish naturalist would succumb immediately to the even more pernicious idea that during the earth's past history, species likewise developed out of one another. "The doctrine of Transmutation of forms during the Embryonal phases," he cautioned, "is closely allied to that still more objectionable one, the transmutation of Species." He thought both propositions would be "crushed in an instant when disrobed of the figurative expressions in which they are often enveloped; and examined by the light of a severe logic."[40] Well, one of Owen's friends did examine these ideas and did not blush to apply a creative logic that confirmed them both.

Darwin had been working on a theory of species development since shortly after his return from the *Beagle* voyage. In the initial entry of his first transformation notebook, this underlaborer in the natural history field made the precise linkage that Owen execrated. On the first page of his "Notebook B"—begun in July 1837, a short time after Owen's lecture series—Darwin considered two kinds of generation. One sort produced identical individuals asexually through budding or division. The other sort was

<hr />

39. The *Manchester Spectator* (8 December 1849) sounded a cautionary note about the quite obvious dangers to the scientifically ignorant in Owen's apparent adoption of Germanic *Naturphilosophie*. Owen responded (22 December 1849) to the *Spectator* article, saying that his language in his conclusion was "figurative," something allowed at the end of a lecture. He contended that the unity of plan displayed in organisms "testifies to the oneness of their Creator."

40. Richard Owen, *The Hunterian Lectures in Comparative Anatomy, May and June 1837*, ed. with introduction and commentary by Phillip Sloan (Chicago: University of Chicago Press, 1992), p. 192 (MS 98–99).

the "ordinary kind," as he termed it, "the new individual passing through several stages (typical, ⟨of the⟩ or shortened repetition of what the original molecule has done)."[41] Darwin here proposed that during ontogeny, the individual would pass through the same morphological stages, by way of a shortened repetition, that the first living speck of life went through in its evolutionary trajectory. He continued to work out the details of this very German idea of evolutionary recapitulation through the period leading up to the composition of the *Origin of Species*.[42]

Darwin's friend and future antagonist Owen had initially brought several objections to recapitulation theory. In conformity to Karl Ernst von Baer's similar strictures, Owen contended that the fetus of a more developed animal never displayed during gestation the individual adult form of lower animals. Von Baer had epitomized this objection by postulating that fetal development passed from a more general morphological state (for instance, the general mammal form) to the more individual structures of the species and particular individual. Hence the fetus at early stages of development could only represent generalized animal patterns, not the adult forms of definite primitive species. Darwin worked through objections of this sort in the many scattered manuscript notes that he left on the theory of recapitulation prior to the publication of the *Origin*. Perhaps his most pointed rejoinder came in a note he penciled on the back flyleaf of his copy of Owen's *On the Nature of Limbs*. There Darwin wrote: "I look at Owen's Archetypes as more than idea, as a real representation as far as the most consummate skill & loftiest generalization can represent the parent form of the Vertebrata."[43] Darwin thus suggested that the generalized archetype of the vertebrates did not lie hidden away as an idea in the mind of God; rather, it was the form of a creature that walked the earth many generations ago. Embryos of descendants would, consequently, pass through the stage of a generalized vertebrate; but that generalized vertebrate would have been a real, individ-

41. Charles Darwin, "Notebook B," in *Charles Darwin's Notebooks, 1836–1844*, ed. Paul Barrett et al. (Ithaca: Cornell University Press, 1986), p. 170 (B, 1). (The words in wedge quotations were deleted in the manuscript.) Darwin here was reflecting on his grandfather's book *Zoonomia*, and likely the passage in which Erasmus Darwin, after considering the morphological similarity of creatures, asked: "Would it be too bold to imagine, that in the great length of time, since the earth began to exist, . . . that all warm-blooded animals have arisen from one living filament, which THE GREAT FIRST CAUSE endued with animality, with the power of acquiring new parts, directed by irritations, sensations, volitions, and associations; and thus possessing the faculty of continuing to improve by its own inherent activity, and of delivering down those improvements by generation to its posterity, world without end!" See Erasmus Darwin, *Zoonomia, or The Laws of Organic Life*, 2 vols. (London: Johnson, 1794–96), 1:505.

42. I have traced this history in my *Meaning of Evolution*, pp. 91–166.

43. This inscription is on the back flyleaf of Richard Owen's *On the Nature of Limbs*, held in the Manuscript Room of Cambridge University Library.

ual creature. In the *Origin of Species*, especially in the later editions, Darwin made perfectly clear the central role archetype and recapitulation theory played in his conception of evolution.

As an instance of the molding force exercised by archetype theory on the deep structure of the *Origin*, one might point to Darwin's chronic presumption that transformation of species occurred without common descent. In his "Essay of 1844," for example, he assumed that evolution would occur within the archetypes of, say, the articulata, radiata, mollusca, and vertebrata, but allowed that no common ancestor would be found for these branches of the animal kingdom. He thus concluded that for the animal and plant kingdoms, "all the organisms yet discovered are descendants of probably less than ten parent forms."[44] In the *Origin*, Darwin advanced this same conviction that "animals have descended from at most only four or five progenitors, and plants from an equal or lesser number." This hypothesis, he maintained, was sufficient for his general theory. He did venture, however, that analogy suggested that "probably all the organic beings which have ever lived on this earth have descended from some one primordial form, into which life was first breathed."[45]

The principal of recapitulation also sunk deeply into the foundations of Darwin's theory. Of the many expressions of the principle in the *Origin*, perhaps the most straightforward comes in the penultimate chapter of the final edition: "As the embryo often shows us more or less plainly the structure of the less modified and ancient progenitor of the group, we can see why ancient and extinct forms so often resemble in their adult state the embryos of existing species of the same class."[46]

Thus central and diverting blooms of Darwin's theory of evolution opened from ideas initially cultivated in Romantic *Naturphilosophie*. But these intricate parts of the composition could not have taken their form if the very root of his theory, his idea of nature, did not also draw from that quite fertile soil.

Nature as Organic and Value Laden

No phrase comes so trippingly to the lips of contemporary biologists as "the *mechanism* of natural selection." Almost reflexively, we think of natural selection, paradoxically, in nonorganic terms. And the nature that selection

44. Charles Darwin, "Essay of 1844," in *The Foundations of the Origin of Species: Two Essays Written in 1842 and 1844 by Charles Darwin*, ed. Francis Darwin (Cambridge: Cambridge University Press, 1909), p. 252.

45. Charles Darwin, *On the Origin of Species* (London: Murray, 1859), p. 484.

46. Charles Darwin, *On the Origin of Species by Charles Darwin: A Variorum Text*, ed. Morse Peckham (Philadelphia: University of Pennsylvania Press, 1959), p. 704.

creates, that too appears to the contemporary scientist as something best fixed, at least in principle, in the hard, mechanistic language of chemistry and physics. Yet the belief that nature was nothing but a vast machine and that natural selection operated according to principles of a Manchester spinning loom—all of this remained quite distant to the mind that originally composed the *Origin of Species*. Darwin never referred to or conceived natural selection as operating in mechanical fashion, and the nature to which selection gave rise was perceived in its parts and in the whole as a teleologically self-organizing structure.[47]

The most colorful instance of Darwin considering the integral operations of the web of life—the mutual ends-means relations existing among different organisms—occurred when he pondered, almost as a whimsical question, how cats in a district might control the growth of red clover, to wit: cats prey on mice who destroy the nests of humble bees that pollinate the clover.[48] Life, to use his very rich metaphor, could be compared to an entangled bank, in which bushes, plants, snails, birds—all exert finely balanced forces upon each other to produce "the proportional numbers and kinds" of life-forms exhibited in the complex. These were the sorts of entangling relationships that Humboldt insisted upon, and that Darwin, under the genial guidance of his predecessor, had described in great detail in his *Journal of Researches*. The deeply nonmechanistic character of Darwin's theory is, however, most conspicuously displayed in the principle of natural selection itself.

Consider how Darwin compared in the *Origin* the selection that human beings practiced with that of nature, natural selection:

> Man can act only on external and visible characters: nature cares nothing for appearances, except in so far as they may be useful to any being. She can act on every internal organ, on every shade of constitutional difference, on the whole machinery of life. Man selects only for his own good; Nature only for that of the being which she tends. . . . It may be said that natural selection is daily and hourly scrutinizing, throughout the world, every variation, even the slightest; rejecting that which is bad, preserving and adding up all that is good; silently and insensibly working, whenever and wherever opportunity offers, at the improvement of each organic being in relation to its organic and inorganic conditions of life.[49]

47. The term "machine" in any of its forms appeared only once in the *Origin of Species* and even then in a context that vitalized its significance. See Darwin, *Origin of Species*, p. 83.

48. Ibid., p. 73.

49. Ibid., pp. 83–84. The term "machinery" in this quotation is the only instance of any form of that word appearing in the *Origin*. The context makes clear it has no semantically significant role in his description.

The productions of nature, Darwin resonantly observed, were "far 'truer' in character than man's productions." They plainly bore, he averred, "the stamp of far higher workmanship."[50] But did this biblical echo refer to the stamp of a machine? Most assuredly not, as the archaeology of these passages indicates.

Darwin's description here of the activities of natural selection can be easily traced back to the large manuscript (to be called *Natural Selection*) that he abandoned to undertake an abridged version (*Origin of Species*) after he had received Alfred Russel Wallace's letter outlining a theory close to his own. In that somewhat older version, Darwin described the operations of natural selection this way:

> [Man] selects any peculiarity or quality which pleases or is useful to him, regardless whether it profits the being. . . . See how differently Nature acts! She cares not for mere external appearance; she may be said to scrutinize with a severe eye, every nerve, vessel & muscle. . . . Can we wonder then, that nature's productions bear the stamp of a far higher perfection than man's product by artificial selection. With nature the most gradual, steady, unerring, deep-sighted selection,—perfect adaption to the conditions of existence.[51]

These passages, which describe natural selection as peering into the very fabric of a creature, selecting altruistically that which is good and casting out what is bad, a natural selection that operates perfectly (as the earlier manuscript has it)—these passages hardly describe the operations of Locke's spinning jenny or even the clatter and wheeze of a Manchester mechanical loom. When we probe still further back into Darwin's earliest speculations about transformational forces, then we see even more clearly that his conception of natural selection sprung from the head of a divinized nature. His "Essay of 1842" and "Essay of 1844," which sketched out his theory in detail for the first time, reveal his model for understanding the "truer" work of natural selection. It is crucially important to understand that in these essays Darwin came *to explain to himself* the operations of selection. And in this explanation, he employed an aesthetic device, a poetic creation, by which intuitively to explore the possibilities of his embryonic idea.

50. Ibid., p. 84.

51. Charles Darwin, *Charles Darwin's Natural Selection: Being the Second Part of His Big Species Book written from 1856 to 1858*, ed. R. C. Stauffer (Cambridge: Cambridge University Press, 1975), pp. 224–25. After the sentence "See how differently Nature acts!" Darwin inserted a new and, perhaps, ameliorative emendation into the manuscript: "By nature, I mean the laws ordained by God to govern the Universe."

In these early essays, Darwin, struggling through inchoate notions, grasped in metaphor and images the slowly forming structure of natural selection. These images, I believe, have distinctly shaped the deep logic of his formulation. In his "Essay of 1844," we can see more clearly the concept in gestation:

> Let us now suppose a Being with penetration sufficient to perceive differences in the outer and innermost organization quite imperceptible to man, and with forethought extending over future centuries to which with unerring care and select for any object the offspring of an organism produced under the foregoing circumstances; I can see no conceivable reason why he should not form a new race (or several were he to separate the stock of the original organism and work on several islands) adapted to new ends. As we assume his discrimination, and his forethought, and his steadiness of object, to be incomparably greater than those qualities in man, so we may suppose the beauty and complications of the adaptations of the new races and their differences from the original stock to be greater than in the domestic races produced by man's agency.[52]

Here as well as in the 1842 essay, Darwin worked out for himself the character of natural selection; and that character was cast in the image of a divine Being, whose "forethought" might teleologically produce creatures of great "beauty" and with progressively intricate "adaptations." Natural selection, in its original, metaphorical conception, was hardly machinelike, rather godlike.

Darwin's imaginative construction of natural selection exemplifies his early notions about the role of poetic fancy in science—a notion quite in conformity to that of the Romantics. He came to believe, likely from reading Humboldt and also Wordsworth, that fabricating "castles in the air" must set the path for more rigidly analytic thought about a subject. Such fancy, he believed, was no less difficult than "the closest train of geological thought"; and, indeed, he judged that the capacity for imaginative constructions "makes a discoverer." His own employment of such "castles in the air" not only aided him in the discovery of natural selection, but also deeply structured that discovery.[53]

---

52. Darwin, "Essay of 1844," in *The Foundations of the Origin of Species*, p. 85.

53. Darwin, "Notebook M," in *Charles Darwin's Notebooks*, p. 527 (M 34–35): "I observe a long castle in the air, is as hard work . . . as the closest train of geological thought.—the capability of such trains of thought makes a discoverer, & therefore (independent of improving powers of invention) such castles in the air are highly advantageous, before real train of inventive thoughts are brought into play."

When Darwin's earlier understanding of natural selection is revealed, we can see more clearly the import of those descendant ideas that animate the *Origin*. In the long passages quoted above, natural selection is depicted as operating on the least shade of difference in organic life, penetrating to the very core of that life, working unceasingly, intricately, aesthetically, and teleologically. Moreover, natural selection picks out the bad and preserved the good not selfishly but altruistically ("Man selects only for his own good; Nature only for that of the being which she tends"). The good that nature fostered, in Darwin's conception, contributed to that ever-growing, morally progressive state that the history of evolution exemplified and, ultimately, produced. For "as natural selection works solely by and for the good of each being, all corporeal and mental endowments will tend to progress toward perfection."[54]

The moral character of nature's actions in regard to her own creations is whispered, as in those just-mentioned passages, throughout the *Origin of Species*, the muted attributions slipping out of the deeper recesses of Darwin's basic concept of nature. That moral perception, however, led also to Darwin's more explicit recognition of a great paradox, a paradox that had bedeviled generations of theologians. How can a morally good, infinite being produce evil?—and the destruction wrought by selection certainly appeared evil. Darwin, I think, found the key to the solution of this metaphysical conundrum in a poetic source.

During his *Beagle* voyage, both on ship and as he traveled into the interior of South America, Darwin always had his copy of Milton's *Paradise Lost* with him. Milton, that favorite of both the German and English Romantics, had captured in sublime poetry the solution to the problem of evil. In one passage, as Satan approaches the Garden of Eden, he is stopped by an entangled bank of bushes and undergrowth, not unlike the barriers Darwin constantly met in his own jungle-garden adventures. Milton wrote:

Now to the ascent of that steep savage hill
Satan had journeyed on, pensive and slow,
But further way found none, so think entwined,
As one continued brake, the undergrowth
Of shrubs and tangling bushes had perplexed
All path of man or beast that passed that way . . .
Thence up he flew, and on the Tree of Life,
The middle tree and highest there that grew,
Sat like a cormorant, yet not true life
Thereby regained, but sat devising death

54. Darwin, *Origin of Species*, p. 489.

To them who lived, nor on the virtue thought
Of that life-giving plant, but only used
For prospect what, well used, had been the pledge
Of immortality.[55]

With the fall of Adam and Eve, the progenitors of us all, came, however, the happy possibility of our salvation. So through the pain and desolation visited upon this earth by what seems an imperfect law, the Redeemer comes and with his own death the world is transformed:

From the conflagrant mass, purged and refined,
New Heavens, new Earth, Ages of endless date
Founded in righteousness and peace and love,
To bring forth fruits, joy and eternal bliss.[56]

This orthodox explanation of evil—that it is only apparent and really the guise of the good—echoes through Milton's lyrics: out of death and destruction comes life more abundant, life transformed. And this is exactly the resolution that nature, in Darwin's divinized reconstruction, offers: out of struggle and death comes the greatest perfection, the higher creatures. We should thus be confident, Darwin reassures his reader, that nature ameliorates the struggle for existence and endows it with a redemptive purpose: "When we reflect on this struggle, we may console ourselves with the full belief, that the war of nature is not incessant, that no fear is felt, that death is generally prompt, and that the vigorous, the healthy, and the happy survive and multiply."[57] And in the end, the purpose of nature will be fulfilled, the transformation of the lowly and debased into higher beings. In the last paragraph of the *Origin*, the Edenic entanglements of nature continue to protect against evil; and in Miltonic cadences, Darwin draws the explicit lesson:

It is interesting to contemplate an entangled bank, clothed with many plants of many kinds, with birds singing on the bushes, with various insects flitting about, and with worms crawling through the damp earth, and to reflect that these elaborately constructed forms, so different from each other, and dependent on each other in so complex a manner, have all been produced by laws acting around us. These laws, taken in the largest sense, being Growth with Reproduction . . . a Ratio of Increase so high as to lead to a Struggle for Life, and as a consequence to Natural Selection, entailing Divergence of Character and the Extinction of less-

55. John Milton, *Paradise Lost*, 4.172–201.
56. Ibid., 12.548–51.
57. Darwin, *Origin of Species*, p. 79.

improved forms. Thus, from the war of nature, from famine and death, the most exalted object which we are capable of conceiving, namely, the production of the higher animals directly follows. There is grandeur in this view of life, with its several powers, having been originally breathed into a few forms or into one; and that, whilst this planet has gone cycling on according to the fixed law of gravity, from so simple a beginning end-less forms most beautiful and most wonderful have been, and are being, evolved.[58]

O felix culpa! Or rather, O felix natura!—nature, that thoroughly organic being which embodies aesthetic and moral values.

By the time Darwin began to compose the *Origin of Species*, his images and metaphors had shed much of the rich fabric with which they had origi-nally been adorned. Yet his simpler expressions nonetheless disclose, via a deeper aesthetic logic, a morally saturated nature. That logic thus belies the usual view of Darwin's accomplishment, which was to show, according to Susan Cannon, nature to be "morally meaningless."[59]

Darwin's conception of nature, an intelligent and moral nature, very much like the nature formulated by Schelling and purveyed by Humboldt and Owen—Darwin's conception makes sense of certain aspects of his gen-eral evolutionary theory that seem inexplicable on the assumption of a na-ture clanking along in the manner of a nineteenth-century steam engine. Were natural processes really machinelike, ought not the products be iden-tical—same mold, same brick? But the products of nature, characterized by an underlying theme, to be sure, were infinitely varied, exuding the great abundance of life. Moreover, machines, at least those of Darwin's acquain-tance, could hardly produce traits of near perfection. From the time he read Paley's *Natural Theology*, Darwin never doubted that organs like the eye—Paley's favorite example—were adaptations of extreme perfection, hardly the sort of thing a machine could produce. He later confessed that when-ever he thought of the eye, his blood ran cold. But if the agency producing an eye were virtually a lesser god, then such production might well be intel-ligible. Nature's agency displayed a subtle sense of possibility in Darwin's construction. She very gradually and over long periods of time refined and shaped her creatures. Darwin's friend Huxley (and some present-day Dar-winians) urged that nature be conceived as hopping along in fits and starts, producing stuttering advances in species forms instead of the slow gradual progress Darwin's theory supposed.[60] Huxley's insistence would be appro-

58. Ibid., pp. 489–90.
59. Cannon, *Science in Culture*, p. 276.
60. In his review of the *Origin*, Huxley singled out gradualism as one of the few definite flaws in

priate if Darwin had indeed construed nature as a machine—nineteenth-century variety. But his nature conceptually grew according to the model of the departed divinity; hence, it could move gradually, slowly, and majestically toward the perfection it teleologically sought. There are many other features of Darwin's conception of nature that hardly mesh with the gears of a machine; but perhaps I've suggested enough to make the case that the usual conception of Darwin's accomplishment simply fails to treat seriously the canonic expression of his theory in the *Origin of Species*.

Given Darwin's conception of nature, one in large part, I have argued, expressive of the kind of Romanticism cultivated originally in Germany and imported to England under various guises—given that conception would it be surprising that Darwin did not eviscerate human nature of moral capacity?

Darwin's Theory of Morals

Moral Theory prior to the *Origin of Species*

Darwin not only reinfused nature with value; he refused to leave man morally naked to the world. This, of course, is not the general view of Darwin's construction of morality. Most commonly he is taken to have advocated a kind of selfish utilitarianism disguised in decorous Victorian language. Yet such an evaluation completely misses what Darwin himself thought to be distinctive of his biology of morality, namely, that it overturned utilitarianism.

During his five-year voyage on H.M.S. *Beagle*, Darwin experienced the extremes of moral behavior, from the brutality he frequently observed among the South American gauchos to the nobility of the Indians whom they slaughtered. And he reacted with particular moral revulsion at the institution of slavery. Humboldt had readied him for his encounter with that trade in which both "copper-colored Indians" and "African Negroes" suffered the kind of injury that resulted "in rendering both the conquerors and the conquered more ferocious."[61] But his response to the reality was hardly less genuine for being filtered through the somber lens of his German predecessor. He felt simultaneously depressed and furious as he witnessed African families being separated at slave auction and slaves being beaten and degraded.[62] Though it was only in the tranquility of his study, as he

---

Darwin's theory. See Thomas Henry Huxley, "The Origin of Species" (1860), in his *Darwiniana*, vol. 2 of *Collected Essays* (London: McMillan, 1893), p. 77.

61. Humboldt, *Personal Narrative*, 3:3.

62. Darwin, *Journal of Researches of the Beagle*, pp. 22, 27–28.

composed his *Journal of Researches*—and then recomposed it in a second edition (1845)—that the Humboldtian framework focused his considerations on the peculiar institution. In that second edition, Darwin concluded:

> On the 19th of August we finally left the shores of Brazil. I thank God, I shall never again visit a slave-country. To this day, if I hear a distant scream, it recalls with painful vividness my feelings, when passing a house near Pernambuco, I heard the most pitiable moans, and could not but suspect that some poor slave was being tortured, yet knew that I was as powerless as a child even to remonstrate. . . . It is argued that self-interest will prevent excessive cruelty; as if self-interest protected our domestic animals, which are far less likely than degraded slaves, to stir up the rage of their savage masters. It is an argument long since protested against with noble feeling, and strikingly exemplified, by the ever illustrious Humboldt.[63]

As he reflected on the nature of moral behavior, which formed a topic of his considerations from the very beginning of his work on the species question, Darwin's patience for utilitarian arguments gradually became exhausted. Initially, though, prior to his reflective consideration of Humboldt and prior to the formulation of natural selection, utilitarianism did seem the right way to think about morals.

While an undergraduate at Cambridge, Darwin had to get up William Paley's *Moral and Political Philosophy* for his exams. The utilitarian position worked out by this Anglican divine provided him with a preliminary framework through which to weave his emerging biological theory of behavior.

About three weeks before he read Malthus, who ignited the spark for the idea of natural selection in his imagination, Darwin reread Paley, who also kindled an idea. In his "Notebook M," he considered "Paley's Rule." In *Moral and Political Philosophy*, Paley offered this rule of "expediency" as the central axiom of his ethics:

> Whatever is expedient is right. But then it must be expedient on the whole, at the long run, in all its effects collateral and remote, as well as in those which are immediate and direct.[64]

Darwin, as was his habit, gave this rule a biological interpretation:

> Sept 8th. I am tempted to say that those actions which have been found necessary for long generation, (as friendship to fellow animals in social

---

63. Charles Darwin, *The Voyage of the Beagle*, ed. Leonard Engel (New York: Doubleday Anchor, 1962), pp. 496, 497.

64. William Paley, *Moral and Political Philosophy*, in *The Works of William Paley* (Philadelphia: Woodward, n.d.), p. 40.

animals) are those which are good & consequently give pleasure, & not as Paley's rule is then that on long run *will* do good.—alter *will* in all such cases to *have* & *origin* as well as rule will be given.[65]

Darwin constructed this interpretation of Paley's rule prior to having formulated his principle of natural selection. Up to this time, he was persuaded that habits, practiced over several generations, could become instinctive, that is, innate. Continued exercise of instincts—such as that of ancient birds' swimming out on a pond—might in the course of generations alter anatomy (e.g., the birds, by stretching their toes in swimming, might produce webbing, and so ducks would be born). This habit-instinct device became for Darwin, prior to reading Malthus, the principal explanation of the alteration of species over time.[66] In this light, his interpretation of Paley's rule becomes a bit more clear. He suggested that those useful and expedient habits that have been necessary to preserve animal groups, allowing them over long periods to propagate and protect their young (habits such as friendship, parental nurture, and the like) were what we had come to call morally good. The continued practice of such useful social behaviors would produce moral instincts that conformed to a temporally readjusted rule of expediency: what has been good will become interred in an animal's bones, and thus will continue to be what animals and their offspring, including humans, regard as good.

In late September 1838, Darwin read Thomas Malthus's *Essay on the Principle of Population* (1798), which allowed him to bring together several inchoate ideas about the causes of species change. These developing ideas would eventually travel under the rubric of "natural selection." The clear formulation of the principle required Darwin also to adjust his account of the origins of moral behavior. He was aided in this task by his conversations with James Mackintosh—the brother-in-law of his uncle Josiah Wedgwood—and by the reading of Mackintosh's *Dissertation on the Progress of Ethical Philosophy*.[67]

Darwin began reading Mackintosh's *Dissertation* on and off from summer of 1838 to spring of 1839. In this historical and critical treatise, Mackintosh provided a survey of ethical theories up to his own time. He objected particularly to Paley's utilitarianism, since it allotted human beings a nature that responded only to the urgings of pleasure and pain for self. Mackintosh

---

65. Darwin, "Notebook M," in *Charles Darwin's Notebooks*, p. 552 (MS 132e).

66. I have given a more extensive account of Darwin's pre-Malthusian theories of species change in my *Darwin and the Emergence of Evolutionary Theories of Mind and Behavior* (Chicago: University of Chicago Press, 1987), pp. 83–98.

67. James Mackintosh, *Dissertation on the Progress of Ethical Philosophy* (Edinburgh: Adam and Charles Black, 1836).

believed human beings to be endowed with a more generous nature, one that might act altruistically for the good of others. In this conviction, he aligned himself with the views of Shaftesbury, Butler, and Hutchinson. He argued that human nature came equipped with a moral sense for right conduct. That sense might be educated through experience, but its roots were innately imbedded in the human constitution. This meant that individuals could be and were motivated by urges other than those of selfish pleasure. They rather spontaneously sought to improve the welfare of others and approved of such behavior when exhibited by acquaintances. Mackintosh did not completely disavow Paley's or Bentham's utilitarianism. He granted that in a cool hour, when we considered the *criterion* for right behavior, we would recognize the utility of acts that promoted the common welfare and the disutility of behavior that compromised it. Mackintosh insisted, however, that the *moral sense* for right conduct, that is, our immediate perception of what we ought to do in a situation, did not depend on any rational calculation of pleasures and pains, utilities and disutilities. We simply recognized innately what behaviors were morally required in a situation.

Mackintosh's theory of moral behavior had two glaring difficulties, at least to Darwin's mind. He had no good explanation of the origins of the moral sense, nor could he adequately account for the conjunction of the moral sense and the moral criterion. Just why was it that what a person might do spontaneously and without reflection would conform to what appeared, after sufficient reflection, to be the most useful act that could have been chosen in the situation? To these difficulties Darwin would suggest an ingenious answer.

Mackintosh's attack against the belief that human beings could only be motivated by selfishness, even in matters of morals, made a considerable impression on Darwin. This is evidenced in three different sets of documents: various passages in his "Notebook N" (kept between fall of 1838 and spring of 1840); remarks left in a bundle of papers he labeled "Old and Useless Notes"; and a nineteen-page manuscript on Mackintosh's views that he composed in spring of 1839.[68] In these various jottings, Darwin gradually sketched out a theory of moral behavior and conscience that marked another stage in his developing considerations about these subjects. The manuscripts show him amalgamating his earlier Paleyesque ideas to Mackintosh's scheme and thus laying the foundation for the ethical theory presented in his *Descent of Man*.

---

68. These notebook and manuscript pages have been transcribed and published in *Charles Darwin's Notebooks, 1836–1844*, pp. 563–96, 599–629. The Mackintosh manuscript is included among the "Old and Useless Notes," pp. 618–29.

Fundamental to Mackintosh's moral-sense theory was the proposal that a certain kind of knowledge lay buried in the human soul, a proposal antithetic to the main empiricist stream of British thought. This conception of innate knowledge received a theoretical boost, at least for Darwin, by an essay that appeared in the *Westminster Review* in 1840. The essay summarized and evaluated the general philosophy of Samuel Taylor Coleridge, who borrowed many of his ideas from Schelling. The appreciation by the anonymous author (none other than John Stuart Mill), though critical, recognized that the Romantic philosophy of Coleridge pervaded the minds and hearts of a significant portion of British intellectuals. That philosophy maintained that the fundamental principles of morals (along with those of religion, mathematics, and the basic laws of physics) came to reason not through experience but through deeper channels of the soul.[69] Darwin's own response to the essay shows him, typically, finding a biological interpretation of a theory with which he was in sympathy. When Mill, in his account of Coleridge, objected that only experience could be the object of our knowledge, Darwin reflected, "Is this not almost a question whether we have any instincts, or rather the amount of our instincts—surely in animals according to usual definition, there is much knowledge without experience, so there *may* be in men—which the reviewer seems to doubt."[70] Instinct, Darwin would insist, formed the basis of our moral sense and anything we might call innate knowledge.

The theory of moral behavior that Darwin developed in his early notebooks—and, with a few modifications, presented much later in the *Descent of Man*—proposed that our judgments about appropriate behavior, the sort of behavior we call moral, stemmed from a kind of innate knowledge, from an instinct for right action. In his theory Darwin distinguished two general kinds of instinct: impulsive instincts, really emotional reactions—for example, a stab of lust or a flash of anger that impels an immediate response; and the more calm and persistent social instincts, which though powerful in the longer term do not have that same immediate force. But it was instincts of this latter sort, Darwin reflected, that kept a mother bird patiently tending her brood. Sometimes, however, the good mother might, while out forging for her nestlings, catch sight of the migrating flock and spontaneously fly away to better climes, without adverting to her young. In a sun-

69. Mill's essays on Bentham and Coleridge are conveniently reprinted with an splendid introduction in John Stuart Mill, *Mill on Bentham and Coleridge*, introduced by F. R. Leavis (London: Chatto & Windus, 1950).

70. Darwin, "Old and Useless Notes," in *Charles Darwin's Notebooks*, p. 610.

nier environment, if this reprobate mother had mind enough to recall her starving chicks, her social instincts would again take hold. She would feel that pull to aid her offspring, though the instinct would remain unsatisfied. A rational animal in these circumstances, Darwin supposed, would confess a troubled conscience. All that was required to turn an animal into a moral creature, according to this scheme, was an intellect approaching that of man. On 3 October 1838, just a few days after he had read Malthus, Darwin reformulated his theory of moral conscience and relished its implications:

> Dog obeying instinct of running hare is stopped by fleas, also by greater temptation as bitch. . . . Now if dogs mind were so framed that he constantly compared his impressions, & wished he had done so & so for his interest, & found he disobeyed a wish which was part of his system, & constant, for a wish which was only short & might otherwise have been relieved, he would be sorry or have troubled conscience.—Therefore I say grant reason to any animal with social & sexual instincts & yet with passion he *must* have conscience—this is capital view. Dogs conscience would not have been same with mans because original instincts different.[71]

In his manuscript on Mackintosh, Darwin further developed this account of conscience. He considered that the useful habits members of a species would develop in social circumstances—cooperative foraging, group defense, parental nurturing, and the like—would continue to be practiced over generations. Those other behaviors directed to particular and individual desires would, however, not be constantly practiced from one generation to the next and would, therefore, not become deeply ingrained in the heritable substance of the animal. Those remaining practices directed to the common good, on the other hand, would seep into the heritable core of individuals and be passed on as instincts. In the human species, characterized by sufficient memory for instincts to remain active, even in the absence of their eliciting situations, these social instincts would become moral motives. Human moral sense would thus evolve along with other species-characteristic traits. When individuals reflected on their behaviors and moral impulses, they would quite naturally feel a particular kind of quiet pleasure in the satisfaction of such instincts and, simultaneously, perceive the utility in their exercise. Rational creatures would thus take as criterion the utility subsequently recognized in such reflections. In this fashion, Darwin had solved Mackintosh's residual problem: he found a

71. Charles Darwin, "Notebook M," in *Charles Darwin's Notebooks*, p. 536 (MS pp. 2–3).

biological explanation for the agreement between spontaneous moral behavior and reflective judgment about the utility of that behavior for the
community at large.

Darwin's account of moral behavior in his early notebooks relied entirely
on the principle of the inheritance of acquired characteristics. But during
the roughly two decades prior to the publication of the *Origin of Species*, he
slowly brought his device of natural selection to explain all sorts of species
characteristic traits, including instinct. However, a stubborn problem surfaced when he applied his device to those instincts promoting moral behavior. Moral behavior, unlike other traits that might be selected for,
conferred benefit not on its agent but on its recipient. The social and moral
instincts were fundamentally altruistic. Yet natural selection seemed to operate only on traits that gave their agent an advantage, not their recipient.
How to explain it?

In the mid-1840s, the problem of the social instincts and their natural-
selectionist explanation became quite poignant. It captured Darwin's attention when he sought to explain the distinctive instincts and anatomical
traits of the social insects, especially those of the workers in beehives and
ant nests. Soldier bees, for example, would guard the hive and even give up
their lives in its defense—altruistic behavior that had a positive disadvantage to the individual. But these casts of insects displayed another feature
that seemed to make a selectionist account of their behavior unworkable—
they were neuters and left no offspring to inherit any advantageous traits.
As I have shown elsewhere, Darwin worried about this difficulty for some
time, feeling that it would undo his general theory of evolution altogether.[72] He attempted various resolutions, initially without success; he
suffered the chilling feeling that he himself might have exposed the fatal
flaw in his own theory. It was only during the first months of 1858, while la-
boring over the manuscript that he intended to publish under the title of
Natural Selection, that he fixed upon a solution that would be highlighted in
the *Origin of Species*, which became the more compact successor of that earlier manuscript. In those first months of 1858, Darwin sketched out his new
theory:

> In the eighth chapter, I have stated that the fact of a neuter insect often
> having a widely different structure & instinct from both parents, & yet
> never breeding & so never transmitting its slowly acquired modifications

72. See Richards, *Darwin and the Emergence of Evolutionary Theories of Mind and Behavior*,
pp. 127–56.

to its offspring, seemed at first to me an actually fatal objection to my whole theory. But after considering what can be done by artificial selection, I concluded that natural selection might act on the parents & continually preserve those which produced more & more aberrant offspring, having any structure or instincts advantageous to the community.[73]

In the *Origin of Species*, Darwin elaborated this solution, namely, that natural selection operated not on the individual workers to provide their unusual traits but on the whole hive or community, which would contain relatives of the workers. (And in the fifth edition of the *Origin*, he extended the idea of group selection to any assemblage of social animals, including human beings.)[74] Thus the altruistic behavior of a soldier bee in sacrificing its life for the nest could be explained as the result of community selection. Natural selection could indeed, then, be brought to account for moral behavior. And in the *Descent of Man*, this is precisely what Darwin did, and then some.

Ethical Theory in the *Descent of Man*

Darwin had originally intended to consider human evolution in the *Origin of Species*. The haste with which he composed the book, however, and, more importantly, his desire to avoid unnecessary provocation counseled postponement. In the wake of the success of the *Origin*, when he decided to write on matters of domestication, variation, and heredity, he mischievously mentioned to Wallace that he would include in his contemplated volume an essay on man, since that creature seemed "an eminently *domesticated* animal."[75] Darwin's *On the Variation of Animals and Plants under Domestication* (1867) did not, however, carry the intended essay. Instead, Darwin had decided, as the result of a dispute with Wallace over sexual selection, to treat human evolution in a volume dedicated to that subject. As late as 1869, he seemed to have had no serious intention of discussing hu-

---

73. Darwin, *Charles Darwin's Natural Selection*, p. 510.

74. In the fifth edition of the *Origin* (1869), Darwin quietly extended the idea of selection to the entire community of individuals. What is noteworthy about this extension is that he did not stipulate that the members of the community should be closely related. He wrote: "In social animals it [natural selection] will adapt the structure of each individual for the benefit of the community; if this in consequence profits by the selected change." In the sixth edition, he directly and simply advanced this proposition: "Natural selection will modify the structure of the young in relation to the parent, and of the parent in relation to the young. In social animals it will adapt the structure of each individual for the benefit of the whole community; if the community profits by the selected change." See *The Origin of Species by Charles Darwin: A Variorum Text*, p. 172.

75. Darwin to Alfred Wallace (March 1867), in *Alfred Russel Wallace: Letters and Reminiscences*, ed. James Marchant, 2 vols. (London: Castile, 1916), 1:181.

Figure 14.4
Charles Darwin.
Photograph taken in
1860, just after the
publication of the *Origin
of Species*. (Courtesy
Down House and Royal
College of Surgeons.)

man morality in the planned book; he wished only to consider those traits of males and females that could be attributed to sexual selection, for instance, the male's greater musculature, secondary sexual characteristics, and other minor traits. However, several of his friends, notably Charles Lyell (1797–1875) and Asa Gray (1810–1888), had during the mid-1860s publicly suggested that natural selection could not explain the distinctive features of human beings, especially their moral nature.[76]

The greatest challenge in this regard, however, came from the cofounder of evolution by natural selection, Alfred Russel Wallace (1823–1913). Darwin's friend had undergone a kind of spiritualist conversion, and in the *Quarterly Review* (1869), Wallace pressed the objection that while the animal kingdom arose through the power of natural selection, human mental and moral traits could not have:

> Neither natural selection or the more general theory of evolution can give any account whatever of the origin of sensational or conscious life. . . . But the moral and higher intellectual nature of man is as unique

76. I have discussed the theological backsliding of Darwin's friends in my *Darwin and the Emergence of Evolutionary Theories of Mind and Behavior*, pp. 157–84.

a phenomenon as was conscious life on its first appearance in the world, and the one is almost as difficult to conceive as originating by any law of evolution as the other.[77]

Darwin felt something akin to despair over his friend's abandonment of natural selection in the case of man. But Wallace's new attitude, along with the stated reservations of Lyell and Gray, provided just the stimulus to alter Darwin's intentions for the new book. He now decided hastily to resurrect his early notes on human moral evolution but to reformulate those youthful ideas in light of his theory of community selection. The result was an articulated theory of moral conscience and a firm hypothesis as to the origins of that faculty. Darwin's labors yielded two volumes, which appeared in 1871 under the title *The Descent of Man and Selection in Relation to Sex.*

The theory of conscience that Darwin presented in the *Descent* bore the distinctive marks of his earlier considerations. He maintained that human moral judgment lay anchored in social instinct, the kind of instinct that would have urged the mother bird to forage for her nestlings. Human conscience, of course, had to be more than simple social instinct. Darwin argued that faculty would have arisen through four overlapping stages: first, protohumans had to develop a set of social instincts strong enough to bind them together into a society; second, members of this society had to have acquired sufficient intellect to recall a social instinct that might have been momentarily swamped by a more insistent urge; third, language would be required to codify and communicate the needs of other society members; and finally, members of this community would have to develop habits of attending to the needs of others—and with this stage they would have attained the truly human state, that of a moral creature.[78]

Darwin had essentially planted Mackintosh's ethical theory in biological ground. Moral sense consisted of the variety of social instincts, that is, altruistic instincts, which had been evolutionarily acquired by a social group, instincts that its members could reflect upon, codify through language, and stabilize in habit. Those altruistic instincts, he maintained, would have arisen through community selection. He put it this way:

> It must not be forgotten that although a high standard of morality gives but a slight or no advantage to each individual man and his children over the other men of the same tribe, yet that an advancement in the standard of morality and an increase in the number of well-endowed men will cer-

77. Alfred Russel Wallace, "Review of *Principles of Geology* and *Elements of Geology* by Charles Lyell," *Quarterly Review* 120 (1869): 359–94; quotation from p. 391.

78. Darwin, *Descent of Man*, 1:72–73.

tainly give an immense advantage to one tribe over another. There can
be no doubt that a tribe including many members who, from possessing in
a high degree the spirit of patriotism, fidelity, obedience, courage, and
sympathy, were always ready to give aid to each other and to sacrifice
themselves for the common good, would be victorious over most other
tribes, and this would be natural selection.[79]

Thus an idea first developed to solve the problem of the evolution of the social insects became adapted to solving a crucial problem of the evolution of
social human beings. But Darwin, shrewd as he was, anticipated a difficulty
with his solution: How did these moral traits arise *within* one tribe in the
first place? After all, as he noted, it is not likely that parents of an altruistic
temper would raise more children than those of a selfish attitude. Moreover,
those who were inclined to self-sacrifice might leave no offspring at all.[80]
Darwin employed his device of use-inheritance to explain the origin of
such social behaviors within a given tribe. He proposed two related sources
for such behaviors. The first is the prototype of contemporary theories of reciprocal altruism. He observed that as the reasoning powers of members of
a tribe improved, each would come to learn from experience "that if he
aided his fellow-men, he would commonly receive aid in return." From this
"low motive," as he regarded it, each might develop the habit of performing
benevolent actions; and this habit might be inherited and thus furnish suitable material on which natural selection might operate.[81] The second
source relied on the rather standard assumption that "praise and blame" of
certain social behaviors would feed our animal need to enjoy the admiration of others and to avoid feelings of shame and reproach. This kind of social control would also lead to heritable habits.[82]

Darwin conjectured that our ancestors lived in small tribal communities
that would compete with one another, not unlike groups of social-insect
hives. Those communities would reap the propagative advantage if some
members exhibited altruistic impulses that directed their behavior to the
welfare of the whole. And over generations, he believed, this process would
further inculcate altruism in successful tribal communities. During the
course of ages, intellectual acquisitions, learned customs, and advances in
knowledge that a group might enjoy would focus the altruistic instincts of
members on actions that would be ever more effective in producing real
benefit for others (for instance, the discovery of the value of inoculation).

79. Ibid., 1:166.
80. Ibid., 1:163.
81. Ibid., 1:163–64.
82. Ibid., 1:164–65.

Darwin thus wedded notions of cultural progress in learning with his theory of community selection to produce a conception that made human beings intrinsically moral animals, but animals whose ethical prescriptions would be informed by increasing knowledge of what really is in the best interests of their community.

The utilitarians, Darwin observed, had claimed that "the foundation of morality lay in a form of Selfishness; but more recently in the 'Greatest Happiness principle.'"[83] His own theory, by contrast, did not suppose that moral action was motivated by self-interest or executed to achieve the greatest happiness. Rather, human beings, he maintained, acted spontaneously, impelled by their altruistic instincts, to advance the welfare of others without counting the cost to self. They neither sought the greatest happiness of the greatest number, nor even the greatest happiness for themselves; rather they directed their actions to achieve the "greatest good," which Darwin interpreted to be the "vigor and health" of the greatest possible number of community members. He thus concluded that "the reproach of laying the foundation of the most noble part of our nature in the base principle of selfishness is removed."[84]

Darwin, that humble—and, at times, seemingly bumbling—biologist had constructed an ethical theory of elegance, power, and nobility. He held that human behavior, in some of its forms, could and did achieve those ideals that have been enshrined in the Western ethical tradition. He preserved man as an intrinsically moral being, a being whose morality tinctured the very core of its substance. Certainly human beings act selfishly on occasion. Darwin, nonetheless, believed that they could recognize the needs of others and could respond unselfishly to satisfy those needs. His ethical theory, therefore, stands apart from the typical Benthamite systems of his contemporaries. This is because, I believe, from the very beginning Darwin had recognized in nature a source of moral and aesthetic value. Alexander von Humboldt had inculcated him with the kind of moral evaluation of nature that simply could not be reduced to the low utilitarianism infecting most British moral philosophy of the time.

Darwin's theory of the rise of moral behavior had an added benefit, which seems to have struck him only in the composition of the *Descent*. It led to a powerful explanation of the development of human intelligence. His friend Wallace had not only objected to a naturalistic account of morals; he objected also to a naturalistic account of human intelligence—and for comparable reasons. Wallace had argued that man's high intellect

83. Ibid., 1:97.
84. Ibid., 1:98.

was quite superfluous for the simple needs of survival and hence that human faculty had to have a source other than natural selection.[85] But now Darwin had a powerful account for the origin of traits that seemed of little use to their possessor, namely, community selection; and this account he now brought to the explanation of human intellect: if in a tribal group, a genius by chance appeared, that primitive Newton would benefit his whole community; community members would learn his tricks, thus giving them an advantage in competition with other tribes; and since his tribe would include his relatives, who would bear seeds of his intellectual talents, improved mind would rise among succeeding generations.[86]

Darwin's account of morals and intelligence had a common root in his theory of community selection. Community selection operated to produce community mind. This sort of explanation resonates with theories found among the German Romantics, particularly in Schilling's thesis that absolute mind produced individual mind and its moral structures. To argue for a direct descent of ideas here would certainly press my case for the influence of German Romantics on Darwin beyond the endurance of the most tolerant of readers. Yet there seems to be, at least in this instance, something like convergent evolution: the Humboldtian legacy created an environment that favored Darwin's construction of mind resident in nature, but mind not bereft of those capacities and values so esteemed by Romantic writers.

Conclusion

The venerable Darwin who peers out from John Collier's posthumous portrait, done in 1884, has the visage of a terrible Old Testament prophet. Photographs taken during his last years confirm that the artifice embraced the man, not merely the painting. These are the images of Darwin we remember most vividly—hardly the kind of figure one would think of as a Romantic revolutionary. Yet, that he was a revolutionary, there can be no doubt. Nor, I believe, can we deny, at least when the written evidence is carefully considered, the deep Romantic strains of his thought.

Darwin came by his attitudes much as the earlier German Romantics had, through prolonged contact with exotic nature—but nature as filtered through a certain literature. In Darwin's case, the literature was singularly provided by the conceptually and aesthetically lush works of Alexander von Humboldt, who taught him how to experience the sublime and how

85. Wallace, "Review of Lyell," p. 392.

86. Darwin, *Descent of Man*, 1:161. Darwin also argued that the development of language would have a rebounding effect on the brain, so as to improve its structure, an improvement that would, via inherited biological characteristics, be passed to subsequent generations (ibid., 1:58).

Figure 14.5
Charles Darwin.
Posthumous portrait
(1884) by John Collier.
(Courtesy Down House
and Royal College of
Surgeons.)

morally to evaluate the nature he met in the jungles, mountains, and plains of South America. That early experience, formed and shaped under the guiding images provided by Humboldt, settled deeply into the conceptual structure of the *Origin of Species* and the *Descent of Man*. The sensitive reader of Darwin's works, a reader not already completely bent to early-twenty-first-century evolutionary constructions, will feel the difference between the nature that Darwin describes and the morally effete nature of modern theory.

Darwin's nature, like that of other German Romantics, exemplified archetypal patterns beneath the wild frenzy of their variations. These patterns gradually changed not under the aegis of Paley's God and certainly not as the products of a Victorian stamping press; they arose and altered, rather, through the power of a creative nature (*natura naturans*)—ever fruitful and rich in possibilities, realizing those possibilities in the best interests of her creatures. Darwin's nature, like that of the other Romantics, progressively produced organisms of ever-greater value, "the higher creatures," as he labeled them. His nature acted altruistically: unlike human beings, she tended her creatures for their own sake, improving their lot and that of the whole interconnected assembly constituting her frame. The intrinsic moral aspect of nature also imbued her most developed creatures, human beings.

Darwin's early attitudes about nature obviously became subject to conceptual influences other than those of the German Romantics—he was not simply, after all, Werther in his blue frock coat and yellow vest, reading his Homer and suffering unrequited love, albeit in a jungle clearing. But neither was he that unflinching mechanist who deprived nature of her soul of loveliness.

Bibliography

Allen, Danielle. "Burning the Fable of the Bees: The Incendiary Authority of Nature." In *The Moral Authority of Nature*, ed. Lorraine Daston and Fernando Vidal. Chicago: University of Chicago Press, forthcoming.

Ameriks, Karl. *Kant and the Fate of Autonomy: Problems in the Appropriation of the Critical Philosophy*. Cambridge: Cambridge University Press, 2000.

———, ed. *The Cambridge Companion to German Idealism*. Cambridge: Cambridge University Press, 2000.

Amrine, Frederick, Francis Zucker, and Harvey Wheeler, eds. *Goethe and the Sciences: A Reappraisal*. Dordrecht: Reidel, 1987.

Anon. "Review of *Historia insectorum generalis, ofte algemeene verhandeling van de bloedeloose dierkens*." *Philosophical Transactions of the Royal Society* 5 (1670): 2078–80.

Atkins, Stuart. "The Evaluation of Romanticism in Goethe's *Faust*." *Journal of English and Germanic Philology* 54 (1955): 9–38.

———. "*Italienische Reise* and Goethean Classicism." In *Aspekte der Goethezeit*, ed. Stanley Corngold, Michael Curschmann, and Theodore Ziolkowski. Göttingen: Vandenhoeck & Ruprecht, 1977.

Auden, W. H. *Selected Poems*. Ed. Edward Mendelson. New York: Vintage Books, 1979.

Bach, Thomas. "Kielmeyer als 'Vater der Naturphilosophie'? Anmerkungen zu seiner Rezeption im deutschen Idealismus." In *Philosophie des Organischen in der Goethezeit: Studien zu Werk und Wirkung des Naturforschers Carl Friedrich Kielmeyer (1765–1844)*, ed. Kai Kanz, pp. 232–51. Stuttgart: Franz Steiner Verlag, 1994.

Baer, Karl Ernst von. *Über Entwickelungsgeschichte der Thiere*. 2 vols. Königsberg: Bornträger, 1828–37.

Balass, Heinrich. "Kielmeyer als Biologe." *Sudhoffs Archiv für Geschichte der Medizin* 23 (1930): 271–72.

Baumgartner, Hans, ed. *Schelling: Einführung in seine Philosophie*. Munich: Karl Alber, 1975.

Beck, Hanno. *Alexander von Humboldt*. 2 vols. Wiesbaden: Steiner Verlag, 1959–61.

Beddow, Michael. "Don't Fell the Walnut Trees." *Times Literary Supplement*, no. 5054 (11 February 2000): 3–4.

Behler, Ernst. *Frühromantik*. Berlin: Walter de Gruyter, 1992.

———. *German Romantic Literary Theory*. Cambridge: Cambridge University Press, 1993.

———. "The Origins of Romantic Literary Theory." *Colloquia Germanica*, nos. 1–2 (1968): 109–26.

———. *Die Zeitschriften der Brüder Schlegel*. Darmstadt: Wissenschaftliche Buchgesellschaft, 1983.

Beiser, Frederick. *Enlightenment, Revolution, and Romanticism*. Cambridge: Harvard University Press, 1992.

———. *The Fate of Reason: German Philosophy from Kant to Fichte*. Cambridge: Harvard University Press, 1987.

———. *German Idealism*. Cambridge: Harvard University Press. Forthcoming.

Berlin, Isaiah. *The Roots of Romanticism*. Ed. Henry Hardy. Princeton: Princeton University Press, 1999.

———. "European Unity and Its Vicissitudes" and "The Apotheosis of the Romantic Will." In *The Crooked Timber of Humanity*, ed. Henry Hardy. New York: Knopf, 1991.

———. "The Romantic Revolution." In *The Sense of Reality*, ed. Henry Hardy. New York: Farrar, Straus and Giroux, 1997.

Bernhardt Eva. *Goethe's Römische Elegien: The Lover and the Poet*. Frankfurt a. M.: Peter Lang, 1990.

Biermann, Kurt-R. et al. *Alexander von Humboldt: Chronologische Übersicht über wichtige Daten seines Lebens*. Berlin: Akademie-Verlag, 1968.

Blumenbach, Johann Friedrich. *The Anthropological Treatises of Johann Friedrich Blumenbach*, trans. Thomas Bendysche. London: Longman, Green, Longman, Roberts & Green, 1865.

———. *Beyträge zur Naturgeschichte*. Part 1. Göttingen: Johann Christian Dieterich, 1790.

———. *De generis humani varietate nativa*, 2nd ed. Göttingen: Vandenhoek et Ruprecht, 1781; 3rd ed. 1795.

———. *De generis humani varietate nativa*. (1775). In *The Anthropological Treatises of Johann Friedrich Blumenbach*. Trans. Thomas Bendysche. London: Longman, Green, Longman, Roberts & Green, 1865.

———. *Geschichte und Beschreibung der Knochen des menschlichen Körpers*. Göttingen: Johann Christian Dieterich, 1786.

———. *Handbuch der Naturgeschichte*. 12th ed. 2 vols. Göttingen: Johann Christian Dieterich, 1779–80. Göttingen: Dieterich'schen Buchhandlung, 1830.

———. *De Nisu formativo et generationis negotio*. Göttingen: Johann Christian Dieterich, 1787.

———. "Über den Bildungstrieb (Nisus formativus) und seinen Einfluss auf die Generation und Reproduktion," *Göttingisches Magazin der Wissenschaften und Litteratur* 1, no. 5 (1780): 247–66.

———. *Über den Bildungstrieb und das Zeugungsgeschäfte*. Göttingen: Johann Christian Dieterich, 1781. 2nd ed., 1789; 3rd ed., 1791.

———. "Über eine ungemein einfache Fortpflanzungsart." *Göttingisches Magazin der Wissenschaften und Litteratur* 2, no. 1 (1781): 80–89.

Bode, Wilhelm. *Charlotte von Stein*. 3rd ed. Berlin: Mittler und Sohn, 1917.

———. *Goethes Leben: Rom und Weimar, 1787–1790*. Berlin: Mittler & Sohn, 1923.

———, ed. *Goethe in vertraulichen Briefen seiner Zeitgenossen*. 3 vols. Berlin: Aufbau Verlag, 1999.

Boerhaave, Hermann. *Praelectiones academicae*. Notes by Albertus Haller. 6 vols. in 3. Göttingen: Vandenhoeck, 1744.

Boisserée, Sulpiz. *Tagebücher*. Ed. Hans-J. Weitz. 5 vols. Darmstadt: Eduard Roether Verlag, 1978–1995.

Bonnet, Charles. *Considerations sur les corps organisés*. 2 vols. Amsterdam: Marc-Michel Rey, 1762.

———. *La Palingénésie philosophique, ou Idées sur l'état passé et sur l'état futur des êtres vivans*. 2 vols. Geneva: Philibert et Chiroi, 1769.

Börne, Ludwig. *Sämtliche Schriften*. Ed. Inge and Peter Rippmann. 5 vols. Düsseldorf: Joseph Melzer Verlag, 1964.

Bowie, Andrew. *Schelling and Modern European Philosophy*. London: Routledge, 1993.

Boyle, Nicholas. *Goethe: The Poet and the Age.* 2 vols. Oxford: Oxford University Press, 1991–2000.

Brandis, Johann David. *Versuch über die Lebenskraft.* Hannover: Hahn'schen Buchhandlung, 1795.

Bräuning-Oktavio, Hermann. *Oken und Goethe im Lichte neuer Quellen.* Weimar: Arion, 1959.

Breidbach, Olaf. *Matthias Jakob Schleiden, Schellings und Hegels Verhältnis zur Naturwissenschaft.* Weinheim: VCH, Acta Humaniora, 1988.

Broderip, William. "Review of *Journal of Researches* by Charles Darwin." *Quarterly Review* 65 (1839): 194–234.

Broman, Thomas. *The Transformation of German Academic Medicine, 1750–1820.* Cambridge: Cambridge University Press, 1996.

———. "University Reform in Medical Thought at the End of the Eighteenth Century." *Osiris* 2nd ser. 5 (1989): 35–53.

Brown, John. *The Elements of Medicine; or, a Translation of the Elementa Medicinae Brunonis by the author of the Original Work.* 2 vols. London: Johnson, 1788.

———. *Grundsätze der Arzeneylehre.* Trans. M. A. Weikard. 2 vols. Frankfurt a. M.: Andreaischen Buchhandlung, 1795.

Brunschwig, Henri. *Enlightenment and Romanticism in Eighteenth-Century Prussia.* Trans. Frank Jellinek. Chicago: University of Chicago Press, 1974.

Buffon, Georges Louis Leclerc, comte de. *Histoire naturelle.* In *Oeuvres complètes de Buffon.* Ed. Pierre Flourens. 12 vols. Paris: Garnier, 1853–55.

Bulhof, Ilse. *The Language of Science, with a Case Study of Darwin's* The Origin of Species. Leiden: Brill, 1992.

Burdach, Karl Friedrich. *Propädeutik zum Studium der gesammten Heilkunst.* Leipzig: Dyt'schen Buchhandlung, 1800.

Burwick, Frederick, ed. *Approaches to Organic Form.* Dordrecht: Reidel, 1987.

Bush, Werner. "Die 'grosse, simple Linie' und die 'allgemeine Harmonie' der Farben." *Goethe-Jahrbuch* 105 (1988): 144–64.

Buttersack, Felix. "Karl Friedrich Kielmeyer, ein vergessene Genie." *Sudhoffs Archiv für Geschichte der Medizin* 23 (1930): 236–46.

Bynum, W. F., and Roy Porter, eds. *William Hunter and the 18th-Century Medical World.* Cambridge: Cambridge University Press, 1985.

Caneva, Kenneth. "Teleology with Regrets." *Annals of Science* 47 (1990): 291–300.

Cannon, Susan Faye. *Science in Culture: The Early Victorian Period.* New York: Science History Publications, 1978.

Cassirer, Ernst. *Kant's Life and Thought.* Trans. James Haden. New Haven: Yale University Press, 1981.

Coleman, William. "Limits of the Recapitulation Theory: Carl Friedrich Kielmeyer's Critique of the Presumed Parallelism of Earth History, Ontogeny, and the Present Order of Organisms." *Isis* 64 (1973): 341–50.

Conrady, Karl Otto. *Goethe: Leben und Werk.* 2 vols. Frankfurt a. M.: Fischer Verlag, 1988.

Cotta, Heinrich. *Naturbeobachtungen über die Bewegung und Funktion des Saftes in den Gewächsen.* Weimar: Hoffmann, 1806.

Cunningham, Andrew, and Nicholas Jardine, eds. *Romanticism and the Sciences.* Cambridge: Cambridge University Press, 1990.

Cuvier, Georges. *Histoire des progrès des sciences naturelles depuis 1789 jusqu'a ce jour.* 4 vols. Paris: Baudouin Frères, 1829.

———. *Le Règne animal distribué d'aprè son organisation*. 2nd ed. 5 vols. Vols. 4 and 5 by P. A. Latreille. Paris: Déterville, 1829–30.

Damm, Sigrid. *Caroline Schlegel-Schelling in ihren Briefen*. Darmstadt: Luchterhand, 1980.

———. *Christiane und Goethe: Eine Recherche*. Frankfurt a. M.: Insel Verlag, 1999.

Darwin, Charles. *The Autobiography of Charles Darwin, 1809–1882*. Ed. Nora Barlow. New York: W. W. Norton, 1969.

———. *Beagle Diary*. Ed. R. D. Keynes. Cambridge: Cambridge University Press, 1988.

———. *Charles Darwin's Natural Selection: Being the Second Part of His Big Species Book written from 1856 to 1858*. Ed. R. C. Stauffer. Cambridge: Cambridge University Press, 1975.

———. *Charles Darwin's Notebooks, 1836–1844*. Ed. Paul Barrett et al. Ithaca: Cornell University Press, 1986.

———. *The Correspondence of Charles Darwin*. Ed. Frederick Burkhardt et al. 12 vols. to date. Cambridge: Cambridge University Press, 1985–.

———. *The Foundations of the Origin of Species: Two Essays Written in 1842 and 1844 by Charles Darwin*. Ed. Francis Darwin. Cambridge: Cambridge University Press, 1909.

———. *Journal of Researches into the Geology and Natural History of the Various Countries Visited by H.M.S. Beagle* (London: Henry Coburn, 1839).

———. *On the Origin of Species*. London: Murray, 1859.

———. *On the Origin of Species by Charles Darwin: A Variorum Text*. Ed. Morse Peckham. Philadelphia: University of Pennsylvania Press, 1959.

———. *The Voyage of the Beagle*. Ed. Leonard Engel. New York: Doubleday Anchor, 1962.

Darwin, Erasmus. *Zoonomie, oder Gesetze des organischen Lebens*. Trans. J. D. Brandis. 3 vols. in 5. Hannover: Gebrüder Hahn, 1795–99.

———. *Zoonomia, or The Laws of Organic Life*. 2 vols. London: Johnson, 1794–96.

Daston, Lorraine, and Fernando Vidal. *The Moral Authority of Nature*. Chicago: University of Chicago Press. Forthcoming.

Deibel, Franz. *Dorothea Schlegel als Schriftstellerin im Zusammenhang mit der romantischen Schule*. Berlin: Mayer & Müller, 1905.

Dettelbach, Michael. "Humboldtian Science." In *Cultures of Natural History*, ed. N. Jardine, J. A. Secord, and E. C. Spary, pp. 287–304. Cambridge: Cambridge University Press, 1996.

Dilthey, Wilhelm. *Leben Schleiermachers*. 3rd ed. Vols. 13 and 14 of *Gesammelte Schriften*. Göttingen: Vandenhoeck & Ruprecht, 1991.

Dischner, Gisela. *Caroline und der Jenaer Kreis*. Berlin: Verlag Klaus Wagenback, 1979.

———, ed. *Friedrich Schlegels Lucinde und Materialien zu einer Theories des Müssiggangs*. Hildesheim: Gerstenberg Verlag, 1980.

Duchesneau, Francois. "Vitalism in Late Eighteenth-Century Physiology: The Cases of Barthez, Blumenbach and John Hunter." In *William Hunter and the 18th-Century Medical World*, ed. W. F. Bynum and Roy Porter, pp. 259–95. Cambridge: Cambridge University Press, 1985.

Düntzer, Heinrich, ed. *Zur deutschen Literatur und Geschichte: Ungedruckte Briefe aus Knebels Nachlass*. 2 vols. Nürnberg: Bauer und Raspe, 1858.

Dupre, John. *The Disorder of Things: Metaphysical Foundations of the Disunity of Science*. Cambridge: Harvard University Press, 1993.

Ecker, Alexander. *Lorenz Oken: Eine biographische Skizze*. Stuttgart: Schweizerbart'sche Verlagshandlung, 1880.

Eckermann, Johann Peter. *Gespräche mit Goethe in den letzten Jahren seines Lebens*. 3rd ed. Berlin: Aufbau-Verlag, 1987.

Eichendorff, Joseph von. *Joseph von Eichendorff Werke*. Ed. Hartwig Schultz. 6 vols. Frankfurt a. M.: Deutscher Klassiker Verlag, 1993.

Eissler, Kurt R. *Goethe: A Psychoanalytic Study, 1775–1786*. 2 vols. Detroit: Wayne State University Press, 1963.

Encyclopaedia Britannica. 9th ed. Boston: Little, Brown, 1878.

Engelberg, Günter. *Aus dem Leben des Dr. J. C. Reil: 19 Beilagen des General-Anzeiger Westrhauderfehn* (Westrhauderfehn, 22 November 1958–28 March 1959).

Engelhardt, Dietrich von. "Natur und Geist, Evolution und Geschichte: Goethe in seiner Beziehung zur romantischen Naturforschung und metaphysischen Naturphilosophie." In *Goethe und die Verzeitlichung der Natur*, ed. Peter Matussek, pp. 58–74. Munich: C. H. Beck, 1998.

———. "Schellings philosophische Grundlegung der Medizin." In *Natur und geschichtlicher Prozess: Studien zur Naturphilosophie F. W. J. Schellings*, ed. Hans Jörg Sandkühler, pp. 305–25. Frankfurt a. M.: Suhrkamp Verlag, 1984.

Eschenmayer, Karl. "Spontaneität = Weltseele oder über das höchste Princip der Naturphilosophie." *Zeitschrift für Speculative Physik* 2 (1801): 1-68.

Eulner, Hans-Heinz. "Johann Christian Reil: Leben und Werk." *Nova Acta Leopoldina (Abhandlungen der Deutschen Akademie der Naturforscher Leopoldina)*, n.s. 20 (1960): 7–50.

Fairley, Baker. *A Study of Goethe*. Oxford: Oxford University Press, 1947.

Fichte, Johann Gottlieb. *Fichtes Werke*. Ed. Immanuel Hermann Fichte. 11 vols. Berlin: Walter de Gruyter, 1971 [1834–46].

———. *Fichte: Early Philosophical Writings*. Trans. and ed. Daniel Breazeale. Ithaca: Cornell University Press, 1988.

———. *J. G. Fichte—Gesamtausgabe der Bayerischen Akademie der Wissenschaften*. Ed. Reinhard Lauth and Hans Jacob. Stuttgart: Friedrich Frommann Verlag, 1964–.

———. *J. G. Fichte im Gespräch*. Ed. Erich Fuchs. 6 vols. Stuttgardt: Bad Cannstatt: Frommann-Holzboog, 1980.

———. *J. G. Fichtes Werke*. 6 vols. Leipzig: Fritz Eckardt Verlag, 1911.

———. *Johann Gottlieb Fichte, Briefe*. Ed. Manfred Buhr. 2nd ed. Leipzig: Verlag Philip Reclam jun., 1986.

Fischer, Kuno. *Friedrich Wilhelm Joseph Schelling*. 2 vols. Vol. 6 of *Geschichte der neuern Philosophie*. Heidelberg: Carl Winter's Universitätsbuchhandlung, 1872–77.

Forberg, Friedrich Karl. *Fragmente aus meinen Papieren*. Jena: Voigt, 1796.

Förster, Eckhart. "'To Lend Wings to Physics Once Again': Hölderlin and the 'Oldest System-Programme of German Idealism.'" *European Journal of Philosophy* 3 (1995): 174–98.

Forster, Georg. *A Voyage Round the World, in His Britannic Majesty's Sloop, Resolution, Commanded by Capt. James Cook, during the Years 1772, 3, 4, and 5*. 2 vols. London: B. White, 1777.

Foucault, Michel. *The History of Sexuality*. Trans. Robert Hurley. 3 vols. New York: Vintage Books, 1990.

Frank, Manfred. *Unendliche Annäherung*. Frankfurt: a. M.: Suhrkamp, 1997.

Galen, Claudii. *Claudii Galen Opera Omnia*. Ed. D. Carolus Kühn. 20 vols. Leipzig, 1821–33.

Galvani, Luigi. *De viribus electricitatis in motu musculari commentarius* (Commentary on the forces of electricity in muscle movement). 1791. English trans. in Luigi Galvani, *Commentary on the Effects of Electricity on Muscular Motion*. Trans. Margaret Foley. Ed. I. Bernard Cohen. Norwalk, Conn.: Burndy Library, 1953.

Geiger, Ludwig. *Briefwechsel des jungen Börne und Henriette Herz*. Leipzig: Schulzesche Hof-Buchhandlung, 1905.

Gerabek, Werner. *Friedrich Wilhelm Joseph Schelling und die Medizin der Romantik*. Frankfurt a. M.: Peter Lang, 1995.

Ghiselin, Michael. "Darwin and Evolutionary Psychology." *Science* 179 (1973): 964–68.

Gillispie, Charles Coulston. *The Edge of Objectivity*. Princeton: Princeton University Press, 1960.

Gilman, Sander. *Inscribing the Other*. Lincoln: University of Nebraska Press, 1991.

Ginsborg, Hannah. "Kant on Aesthetic and Biological Purposiveness." In *Reclaiming the History of Ethics: Essays for John Rawls*, ed. Andrews Reath et al. Cambridge: Cambridge University Press, 1997.

Goethe, Johann Wolfgang von. *Briefe an Goethe* (Hamburger Ausgabe). Ed. Karl Robert Mandelkow. 3rd ed. 2 vols. Munich: C. H. Beck, 1988.

———. *Goethe-Briefe*. Ed. Philipp Stein. 8 vols. Berlin: Wertbuchhandel, 1924.

———. *Goethe in vertraulichen Briefen seiner Zeitgenossen*. Ed. Wilhelm Bode and Regine Otto. 2nd ed. 3 vols. Berlin: Aufbau-Verlag, 1982.

———. *Goethes Briefe* (Hamburger Ausgabe). Ed. Karl Robert Mandelkow. 4th ed. 4 vols. Munich: C. H. Beck, 1988.

———. *Goethes Briefe an Charlotte von Stein*. Ed. Julius Petersen. 4 vols. Leipzig: Insel Verlag, 1923.

———. *Goethes Gespräche*. Ed. Flodoard von Biedermann. Expanded by Wolfgang Herwig. 5 vols. Munich: Deutscher Taschenbuch Verlag, 1998.

———. *Goethes Werke* (Weimar Ausgabe). 133 vols. Weimar: Hermann Böhlau, 1888.

———. *Die Leiden des jungen Werther*. Munich: Deutscher Taschenbuch Verlag, 1978.

———. *Sämtliche Werke nach Epochen seines Schaffens* (Münchner Ausgabe). Ed. Karl Richter et al. 21 vols. Munich: Carl Hanser Verlag, 1985–98.

———. *Die Schriften zur Naturwissenschaft*. Ed. Dorothea Kuhn. 21 vols. Weimar: Hermann Böhlaus Nachfolger, 1977.

———. *Werke* (Hamburger Ausgabe). Ed. Erich Trunz et al. 14 vols. Munich: C. H. Beck, 1988.

Goldstein, Jan. *Console and Classify: The French Psychiatric Profession in the Nineteenth Century*. Cambridge: Cambridge University Press, 1987.

Grafton, Anthony. *Defenders of the Text: The Traditions of Scholarship in an Age of Science, 1450–1800*. Cambridge: Harvard University Press, 1991.

Gray, Ronald. *Goethe the Alchemist*. Cambridge: Cambridge University Press, 1952.

———. Letter to the *Times Literary Supplement*, no. 5056 (25 February 2000): 17.

Green, Joseph Henry. *Vital Dynamics: The Hunterian Oration before the Royal College of Surgeons in London, 14th February 1840*. London: William Pickering, 1840.

Gregory, Friedrich. "Hat Müller die Naturphilosophie wirklich aufgegeben?" In *Johannes Müller und die Philosophie*. Berlin: Akademie Verlag, 1992.

———. *Nature Lost? Natural Science and the German Theological Traditions of the Nineteenth Century*. Cambridge: Harvard University Press, 1992.

———. "Theology and the Sciences in the German Romantic Period." In *Romanticism and the Sciences*, ed. Andrew Cunningham and Nicholas Jardine, pp. 69–81. Cambridge: Cambridge University Press, 1990.

Grimm, Jacob, and Wilhelm Grimm. *Deutsches Wörterbuch*. 16 vols. in 32. Leipzig: Verlag von S. Hirzel, 1854–1952.

Haeckel, Ernst. "Ueber die Entwickelungstheorie Darwin's." In *Amtlicher Bericht über die acht und dreissigste Versammlung Deutscher Naturforscher und Ärzte in Stettin*. Stettin: Hessenland's Buchdruckerei, 1864.

Hagner, Michael. *Homo cerebralis: Der Wandel vom Seelenorgan zum Gehirn*. Berlin: Berlin Verlag, 1997.

Hahn, Karl-Heinz. "Zwei ungedruckte Briefe Goethes an Schelling." *Goethe, Neue Folge des Jahrbuchs* 19 (1957): 219–25.

Haller, Albrecht von. *Primae Lineae Physiologiae in usum Praelectionum Academicarum*. 4th ed. Lausanne: Grasset et Socios, 1771.

——. *Sur la formation du coeur dans le poulet; sur l'oeil; sur la structure du jaune &c.* 2 vols. Lausanne: Bousquet, 1758.

——. *Von den empfindlichen und reizbaren Teilen des menschlichen Körpers* (Klassiker der Medizin). Ed. Karl Sudhoff. Leipzig: Verlag von Johann Ambrosius Barth, 1922.

Handwerk, Gary. "Romantic Irony." In *The Cambridge History of Literary Criticism*. Vol. 5: *Romanticism*. Ed. Marshall Brown. Cambridge: Cambridge University Press, 2000.

Hankins, Thomas. "The Ocular Harpsichord of Louis-Bertrand Castel; or, The Instrument that Wasn't." *Osiris* 9 (1994): 141–43.

Hanov, Michael Christoph. *Philosophiae naturalis sive physicae dogmaticae: tomus III, continens geologiam, biologiam, phytologiam generalis*. Halle: Renger, 1766.

Hansen, Frank-Peter. *"Das älteste Systemprogram des Deutschen Idealismus": Rezeptionsgeschichte und Interpretation*. Berlin: Walter de Gruyter, 1989.

Hansen, LeeAnn. "From Enlightenment to *Naturphilosophie*: Marcus Herz, Johann Christian Reil, and the Problem of Border Crossings." *Journal of the History of Biology* 26 (1993): 39–64.

——. "Metaphors of Mind and Society: The Origins of German Psychiatry in the Prussian Reform Era." Paper delivered to the Fishbein Workshop in the History of the Human Sciences. The University of Chicago. 1994.

Hardenberg, Friedrich von [Novalis]. *Novalis Schriften*. Ed. Paul Kluckhohn and Richard Samuel. 2nd ed. 5 vols. Stuttgart: Kohlhammer, 1960–75.

——. *Novalis Schriften*. Ed. Ludwig Tieck and Friedrich Schlegel. 5th ed. 3 vols. Berlin: Reimer, 1837–46.

——. *Novalis Werke*. Ed. Gerhard Schulz. 3rd ed. Munich: Beck, 1989.

Hardy, Thomas. *The Essential Hardy*. Ed. Joseph Brodsky. Hopewell, N.J.: Ecco Press, 1995.

Harvey, William. *Exercitationes de generatione animalium*. London: DuGaidianis, 1651.

Hasler, Ludwig, ed. *Schelling, Seine Bedeutung für eine Philosophie der Natur und der Geschichte*. Stuttgart-Bad Cannstatt: Frommann-Holzboog, 1981.

Haym, Rudolf. *Die romantische Schule: Ein Beitrag zur Geschichte des deutschen Geistes*. Berlin: Rudolph Gaertner, 1870.

Hegel, Georg Wilhelm Friedrich. *Differenz des Fichte'schen und Schelling'schen Systems der Philosophie* (Jena: Seidler, 1801). Translated by H. S. Harris and Walter Cerf as *The Difference between Fichte's and Schelling's System of Philosophy*. Albany: State University of New York Press, 1977.

——. *Hegel's Philosophy of Right*. Trans. T. M. Knox. Oxford: Oxford University Press, 1967.

——. *Sämtliche Werke*. 5th ed. of the Jubiläumsausgabe. Stuttgart-Bad Cannstatt: Friedrich Frommann Verlag, 1977.

Heine, Heinrich. *Sämtliche Schriften*. Ed. Klaus Briegleb. 3rd ed. 6 vols. Munich: Deutscher Taschenbuch Verlag, 1997.

Helmholtz, Hermann von. *Goethes Vorahnungen kommender naturwissenschaftlicher Ideen.*
    Rede, gehalten in der Generalversammlung der Goethe-Gesellschaft zu Weimar den
    11 Juni 1892. Berlin: Verlag von Gebrüder Paetel, 1892.
———. "Ueber Goethes wissenschaftliche Arbeiten. Ein vortrag, gehalten in der
    deutschen Gesellschaft in Königsberg, 1853," *Allgemeine Monatsschrift für Wis-*
    senschaft und Literatur, 1853, pp. 383–98. Republished in *Populäre wissenschaft-*
    liche Vorträge. Vol. 1. Braunschwieg: Friedrich Vieweg und Sohn, 1865.
Henrich, Dieter. *The Course of Remembrance and Other Essays on Hölderlin.* Ed. Eckart
    Förster. Stanford: Stanford University Press, 1997.
Herder, Johann Gottfried von. *Herders Briefe,* ed. Wilhelm Dobbek. Weimar: Volksver-
    lag, 1959.
———. *Italienische Reise: Briefe und Tagebuchaufzeichnungen, 1788–1789,* ed. Albert Meier
    and Heide Hollmer. Munich: Deutscher Taschenbuch Verlag, 1988.
———. *Johann Gottfried Herder, Briefe.* Ed. Wilhelm Dobbek and G. Arnold, 10 vols.
    Weimar: Böhlaus, 1977–96.
———. *Johann Gottfried Herders Sprachphilosophie, aus dem Gesamtwerk ausgewählt.* Ed.
    Erich Heintel. Hamburg: Felix Meiner, 1975.
———. *Johann Gottfried Herder Werke.* Ed. Martin Bollacher et al. 10 vols. Frankfurt
    a. M.: Deutscher Klassiker Verlag, 1985–.
Herder, Maria Caroline von. *Erinnerungen aus dem Leben Joh. Gottfrieds von Herder.* 2
    vols. Tübingen: J. G. Cotta'schen Buchhandlung, 1820.
Hertz, Deborah. *Jewish High Society in Old Regime Berlin.* New Haven: Yale University
    Press, 1988.
Herz, Henriette. *Berliner Salon. Erinnerungen und Portraits.* Frankfurt: Verlag Ullstein,
    [1850] 1986.
Herz, Marcus. *Betrachtungen aus der spekulativen Weltweisheit.* Königsberg: Hohann Jakob
    Kanter, 1771.
———. *Versuch über den Schwindel.* Berlin: Christian Friedrich Voss und Sohn, 1786.
Hiebel, Friedrich. *Novalis: Deutscher Dichter, Europäischer Denker, Christlicher Seher.* 2nd
    ed. Bern: Francke Verlag, 1972.
Historical Institute of the Friedrich-Schiller University. *Geschichte der Universität Jena,*
    1548/58–1958. 2 vols. Jena: Gustav Fischer, 1958.
Hölderlin, Friedrich. *Sämtliche Werke.* 3rd ed. 4 vols. Berlin: Propyläen-Verlag, 1943.
Holstein, Madame de Staël. *De l'Allemagne.* 3 vols. Paris: H. Nicolle, 1810 [London:
    Murray, 1813].
Hufeland, Christian Wilhelm. *Ideen über Pathogenie und Einfluss der Lebenskraft auf*
    Entstehung und Form der Krankheiten. Jena: Akademische Buchhandlung, 1795.
Humboldt, Alexander von. *Ansichten der Natur.* 3rd ed. Stuttgart: Cotta'schen Buch-
    handlung, 1849. Translated by E. C. Otté and Henry Bohn as *Views of Nature, or*
    Contemplations on the Sublime Phenomena of Creation. London: Henry Bohn,
    1850.
———. *Beobachtungen aus der Zoologie und vergleichenden Anatomie: gesammelt auf einer*
    Reise nach den Tropen-Ländern des neuen Kontinents. Tübingen: F. G. Cotta, 1806.
———. *Florae Fribergensis specimen, plantas cryptogamicas praesertim subterraneas exhibens.*
    Berolini: H. A. Rottman, 1793.
———. *Ideen einer Geographie der Pflanzen.* Tübingen: F. G. Cotta, 1807.
———. *Ideen zu einer Physiognomik der Gewächse.* Tübingen: Cotta, 1806.
———. *Die Jugendbriefe Alexander von Humboldts, 1787–1799.* Ed. Ilse Jahn and Fritz
    Lange. Berlin: Akademie-Verlag, 1973.
———. *Kosmos, Entwurf einer physischen Weltbeschreibung.* 5 vols. Stuttgart: Cotta'scher

Verlag, 1845–58. Translated by E. C. Otté as *Cosmos: A Sketch of a Physical Description of the Universe*. 5 vols. New York: Harper, n.d.

———. "Die Lebenskraft oder der rhodische Genius, ein Erzählung." *Die Horen*. Part 5 (1795): 90–96.

———. *Versuche über die gereizte Muskel- und Nervenfaser, nebst Vermuthungen über den chemischen Process des Lebens in der Thier- und Pflanzenwelt*. 2 vols. Berlin: Heinrich August Rottmann, 1797–99.

Humboldt, Alexander von, and Aimé Bonpland. *Personal Narrative of Travels to the Equinoctial Regions of the New Continent, during the Years 1799–1804*. Trans. Helen Williams. 7 vols. London: Longman, Hurst, Rees, Orme, and Brown, 1818–29.

Huxley, Thomas Henry. *Collected Essays*. London: McMillan, 1893.

Immerwahr, Raymond. "The Word 'romantisch' and Its History." In *The Romantic Period in Germany*, ed. Siegbert Prawer, pp. 34–63. London: Weidenfeld and Nicolson, 1970.

Jacobi, Friedrich Heinrich. *Friedrich Heinrich Jacobi Briefwechsel*. Ed. Michael Brüggen, Heinz Gockel, and Peter-Paul Schneider. 3 vols. Stuttgart: Friedrich Frommann Verlag, 1987.

———. *Über die Lehre des Spinoza in Briefen an den Herrn Moses Mendelssohn*. Breslau, 1785. Ed. Marion Lauschke. Hamburg: Felix Meiner Verlag, 2000.

Jackson, Myles. "Natural and Artificial Budgets: Accounting for Goethe's Economy of Nature." *Science in Context* 7 (1994): 409–31.

Jaeger, Georg. "Ehrengedächtniss der Königl. Würtembergischen Staatsraths von Kielmeyer." *Novorum Actorum Academiae Caesareae Leopoldino-Carolinae Naturae Curiosorum* (*Verhandlungen der Kaiserlichen Leopoldinisch-Carolinischen Akademie der Naturforscher*) 13, 2nd part (1845).

Jantzen, Jörg. *Physiologische Theorien*. In *Friedrich Wilhelm Joseph Schelling, Ergänzungsband zu Werke Band 5 bis 9: Wissenschaftshistorischer Bericht zu Schellings naturphilosophischen Schriften, 1797–1800*. Stuttgart: Frommann-Holzboog, 1994.

Jardine, Nicholas. *The Scenes of Inquiry*. Oxford: Clarendon Press, 1991.

Jardine, N., J. A. Secord, and E. C. Spary, eds. *Cultures of Natural History*. Cambridge: Cambridge University Press, 1996.

Johnson, Samuel. *A Dictionary of the English Language*. 1756. Abridged from H. J. Todd's corrected ed., reedited by Alexander Chalmers. Philadelphia: Kimber and Sharpless, 1842.

Kant, Immanuel. *Briefwechsel von Imm. Kant in drei Bänden*. Ed. H. E. Fischer. 3 vols. Munich: Georg Müller, 1912.

———. *Immanuel Kant, Briefwechsel*. Ed. Otto Schöndörffer. Hamburg: Felix Meiner Verlag, 1972.

———. *Immanuel Kant Werke*. Ed. Wilhelm Weischedel, 6 vols. Wiesbaden: Insel Verlag, 1957.

Kanz, Kai. *Kielmeyer-Bibliographie*. Stuttgart: Verlag für Geschichte der Naturwissenschaften und der Technik, 1991.

———. "Zur Frühgeschichte des Begriffs 'Biologie,' die botanische Biologie von Johann Jakob Planer." *Verhandlungen zur Geschichte und Theorie der Biologie*, 5 (2000): 269–82.

———, ed. *Philosophie des Organischen in der Goethezeit: Studien zu Werk und Wirkung des Naturforschers Carl Friedrich Kielmeyer (1765–1844)*. Stuttgart: Franz Steiner Verlag, 1994.

Kauffmann, Doris. *Aufklärung, bürgerliche Selbsterfahrung, und die 'Erfindung' der Psychiatrie in Deutschland 1770–1850*. Göttingen: Vandenhoeck and Ruprecht, 1995.

Kielmeyer, Carl Friedrich von. *Gesammelte Schriften*. Ed. F.-H. Holler. Berlin: W. Keiper, 1938.

———. *Ueber die Verhältnisse der organischen Kräfte*. Introduction by Kai Torsten Kanz. Marburg an der Lahn: Basilisken-Presse, 1993.

———. "Ueber die Verhältnisse der organischen Kräfte unter einander in der Reihe der verschiedenen Organisationen, die Gesetze und Folgen dieser Verhältnisse." *Sudhoffs Archiv für Geschichte der Medizin* 23 (1930): 247–67.

Kluckhohn, Paul. *Das Ideengut der deutschen Romantik*. 3rd ed. Tübingen: Niemeyer Verlag, 1953.

Knight, David. "Romanticism and the Sciences." In *Romanticism and the Sciences*, ed. Andrew Cunningham and Nicholas Jardine, pp. 13–24. Cambridge: Cambridge University Press, 1990.

Knight, Richard Payne. *An Account of the Remains of the Worship of Priapus*. London: Spilsbury, 1786.

Kohn, David. "The Aesthetic Construction of Darwin's Theory." In *Aesthetics and Science: The Elusive Synthesis*, ed. Alfred Tauber. Dordrecht: Kluwer, 1996.

Komar, Kathleen. "Fichte and the Structure of Novalis's 'Hymnen an die Nacht.'" *German Review* 54 (1979): 137–44.

Körner, Josef. *Romantiker und Klassiker: Die Brüder Schlegel in ihren Beziehungen zu Schiller und Goethe*. Berlin: Askanischen Verlag, 1924.

Kramer, Cheryce. "Illenau, Château de Plaisir." Paper delivered at the Fishbein Workshop in the History of Human Sciences. University of Chicago, 1994.

———. "The Psychic Constitution: Psychiatry in Early 19th Century Germany." Master's thesis, University of Cambridge, 1991.

Kuhn, Dorothea. "Uhrwerk oder Organismus, Karl Friedrich Kielmeyers System der organischen Kräfte." *Nova Acta Leopoldina* (*Abhandlungen der Deutschen Akademie der Naturforscher Leopoldina*), n.s. 36, no. 198 (1970): 157–67.

Kuzniar, Alice, ed. *Outing Goethe and His Age*. Stanford: Stanford University Press, 1996.

Larmore, Charles. "Hölderlin and Novalis." In *The Cambridge Companion to German Idealism*, ed. Karl Ameriks, pp. 141–60. Cambridge: Cambridge University Press, 2000.

———. *Romantic Legacy*. New York: Columbia University Press, 1996.

Larson, James. *Interpreting Nature: The Science of Living Form from Linnaeus to Kant*. Baltimore: Johns Hopkins University Press, 1994.

Lauth, Reinhard. *Die Enstehung von Schellings Identitätsphilosophie in der Auseinandersetzung mit Fichtes Wissenschaftslehre* (*1795–1801*). Freiburg: Verlag Karl Alber, 1975.

Lenoir, Timothy. "The Eternal Laws of Form: Morphotypes and the Conditions of Existence in Goethe's Biological Thought." In *Goethe and the Sciences: A Reappraisal*, ed. Frederick Amrine, Francis Zucker, and Harvey Wheeler. Dordrecht: Reidel, 1987.

———. "The Göttingen School and the Development of Transcendental Naturphilosophie in the Romantic Era." *Studies in History of Biology* 5 (1981): 111–205.

———. "Kant, Blumenbach, and Vital Materialism in German Biology." *Isis* 71 (1980): 77–108.

———. *The Strategy of Life: Teleology and Mechanics in Nineteenth-Century German Biology*. 2nd ed. Chicago: University of Chicago Press, 1989.

Levere, Trevor. "Romanticism, Natural Philosophy, and the Sciences: A Review and Bibliographic Essay." *Perspectives on Science* 4 (1996): 463–88.

Linné, Carl von. *Bibliotheca botanica*. Munich: Fritsch, [1736] 1968.

Lovejoy, Arthur. "On the Meaning of 'Romantic' in Early German Romanticism." *Modern Language Notes* 31 (1916): 385–96; 32 (1917): 65–77.

Löw, Reinhard. *Philosophie des Lebendigen: Der Begriff des Organischen bei Kant, sein Grund und seine Aktualität*. Frankfurt a. M.: Suhrkamp Verlag, 1980.

Lyell, Charles. *Principles of Geology*. 3 vols. London: Murray, 1830–33.

Mackintosh, James. *Dissertation on the Progress of Ethical Philosophy*. Edinburgh: Adam and Charles Black, 1836.

Maisak, Petra. "Wir Passen Zusammen als Hätten Wir Zusammen Gelebt." In *J. H. W. Tischbein, Goethes Maler und Freund*, ed. Hermann Mildenberger. Neumünster: Karl Wachholtz Verlag, 1986.

Mandelkow, Karl, ed. *Goethe im Urteil seiner Kritiker*. 5 vols. Munich: C. H. Beck, 1975.

Marchant, James, ed. *Alfred Russel Wallace: Letters and Reminiscences*. 2 vols. London: Castile, 1916.

Martin, Wayne. *Idealism and Objectivity: Understanding Fichte's Jena Project*. Stanford: Stanford University Press, 1997.

Matussek, Peter, ed. *Goethe und die Verzeitlichung der Natur*. Munich: C. H. Beck, 1998.

Mayr, Ernst. *The Growth of Biological Thought*. Cambridge: Harvard University Press, 1982.

McLaughlin, Peter. "Blumenbach und der Bildungstrieb." *Medizin historisches Journal* 17 (1982): 357–72.

Meckel, Johann Friedrich. *Beyträge zur vergleichenden Anatomie*. 2 vols. in 3. Leipzig: Carl Heinrich Reclam, 1812.

Michler, Markwart. *Medizin zwischen Aufklärung und Romantik: Melchior Adam Weikard (1742–1803) und sein Weg in den Brownianismus*. In *Acta historica Leopoldina*, no. 24. Halle: Deutsche Akademie der Naturforscher Leopoldina, 1995.

Mill, John Stuart. *Mill on Bentham and Coleridge*. Intro. by F. R. Leavis. London: Chatto & Windus, 1950.

Mocek, Reinhard. *Johann Christian Reil (1759–1813)*. Frankfurt a. M.: Peter Lang, 1995.

Mommsen, Wilhelm. *Die Politischen Anschauungen Goethes*. Stuttgart: Deutsche Verlags-Anstalt, 1948.

Müller-Sievers, Helmut. "Skullduggery: Goethe and Oken, Natural Philosophy and Freedom of the Press." *Modern Language Quarterly* 59 (1998): 231–59.

Nagel, Thomas. "How to Be Free and Happy." *New Republic* 224 (7 May 2001): 30–34.

Neuburger, Max. *Johann Christian Reil: Gedenkrede*. Stuttgart: Verlag von Ferdinand Enke, 1913.

Neuhouser, Frederick. *Fichte's Theory of Subjectivity*. Cambridge: Cambridge University Press, 1990.

Newton, Isaac. *Mathematical Principles of Natural Philosophy*. 1729. Trans. Andrew Motte. Ed. Florian Cajori. 2 vols. Berkeley: University of California Press, 1962.

Nisbet, Hugh. "Historisierung: Naturgeschichte und Humangeschichte bei Goethe, Herder und Kant." In *Goethe und die Verzeitlichung der Natur*, ed. Peter Matussek, pp. 15–43. Munich: C. H. Beck, 1998.

Nordenskiöld, Erik. *The History of Biology*. New York: Tudor, 1936.

O'Brien, William Arctander. *Novalis: Signs of Revolution*. Durham, N.C.: Duke University Press, 1995.

Oken, Lorenz. *Über die Bedeutung der Schädelknochen: Ein Programm beim Antritt der Professur an der Gesammt-Universität zu Jena*. Jena: Göpferdt, 1807.

———. *Die Zeugung*. Bamberg: Goebhardt, 1805.

———. Review of *Ueber die Bedeutung der Schädelknochen*, *Isis* 1 (1817): 1204–8.

———. [Discussion of *Ueber die Bedeutung der Schädelknochen*], *Isis* 1 (1818): 510–11.

Oken, Lorenz, and Dietrich Georg Kieser. *Beiträge zur vergleichenden Zoologie, Anatomie und Physiologie*. 2 vols. Bamberg: Goebhardt, 1806–7.

Owen, Richard. *The Hunterian Lectures in Comparative Anatomy, May and June 1837*. Ed. with introduction and commentary by Phillip Sloan. Chicago: University of Chicago Press, 1992.

———. *On the Nature of Limbs, a Discourse delivered on Friday, February 9, at an Evening Meeting of the Royal Institution of Great Britain*. London: John Van Voorst, 1849.

———. *Report on the Archetype and Homologies of the Vertebrate Skeleton*. In *Report of the Sixteenth Meeting of the British Association for the Advancement of Science; held at Southampton in September 1846*. London: Murray, 1847.

Paley, William. *Moral and Political Philosophy*. In *The Works of William Paley*. Philadelphia: Woodward, n.d.

Pander, Christian, and Eduard d'Alton. *Das Riesen-Faultier Bradypus Giganteus*. Bonn: Weber, 1821.

Panofsky, Erwin. *Meaning in the Visual Arts*. Chicago: University of Chicago Press, [1955] 1982.

Park, Katherine. "Nature in Person: Medieval and Renaissance Allegories and Emblems." In *The Moral Authority of Nature*, ed. Lorraine Daston and Fernando Vidal. Chicago: University of Chicago Press, forthcoming.

Pera, Marcello. *The Ambiguous Frog: The Galvani-Volta Controversy on Animal Electricity*. Trans. Jonathan Mandelbaum. Princeton: Princeton University Press, 1992.

Peter, Klaus. *Friedrich Schlegel*. Stuttgart: Metzlersche Verlagsbuchhandlung, 1978.

Pikulik, Lothar. *Frühromantik: Epoche, Werke, Wirkung*. Munich: C. H. Beck, 1992.

Pinel, Philippe. *Philosophisch-Medicinische Abhandlung über Geistesverirrungen oder Manie*. Trans. Michael Wagner. Vienna: Carl Schaumburg und Compagnie, 1801.

———. *Traité médico-philosophique sur l'aliénation mentale, ou la Manie*. Paris: Richard, Caille et Ravier, 1801.

Pippin, Robert. "Fichte's Alleged Subjective, Psychological, One-Sided Idealism." In *The Reception of Kant's Critical Philosophy*, ed. Sally Sedgwick, pp. 147–70. Cambridge: Cambridge University Press, 2000.

———. "Fichte's Contribution." *Philosophical Forum* 19 (1987–88): 74–96.

———. *Idealism as Modernism*. Cambridge: Harvard University Press, 1997.

Planer, Johann. *Versuch einer teutschen Nomenclatur der Linneischen Gattungen*. Erfurt: Müller, 1771.

Plitt, Gustav. *Aus Schellings Leben. In Briefen*. 3 vols. Leipzig: S. Hirzel, 1869–70.

Prawer, Siegbert, ed. *The Romantic Period in Germany*. London: Weidenfeld and Nicolson, 1970.

Reil, Johann Christian. *Entwurf einer allgemeinen Pathologie*. 3 vols. Halle: Curtschen Buchhandlung, 1815–16.

———. *Kleine Schriften wissenschaftlichen und gemeinnützigen Inhalts*. Halle: Curtschen Buchhandlung, 1817.

———. "An die Professoren Herrn Gren und Herrn Jakob in Halle." *Archiv für die Physiologie* 1 (1795): 3.

———. "Rezensionen." *Archiv für die Physiologie* 1 (1796): 178–92; 2 (1797): 149–52.

————. *Rhapsodieen über die Anwendung der psychischen Curmethode auf Geisteszerrüttungen.* Halle: Curtschen Buchhandlung, 1803.

————. "Ueber den Begriff der Medicin und ihre Verzweigungen, besonders in Beziehung auf die Berichtigung der Topik der Psychiaterie." *Beyträge zur Beförderung einer Kurmethode auf Psychischem Wege* 1 (1808): 161–279.

————. *Ueber die Erkenntniss und Cur der Fieber.* 5 vols. Halle: Curtschen Buchhandlung, 1799–1815.

————. "Ueber das polarische Auseinanderweichen der ursprünglichen Naturkräfte in der Gebärmutter zur Zeit der Schwangerschaft, und deren Umtauschung zur Zeit der Geburt, als Beytrag zur Physiologie der Schwangerschaft und Geburt." *Archiv für die Physiologie* 7 (1807): 402–501.

————. "Von der Lebenskraft," *Archiv für die Physiologie* 1, no. 1 (1795): 66–67. [The title page of the first volume carries the date 1796.]

Reinhold, Karl Leonhard. *Versuch einer neuen Theorie des menschlichen Vorstellungsvermögens.* Jena: C. Widtmann and I. M. Mauke, 1789.

Richards, Robert. *Darwin and the Emergence of Evolutionary Theories of Mind and Behavior.* Chicago: University of Chicago Press, 1987.

————. "The Linguistic Creation of Man: Charles Darwin, August Schleicher, Ernst Haeckel, and the Missing Link in 19th-Century Evolutionary Theory." In *Experimenting in Tongues: Studies in Science and Language,* ed. Matthias Doerres. Stanford: Stanford University Press, 2002.

————. *The Meaning of Evolution: The Morphological Construction and Ideological Reconstruction of Darwin's Theory.* Chicago: University of Chicago Press, 1992.

Riemer, Friedrich Wilhelm. *Mitteilungen über Goethe.* Ed. Arthur Pollmer. Leipzig: Insel Verlag, [1841] 1921.

Risse, Günter. "Kant, Schelling, and the Early Search for a Philosophical 'Science' of Medicine in Germany." *Journal of the History of Medicine* 27 (1972).

Robinson, Henry Crabb. *Diary, Reminiscences, and Correspondence.* Ed. Thomas Sadler. 3 vols. London: Macmillan, 1869.

Roe, Shirley. *Matter, Life, and Generation: 18th-Century Embryology and the Haller-Wolff Debate.* Cambridge: Cambridge University Press, 1981.

Roose, Theodor Georg. *Grundzüge der Lehre von der Lebenskraft.* Braunschweig: Thomas, 1797.

Rupke, Nicholaas. *Richard Owen, Victorian Naturalist.* New Haven: Yale University Press, 1994.

Salmen, Walter. *Johann Friedrich Reichardt: Komponist, Schriftsteller, Kapellmeister und Verwaltungsbeamter der Goethezeit.* Freiburg i. Br.: Atlantis Verlag, 1963.

Samuel, Richard. "Friedrich Schlegel's and Friedrich von Hardenberg's Love Affairs in Leipzig." In *Festschrift for Ralph Farrell,* ed. Anthony Stephens et al., pp. 47–56. Bern: Peter Lang, 1977.

Sandkühler, Hans Jörg. *Natur und geschichtlicher Prozess: Studien zur Naturphilosophie F.W.J. Schellings.* Ed. Hans Jörg Sandkühler. Frankfurt a. M.: Suhrkamp Verlag, 1984.

Schaffer, Simon. "Genius in Romantic Natural Philosophy." In *Romanticism and the Sciences,* ed. Andrew Cunningham and Nicholas Jardine, pp. 82–100. Cambridge: Cambridge University Press, 1990.

Scharf, Joachim-Hermann. "Johann Christian Reil als Anatom." *Nova Acta Leopoldina* (*Abhandlungen der Deutschen Akademie der Naturforscher Leopoldina*), n.s. 20 (1960): 51–97.

Schelling, Friedrich Wilhelm Joseph. *Briefe und Dokumente*. Ed. Horst Fuhrmans. 3 vols. to date. Bonn: Bouvier Verlag, 1962–.

———. *Friedrich Wilhelm Joseph Schelling Historisch-Kritische Ausgabe*. Ed. Hans Baumgartner et al. 5 vols. to date. Stuttgart: Frommann-Holzboog, 1976–.

———. *Sämtliche Werke*. Ed. K. F. A. Schelling. 14 vols. Stuttgart: Cotta'scher Verlag, 1857.

———. *Schelling als Persönlichkeit: Briefe, Reden, Aufsätze*. Ed. Otto Braun. Leipzig: Fritz Eckardt Verlag, 1908.

———. *Schellings Werke*. Ed. Manfred Schröter. 3rd ed. 12 vols. Munich: C. H. Beck, 1927–59.

Schiller, Friedrich. *Friedrich Schiller: Sämtliche Gedichte*. Frankfurt a. M.: Insel Verlag, 1994.

———. *Schiller Werke und Briefe*. Ed. Otto Dann et al. 12 vols. Frankfurt a. M.: Deutscher Klassiker Verlag, 1988–.

———. *Schillers Werke* (Nationalausgabe). Ed. Julius Petersen et al. 55 vols. Weimar: Böhlaus Nachfolger, 1943–.

Schlegel, August Wilhelm. *An das Publicum. Rüge einer in der Jenaischen Allg. Literatur-Zeitung begangen Ehrenschändung*. Tübingen: J. G. Cotta'schen Buchhandlung, 1802.

———. *Briefe von und an August Wilhelm Schlegel*. Ed. Josef Körner. 2 vols. Zurich: Amalthea, 1930.

Schlegel, Friedrich. *Friedrich Schlegel, 1794–1802: Seine prosaischen Jugendschriften*. Ed. Jacob Minor. 2 vols. Vienna: Carl Konegen, 1882.

———. *Friedrich Schlegels Briefe an seinen Bruder August Wilhelm*. Ed. Oskar Walzel. Berlin: Speyer & Peters, 1890.

———. *Kritische Friedrich-Schlegel-Ausgabe*. Ed. Ernst Behler et al. 35 vols. to date. Paderborn: Verlag Ferdinand Schöningh, 1958–.

———. *Kritische Schriften und Fragmente*. Ed. Ernst Behler and Hans Eichner. 6 vols. Paderborn: Ferdinand Schöningh, 1988.

———. *Transcendentalphilosophie*. Ed. Michael Elsässer. Hamburg: Felix Meiner Verlag, 1991.

Schleiden, Matthias Jakob. *Schellings und Hegels Verhältniss zur Naturwissenschaft. (Als Antwort auf de Angriffe des Herrn Nees von Esenbeck in der Neuen Jenaer Lit.-Zeitung, Mai 1843, insbesondere für die Leser dieser Zeitschrift.)* Leipzig: Wilhelm Engelmann, 1844.

Schleiermacher, Friedrich. *Friedrich Daniel Ernst Schleiermacher, Kritische Gesamtausgabe*. Ed. Hans-Joachim Birkner et al. 11 vols. to date. Berlin: Walter de Gruyter, 1980–.

———. *Der Christliche Glaube*. 2nd ed. 2 vols. Berlin: Walter De Gruyter, [1830] 1960.

———. *Monologen*. Berlin: Reimer, 1800.

———. *Monologen, Neujahrspredigt von 1792, und Über den Wert des Lebens*. Ed. Friedrich Schiele. Hamburg: Felix Meiner Verlag, 1978.

———. *Schleiermacher als Mensch: Sein Werden und Wirken, Familien- und Freundesbriefe*. Ed. Heinrich Meisner. Gotha: Friedrich Andreas Perthes, 1922.

———. *Schleiermacher/Briefe*. Jena: Eugen Diederichs, 1906.

Schrenk, Martin. *Über den Umgang mit Geisteskranken*. Berlin: Springer, 1973.

Schüddekopf, Carl, and Oskar Walzel. *Goethe und die Romantik: Briefe mit Erläuterungen*. Vol. 13 of *Schriften der Goethe-Gesellschaft* (1898): L–LI.

Schulz, Gerhard. *Novalis, mit Selbstzeugnissen und Bilddokumente*. 3rd ed. Reinbek bei Hamburg: Rowohlt, 1993.

Sedgwick, Sally, ed. *The Reception of Kant's Critical Philosophy: Fichte, Schelling, and Hegel.* Cambridge: Cambridge University Press, 2000.

Seidel, Siegfried. *Der Briefwechsel zwischen Schiller und Goethe.* 3 vols. Munich: Verlag C. H. Beck, 1984.

Shaffer, Elinor. "Romantic Philosophy and the Organization of the Disciplines: The Founding of the Humboldt University of Berlin." In *Romanticism and the Sciences,* ed. Andrew Cunningham and Nicholas Jardine, pp. 38–54. Cambridge: Cambridge University Press, 1990.

Shaw, George Bernard. *Back to Methuselah.* 1921. London: Penguin Books, 1961.

Sheehan, James J. *German History, 1770–1866.* Oxford: Clarendon Press, 1989.

Shelley, Mary. *Frankenstein; or the Modern Prometheus.* In *The Complete Frankenstein,* ed. Leonard Wolf. New York: Penguin, 1993.

Slatkin, Laura. "Measuring Authority, Authoritative Measures: Hesiod's Works and Days." In *The Moral Authority of Nature,* ed. Lorraine Daston and Fernando Vidal. Chicago: University of Chicago Press, forthcoming.

Sloan, Phillip. "Buffon, German Biology, and the Historical Interpretation of Biological Species." *British Journal for the History of Science* 12 (1979): 109–53.

———. "Darwin, Vital Matter, and the Transformism of Species." *Journal of the History of Biology* 19 (fall 1986): 369–445.

Snelders, H. A. M. *Weltenschap en Intuïtie: Het Duitse romantisch-speculatief natuuronderzoek rond 1800.* Baarn: Uitgeverij Ambo, 1994.

———. "Romanticism and Naturphilosophie." *Studies in Romanticism* 9 (1970): 193–215.

Snow, Dale. *Schelling and the End of Idealism.* Albany: State University of New York Press, 1996.

Soemmerring, Samuel Thomas. *Abbildung des menschlichen Hörorgrans.* Frankfurt a. M.: Varrentrapp und Wenner, 1806.

———. *Samuel Thomas Soemmerring Werke.* Ed. Jost Benedum and Werner Kümmel. 21 vols. Stuttgart: Gustav Fischer Verlag, 1997–.

———. *Vom Baue des menschlichen Körpers. Erster Theil. Knochenlehre,* 2nd ed. Frankfurt a. M.: Varrentrapp und Wenner, 1800.

Spinoza, Benedicti De. *Opera.* Ed. J. Van Vloten and J. Land. 3 vols. Hague: Martinus Nijhoff, 1890.

Spix, Johann Baptist von. *Cephalogenesis; sive, Capitis ossei structura, formatio et significatio per omnes animalium classes, familias, genera ac aetates digesta.* Monachii: F. S. Hübschmannii, 1815.

Staiger, Emil, ed. *Der Briefwechsel zwischen Schiller und Goethe.* Frankfurt a. M.: Insel Verlag, 1966.

Steffens, Henrik. *Johann Christian Reil: Eine Denkschrift.* Halle: Curtschen Buchhandlung, 1815.

———. *Schriften, Alt und Neu.* 2 vols. Breslau: Josef Max, 1821.

———. *Was Ich Erlebte.* 10 vols. Breslau: Josef Max, 1840–44.

Steiger, Robert. *Goethes Leben von Tag zu Tag.* 8 vols. to date. Zurich: Artemis Verlag, 1982–.

Steigerwald, Joan. *Lebenskraft in Reflection: German Perspectives of the Late Eighteenth and Early Nineteenth Centuries.* Forthcoming.

Stephens, Anthony, et al., eds. *Festschrift for Ralph Farrell.* Bern: Peter Lang, 1977.

Stephenson, R. H. *Goethe's Conception of Knowledge and Science.* Edinburgh: Edinburgh University Press, 1995.

Sternberg, A. von. *Erinnerungsblätter.* Berlin: Schindler, 1855.

Stoljar, Margaret. *Athenaeum: A Critical Commentary*. Bern: Verlag Herbert Lang, 1973.

Sturma, Dieter. "The Nature of Subjectivity: The Critical and Systematic Function of Schelling's Philosophy of Nature." In *The Reception of Kant's Critical Philosophy: Fichte, Schelling, and Hegel*. Ed. Sally Sedgwick, pp. 216–31. Cambridge: Cambridge University Press, 2000.

Strickland, Stuart. "Circumscribing Science: Johann Wilhelm Ritter and the Physics of Sidereal Man." Ph.D. diss., Harvard University, 1992.

Swammerdam, Jan. *Historia Insectorum Generalis, Ofte Algemeene Verhandeling Van De Bloedeloose Dierkens*. Utrecht: Van Drevnen, 1669.

———. *Historia insectorum generalis*. Trans. H. Henninius. Holland: Luchtmans, 1685.

Tauber, Alfred, ed. *Aesthetics and Science: The Elusive Synthesis*. Dordrecht: Kluwer, 1996.

Tilliette, Xavier, ed. *Schelling im Spiegel seiner Zeitgenossen*. 6 vols. Torino: Bottega D'Erasmo, 1974–81.

Tobin, Robert. "In and Against Nature: Goethe on Homosexuality and Heterotextuality." In *Outing Goethe and His Age*, ed. Alice Kuzniar, pp. 94–110. Stanford: Stanford University Press, 1996.

Treviranus, Gottfried Reinhold. *Biologie, oder Philosophie der lebenden Natur*. Göttingen: Johann Friedrich Röwer, 1802–22.

Tsouyopoulos, Nelly. *Andreas Röschlaub und die romantische Medizin*. Stuttgart: Gustav Fischer Verlag, 1982.

Unseld, Siegfried. *Goethe and His Publishers*. Trans. Kenneth Northcott. Chicago: University of Chicago Press, 1996.

Vesalius, Andreas. *De humani corporis fabrica*. Basel: Ioannus Oporinus, 1543.

Vicq-d'Azyr, Félix. *Oeuvres de Vicq-D'Azyr*. Ed. J. L. Moreau. 6 vols. Paris: L. Duprat-Duverger, 1805.

Vorländer, Karl. *Kant-Schiller-Goethe*. 2nd ed. Aalen: Scientia Verlag, [1923] 1984.

Wagner, Rudolph. *Samuel Thomas von Soemmerrings Leben und Verkehr mit seinen Zeitgenossen*. 2 vols. Stuttgart: Gustav Fischer Verlag, [1844] 1986.

Wahl, Hans, ed. *Briefwechsel des Herzogs-Grossherzogs Carl August mit Goethe*. 2 vols. Berlin: Ernst Siefgried Mittler und Sohn, 1915.

Waitz, Georg, ed. *Caroline: Briefe an ihre Geschwister, ihre Tochter Auguste, die Familie Gotter, F. L. W. Meyer, A. W. und Fr. Schlegel, J. Schelling u. a.* 2 vols. Leipzig: S. Hirzel, 1871.

Waitz, Georg, and Erich Schmidt, eds. *Caroline: Briefe aus der Frühromantik*. 2 vols. Leipzig: Insel Verlag, 1913.

Wallace, Alfred Russel. "Review of *Principles of Geology* and *Elements of Geology* by Charles Lyell." *Quarterly Review* 120 (1869): 359–94.

Walzel, Oskar. *German Romanticism*. Trans. Alma Lussky. New York: Putnam's Sons, 1932.

———, ed. *Friedrich Schlegels Briefe an seinen Bruder August Wilhelm*. Berlin: Verlag von Speyer & Peters, 1890.

Webster, Charles, ed. *Biology, Medicine and Society, 1840–1940*. Cambridge: Cambridge University Press, 1981.

Weindling, Paul. "Theories of the Cell State in Imperial Germany." In *Biology, Medicine and Society, 1840–1940*, ed. Charles Webster, pp. 99–156. Cambridge: Cambridge University Press, 1981.

Wellbery, David. *The Specular Moment: Goethe's Early Lyric and the Beginnings of Romanticism*. Cambridge: Harvard University Press, 1996.

Wellek, René. *Discriminations: Further Concepts of Criticism*. New Haven: Yale University Press, 1970.

Wells, George. *Goethe and the Development of Science, 1750–1900*. Alphen aan den Rijn: Sijthoff & Noordhoff, 1978.

Wenzel, Manfred, ed. *Goethe und Soemmerring: Briefwechsel 1784–1828*. Stuttgart: Gustav Fischer Verlag, 1988.

Werner, Petra. *Übereinstimmung oder Gegensatz? Zum widersprüchlichen Verhältnis zwischen A. v. Humboldt und F. W. J. Schelling*. Berlin: Alexander-von-Humboldt-Forschungsstelle, 2000.

Wessel, Leonard. "The Antinomic Structure of Friedrich Schlegel's 'Romanticism.'" *Studies in Romanticism* 12 (1973): 648–69.

Wetzels, Walter. "Aspects of Natural Science in German Romanticism." *Studies in Romanticism* 10 (1971): 44–59.

———. "Organicism and Goethe's Aesthetics." In *Approaches to Organic Form*, ed. Frederick Burwick, pp. 71–85. Dortrecht: Reidel, 1987.

Whewell, William. *History of the Inductive Sciences*. 3rd ed. 3 vols. London: Parker and Son, 1857.

White, Alan. *Schelling: Introduction to the System of Freedom*. New Haven: Yale University Press, 1983.

Whytt, Robert. *Observations on the Sensibility and Irritability of the Parts of Men and other Animals Occasioned by the Celebrated M. De Haller's late Treatise on those Subjects*. In *The Works of Robert Whytt, M.D. Published by his Son*. Edinburgh: Becket and DeHondt, 1768.

Wiesing, Urban. "Der Tod der Auguste Böhmer: Chronik eines medizinischen Skandals, seine Hintergründe und seine historische Bedeutung." *History and Philosophy of the Life Sciences* 11 (1989): 275–95.

Williams, John. *The Life of Goethe: A Critical Biography*. Oxford: Blackwell, 1998.

Wilson, Edward O. *Consilience*. New York: Knopf, 1998.

Wilson, W. Daniel. *Das Goethe-Tabu: Protest und Menschenrechte im klassischen Weimar*. Munich: Deutscher Taschenbuch Verlag, 1999.

Winckelmann, Johann Joachim. *Gedanken zur Nachahmung der griechischen Werke in der Mahlerey und Bildhauerkunst*. 2nd ed. Stuttgart: Reclam, [1756] 1995.

———. *Geschichte der Kunst des Altertums*. Darmstadt: Wissenschaftliche Buchgesellschaft, [1764] 1972.

Witte, Bernd, et al. *Goethe Handbuch*. 4 vols. Stuttgart: J. B. Metzler, 1996–98.

Wolff, Caspar Friedrich. *Theoria generationis*. Halle: Hendelianis, 1759.

———. *Theorie von der Generation*. Berlin: Birnstiel, 1764.

Wordsworth, William. *The Essential Wordsworth*. Ed. Seamus Heaney. Hopewell, N.J.: Echo Press, 1988.

Zapperi, Roberto. *Das Inkognito: Goethes ganz andere Existenz in Rom*. Trans. Ingeborg Walter. Munich: Beck, 1999.

Ziolkowski, Theodore. *German Romanticism and Its Institutions*. Princeton: Princeton University Press, 1990.

———. *Das Wunderjahr in Jena, Geist und Gesellschaft 1794/95*. Stuttgart: Cotta'sche Buchhandlung Nachfolger, 1998.

Index

absolute self, Schelling on, 131–34, 142n, 143n, 153, 154–55
adequate ideas (Spinoza), Goethe's use of, 380, 415
aesthetic theory, Karl Friedrich Schlegel on, 105–12
aesthetics and science, complementarity of, 12, 329
alchemy, Goethe on, 339
Allgemeine Literatur-Zeitung, 140–41, 422
"Das älteste Systemprogramm des deutschen Idealismus," 124–25, 124n
altruism, Darwin on, 546–47, 549–51
Ameriks, Karl, 429n
archetype, theory of
    Cuvier, Georges, 517
    Darwin, Charles, 10, 517, 532–33
    Goethe, Johann Wolfgang von, 8–10, 34, 408, 415–16, 421, 440–53, 456–57, 490, 507, 517–18
    Green, Joseph Henry, 527–28
    Herder, Johann Gottfried, 370
    Humboldt, Alexander von, 520–21
    Kant, Immanuel, 8–9, 65, 68, 232–33
    Owen, Richard, 443, 517–18
    Schelling, Friedrich Wilhelm Joseph, 8–10, 145, 183–84, 186, 188–89, 191–92, 302, 311–12
Archiv für die Physiologie, 255
Aristotle, on scientific explanation, 307–8
Arnim, Achim von, 17n
art, Schelling's philosophy of, 160–62, 186–90
atheism controversy, Fichte's involvement with, 46, 89–90
Auden, W. H., 404

Baer, Karl Ernst von, 3, 191, 312, 494
Baruch, Löb. *See* Börne, Ludwig
Batsch, August Johann Georg Carl, 1–2, 375, 413, 423
Behler, Ernst, on meaning of Romanticism, 22n
Behrisch, Ernst Wolfgang, 335, 337
Beiser, Frederick, ethical interpretation of Fichte, 81n
Berg, Franz, 174–75, 175n
Berlin salon of Marcus and Henriette Herz, 91–92
Berlin, Isaiah, on Romanticism, 6, 6n, 201–3
Berlin, university of, 198, 282

Bildungstrieb
    Blumenbach, Johann Friedrich, 218–22, 225–29, 257
    Goethe, Johann Wolfgang von, 447–48
    Herder, Johann Gottfried, 226
    Kant, Immanuel, 220, 227, 231–38
    Kielmeyer, Carl Friedrich, 245
    Reil, Johann Christian, 255, 281
    Schelling, Friedrich Wilhelm Joseph, 293–94, 296–97, 304
biographical explanation, viii, 5, 14, 23, 390n
    Kant on, 229–31, 231n, 309
biology
    foundational science, 114, 116, 157–59, 310
    Romantic, 6–8, 12–14, 512
    term, meaning of, 4n
Blumenbach, Johann Friedrich, 5, 334
    Bildungstrieb, 218–22, 225–29, 257
    embryogenesis, 218–20
    Goethe, Johann Wolfgang von, 368, 373–74
    intermaxillary bone, 368, 373–74
    Kielmeyer, Carl Friedrich, 239, 245
    works
        Beyträge zur Naturgeschichte, 222
        De generis humani varietate native, 217, 221, 373–74
        Geschichte und Beschreibung der Knochen des menschlichen Körpers, 373
        Handbuch der Naturgeschichte, 217–18, 219, 226, 228–29
        Über den Bildungstrieb, 216–22
body as republic, 260, 260n
Boerhaave, Hermann, 212–13
Böhmer, Auguste (daughter of Caroline Böhmer Schlegel Schelling), 36–37, 171–72
    Schelling, Friedrich Wilhelm Joseph, 166–76, 179, 173n
    Schlegel, Karl Friedrich, 105–6, 107–8
Böhmer, Georg (first husband of Caroline Böhmer Schlegel Schelling), 36–37
Bonaparte, Napoléon, 91, 278, 284
Bonnet, Charles, 213–16
Bonpland, Amié, 519
Börne, Ludwig (Löb Baruch), 275n, 276–77
Bowie, Andrew, 143n
Boyle, Nicholas, 344–45, 403n
Brandis, Johann David, 261n
Breidbach, Olaf, 128n
Brentano, Bettina, 331n

Brentano, Clemens, 17n
Brion, Friederike (Goethe's early love), 342–45
Brown, John
  Elements of Medicine, 296, 315
  irritability and sensibility, 315
  medical theory of, 315
Brunswick, duke of, 38
Buff, Charlotte (Goethe's early love), 346–49
Buffon, Georges-Louis Leclerc, comte de, 221
Burdach, Karl Friedrich, 3, 191, 453
Burschenschaften, Oken's relations with, 500–1
Bury, Friedrich, 384
Bush, Werner, 391n, 393n
Büttner, Christian Wilhelm, 217, 413

Camper, Pieter, on intermaxillary bone, 368,
  372, 374
Carl August, duke of Saxe-Weimar-Eisenach
  atheism controversy, 46
  Fichte, dismissal of, 90
  French Revolution, 38
  Goethe, Johann Wolfgang von, 355
  Napoleonic wars, 476
  Oken, Lorenz, 501
Carus, Carl Gustav, 129, 191, 528
cat-piano, Reil on, 253, 271, 271n
classic, definition of, 3, 3n, 458
Coleman, William, on Kielmeyer, 240n
Coleridge, Samuel Taylor, 544
Collingwood, R. G., 390n
compensation, Goethe's law of, 416
Cook's voyage, 37
Cotta (publisher), 177, 476–77
Crancé, Jean-Baptiste Dubois, 40
Cranz, Wilhelm Julius (son of Caroline Böhmer
  Schlegel Schelling), 43
Custine, Adam-Philippe de, 40
Cuvier, Georges
  archetype, 517
  embryogenesis, 216
  Goethe, Johann Wolfgang von, 444, 444n
  Karlsschule, 239
  Kielmeyer, Carl Friedrich, 238, 248–49
  Naturphilosophie, 248n

D'Alton, Eduard, 483–86
Darwin, Charles
  altruism, 546–47, 549–51
  archetype, theory of, 10, 517, 532–33
  Beagle voyage, 523–26, 540–41
  Cambridge, 522
  Coleridge, Samuel Taylor, 544
  community selection, 547, 549–50
  evolution, 5, 192, 303, 305
  Forster, Georg, 37, 37n
  Goethe as predecessor, 210–11, 210n, 435n
  Gray, Asa, 548
  group selection, 547n
  Henslow, John Stevens, 522

Humboldt, Alexander von, influence of, 514,
  521, 521n, 522–26, 552–53
Humboldt's *Cosmos*, 512
Humboldt's *Personal Narrative*, 522–23
  instinct, 544–46
  Lyell, Charles, 548
  Mackintosh, James, 542
  Mackintosh's *Dissertation*, 542–44, 546
  Malthus, Thomas, 542
  Mill, John Stuart, 544
  Milton, John, 537–38
  mind, evolution of, 551–52
  moral theory, 540–52
  natural selection, 533–37
  nature as moral, 34, 537, 539
  nature, not machine, 533–34, 539–40
  nature as value laden, 533–40
  Paley, William, 539, 541–42
  Paley's rule, 542–42
  recapitulation, 531–33
  Romanticism, 6, 514–15, 526, 552–54
  slavery, 540–41
  utilitarianism, 543–551
  Wallace, Alfred Russel, 547–549
  works
    Descent of Man, 514, 543–44, 549
    "Essay of 1842," 535–36
    "Essay of 1844," 535–36
    Journal of Researches, 525–26
    Origin of Species, 6, 34, 514, 534–35, 538–
    39, 547
    *Variation of Animals and Plants under Do-
    mestication*, 547–48
Darwin, Erasmus, 246, 300–1
  evolution, 311
  Goethe, Johann Wolfgang von, 466, 481
  irritability and sensibility, 314–15
  Schelling, Friedrich Wilhelm Joseph, 291,
  300–1, 305
  Zoonomia, 300–1, 314
Dettelbach, Michael, 519n
Dilthey, Wilhelm, on Schleiermacher, 95, 104
Döllinger, Ignaz, 3, 191
Dunn, Joanna, 93n
dynamic evolution. *See* evolution, dynamic

Eichendorff, Joseph von, 18, 274
Eisenstuck, Julie (friend of Hardenberg), 26–27,
  26n
Eisenstuck, Laura (friend of Karl Friedrich
  Schlegel), 26–27, 26n
Eissler, Kurt, 321n
electrophysiology, Humboldt on, 318–21
embryogenesis, 13
  Blumenbach, Johann Friedrich, 218–20
  Boerhaave, Hermann, 212–13
  Bonnet, Charles, 213–14, 216
  Cuvier, Georges, 216
  Haller, Albrecht von Haller, 212–16

Kielmeyer, Carl Friedrich, 245
Needham, John, 216, 219
Swammerdam, Jan, 211–12
Wolff, Caspar Friedrich, 215–16, 219, 220
embryology, 211–16
Engelhardt, Dietrich von, 299, 463n
epigenesis vs. evolution, 212, 416
Eschenmayer, Karl, 181
evolution
　Bonnet, Charles, 215
　D'Alton, Eduard, 483–86
　Darwin, Charles, 303, 305, 531–54
　Darwin, Erasmus, 311
　dynamic, Schelling on, 10, 144–45, 144n,
　　208–11, 298–306, 311–12
　Goethe, Johann Wolfgang von, 208–11, 306,
　　369–70, 406, 466, 477–86, 478–79,
　　490–91
　Huxley, Thomas Henry, 215
　Kant, Immanuel, 232–33
　Kielmeyer, Carl Friedrich, 244–45, 244–48
　Owen, Richard, 531
　Pander, Christian, 483–86
　varieties of, 301
　word, meaning of, 211–12
experimental science
　Goethe, Johann Wolfgang von, 413–14,
　　434–40
　Schelling, Friedrich Wilhelm Joseph, 115,
　　140–45

Falk, Johannes Daniel, 479–80
Fichte, Johann Gottlieb
　Anstoss, 79–82, 80n
　atheism controversy, 89–90
　Carl August's dismissal of, 90
　early life of, 60–62
　Forberg's evaluation of, 87, 87n
　French Revolution, 85–86
　Goethe's evaluation of, 460–61
　Hardenberg's criticism of, 31–32, 31n, 83
　Heine's evaluation of, 83n, 90
　idealism, interpretation of, 79–84
　imagination, productive, 72, 72n
　intellectual intuition, 76–78, 76n, 77n
　intellectus archetypus, 81
　Jacobi, Friedrich, 78–79
　Jena, dismissal from, 89–90
　Kant, attitudes about, 61–62
　Kantianism, 79
　Kielmeyer's criticism of, 249
　philosopher, character of, 13–14, 13n, 82
　politics, 52–53, 53n, 85–86, 88
　professor at Jena, 46, 86–90
　Reinhold, Karl Leonhard, 74–75
　representation, 72–79
　Schelling, break with, 176–80
　Schelling, relations with, 147–48, 166, 176–
　　80, 183, 183n

Schiller, Friedrich, 83
Schlegel's evaluation of, 35, 59–60, 60n, 83,
　84–86, 88
Schleiermacher's evaluation of, 97–98
Schmid, Christian, 78
scholar, obligations of, 87–88
self-consciousness, 72–79
self-positing, 74
striving, 80
student fraternities, 88–89
thing-in-itself, 76–78, 76n
transcendental philosophy, 13, 72–79
works
　Begriff der Wissenschaftslehre, 120
　*Beitrag zur Berichtigung der Urtheile des Pub-
　　likums über die französische Revolution*,
　　120
　Bestimmung des Menschen, 82, 178, 182
　Erste Einleitung in die Wissenschaftslehre,
　　80n, 82n
　*Grundlage der gesammten Wissenschafts-
　　lehre*, 60, 73–74, 80, 83, 86n
　System der Sittenlehre, 124
　Versuch einer Kritik aller Offenbarung, 61,
　　120
　*Versuch eines erklärenden Auszugs aus Kants
　　Kritik der Urteilskraft*, 72n
　*Vorlesungen über die Bestimmung des
　　Gelehrten*, 54, 72n, 86–87
　Zweite Einleitung in die Wissenschaftslehre,
　　74n, 76–78
Fischer, Kuno, 44n, 129, 298
Flourens, Pierre, 220
Forberg, Friedrich Karl, on Fichte, 87, 87n
Forster, Georg
　Cook's voyage, 37
　Darwin, Charles, 37, 37n
　Heyne, Therese, 37, 40
　Humboldt, Alexander von, 37
　Mainz Revolution, 37–40
　politics, 54
　Schelling, Caroline Böhmer Schlegel, 40
fossils, Goethe on, 367n
Foucault, Michel, 511
fragments, romantic genre, 85, 85n
Francis II (emperor), 38
Frank, Manfred, 19n
freedom and determinism
　Kant, Immanuel, 64, 64n, 69
　Schelling, Friedrich Wilhelm Joseph, 123–24
French Revolution. *See also* Mainz Revolution
　Carl August of Saxe-Weimar-Eisenach, 38–39
　Fichte, Johann Gottlieb, 85–86
　Goethe, Johann Wolfgang von, 38–43
　Schelling, Friedrich Wilhelm Joseph, 118–
　　19, 119n
Schlegel, Karl Friedrich, 84n
Friedrich the Great, 51
Friedrich Wilhelm II, 38, 51–52, 91

Friedrich Wilhelm III, 57–58, 58n, 282
Friedrich, Caspar David, 17n

Galen, on intermaxillary bone, 371
Gall, Franz Joseph, 277–78, 474
Galvani, Luigi, 91, 317–18
Gellert, Christian Fürchtegott, 334
generation, spontaneous, Goethe on, 375
genius, theory of
    Goethe, Johann Wolfgang von, 425, 431–33, 470
    Kant, Immanuel, 14, 70
    Schelling, Friedrich Wilhelm Joseph, 14, 70n, 114, 161–64
    Winckelmann, Johann Joachim, 391–93
Ghiselin, Michael, 515n
Giebichenstein (estate of Reichardt), 274–75, 278
Gilman, Sander, 400n
Ginsborg, Hannah, 65n
Goethe, Cornelia (sister of Goethe), 331, 331n, 333
Goethe, Friedrich Georg (grandfather of Goethe), 330
Goethe, Johann Caspar (father of Goethe), 330–32
Goethe, Johann Wolfgang von
    adequate ideas (Spinoza), 380, 415
    aesthetic judgment, 426–27
    alchemy, 339
    anatomy, study of, 367–75, 391
    archetype, animal, 440–53
    archetype, theory of, 408, 415–16, 421, 453, 456–57, 490, 507, 517–18
    artists, Roman community of, 384–85
    Auden, W. H., 404
    Behrisch, Ernst Wolfgang, 335, 337
    Bildungstrieb, 447–48
    Blumenbach, Johann Friedrich, 368, 373–74
    boyhood, 330–34
    Brion, Friederike (early love), 342–45
    Buff, Charlotte (early love), 346–49
    Bury, Friedrich, 384
    Carl August invites to Weimar, 355
    compensation, law of, 416
    Cotta (publisher), 476–77
    Cuvier, Georges, 444, 444n
    D'Alton, Eduard, 483–86
    Darwin, Erasmus, 466, 481
    Darwin's use of his theory of morphology, 435n
    Darwinian evolution, 328
    death of, 507–8
    desire, poetry of, 344
    divorce for Schlegels, 176n
    epigenesis vs. evolution, 416
    evolution, 208–11, 306, 369–70, 406, 466, 477–86, 490–91
    evolution, constituents of theory, 478–89

experimental science, 413–14, 434–40
Falk, Johannes Daniel, 479–80
Faustine, 389–90, 390n
female ideal, 397–400, 401
Fichte's philosophy, 460–61
fossils, 367n
French Revolution, 38–43
Gall, Franz Joseph, 277–78, 474
Gellert, Christian Fürchtegott, 334
generation, spontaneous, 375
genius, theory of, 425, 431–33, 470
geology, 365–67
Goethe, Cornelia, 331n
Gottsched, Johann Christoph, 334
Gray, Ronald on, 408n, 419n
Gretchen (first love), 333–32
Gröschen, Georg Joachim (publisher), 381n
Haugwitz, Christian August von, 353
Heine on, 462, 504
Heinse, Johann Jakob Wilhelm, 349
Helmholtz's evaluation of, 328–29, 435
Herder, Johann Gottfried, early relationship with, 340–42
Herder, Karoline, 410–11, 420, 492
Homer, 395
homoeroticism, 387n
Homunculus, 207–8
Die Horen, 423, 425
Humboldt, Alexander von, 446–47, 449
Humboldt, Wilhelm von, 446
idealism, 465, 467–71, 480, 491
imagination, scientific, 438–39, 438n
intellectual character of, 422, 425–26
intellectus archetypus, 424, 489–90
intermaxillary bone, 368–76, 505
Italian itinerary, 383–84
Italian journey, first, 382–404
Italian journey, second, 419–21
Jacobi, Friedrich Heinrich, 378–80
Jena Natural History Society, 1
Jerusalem, Karl Wilhelm, 348
Kant's Kritik der reinen Vernunft, 428
Kant's Kritik der Urteilskraft, 426, 429–30, 434, 446–47, 488–89
Kant's teleology, 444–43, 450–51
Kantianism, 427–30
Kauffmann, Angelika, 384
Kestner, Johann Georg, 346–49
Kielmeyer, relations with, 239
Klettenberg, Susanna Katharina von, 339
Klopstock, Friedrich, 332–33
Knebel, Karl Ludwig, 375
Kniep, Christoph Heinrich, 391, 395–96
Kotzebue, Amalia, 356
La Roche, Maximiliane (early love), 348
La Roche, Sophie von, 348
Lichtenberg, Georg Christoph, 350, 441n
Linné, Carl von (Linnaeus), 375
Lips, Johann Heinrich, 384, 504

literature as self-reflection, 325–26, 326n, 338
Loder, Justus Christian, 367–68, 371–72
Mainz Revolution, 40–42
marriage, 476
Megatherium, 483–86
memory, reconstructions of, 325–26, 333
Merck, Johann Heinrich, 350, 372, 374
Meyer, Johann Heinrich, 384, 391
model for the Romantics, 107
Moritz, Carl Philipp, 385, 385n
morphology, science of, 453–57
Napoléon, 56n
Napoleonic wars, 474–76
nature
    as female, 326–27, 353–54
    ideas in, 402
    rights of, 404–5, 404n, 422
    in *Werther*, 346–48
Naturphilosophie, 464–65, 467–68
Nees von Esenbeck, Christian, 396
Newton, Isaac, 415, 434, 437
Oken, Lorenz, 463, 492, 497–502, 505
optics, 441, 441n
Palermo, gardens of, 395
Pander, Christian, 483–86
Panofsky, Erwin, 382n
plant development, 417–18
plant, primal, 2
poetry
    "An den Mond," 363–64
    "Bin so in Lieb," 338
    "Kennest du das Land," 381
    "Leipzig Menchen," 334–35
    "Lili" poems, 352–53
    "Maifest," 343
    "Metamorphose der Pflanzen," 466–67
    "Metamorphose der Tiere," 467
    "Natur und Kunst," 469–71
    Reineke Fuchs, 441
    Römische Elegien, 389–90, 389n, 390n, 398–400, 421–22
    "Saale Fluss," 363
    "Unbeständigkeit," 338–39
    "Wandrers Nachtlied I," 362
    "Wandrers Nachtlied II," 362
    "Weltseele," 465–66
    "Werner," 367
    "Willkommen und Abschied," 344
polarity, 417, 429
political attitudes of, 55–56
Priapus, 387, 389n
psychoanalytic interpretation of, 331n, 344n
puppet show, 332
Reiffenstein, Johann Friedrich, 384
Reil, Johann Christian, 474
Reinhold, Karl Leonhard, 427–29
religious beliefs, 336
Riemer, Friedrich, 456–58, 476

Riese, Johann, 334
Romantic circle, 460–63
Romantic, definition of, 3, 430, 458
Romanticism, 2, 3, 14, 329–30, 430–31, 433–34, 457–60, 506
Salzmann, Johann David, 343
Schelling, Caroline Böhmer Schlegel, 461–63
Schelling, relationship with, 116, 140n, 142n, 147–48, 161, 165–66, 172, 172n, 191, 462–71
Schelling's *Einleitung*, 464–65
Schelling's *System*, 467–70
Schelling's *Weltseele*, 465–66
Schiller, death of, 471–73
Schiller, encounter with, 2, 329, 329n, 423–25
Schiller, friendship with, 421–27
Schiller's criticism of, 423
Schlegel, August Wilhelm, 461
Schlegel, Karl Friedrich, 59, 59n, 461–62, 473, 473n
Schlosser, Johann Georg, 331, 335, 336
Schönemann, Lili (early love), 351–53
Schönkopf, Anna Katharina (early love), 336–38
Schröter, Corona, 357
Schubert, Gotthilf Heinrich von, 463
    as a scientist, 407–9, 408n, 457
Shakespeare, William, 345
Soemmerring, Samuel Thomas von, 368, 372, 374, 443
Spinoza, Baruch de, 376–80, 439, 439n, 490, 506
Steffens, Henrik, 463, 472–73
Stolberg-Stolberg, Christian zu, 353
Stolberg-Stolberg, Friedrich Leopold zu, 353
Strasbourg, 339–45
Sturm und Drang, 346
teleology, 444–45, 445n, 450–51
theory, use of, 439n
Tischbein, Johann Heinrich Wilhelm, 384–85, 391
university education, 334–39
Urpflanze, 376, 395–96, 401–2, 414–15, 424
vertebral theory of skull, 328, 420–21, 491–502
Vesuvius, 394
Voigt, Christian Gottlob, 41
Voigt, Friedrich Siegmund, 499–500
Voigt, Johann Carl Wilhelm, 365
Von Stein, Charlotte, 357–67, 377, 380–81, 410–411
Voss, Johann Heinrich, 472
Vulpius, August Walther (son), 412, 415–16
Vulpius, Christiane (mistress and wife), 390, 410–13, 419–20
Waitz, Johann Christian Wilhelm, 372
Weimar administrator, 355–56
Wetzlar, 346

Goethe, Johann Wolfgang von (*continued*)
    Whewell's evaluation of, 434–35
    Wieland, Christoph Martin, 427
    Winckelmann, Johann Joachim, 384, 401
    Wolf, Friedrich August, 473
    women, symbol of creativity, 326–27
    works
        Beiträge zur Optik, 441
        "Briefe aus der Schweiz," 353–54
        Campagne in Frankreich, 38–39
        Claudine von Villa Bella, 352, 391
        Dichtung und Wahrheit, 325–26, 330–34,
            487
        Egmont, 390–91, 422–23
        "Einfache Nachahmung der Nature,
            Manier, Styl," 402–3
        *Erster Entwurf einer allgemeinen Einleitung
            in die vergleichende Anatomie*, 442, 446–
            49
        Erwin und Elmira, 391
        Farbenlehre, 477, 486–87
        Faust, 207–8, 391, 397–98, 398n, 487
        Die Geschwister, 356
        Götz von Berlichingen mit der eisernen Hand,
            345–46
        "In wiefern die Idee: Schönheit sei Voll-
            kommenheit mit Freiheit," 426
        Iphigenie auf Tauris, 2, 357, 390
        Italienische Reise, 382–83, 382n, 397, 488,
            502
        "Klassiker und Romantiker in Italien, sich
            heftig bekämpfend," 458–59
        Die Leiden des jungen Werthers, 2, 21n, 135,
            135n, 332–33, 339, 346–50
        Die Metamorphose der Pflanzen, 407, 416–
            19, 476, 495
        Stella, 352
        Torquato Tasso, 2, 391
        Urfaust, 351
        Venezianische Epigramme, 420, 422
        "Der Versuch als Vermittler von Objekt
            und Subjekt," 435–40
        "Versuch einer allgemeinen Knochen-
            lehre," 442, 443–44
        "Versuch über die Gestalt der Tiere," 440,
            442–43, 498
        *Vorträge über die drei ersten Kapitel des En-
            twurfs*, 442, 449–53
        Wilhelm Meisters Lehrjahre, 2, 108, 109,
            332, 339
        Wilhelm Meisters theatralische Sendung, 356
        Winckelmann und sein Jahrhundert, 473–74
        Xenien, 422
        Zur Morphologie, 372, 488, 491–92, 498, 502
        "Zwischenknochen," 371
Goethe, Katharina Elisabeth Textor (mother of
    Goethe), 331, 331n
Goldhagen, Johann Friedrich, 254, 261n, 263,
    263n

Goldstein, Jan, 269
Gotter, Pauline (second wife of Schelling), 198,
    198n
Gottsched, Johann Christoph, 334
Grappengiesser, Karl, 284
Gray, Asa, 548
Gray, Ronald, 408n, 419n
Green, Joseph Henry, 10, 191–92, 527–28
Gregory, Frederick, 94n, 127n, 187n
Gretchen (Goethe's first love), 333–32
Grimm, Jacob, 20n
Grimm, Wilhelm, 20n
Gröschen, Georg Joachim (publisher of
    Goethe), 381n
group selection, Darwin on, 547n
Grunow, Elenore (friend of Schleiermacher),
    100, 100n, 113n
Guercino (Giovanni Francesco Barbieri), 382,
    382n

Haeckel, Ernst, 5–6, 192, 247, 298
Halle, university of, 282
Haller, Albrecht von
    embryogenesis, 212–16
    irritability and sensibility, 313
Hamilton, Lady. *See* Lyon, Emma
Hamilton, Sir William, 387
Hanov, Michael Christoph, 4n
Hardenberg, Friedrich von (Novalis)
    death of, 35–36, 194
    Eisenstuck, Julie, 26–27, 26n
    erotic interests, 29n, 32n, 33
    Fichte, criticism of, 31–32
    geology, 34–35
    Kühn, Sophie von, 28–30
    personality, romantic, 25–28, 35–36
    philosophical study, 31–32
    poetry as way to infinite, 4, 32
    political writings, 57–58
    Reinhold, Karl Leonhard, 26n
    Romantic circle, early, 17n
    Romanticism, 19, 19n, 200
    Schelling, Friedrich Wilhelm Joseph, 31–32
    Schiller, Friedrich, 26n
    Schlegel, friendship with, 25–27
    Schleiermacher, evaluation of, 103
    Tieck, Ludwig, 36
    Werner, Abraham, 34–35
    Woltmann, Karl, 30
    works
        Blütenstaub, 106
        Die Christenheit oder Europe, 58
        Glauben und Liebe, 57–58
        Hymnen an die Nacht, 30–34, 30n, 106
        Novices at Sais, 6n
Hardenberg, Heinrich Freiherr von (father of
    Friedrich von Hardenberg), 27, 27n
Harvey, William, 212
Haugwitz, Christian August von, 353

Haym, Rudolf, 18n, 286
Hegel, Georg Friedrich
    Berlin, university of, 198
    *Differenz des Fichte'schen und Schelling'schen
        Systems der Philosophie*, 117n, 190–91
    Fichte's idealism, 83n, 90
    philosophy professor at Jena, 46
    Schelling, relationship with, 117, 119, 124
Heine, Heinrich
    Goethe, evaluation of, 462, 504
    Romanticism, 18n
    Schelling, evaluation of, 180n
    Schlegel, August Wilhelm, evaluation of, 195
Heinse, Johann Jakob Wilhelm, on *Werther*, 349
Helmholtz, Hermann von, on Goethe, 328–29,
    435
Henrich, Dieter, 118n
Henslow, John Stevens, 522
Herder, Johann Gottfried
    archetype, 370
    Bildungstrieb, 226
    development, 223–25, 369–70
    Goethe, early relationship with, 340–42
    *Ideen zur Philosophie der Geschichte der Men-
        schheit*, 135, 223–25, 368–70
    Italian journey, 409–10
    Kant's criticism of, 224–25, 225n
    Kauffmann, Angelika, 384
    Kielmeyer's reference to, 245
    language, origins of, 341–2
    Schelling's reference to, 120, 135–36
    Strasbourg, 340–42
Herder, Karoline Flachsland (wife of Johann
    Gottfried Herder), 341, 341n, 410–11,
    420, 492
Herz, Henriette
    Berlin salon, 91–92
    Börne, Ludwig, 276–77
    Reil, relationship with, 275–77
    Schlegel on, 95n
    Schleiermacher, friendship with, 95, 113n
Herz, Marcus
    Berlin salon, 91–92
    Betrachtungen aus der spekulativen Weltweisheit,
        255n
    insanity, 263
    Kant, student of, 255n
    Reil, relationship with, 254
Heyme, Therese (wife of Georg Forster and
    friend of Caroline Böhmer Schlegel
    Schelling), 37, 40
Heyne, Christian Gottlob, 24, 217, 334
historiography, 5–7, 18–19
Hoffmann, E. T. A., 17n
Hölderlin, Friedrich
    Romanticism, 19n
    Schelling, relationship with, 117, 119, 124
    homoeroticism, 387n, 388n
Die Horen (Schiller's journal), 423, 425

Hufeland, Christoph Wilhelm, 46, 261n
Humboldt, Alexander von
    aesthetic judgment, 521n
    Americas, travel to, 519
    archetype, 520–21
    Bonpland, Amié, 519
    Darwin, influence on, 514, 521, 521n, 522–
        26, 552–53
    electrophysiology, 318–21
    Forster, Georg, 37
    Goethe, relationship with, 446–47, 449
    Jena, 446–47, 449
    Kielmeyer, evaluation of, 238
    Lebenskraft, 257, 275n, 293n, 320–21
    life, theory of, 293n, 316–17, 320–21
    Naturphilosophie, 134n
    plants, 519–20
    Schelling, relationship with, 129, 134, 134n,
        139
    works
        Ansichten der Natur, 293
        Florae Fribergensis, 316
        Ideen einer Geographie der Pflanzen, 134n
        Kosmos, 134n, 514
        "Die Lebenskraft order der rhodische Ge-
            nius," 447
        Personal Narrative, 514
        *Versuche über die gereizte Muskel- und Ner-
            venfaser*, 294–95, 319–21
Humboldt, Wilhelm von
    Berlin, university of, 282
    Jena, 446
    Schelling, Caroline Böhmer Schlegel, 42
Hume, David, 308
Huxley, Thomas Henry, 215
Huygens, Christiaan, 328

idealism
    Fichte, Johann Gottlieb, 72–84
    Goethe, Johann Wolfgang von, 465, 467–71,
        480, 491
    interpretations of Fichte's philosophy, 79–84
    Kant, Immanuel, 62n, 63
    Kielmeyer, Carl Friedrich, 248–51
    Schlegel, Karl Friedrich, 110
    Schleiermacher, Friedrich Daniel, 100n, 105
    identity, Schelling's philosophy of, 176–92,
        291
    intellectual love of God, according to Spinoza,
        98, 98n, 377
intellectus archetypus
    Fichte, Johann Gottlieb, 81
    Goethe, Johann Wolfgang von, 424, 489–90
    Kant, Immanuel, 68, 81, 233, 309, 424
    intermaxillary bone
    Blumenbach, Johann Friedrich, 368, 373–74
    Camper, Pieter, 368, 372, 374
    Galen, 371
    Goethe, Johann Wolfgang von, 368–76, 505

intermaxillary bone (*continued*)
Oken, Lorenz, 496
Soemmerring, Samuel Thomas von, 368, 372, 374
Vesalius, Andreas, 371–72, 372n
Vicq-d'Azyr, Félix, 374
intuition of the universe, according to Schleiermacher, 99–102
intuition, intellectual
Fichte, Johann Gottlieb, 76–78, 76n, 77n
Kant, Immanuel, 76, 76n, 77n
Schelling, Friedrich Wilhelm Joseph, 132, 151–55, 154n, 162, 183, 187
irony, Romantic, 112, 112n, 273
irritability and sensibility
Brown, John, 315
Darwin, Erasmus, 314–15
Haller, Albrecht von, 313
Whytt, Robert, 314
Isis (Oken's journal), 500

Jacobi, Friedrich Heinrich
Fichte, Johann Gottlieb, 78–79, 378
Spinoza controversy, 378–79
Über die Lehre des Spinoza, 378–79
Jena, city of, 45, 45n
Jenaische Allgemeine Literatur-Zeitung, 177n
Jerusalem, Karl Wilhelm, 348
judgment
a priori principle of, according to Kant, 68, 68n, 71
aesthetic
Goethe, Johann Wolfgang von, 426–27
Humboldt, Alexander von, 521n
illustrations, 12
Kant, Immanuel, 12–13, 68n, 69–71, 432n
Schelling, Friedrich Wilhelm Joseph, 161–62, 186–87
Schiller, Friedrich, 13, 427n
determinative, according to Kant, 67, 489
reflective, according to Kant, 67, 488–89
regulative, according to Kant, 10–11, 67, 230–31
teleological, according to Kant, 64–69

Kant, Immanuel
aesthetic judgment, 12–13
archetype, 8–9, 65, 68, 232–33
beauty, 69–71
Bildungstrieb, 220, 227, 231–38
biology as a science, 229–31, 231n, 309
ego, 62–64
evolution, 232–33
Fichte, evaluation of, 89
freedom and determinism, 64, 64n, 69
genius, theory of, 14, 70
God, postulation of, 90
Herder, criticism of, 224–25, 225n

idealism, 62n, 63
intellectual intuition, 76, 76n, 77n
intellectus archetypus, 68, 81, 233, 309, 424
judgment
a priori principle of, 68, 68n, 71
aesthetic, 68n, 69–71, 432n
determinative, 67, 489
moral, 12
reflective, 67, 488–89
regulative, 10–11, 67, 230–31
teleological, 64–69
Kielmeyer's criticism of, 241–43, 248–50
matter, concept of, 139n
mechanism, 67, 236, 290, 309–10
organism, 66–67, 70, 138, 290
political trouble, 52
purpose and purposiveness, 65–69, 70
rationalism, 18
Reinhold, Karl Leonhard, 74–75
teleology, 3, 5, 10–12, 309
thing-in-itself, 63, 75–78
works
Idee zu einer allgemeinen Geschichte in weltbürgerlicher Absicht, 135n
Kritik der praktischen Vernunft, 63–64
Kritik der reinen Vernunft, 62–63
Kritik der Urteilskraft, 12, 64–71, 225, 229–37
Metaphysiche Anfangsgründe der Naturwissenschaft, 130n, 429
Prolegomena zu einer jeden künftigen Metaphysik, 63
Zum ewigen Frieden, 52
Kantianism
Fichte, Johann Gottlieb, 79
Goethe, Johann Wolfgang von, 427–30
Schiller, Friedrich, 1, 427, 431
Kanz, Kai, 245n
Karl Eugen, duke of Württemberg, 119, 238, 239, 278
Karlsschule, 239–40
Kauffmann, Angelika, 384
Kestner, Johann Georg, 346–49
Kielmeyer, Carl Friedrich
Bildungstrieb, 245
Blumenbach, relationship with, 239, 245
Cuvier, relationship with, 238
embryogenesis, 245
evolution, 244–48
Fichte, criticism of, 249
Goethe, relationship with, 239
Herder, 245
Humboldt, Alexander, relations with, 238
idealism, analysis of, 248–51
Kant, Immanuel, 241–43, 248–50
organic forces, 241–46
organism, definition of, 241
recapitulation, 247
Schelling, criticism of, 248–50

Schelling, relationship with, 139, 238–39, 299, 304
Schelling's evolutionism, 248
"Ueber die Verhältnisse der organischen Kräfte," 238, 241–46
Klettenberg, Susanna Katharina von, 339
Klopstock, Friedrich, 332–33
Knebel, Karl Ludwig, 375
Kniep, Christoph Heinrich, 391, 395–96
Knight, David, 515n
Knight, Richard Payne, 387n
Kohn, David, 515n
Kotzebue, Amalia, 356
Kramer, Cheryce, 253n, 261n, 270n
Kritisches Journal der Philosophie, 177–78
Kühn, Sophie von (beloved of Hardenberg), 28–30

La Mettrie, Julien Offray de, 308
La Roche, Maximiliane (Goethe's early love), 348
La Roche, Sophie von, 348
Lamarck, Jean-Baptiste de, 246, 484–85
language, origins of, Herder on, 341–2
Laplace, Pierre-Simon de, 11n
Larmore, Charles, 31n, 72n
Lauth, Reinhard, 177n
Lebenskraft
  Humboldt, Alexander von, 257, 275n, 293n, 320–21
  Reil, Johann Christian, 255–61, 279
  Schelling, Friedrich Wilhelm Joseph, 293
Leibniz, Gottfried Wilhelm, 18
Leipzig, battle of, 284
Lenoir, Timothy, 3–4, 210, 227–228, 235, 240n
Leopold II (emperor), 38
Lessing, Gotthold Ephraim, 2, 336, 378–79
Levere, Trevor, 7n
Lichtenberg, Georg Christoph, 350, 441n
Linnaeus. *See* Linné, Carl von
Linné, Carl von (Linnaeus), 4n, 375
Lips, Johann Heinrich, 384, 504
Loder, Justus Christian, 46, 367–68, 371–72
love
  Karl Friedrich Schlegel on, 34n
  Schleiermacher's religion of, 100–2, 105
  way to the infinite, 34
Lovejoy, Arthur, 22n
Luise of Hesse-Darmstadt (wife of Carl August), 355
Lyell, Charles, 548
Lyon, Emma (Lady Hamilton), 387

Mackintosh, James, 542
Mainz Revolution, 36–43
Malthus, Thomas, 542
Marcus, Adalbert Friedrich, 174
marriage, Schlegel's theory of, 93n
Martin, Wayne, 81n

materialism, Reil on, 258
matter
  Kant on, 139n
  Schelling on, 130–32
Mayr, Ernst, 115n
mechanism, xvii, 11, 308
  Cartesian, 308
  Hume, David, 308
  Kant, Immanuel, 67, 236, 290, 309–10
  La Mettrie, Julien Offray de, 308
  Newtonian, 308–10
  Schelling, Friedrich Wilhelm Joseph, 290, 292–93
Meckel, Johann Friedrich, 497
Meckel, Philipp Friedrich, 254
Megatherium, Goethe on, 483–86
memory, Goethe's reconstructions of, 325–26, 333
Mendelssohn, Moses, 378–79
Merck, Johann Heinrich, 350, 372, 374
Mereau, Sophie, 86
Meyer, Johann Heinrich, 384, 391
Michaelis, Johann David (father of Caroline Böhmer Schlegel Schelling), 36, 334
Michaelis, Philip (brother of Caroline Böhmer Schlegel Schelling), 42
Mill, John Stuart, 544
Milton, John, 537–38
mind
  and brain, Reil on, 265–66, 266n
  evolution of, according to Darwin, 551–52
  historical development of, according to Schelling, 135–36
Mocek, Reinhardt, 285–88
monism, 10–11
moral cure, Pinel on, 268–70
Moravian Brotherhood, 94
Moritz, Carl Philipp, 91, 385, 385n
morphology, Goethe's science of, 453–57
Müller-Sievers, Helmut, 497n
mythology, Schelling on, 119–20

Napoléon. *See* Bonaparte, Napoléon
Napoleonic wars, 474–76
natural selection, Darwin on, 533–37
nature
  as female, according to Goethe, 326–27, 353–54
  in Goethe's *Werther*, 346–48
  ideas in, according to Goethe, 402
  not machine, according to Darwin, 533–34, 539–40
  as moral, according to Darwin, 34, 537, 539
  poetry of, according to Schelling, 155, 159–64
  as product and productivity, according to Schelling, 142–45
  rights of, according to Goethe, 404–5, 404n, 422
  and the self, 13, 134–35
  value laden, according to Darwin, 533–40

Naturphilosophie, 516–18
  Cuvier, Georges, 248n
  definition of, 6–11
  Goethe, Johann Wolfgang von, 464–65,
    467–68
  Humboldt, Alexander von, 134n
  Reil, Johann Christian, 280–81, 285–88
  Schelling, Friedrich Wilhelm Joseph, 115n,
    116, 122, 128–51, 297
Needham, John, on embryogenesis, 216,
  219
Nees von Esenbeck, Christian, 129, 396
Neptunism, 34–35, 366–67
Neuhouser, Frederick, 81n
Newton, Isaac
  Goethe on, 415, 434, 437
  Schelling on, 130, 163
Ney, Michel, 475
Niethammer, Friedrich, 126, 147
Novalis. *See* Hardenberg, Friedrich von

Oken, Lorenz
  Baer, Karl Ernst von, 494
  Burschenschaften, 500–1
  Carl August, 501
  Goethe, Johann Wolfgang von, 463, 492,
    497–502, 505
  intermaxillary bone, 496
  Isis, 500
  Owen, Richard, 496, 528
  recapitulation, 493–94
  Schelling, Friedrich Wilhelm Joseph, 492,
    494–95, 496–97, 500, 505
  vertebral theory of skull, 421, 495–98
  works
    Über die Bedeutung der Schädelknochen,
      495–96, 501
    Die Zeugung, 492–93
optics, Goethe on, 441, 441n
organic forces, Kielmeyer on, 241–46
organicism, Schelling on 137–40, 157–59, 185–
  86, 239, 289–90, 292–94, 310
organism
  Kant's definition of, 66–67, 70, 138, 290
  Kielmeyer's definition of, 241
Owen, Richard, 10, 192
  archetype, 443, 517–18
  Carus, Carl Gustav, 528
  evolution, 531
  Oken, Lorenz, 496, 528
  recapitulation, 531
  works
    On the Nature of Limbs, 528, 530, 532
    Report on the Archetype, 528

Paley, William, 539, 541–42
Palus, H. E. G., 119
Pander, Christian, 483–86
Panofsky, Erwin, 382n

personality
  in philosophy, according to Fichte, 13–14, 13n
  Romantic, 13–14
Pinel, Philippe
  moral cure, 268–70
  *Traité médico-philosophique sur l'aliénation men-
    tale*, 261, 268–70
Pippin, Robert, 64n, 74n
Planer, Johann, 4n
plant development
  Goethe on, 417–18
  Humboldt on, 519–20
poetry as way to infinite
  according to Hardenberg, 32
  according to Schleiermacher, 97, 101, 101n
polar forces
  Goethe, Johann Wolfgang von, 417, 429
  Schelling, Friedrich Wilhelm Joseph, 130–32
politics
  Fichte, Johann Gottlieb, 52–53, 53n, 85–86,
    88
  Forster, Georg, 54
  Goethe, Johann Wolfgang von, 55–57
  Hardenberg, Friedrich von (Novalis), 57–59
  Prussia, 51–52, 57–58
  Romanticism, 51–59
  Schiller, Friedrich, 13, 51–55
  Schlegel, Karl Friedrich, 85
Priapus, Goethe on, 387, 389n
Prussia's political situation, 51–52, 57–58
psychiatry, word, coined by Reil, 263

rationalism
  Kant, Immanuel, 18
  Leibniz, Gottfried Wilhelm, 18
  Wolff, Christian, 18
reason, Enlightenment view of, xvii
recapitulation, 11
  Darwin, Charles, 531–33
  Haeckel, Ernst, 247
  Kielmeyer, Carl Friedrich, 247
  Oken, Lorenz, 493–94
  Owen, Richard, 531
Rehberg, Caroline (friend of Karl Friedrich
  Schlegel), 24, 24n
Reichardt, Johann Friedrich
  Gall, Franz Joseph, 277
  Giebichenstein, 274–75, 278
  musical performances, 273–75
Reiffenstein, Johann Friedrich, 384
Reil, Anna (mother of Johann Christian Reil),
  253
Reil, Johann Christian, 3, 5, 191
  Berlin, university of, 282
  Bildungstrieb, 255, 281
  body as republic, 260, 260n
  cat-piano, 253, 271, 271n
  corsets, 275n
  death of, 285

force, definition of, 257
Gemeingefühl, 262
Goethe, relationship with, 474
Herz, Henriette, 275–77
insane, study of, 252, 261–73
Kant, criticism of, 259–60
Kantian dualism, 256
Lebenskraft, 255–61, 279
madhouse, 264
materialism, 258
medicine as Wissenschaft, 255
mind and brain, 265–66, 266n
Napoléon, dedication to, 261
Naturphilosophie, 280–81, 285–88
Pinel, use of, 268–70
pregnancy, study of, 252, 278–81
psychiatry, word, 263
Schelling, disciple of, 252, 263, 267, 280–81, 286–88
Schleiermacher, relationship with, 253
self, 264
self-consciousness, 267–68
Steffens's depiction of, 276n
Steffens, relationship with, 253
works
    Fieberhafte Nervenkrankheiten, 261
    "Gebärmutter," 278–81, 285–86
    Pepinieren zum Unterricht Artzlicher Routiniers, 286
    Rhapsodieen über die Anwendung der psychischen Curmethode auf Geisteszerrüttungen, 263–73
    Ueber die Erkenntniss und Cur der Fieber, 261
    "Von der Lebenskraft, 255–61, 265, 279, 287
Reil, Johann Julius (father of Johann Christian Reil), 253
Reinhold, Karl Leonhard
    Fichte, influence on, 74–75
    Goethe, association with, 427–29
    Kant, interpreter of, 74–75, 428–29, 428n
    philosophy professor at Jena, 45–46
    representation, theory of, 74–75, 74n, 75n, 428–29
    Schelling on, 126n
representation
    Fichte, Johann Gottlieb, 72–79
    Reinhold, Karl Leonhard, 74–75, 74n, 75n, 428–29
Rhenish Republic, 40
Riemer, Friedrich Wilhelm (secretary to Goethe), 457–58, 476
Riese, Johann, 334
Rist, Johann Georg, 86n
Ritter, Johann Wilhelm, 35, 179, 179n
Robinson, Henry Crab, 150
Rockenthien, Johann von (stepfather of Sophie von Kühn), 28
Roller, Christian, 270n

Romantic
    biology, 6–8, 12–14, 512
    Darwin as, 514–15, 552–54
    Hardenberg as, 25–28, 35–36
    irony, 112, 112n
    movement, 460–63, 516–22
    personality, 13–14
    poetry, 22, 107–12, 200–2
    Schleiermacher as, 104–5
    word, 20–21
Romanticism
    Behler, Ernst, 22n
    Berlin, Isaiah, 6, 6n, 201–3
    Darwin, Charles, 6, 526
    Eichendorff, Joseph von, 18
    Frank, Manfred, 19n
    Frühromantiker, 17n
    Goethe, Johann Wolfgang von, 2, 3, 14, 21–22, 22n, 329–30, 430–31, 433–34, 457–60, 506
    Goethe's definition of, 3, 430, 458
    Grimm, Jacob, 20n
    Grimm, Wilhelm, 20n
    Haeckel, Ernst, 6
    Halle, 273–78
    Hardenberg, Friedrich von (Novalis), 19, 19n, 200
    Heine, Heinrich, 18n
    Hölderlin, Friedrich, 19n
    Lovejoy, Arthur, 22n
    meaning of, xvii, 3, 6–7, 17–19, 20–23, 22n, 109–12, 199–203
    politics of, 51–59
    Schelling, Caroline Böhmer Schlegel, 19n
    Schlegel, Dorothea Veit, 19n
    Schlegel, Karl Friedrich, 19n, 20–23, 109–112
    Roose, Theodor Georg, 4n
    Röschlaub, Andreas, 174, 286

salons in Germany, 91–92
Salzmann, Johann David, 343
Schellhorn, Cornelia (grandmother of Goethe), 330
Schelling, Caroline Böhmer Schlegel (née Michaelis)
    Böhmer, Auguste (daughter), 36–37, 169–72
    Böhmer, Georg (first husband), 36–37
    death of, 197–98
    early life, 36–37
    Forster, Georg, 40
    Goethe, relationship with, 461–63
    Heyme, Therese, 37, 40
    Humboldt, Wilhelm von, 42
    jailed, 42
    Mainz Revolution, 36–37, 40–43
    Michaelis, Philip (brother), 42
    political figure, 51–54
    Romanticism, 19n

Schelling, Caroline Böhmer Schlegel (continued)
    Schelling, relationship with, 116, 146, 149,
        166–76, 179–80
    Schlegel, August Wilhelm, 42–43, 45
    Schlegel, Dorothea Veit, evaluation by, 169–
        70
    Schlegel, Karl Friedrich, 43–45, 43n, 107–8
    Tatter, Georg, 37
Schelling, Friedrich Wilhelm Joseph
    the absolute, 182–92
    absolute self, 131–34, 142n, 143n, 153, 154–
        55
    aesthetic judgment, 161–62, 186–87
    Allgemeine Literatur-Zeitung, criticism by,
        140–41
    archetype, 8–10, 145, 183–4, 186, 188–89,
        191–92, 302, 311–12
    art, philosophy of, 160, 186–90
    artistic activity, theory of, 161–62
    Bildungstrieb, 293–94, 296–97, 304
    biology, 114, 116, 157–59
    Böhmer, Auguste, 166–76, 179, 173n
    Carus, Carl Gustav, 129
    chemistry, 130–32
    criticism and dogmatism, 122–23
    Darwin, Erasmus, 291, 300–1, 305
    Dresden, 147–48
    early life, 116–28
    empirical ego, 133–35
    evolution, dynamic, 144–45, 144n, 208–11,
        298–306, 311–12
    evolution, genealogical theory of, 301
    excitability, 295–97
    experimental science, 115, 140–45
    Fichte, break with, 176–80
    Fichte, early influence of, 120–23
    Fichte, relations with, 147–48, 166, 176–80,
        183, 183n
    forces, dynamic, 295
    freedom and determinism, 123–24
    French Revolution, 118–19, 119n
    genius, theory of, 70n, 114, 161–64
    Goethe, relationship with, 116, 140n, 142n,
        147–48, 161, 165–66, 172, 172n, 191,
        462–71
    Hardenberg, relationship with, 31–32
    Hegel, Georg Friedrich, 117, 119, 124
    Heine's evaluation of, 180n
    Herder, Johann Gottfried, 120, 135–36
    Hölderlin, Friedrich, 117, 119, 124
    Humboldt, Alexander, relationship with, 129,
        134, 134n, 139
    idealism of nature, 176
    idealism as realism, 165
    identity, philosophy of, 176–92, 291
    intellectual intuition, 132, 151–55, 154n,
        162, 183, 187
    Jena, call to, 147–48
    Kantian revolution, 118, 119n

Kantians, criticism of, 126–27
Kielmeyer, relationship with, 139, 238–39,
    299, 304
last years, 196–98
Lebenskraft, 293
life and death, nature of, 158–59, 292
matter, theory of, 130–32
mechanism, 290, 292–93
medicine, 125, 128n
mind, historical development of, 135–36
mysticism, charged with, 115n, 128n
mythology, 119–20
nature
    poetry of, 159–64
    as product and productivity, 142–45
    and self, 134–35, 155
nature philosophy, task of, 122, 133
Naturphilosophie, 115n, 116, 122, 128–51,
    297
Nees von Esenbeck, Christian Gottfried, 129
Newton, Isaac, 130, 163
objective self, 131–34, 142n, 143n, 152, 153,
    154–55
organicism, 137–40, 157–59, 185–86, 239,
    289–90, 292–94, 310
philosophy into poetry, 160
polar forces, 130–32
professor of philosophy at Jena, 46
Reinhold, Karl Leonhard, 126n
Robinson, Henry Crab, 150
Romanticism, early, 17n
Schelling, Caroline Böhmer Schlegel, rela-
    tionship with, 116, 146, 149, 166–76,
    179–80
Schiller, relationship with, 147–48
Schlegel, Karl Friedrich, relations with, 35,
    52–54, 148–49, 148n
Schleiden, Matthias Jakob, 128–29
scientific study, 125–26
self-positing, 151–53
seminary studies, 116–23
sex and love, 145–46, 303, 303n
species, nature of, 301–5
Spinoza, Baruch de, 116, 121, 181–84, 186,
    187
Steffens, Henrik, relationship with, 150–51
teleology, 11
thing-in-itself, 131, 154, 155
transcendental philosophy, task of, 122, 161
tutor, 123–28
unconscious activity of self, 153–54, 161–62
works
    Abhandlungen zur Erläuterung des Idealismus
        der Wissenschaftslehre, 126n, 138n
    Antiquissimi de prima malorum humanorum
        origine, 119n
    Bruno, 180, 182–86
    Darstellung meines Systems der Philosophie,
        178, 180, 181–82

"Das älteste Systemprogramm des deutschen Idealismus," 124–25, 124n
Einleitung zu dem Entwurf eines Systems der Naturphilosophie, 127, 140–45, 145–46, 149–50, 291
Epikurisch Glaubensbekenntniss Heinz Widerporstens, 103–4, 104n
Erster Entwurf eines Systems der Naturphilosophie, 127, 140–45, 149, 291, 295, 306
Ideen zu einer Philosophie der Natur, 31n, 127, 129–37, 140–41, 147
"Neue Deduktion des Naturrechts," 123
Philosophische Briefe über Dogmatismus und Kritizismus, 122
Philosophie der Kunst, 120n, 189–90
System des transscendentalen Idealismus, 106, 140–45, 150, 151–66, 168, 178n, 186–87, 287
Über die Möglichkeit einer Form der Philosophie überhaupt, 212
"Über Mythen, historische Sagen und Philosopheme des ältesten Welt," 120
Ueber den wahren Begriff der Naturphilosophie, 180, 181
Vom Ich als Prinzip der Philosophie, 121
Weltseele, 127, 138–39, 138n, 287, 291, 294–95, 298–99, 304
Schelling, Joseph Friedrich (father of Friedrich Wilhelm Joseph Schelling), 116–117
Schelling, Karl (brother of Friedrich Wilhelm Joseph Schelling), 238
Schiller, Charlotte (wife of Friedrich Schiller), 421
Schiller, Friedrich
    aesthetic judgment, 13, 427n
    Fichte, Johann Gottlieb, 83
    Goethe, criticism of, 423
    Goethe, encounter with, 2, 329, 329n, 423–25
    Goethe, friendship with, 421–27
    Hardenberg, relationship with, 26n
    history professor at Jena, 46
    Die Horen, 423, 425
    intellectual character of, 422, 425–26
    Karlsschule, 239
    moral judgment, 13
    naive and sentimental poets, 431–34
    politics, 13, 51–55
    Schelling, relationship with, 147–48
    Schlegel, Karl Friedrich, relationship with, 49–51, 49n
    science, Goethe's, 438n
    works
        Die Bestimmung des Gelehrten, 54, 86–88
        Geschichte des Abfalls der vereinigten Niederlande, 410n
        "Die Götter Griechenlandes," 409
        Die Räuber, 1, 409, 422

Über Anmut und Würde, 1, 423, 426
Über die ästhetische Erziehung des Menschen, 54–55, 163
Über naive und sentimentalische Dichtung, 106, 422, 430–34
Xenien, 422
Schlegel, August Wilhelm, 4
    Böhmer, Auguste, death of, 171–72
    early life, 23–25
    Goethe, relationship with, 461
    Heine's evaluation of, 195
    Lucinde, reaction to, 112n
    Schelling, Caroline Böhmer Schlegel divorce from, 175–76, 176n
    early relationship with, 42–43
    marriage to, 45
    Shakespeare, William, 32
    university study, 23–27
Schlegel, Dorothea Veit (mistress and wife of Karl Friedrich Schlegel), 17n
    Lucinde, 92–93
    maternal quality of, 93n
    Mendelssohn, Moses (daughter of), 91
    Romanticism, 19n
    Schelling, Caroline Böhmer Schlegel, judgment about, 169–70
    Schlegel, Friedrich, 92–93
Schlegel, Friedrich. *See* Karl Friedrich Schlegel
Schlegel, Johann Adolf (father of August Wilhelm and Karl Friedrich Schlegel), 23
Schlegel, Johanna Christiane, née Hübsch (mother of August Wilhelm and Karl Friedrich Schlegel), 23
Schlegel, Karl Friedrich, 4
    aesthetic theory, 105–12
    Böhmer, Auguste, love for, 105–6, 107–8
    Catholicism, conversion to, 58–59, 58n
    early life, 20
    Eisenstuck, Laura, 26–27, 26n
    Fichte, evaluation of, 35, 59–60, 60n, 83, 84–86, 88
    fragments, 85, 85n
    French Revolution, 84n
    Goethe, as model of Romantic writer, 107
    Goethe, relationship with, 59, 59n, 461–62, 473, 473n
    Hardenberg, friendship with, 25–27
    Herz, Henriette, 95n
    idealism, 110
    irony, Romantic, 112, 112n, 273
    last years, 195–96
    love, theory of, 34n
    marriage, views on, 93n
    political writings, 52–54, 53n, 58, 85
    Privatdozent in Jena, 187–88, 187n
    Rehberg, Caroline, 24, 24n
    Romantic poetry, 22, 107–12
    Romanticism, meaning of, 19n, 20–23, 109–112

Schlegel, Karl Friedrich (*continued*)
    Schelling, Caroline Böhmer Schlegel, love of,
        43–45, 43n, 107–8
    Schelling, relations with, 35, 52–54, 148–49,
        148n
    Schiller, antagonism toward, 49–51, 49n
    Schiller's *Naive und sentimentalische Dichtung*,
        influence of, 21n, 22n, 107
    Schleiermacher, friendship with, 96–97
    Spinoza, Baruch de, 110–11
    transcendental philosophy, 187, 187n
    university studies, 23–24
    works
        Athenaeum, 20, 85, 106
        Athenaeum Fragmente, 22, 85n, 106, 108
        "Georg Forster," 53–54
        Gespräch über die Poesie, 85n, 106, 108–12
        Kritische Fragmente, 53
        Lucinde, 24, 24n, 43–44, 110, 112–113,
            112n, 113n, 146
        Über das Studium der griechischen Poesie,
            21n, 49, 106
        "Versuch über den Begriff der Repub-
            likanismus," 52
Schleicher, August, 342n
Schleiden, Matthias Jakob, 128–29
Schleiermacher, Friedrich Daniel, 4
    early life, 94–97
    Fichte, Johann Gottlieb, 97–98, 178
    Grunow, Elenore, 100, 100n, 113n
    Halle, 194–95, 275
    Hardenberg, Friedrich von, on, 103
    Herz, Henriette, friendship with, 95, 113n
    idealism, 100n, 105
    intuition of *Universum*, 99–102
    Kant, evaluation of, 98–99
    love and religion, 100–2, 105
    miracles, 102n
    Moravian Brotherhood, 94
    poets and artists, 97, 101, 101n
    Reil, relationship with, 253
    Religion, evaluation of, 103–4
    as a Romantic, 104–5
    Romanticism, early, 17n
    Schelling, Friedrich Wilhelm Joseph, 103–4
    Schlegel, Karl Friedrich, friendship with, 96–
        97
    science and religion, 102–3
    Spinoza, Baruch de, 98
    works
        Der christliche Glaube, 102, 102n
        Monologen, 100n, 105, 275
        Über die Religion, 97–105, 275
        *Vertraute Briefe über Friedrich Schlegels Lu-
            cinde*, 112–13, 113n
Schlosser, Johann Georg, 331, 335, 336
Schmid, Christian, 78
Schönemann, Lili (early love of Goethe), 351–
    53

Schönkopf, Anna Katharina (early love of
    Goethe), 336–38
Schröter, Corona (actress at Weimar), 357
Schubert, Gotthilf Heinrich von, 463
Schütz, Christian Gottfried, 175
science and religion, Schleiermacher on, 102–3
sex and love, Schelling on, 145–46, 303, 303n
Shakespeare, William
    Goethe, Johann Wolfgang von, 345
    Schlegel, August Wilhelm, 32
Shaw, George Bernard, 515–16
slavery, Darwin on, 540–41
Snelders, H. A. M., Schelling as mystic, 115n,
    128n
Snow, Dale, 116n
Soemmerring, Samuel Thomas von, 368, 372,
    374, 443
species, nature of, 301–5
Spinoza, Baruch de,
    Ethica, 376–80
    Goethe, Johann Wolfgang von, 439, 439n,
        490, 506
    intellectual love of God, 98, 98n, 377
    monism, 11
    poetry, 110–11
    Schelling, Friedrich Wilhelm Joseph, 116,
        121, 181–84, 186, 187
    Schlegel, Karl Friedrich, 110–11
    Schleiermacher, Friedrich Daniel, 98
Spinoza controversy, 378–79
Spix, Johann Baptist von, 497
Steffens, Henrik
    Fichte, student of, 86n
    Gall, criticism of, 278
    Goethe, relationship with, 463, 472–73
    Jena, life in, 193–94
    political attitude of, 55n
    Reichardt (son-in-law of), 275, 275n
    Reil, relationship with, 253
    Schelling, relations with, 150–51
Steigerwald, Joan, 313n
Stein, Charlotte von. *See* Von Stein, Charlotte
Stolberg-Stolberg, Christian zu, 353
Stolberg-Stolberg, Friedrich Leopold zu, 353
Sturm und Drang, 346
Swammerdam, Jan, 211–12

Tatter, Georg (friend of Caroline Böhmer
    Schlegel Schelling), 37
teleology
    Goethe, Johann Wolfgang von, 11, 444–45,
        445n, 450–51
    Kant, Immanuel, 3, 5, 10–12, 309
    Schelling, Friedrich Wilhelm Joseph, 11
teleomechanism, 3–4, 210, 227–28, 235, 237
Textor, Johann Wolfgang (grandfather of Johann
    Wolfgang von Goethe), 331
thing-in-itself
    Fichte, Johann Gottlieb, 76–78, 76n

Kant, Immanuel, 63, 75–78
Schelling, Friedrich Wilhelm Joseph, 131, 154, 155
Tieck, Ludwig, 17n, 36, 91
Tischbein, Johann Heinrich Wilhelm, 384–85, 391
transcendental philosophy
  Fichte, Johann Gottlieb, 13, 72–79
  Schelling, Friedrich Wilhelm Joseph, 122, 161
  Schlegel, Karl Friedrich, 187, 187n
Treviranus, Gottfried Reinhold, 4n
Tritschler, Johann, 246

Uhland, Johann Ludwig, 17n
unconscious activity of self, Schelling on, 153–54, 161–62
Urpflanze, Goethe on, 376, 395–96, 401–2, 414–15, 424
utilitarianism, Darwin on, 543–551

Veit, Dorothea. *See* Schlegel, Dorothea Veit
vertebral theory of skull
  Goethe, Johann Wolfgang von, 328, 420–21, 491–502
  Oken, Lorenz, 421, 495–98
Vesalius, Andreas, 371–72, 372n
Vicq-d'Azyr, Félix, 374
Virchow, Rudolf, 261
Voigt, Christian Gottlob, 41, 147–48
Voigt, Friedrich Siegmund, 499–500
Voigt, Johann Carl Wilhelm, 365
Volta, Alessandro, 317–18
Von Stein, Charlotte, 357–67, 377, 380–81
  "An den Mond nach meiner Manier," 364
  description of, 357–58
  Goethe, relationship with, 357–67, 377, 380–81, 410–11
Voss, Johann Heinrich, 472
Vulcanism, 35, 366–67
Vulpius, August Walther (son of Goethe), 412, 415–16

Vulpius, Christian August, 410
Vulpius, Christiane (mistress and wife of Goethe), 390, 410–13, 419–20

Wackenroder, Wilhelm, 17n
Wagnitz, Heinrich, 263
Waitz, Johann Christian Wilhelm, 372
Wallace, Alfred Russel, 547–549
Weimar, city of, 355n
Wellbery, David, 350n
Wells, George, 478n, 483n
Werner, Abraham Gottlob, 34–35, 365–67
Wetzels, Walter, 115n
Whewell, William, 434–35
Whytt, Robert, 314
Wieland, Christoph Martin, 427
Wilson, E. O., 3
Wilson, W. Daniel, 90n
Winckelmann, Johann Joachim, 106
  artistic genius, nature of, 391–93
  beauty, 401
  environmentalism, 403n
  Goethe, Johann Wolfgang von, 384, 401
  homoeroticism, 388n
  works
    Gedanken zur Nachahmung der griechischen Werke, 393n
    Geschichte der Kunst des Altertums, 391–92, 392n
Windischmann, Karl Joseph, 246
Wolf, Friedrich August, 275, 276n, 277–78, 473
Wolff, Caspar Friedrich, 215–16, 219, 220
Wolff, Christian, 4n, 18, 94
Wöllner, Johann Christoph von, 52
Woltmann, Karl, 30

Zeitschrift für Speculative Physik, 181
Zimmerman, Georg, 357
Ziolkowski, Theodore, 258n
Zwischenkiefer. *See* intermaxillary bone